T0192453

Lecture Notes in Physics

Volume 977

The Lecture Notes in Physics

The series Lecture Notes in Physics (LNP), founded in 1969, reports new developments in physics research and teaching-quickly and informally, but with a high quality and the explicit aim to summarize and communicate current knowledge in an accessible way. Books published in this series are conceived as bridging material between advanced graduate textbooks and the forefront of research and to serve three purposes:

- to be a compact and modern up-to-date source of reference on a well-defined topic;
- to serve as an accessible introduction to the field to postgraduate students and nonspecialist researchers from related areas;
- to be a source of advanced teaching material for specialized seminars, courses and schools.

Both monographs and multi-author volumes will be considered for publication. Edited volumes should however consist of a very limited number of contributions only. Proceedings will not be considered for LNP.

Volumes published in LNP are disseminated both in print and in electronic formats, the electronic archive being available at springerlink.com. The series content is indexed, abstracted and referenced by many abstracting and information services, bibliographic networks, subscription agencies, library networks, and consortia.

Proposals should be sent to a member of the Editorial Board, or directly to the responsible editor at Springer:

Dr Lisa Scalone
Springer Nature
Physics
Tiergartenstrasse 17
69121 Heidelberg, Germany
lisa.scalone@springernature.com

More information about this series at http://www.springer.com/series/5304

Edward Shuryak

Nonperturbative Topological Phenomena in QCD and Related Theories

 Springer

Edward Shuryak
Physics and Astronomy
Stony Brook University
Stony Brook, NY, USA

ISSN 0075-8450   ISSN 1616-6361   (electronic)
Lecture Notes in Physics
ISBN 978-3-030-62989-2   ISBN 978-3-030-62990-8   (eBook)
https://doi.org/10.1007/978-3-030-62990-8

This Springer imprint is published by the registered company Springer Nature Switzerland AG.
The registered company address is: Gewerbestrasse 11, 6330 Cham, Switzerland

Preface

The book is a summary of various lectures on nonperturbative QCD which I have given during the last three decades at Stony Brook. It is about the *topological objects* present in gauge theories, their semiclassical theory, and applications to multiple physical phenomena. The overall content of the book can be grasped from the list of chapters and sections below, so let me mention here, at the start, what is different in this book relative to others. There are good classic reviews, books, and lecture notes on *magnetic monopoles* and *instantons*. Yet those about *instanton-dyons, sphalerons*, and even such traditional object as *QCD flux tubes* are yet to be found. Even for subjects on which there is extensive pedagogic-style literature, they all have rather different focus. Usually these objects—topological solitons, as one can call them collectively—are treated *individually*. (It can be compared to a visit to a zoo: here is a *lion*, and here is a *gazelle*, etc.)

Of course, we will have similar individual discussion of all these objects below as well, but the focus will be on their *ensembles* and *phenomena* which such objects generate collectively. (Think of it as an actual trip to African savannas.) For example, at some settings, such solitons can exist in finite cluster or groups and in another in infinite (scaled as volume) "condensates." Under certain conditions, *monopoles* undergo Bose-Einstein condensation, related to "confinement-deconfinement" phase transitions. *Instantons* lead to quark pairing and condensation, breaking spontaneously the chiral symmetry. This leads to effective quark mass (and therefore large fraction of the nucleon's mass—as well as our own!). *Sphalerons* lead to chiral imbalance in heavy ion collisions and in early Universe, producing also sounds, gravity waves, and perhaps even the baryon asymmetry.

The semiclassical models of all these phenomena have also strong roots in the first-principle numerical approach to gauge theories, known as *lattice gauge theories*. For all of these objects, observations of them "on the lattice," and verification of their collective effects, would also be the important part of many chapters below.

Crucial feature of modern field/string theories is a notion of *"dualities"*, namely, existence of *different yet physically equivalent* descriptions. We will discuss three of those: (1) the famed electric-magnetic duality; (2) the "Poisson duality," e.g., between monopoles and instanton-dyons; as well as (3) the holographic gauge-string or AdS/CFT duality. Discovery of each duality is always a surprise, complemented with original disbelief, then some confusion, and finally multiple tests. This makes internal logic of a book a bit complex, since some chapters are "dual" to others. Let me illustrate the situation for "Poisson duality," much less known than the others. It tells us that the partition function (and of course everything else, stemming from it) using *monopoles* looks very different from that using *instanton-dyons*: but the results are the same. Therefore, one should either use one or the other formulation, not their sum or other combination: they are just two different ways of organizing evaluation of the same path integral!

Stony Brook, NY, USA   Edward Shuryak
August 2020

Acknowledgments

The book describes development during the last 50 years, and clearly there were too many people contributing to mention them. Let me only emphasize theory of instanton-dyons, which would not be possible without contributions by Pierre van Baal and Dmitry Diakonov, who are no longer with us. Many other colleagues were explaining to me issues described above, bit by bit. I am indebted to my long-time collaborators at Stony Brook, especially to Jac Verbaarschot, Ismail Zahed, and to my former students Pietro Faccioli, Jinfeng Liao, Shu Lin, Rasmus Larsen, and Adith Ramamurti.

Contents

Notations and Units

Some Abbreviations Used

AdS	Anti-de Sitter spacetime
CFL=CSC3	Color-flavor locked phase, or color superconducting phase with three flavors
CSC2	Color superconducting phase with two flavors
CFT	Conformal field theory
DIS	Deep inelastic scattering
EoS	Equation of state
N_c and N_f	Numbers of quark colors and flavors
NJL	Nambu-Jona Lasinio model
MFA and RPA	The Mean Field and Random Phase Approximations
OPE	Operator product expansion
QCD	Quantum chromodynamics, pQCD is its perturbative version
QED	Quantum electrodynamics
QGP	Quark-gluon plasma
RG	Renormalization group
RILM and IILM	Random Instanton Liquid Model and Interacting Instanton Liquid Model
SUSY	Supersymmetric
SYM	Supersymmetric Yang-Mills theory
UV and IR	Ultraviolet and infrared limits, meaning the limits of large and small momentum scales
VEV	Vacuum expectation value
ZMZ	Zero mode zone

Units

We use standard "natural units" of high energy/nuclear physics in which the speed of light and the Plank constant $\hbar = c = 1$. Thus length and time has the same dimension, the inverse of momentum and energy. Transition between units occurs by a convenient substitution of 1 according to

$$1 = 0.19732 \, \text{fm} \cdot \text{GeV}$$

and then cancellation of femto-meters ($fm = 10^{-15}$ m, also known as *fermis*) or GeV (10^9 eV or Giga-electron-volts) as needed.

Discussion of the temperature T uses also Boltzmann constant $k_B = 1$, so it is measured in energy units (e.g., GeV).

Space-Time and Other Indices, Standard Gamma Matrices

We follow standard physics convention that an index appearing twice on one side of the equation is a dummy variable with the summation implied, e.g., $a_m b_m \equiv \sum_m a_m b_m$.

We use Latin letters $a, b \ldots$ to count color generators, 1–8 or from 1 to $N_c^2 - 1$, and $i, j..$ to count colors, 1–3 or 1-N_c. We use letters l, m, n also to count spatial vectors 1–3.

Greek letters are generally used for space-time. Standard Minkowski metrics $g_{\mu\nu}^M = diag(1, -1, -1, -1)$ is implied in sums,

$$a_\mu b_\mu \equiv \sum_{\mu\nu} g_{\mu\nu} a_\mu b_\nu$$

Transition to Euclidean time is done with

$$x_0^M = -i x_4^E \qquad x_m^M = x_m^E$$

The Euclidean metrics is just $g_{\mu\nu}^E = \delta_{\mu\nu} = diag(1, 1, 1, 1)$.

Pauli matrices τ_{ij}^m, all indices 1...3, are twice the generators of the SU(2) rotations. They satisfy the basic relation

$$\tau^a \tau^b = \delta^{ab} + i\epsilon^{abc} \tau^c$$

Color SU(3) generators are half of the Gell-Mann matrices $T^a = t^a/2$, $a = 1\ldots 8$. We also use a notation λ^a for the same set of matrices and also use those for SU(3) flavor. Their product can be written in a form similar to that for Pauli matrices

$$t^a t^b = \frac{2}{3}\delta^{ab} + t^c(d^{abc} + if^{abc})$$

where d, f are some standard numerical tensors of the SU(3) group.

Angular Momentum in Four Dimensions and t'Hooft η Symbol

Angular momentum algebra related with three-dimensional rotation group $O(3)$ is assumed to be familiar from quantum mechanics textbooks. Just reminding, classically one defines the angular momentum as $\vec{x} \times \vec{p}$, so its quantum version is[1]

$$\hat{L}_i = \epsilon_{ijk} x_j \left(-i \frac{\partial}{\partial x_k} \right)$$

Its square $\hat{L}_i^2$ enters Laplace-Beltrami operator of the Laplacian, producing the so-called centrifugal potential

$$V_l = \frac{\hat{L}^2}{r^2} = \frac{l(l+1)}{r^2}$$

In four dimensions the group of rotations $O(4)$ has six generators. There exists a standard formalism with total angular momentum and different angular functions for each dimension in math literature, but we would not follow it here. Instead we will view these six generators as *two pairs* of $O(3)$ or $SU(2)$ algebras, much more familiar to us. Often these two are called left- and right-handed sets of angular momenta.

In taking this road, one has to face a certain dilemma: one can either keep (1) the standard normalization of the angular momentum or (2) the standard form of the $SU(2)$ commutation relation. Following t' Hooft, we use the latter choice, defining these *two pairs* of angular momenta operators by

$$L_1^a = -\frac{i}{2} \eta_{a\mu\nu} x^\mu \frac{\partial}{\partial x_\nu} \tag{1}$$

$$L_2^a = -\frac{i}{2} \bar{\eta}_{a\mu\nu} x^\mu \frac{\partial}{\partial x_\nu} \tag{2}$$

which do not have standard normalization (due to extra 1/2) but commute in the standard way, namely,

$$\left[L_p^a L_q^b \right] = i \delta_{pq} \epsilon_{abc} L_p^c$$

Note that δ_{pq} indicates that two sets are mutually commuting. Note extra $1/2$ in definition of $L_{1,2}$ compared to standard quantum mechanics in 3-d: it leads to

---

[1] The antisymmetric d-dimensional tensor ϵ in d dimensions is known as Levi-Civita tensor in mathematics. Often it is substituted by the so-called wedge symbol $\hat{\ }$.

extra factor 4 in angular part of the Laplacian. With it the radial equation adds the centrifugal potential of the form $+2(L_1^2 + L_2^2)/r^2$.

These notations imply that some simple object look a bit unusual. For example, the 4-vector is not simply $l = 1$ object, as it is in 3-d, but a tensor made of left and right spinors $(l_1, l_2) = (1/2, 1/2)$ representation of $O(4)$. The corresponding centrifugal term in the vector channel obtains the coefficient of the centrifugal potential $2l_1(l_1 + 1) + 2l_2(l_2 + 1) = 3$.

(One more reason physicists are using such notations is that the $O(4)$ group is obviously a close relative of the Lorentz group, of 1+3 Minkowski space-time. One of the $SU(2)$ is rotations, and another (modified) is that of the boosts.)

Before defining 't Hooft symbol $\eta_{a\mu\nu}$, we elevate three Pauli matrices to four quaternions (invented by Hamilton)

$$\tau_\mu^\pm = (\vec{\tau}, \mp i), \tag{3}$$

where $\tau^a \tau^b = \delta^{ab} + i\epsilon^{abc}\tau^c$ and

$$\tau_\mu^+ \tau_\nu^- = \delta_{\mu\nu} + i\eta_{a\mu\nu}\tau^a, \tag{4}$$

$$\tau_\mu^- \tau_\nu^+ = \delta_{\mu\nu} + i\bar{\eta}_{a\mu\nu}\tau^a, \tag{5}$$

with the η-symbols given by

$$\eta_{a\mu\nu} = \epsilon_{a\mu\nu} + \delta_{a\mu}\delta_{v4} - \delta_{av}\delta_{\mu4}, \tag{6}$$

$$\bar{\eta}_{a\mu\nu} = \epsilon_{a\mu\nu} - \delta_{a\mu}\delta_{v4} + \delta_{av}\delta_{\mu4}. \tag{7}$$

The η-symbols are (anti) self-dual in the vector indices

$$\eta_{a\mu\nu} = \frac{1}{2}\epsilon_{\mu\nu\alpha\beta}\eta_{a\alpha\beta}, \qquad \bar{\eta}_{a\mu\nu} = -\frac{1}{2}\epsilon_{\mu\nu\alpha\beta}\bar{\eta}_{a\alpha\beta} \qquad \eta_{a\mu\nu} = -\eta_{a\nu\mu}. \tag{8}$$

We have the following useful relations for contractions involving η symbols

$$\eta_{a\mu\nu}\eta_{b\mu\nu} = 4\delta_{ab}, \tag{9}$$

$$\eta_{a\mu\nu}\eta_{a\mu\rho} = 3\delta_{v\rho}, \tag{10}$$

$$\eta_{a\mu\nu}\eta_{a\mu\nu} = 12, \tag{11}$$

$$\eta_{a\mu\nu}\eta_{a\rho\lambda} = \delta_{\mu\rho}\delta_{v\lambda} - \delta_{\mu\lambda}\delta_{v\rho} + \epsilon_{\mu\nu\rho\lambda}, \tag{12}$$

$$\eta_{a\mu\nu}\eta_{b\mu\rho} = \delta_{ab}\delta_{v\rho} + \epsilon_{abc}\eta_{cv\rho}, \tag{13}$$

$$\eta_{a\mu\nu}\bar{\eta}_{b\mu\nu} = 0. \tag{14}$$

The same relations hold for $\bar{\eta}_{a\mu\nu}$, except for

$$\bar{\eta}_{a\mu\nu}\bar{\eta}_{a\rho\lambda} = \delta_{\mu\rho}\delta_{\nu\lambda} - \delta_{\mu\lambda}\delta_{\nu\rho} - \epsilon_{\mu\nu\rho\lambda}. \tag{15}$$

Some additional relations are

$$\epsilon_{abc}\eta_{b\mu\nu}\eta_{c\rho\lambda} = \delta_{\mu\rho}\eta_{a\nu\lambda} - \delta_{\mu\lambda}\eta_{a\nu\rho} + \delta_{\nu\lambda}\eta_{a\mu\rho} - \delta_{\nu\rho}\eta_{a\mu\lambda}, \tag{16}$$

$$\epsilon_{\lambda\mu\nu\sigma}\eta_{a\rho\sigma} = \delta_{\rho\lambda}\eta_{a\mu\nu} + \delta_{\rho\nu}\eta_{a\lambda\mu} + \delta_{\rho\mu}\eta_{a\nu\lambda}. \tag{17}$$

About the Author

Edward Shuryak was born in 1948 and grew up in Odessa, Ukraine. Winning the second place in Siberian Mathematics Olympiad, he was admitted into a special high school in 1964 and then into Novosibirsk State University, where he graduated in 1970 in physics. Under the supervision of S.T. Belyaev, Shuryak received his Ph.D. in 1974 from the Budker Institute of Nuclear Physics, where he continued on as a researcher while simultaneously teaching at Novosibirsk State University. He became a full professor in 1982, the year in which he also gave the first series of lectures at CERN about quark-gluon plasma, a new form of matter for which he proposed the name in a 1978 paper. He moved to the United States in 1990 and became professor of physics at Stony Brook University, leading the Nuclear Theory Center. In 2004, Shuryak was promoted to distinguished professor, the highest academic appointment rank in the State University of New York system. Shuryak is the author or co-author of nearly 400 papers which in total have been cited more than 32,000 times; five of these papers have been cited more than 1000 each, and another nine have been cited more than 500 times each. His H-index is 88, according to Google Scholar. The outstanding scientific achievements of Edward Shuryak have been recognized internationally. He was elected as fellow of American Physical Society in 1996, "for his seminal contributions to the study of the quark-gluon plasma." He was the 2004 recipient of the Dirac Medal from University of New South Wales in Australia and the 2005 recipient of the A. von Humboldt Prize from Germany. More recently, he was awarded the 2018 Herman Feshbach Prize in Theoretical Nuclear Physics, "for his pioneering contributions to the understanding of strongly interacting matter under extreme conditions, and for establishing the foundations of the theory of quark-gluon plasma and its hydrodynamical behavior."

Introduction

1

1.1 What Are the "Nonperturbative Topological Phenomena"?

In **quantum field theory (QFT)** courses, one first learns about *weakly coupled* field theories, such as quantum electrodynamics and *perturbative* QCD.[1] They start with harmonic oscillator quantization, define "quanta" of the fields, and their weak interaction, described by *Feynman diagrams*.[2] Expanding the partition sum of a theory, represented by some path integral, in powers of the coupling one get results which schematically looks like this

$$Amplitude(p) = \sum_n (\alpha_s)^n C_n(p) \qquad (1.1)$$

where $\alpha_s \equiv g^2/4\pi$ is assumed to be small enough to make series convergent in some "practical sense."[3] The coefficients C_n are calculated functions of the kinematical parameters, e.g., products of 4-momenta of in- and outgoing quanta. This approach

---

[1]It is a big field now, but for readers who need concise and compact introduction, I recommend Feynman lectures (Feynman and Cline 2020) published recently with 30+ year delay.

[2]Of course, Feynman diagrams have now much wider range of applications than QFTs. Readers interested in the easiest introduction to Feynman diagrams in quantum/statistical mechanics can find it in my textbook "Manybody theory in the nutshell," Princeton University Press, where they are explained using basic toy models from quantum and statistical mechanics.

[3]While these series are known to be divergent or asymptotic, several of its terms approach certain limit at sufficiently small α_s, before diverging away from it in higher order. General issues related with perturbative series we will discuss in connection to the so-called transseries later.

© The Author(s), under exclusive license to Springer Nature Switzerland AG 2021
E. Shuryak, *Nonperturbative Topological Phenomena in QCD and Related Theories*, Lecture Notes in Physics 977,
https://doi.org/10.1007/978-3-030-62990-8_1

was created in the 1950s in quantum electrodynamics, QED. The corresponding coupling[4]

$$e^2/\hbar c \approx 1/137 \ll 1$$

is small, serving as a natural small parameter of the perturbation theory.

In non-Abelian gauge theories such as QCD, the coupling constant $\alpha_s(\mu) = g^2(\mu)/4\pi$ is *running*, it depends on the momentum scale μ involved. The coupling is small at large momentum transfer ("hard") processes. Perturbative QCD (pQCD) also uses Feynman diagrams: but we will not discuss any of that in these lectures. (The most important perturbatively calculated function is the *beta* function of the renormalization group, which is discussed in the Appendix. Another perturbatively calculated quantity we will need is the *effective action*, created by quantum fluctuations for constant classical $\langle A_0 \rangle \neq 0$, also given in Appendix.)

These lectures are not about Feynman diagrams and perturbation theory, but about *nonperturbative phenomena*. Unfortunately, the word "nonperturbative" (appearing in the book's title) is used in literature with several different meanings. The weakest of them is a situation in which the perturbative series are re-summed. Sometimes the sum is simpler than each diagram by itself: this, for example, happens if the answer includes *non-integer* power of α_s. The most important re-summation will be that given by renormalization group (RG) (Gell-Mann-Low beta function).

However, in this book we adopt stronger meaning of the term, namely, we will call "nonperturbative" any phenomena which are *invisible*[5] *in the perturbative context*. What it means is exemplified by a function of the coupling like this

$$f(\alpha_s) \sim \exp\left(-\frac{const}{\alpha_s}\right)\left(\sum_{n=0}^{\infty}(\alpha_s)^n B_n\right) \qquad (1.2)$$

Attempting to expand it in powers of the coupling, one finds that all coefficients of its Taylor expansion are zero. The exponential term and the perturbative series build on it are derived by the *semiclassical* methods, which we will study. All exponential, perturbative, and logarithmic (not shown) terms together form the so-called *transseries*.[6] Relations between coefficients B_n and C_n of (1.1), if found, are called *resurgence relations*.

The *"topological"* phenomena are related to existence of *topological solitons*, made of so strong fields, that the interaction (non-quadratic) parts of Lagrangian is

---

[4]We will use now standard high energy physics units $\hbar = c = 1$, showing them explicitly only in few cases such as this one. For more on units and notations, see Appendix.

[5]Stickily speaking, they are visible via the so-called Dyson phenomena explaining why the perturbative series are asymptotic (badly divergent) series.

[6]A curious reader may wonder if exponent, log, and powers do form together a sufficiently complete set. The answer to it is affirmative.

as large as quadratic ones. In the case of gauge theories, it means $A_\mu^a = O(1/g)$. Specifically, we will discuss:

(i)   **instantons** and their constituents, **instanton-dyons**
(ii) **sphalerons**, unstable magnetic balls, and their explosions
(iii) magnetic **monopoles**
(iv) confining electric flux tubes, also known as the **QCD strings**

As interesting as those objects are by themselves, as some mathematical curiosities, we will be mostly interested in "what they can do to help us to understand the world around us," to mention a standard textbook-style sentence. Therefore, let me mention on the onset *why* we are going to study them.

Instantons are four-dimensional solitons, historically important as evidence of tunneling phenomena existing not only in quantum mechanical systems (like nuclear α decays) but also in QFT as well, in gauge theories especially. They also were the first objects for which semiclassical approximation was used in the QFT context.[7]

Instantons also are very important in QCD because their *fermionic zero modes* generate nontrivial interaction between light quarks (in QCD) or quarks and leptons (in electroweak theory). If instantons are present in large enough density, this effective multi-quark forces are strong enough to form the so-called *quark condensate*[8]

$$\langle \bar{q}q \rangle \neq 0$$

This does happen in the QCD vacuum we leave in, making near-massless u, d, s quarks look like objects with an effective mass $\sim 400\,\text{MeV}$. So instantons are responsible for significant fraction of the nucleon mass (and thus our mass as well!): this alone means that they deserve to be studied. Quark condensate "melts" at $T > T_\chi$, the so-called temperature of the chiral symmetry restoration. In QCD with physical quarks, both are close

$$T_\chi \approx T_c \approx 155\,\text{MeV}$$

Another important phenomenon in QCD at finite temperatures is *deconfinement*: both transitions seem to happen at the same temperature. This however seems to be occasional.[9]

---

[7]More generic semiclassical objects called **fluctons** were also developed and will be discussed in Sect. 5.1. We did not list them above since they are not generally topological, and their role in QFTs is not worked out in detail so far.

[8]Here and below angular brackets indicate vacuum expectation value, or VEV.

[9]We think so because some deformations of QCD do split them significantly and even make them of different transition order.

Monopoles are 3-d solitons, made of the gauge and scalar fields, with nonzero magnetic charges. They are bosons, and their ensemble can undergo Bose-Einstein condensation. In QCD this happens below some temperature T_c. Their condensate expels the color-electric field into confining electric flux tubes. That is why T_c is called the *deconfinement* temperature: above it the matter is quark-gluon plasma (QGP).

Sphalerons, like monopoles, are magnetic 3-d solutions to Yang-Mills (plus scalar) equations. Unlike monopoles, their magnetic field goes in circles, and so they lack magnetic charge. Although they satisfy Yang-Mills equation, they are not minima of the action but only saddle points. Therefore they are *unstable* and, with small perturbation, can *explode* along the so-called *sphaleron path*. The explosion itself allows for analytic solution and we will study it in sphaleron chapter. While sphalerons are important both in QCD and electroweak theory, we will discuss them mostly in the context of cosmological electroweak phase transitions (EWPT), since they violate the baryon/lepton number conservation and are suspect to generate the observed baryon asymmetry of the Universe.

Certain properties of hadronic amplitudes and spectrum lead to the idea that quarks in the QCD vacuum are connected by the **QCD strings**. Studies of effective string description led to creation of the *string theory*: but we will not go in this direction. Instead we will discuss their structure, interactions, and in general their role in strong interaction physics. The main applications we will consider are *inter-quark confining potentials*, closely related to *Pomerons and Reggeons*, the somewhat mysterious objects which emerged in the 1960s from phenomenology of hadronic scattering. We will of course have a look at modern lattice and experimental data, as well as some modern derivation of the Pomeron amplitude.

Let us now return to "perturbative" phenomena, for a moment. A re-summation of certain sequence of diagrams can lead to properties of new quasiparticles, collective excitations, etc. Even more advance theory can be based on the renormalization group: the calculations may start in weak coupling and lead to fixed points. They may also indicate coupling flow to large values (strong coupling) and explain existence of qualitatively new phases of the theory. A classic example is the BCS theory of superconductivity: weak phonon-induced attraction between electrons gets stronger near the Fermi surface, till finally it becomes strong enough to form Cooper pairs.

"QCD-like gauge theories" we will discuss also possess the *asymptotic freedom* phenomenon. In the UV (high momenta, small distances), the coupling is weak $g(p \to \infty) \to 0$. To one-loop accuracy their coupling "runs" according to

$$\frac{8\pi^2}{g^2(p)} = \left(\frac{11 N_c}{3} - \frac{2 N_f}{3} \right) \log \left(\frac{p}{\Lambda_{QCD}} \right) \tag{1.3}$$

The coefficient, known as the first (one-loop) coefficient in the r.h.s.

$$b = \left(\frac{11 N_c}{3} - \frac{2 N_f}{3} \right)$$

of the beta function depends on the number of colors N_c and light quark flavors N_f. It will always be positive in this course, so the number of fermions would be limited by $11N_c - 2N_f > 0$.

Substitution of this expression into perturbative series, one gets series in inverse power of $\log(p)$, small at large p^2. Since log is not a very strong function, one would need to reach rather far in momentum scale, compared to the basic scale $\Lambda_{QCD} \approx 1/fm \sim 0.2\,\text{GeV}$ to make coupling really weak. To give an idea, $\alpha_s(100\,\text{GeV}) \approx 0.11$ and $\alpha_s(1\,\text{GeV}) \approx 1/3$.

Now we return to nonperturbative phenomena. Substituting it into the exponential function mentioned above, one obtains certain *powers* of the momentum scale

$$\exp\left[-C\frac{8\pi^2}{g^2(p)}\right] = \left[\frac{\Lambda_{QCD}}{p}\right]^{C \cdot b} \tag{1.4}$$

So, one may reformulate our definition of the nonperturbative effects as those depending on the relevant momentum scale as its *inverse powers*. Furthermore, one might expect that those powers would be some integers: and indeed, one expects such effects to be present, they go under the term "infrared renormalons" and have large literature devoted to them. Nevertheless, the origin of those effects remains rather obscure and will not be discussed.

Instead, we will focus on the *topological nonperturbative effects* induced by topological solitons of various kinds. Unlike small perturbative fields—photon or gluon waves—they by no means are small perturbations around "classical vacuum," the zero fields. Their masses or actions in the weak coupling are large $O(1/g^2)$. Account for such effects complement/generalize the perturbative series to the so-called *transseries*.

The main analytic method to be used is *semiclassical* approximation, in various settings. In this lectures we will compare the results with what is obtained in numerical simulations of the gauge theories, which is "based on first principles of the theory," and by definition include all effects, perturbative and nonperturbative. Needless to say, the ultimate judge in physics is experiment, to which we will refer whenever possible.

1.2 Brief History of Non-Abelian Gauge Theories and Quantum Chromodynamics

While most readers can skip this section, some may still need some historic perspective. Maxwellian electrodynamics is clearly the prototype field theory, on which significant part of physics and its applications is firmly based. As shown by Lorentz and Poincare, electrodynamics requires transforms from one moving frame into another by (then unusual) *Lorentz transformations*. Einstein in 1905 extended Lorentz invariance to all fields of physics, including mechanics, explaining that previously used Galilean transformation is but an approximation at small velocities $v \ll c$.

After creation of quantum mechanics was complete, electrodynamics was also extended, from classical field theory to quantum one, known now as *quantum electrodynamics* (QED). Development of its perturbative formulation and working out multiple applications (splitting of atomic levels, anomalous magnetic moments of electron and muons, etc.) was quite a triumph.

QED is an Abelian theory: this follows from the fact that photons, quanta of electromagnetic field A_μ, are not themselves charged. Only other particles, say electrons, can provide a source for electromagnetic fields. It is a "gauge theory": the wave function of charged objects allows transformation of its phase

$$\psi(x) \rightarrow e^{i\phi(x)}\psi(x) \tag{1.5}$$

where $\phi(x)$ is an *arbitrary* function of the space-time coordinates. The invariance is created by appropriate "compensation," subtracting gradient of ϕ from the gauge field. The corresponding group of transformations is Abelian, which means that one can multiply by these phase factors in any order.

In (Yang and Mills 1954) QED was famously generalized to the case in which there are several types of charges, known as *non-Abelian gauge theories*. The local gauge symmetry is not a "symmetry" in the usual sense, because it is indeed quite different from, say, translational, rotational, or flavor symmetries we are used to. All physical observables are singlets under gauge rotations and therefore are not really transformed. Thus there cannot be any relations between them to be deduced from this symmetry. Rather it tells us which degrees of freedom of the fields are unphysical, or *redundant*, variables.

In order to explain its general meaning, it is useful to trace an analogy with the Einstein's general relativity. Special relativity postulates that all inertial reference frames can equally be used. General relativity goes further and suggests that one can make arbitrary coordinate transformations, independently *at any point*. Similarly, in gauge theories, "global symmetries" (rotations in the space of internal quantum numbers), are elevated to local gauge invariance, based on *independent* rotations in space of charges, at different space-time points.

Gauge transformation of the charged fields, now possessing some additional index, is generalized to unitary matrix[10]

$$\psi_i(x) \rightarrow U_{ij}(x)\psi_j(x) \tag{1.6}$$

reflecting the fact that in the space of charges a coordinate system can be chosen arbitrarily and it can be done *independently* at each space-time points. The corresponding groups of U are non-Abelian, since order of transformation is important: in general unitary matrices do not commute with each other. This leads to many

---

[10]Like in other papers and books, we use here and below the Einstein's notations: any repeated index is summed.

consequences, in particularly to non-Maxwellian relation between the vector field A_μ and the field strength $G_{\mu\nu}$ containing the commutator.[11]

Yang and Mills first interpreted the index $i = 1, 2$ as *isospin*, the $SU(2)$ flavor symmetry due to (approximate) similarity of the two lightest quarks, u and d. Therefore their non-Abelian field was $SU(2)$ adjoint vector field identified with the ρ mesons. Subsequent natural application of such $SU(2)$ group was related to the "weak isospin" of weak interactions, with non-Abelian fields being vector W, Z bosons of weak interactions. In both of these cases, great original difficulty was that gauge symmetry requires the vector field quanta to be *massless*. Needless to say, neither ρ mesons nor W, Z bosons are massless. This puzzle was eventually solved, by "soft" spontaneous breaking of the $SU(2)$ gauge symmetry of weak interactions in Weinberg-Salam theory.

Quantum chromodynamics (QCD) is based on the $SU(3)$ group of rotations in space of quark's "color" degree of freedom. This gauge group is *not* broken, so gluons are massless. Yet they are strongly interacting and do not propagate individually due to *color confinement* phenomenon which we will discuss in the next subsection.

Historically, QCD became standard theory of strong interaction after the theoretical discovery of "asymptotic freedom" (Gross and Wilczek 1973; Politzer 1973) ("running" of the coupling toward zero at small distances), nicely correlated in time with the experimental discovery of weakly interacting pointlike quarks inside the nucleons (and other hadrons, of course). We will not discuss perturbative QCD as such, and thus discussion and the relevant expressions can be found in Appendix B.1. Derivation of the two-loop beta function for supersymmetric relatives of QCD, based on instantons, will be given in Sect. 6.2.4.

While the main playground of QED is in the atomic physics, that of QCD is the physics of strong interactions. Instead of atoms one has particles collectively called "hadrons." Some authors emphasize the fact that while atoms are nonrelativistic, hadrons show large variety of regimes including very light and very heavy quarks. We would emphasize even more important distinction between them: in QED the vacuum is "empty," while in QCD it is very nontrivial. It is the understanding of the "vacuum structure" which is necessary in order to understand structure of various hadrons.

We will not go into large field of hadronic models here and just present brief overview of few main directions:

(i) *Traditional quark models* (too many to mention here) are aimed at calculation of static properties (e.g., masses, radii, magnetic moments, etc.). Normally all calculations are done in hadron's rest frame, using certain model Hamiltonians. Typically, chiral symmetry breaking is included via effective "constituent quark" masses, and the Coulomb-like and confinement forces are included via

---

[11]The readers not familiar with this may need to consider more introductory texts.

some potentials. In some models also a "residual" interaction is also included, via some 4-quark terms of the Nambu-Jona-Lasinio type.

(ii) Numerical calculation of *Euclidean-time* two- and three-point correlation functions is another general approach, with a source and sink operators creating a state with needed quantum numbers and the third one in between, representing the observable. Originally started from small-distance OPE and the QCD sum rule method, it moved to intermediate distances (see review, e.g., in (Shuryak 1993)) and is now mostly used at large time separations $\Delta\tau$ (compared to inverse mass gaps in the problem $1/\Delta M$) by the lattice gauge theory (LGT) simulations. This condition ensures "relaxation" of the correlators to the lowest mass hadron in a given channel. We will discuss Euclidean correlation functions and lattice gauge theories in the corresponding chapters below.

(iii) Light-front quantization using also certain model Hamiltonians, aimed at the set of quantities, available from experiment. Deep inelastic scattering (DIS), as well as many other hard processes, uses factorization theorems of perturbative QCD and nonperturbative *parton distribution functions* (PDFs). Hard exclusive processes (e.g., form factors) are described in terms of nonperturbative hadron on-light-front wave functions (LCWFs) (for reviews, see (Brodsky and Lepage 1989; Chernyak and Zhitnitsky 1984). We will discuss it in Chap. 10.

(iv) Relatively new approach is "holographic QCD," describing hadrons as quantum fields propagating in the "bulk" space with extra dimensions. It was originally supposed to be a dual description to some strong-coupling regime of QCD and therefore was mostly used for description of quark-gluon plasma (QGP) phase at high temperatures. Nevertheless, its versions including confinement (via dilaton background with certain "walls" (Erlich et al. 2005)) and quark-related fields (especially in the so-called Veneziano limit in which both the number of flavors and colors are large $N_f, N_c \rightarrow \infty, N_f/N_c = fixed$ (Jarvinen and Kiritsis 2012)) do reproduce hadronic spectroscopy, with nice Regge trajectories. The holographic models also led to interesting revival of baryons-as-solitons-type models, generalizing skyrmions and including also vector meson clouds. Brodsky and de Teramond (2006) proposed to relate the wave functions in extra dimension z to those on the light cone, identifying z with certain combination of the light-cone variables ζ. Needless to say, all of these are models constructed "bottom-up," but with well-defined Lagrangians and some economic set of parameters, from which a lot of (mutually consistent) predictions can be worked out.

While (i) and (iv) remain basically in realm of model building, (ii) remains the most fundamental and consistent approach. Lattice studies, starting from the first principles of QCD, had convincingly demonstrated that they correctly include all nonperturbative phenomena. They do display chiral symmetry breaking and confinement and reproduce accurately hadronic masses. Yet its contact with PDFs and light-front wave functions remains difficult. The light-front direction, on the other hand, for decades relied on perturbative QCD, in denial of most of nonperturbative physics.

1.3 Introduction to Chiral Symmetries and Their Breaking

QCD is a "vectorial" theory, and the quark current interacting with gluons, $J_\mu A^\mu$, has a structure

$$J_\mu = \bar{q}_L \gamma_\mu q_L + \bar{q}_R \gamma_\mu q_R$$

which is not mixing the left and right (L,R) components of fermions.[12]

In QCD we have three "light quarks" u, d, s, with masses[13]

$$m_u = 2.32(10)\,\text{MeV}, m_d = 4.71(9)\,\text{MeV}, \ m_s = 92.03(88)\text{MeV} \qquad (1.7)$$

If those masses be the same, one would have additional "flavor symmetries" of rotation in quark flavor space. If only $m_u = m_d$, this is called the $SU(2)$ isospin symmetry.

Putting quark masses to zero, one finds a theory possessing extra *chiral symmetries*. Since the mass terms in the Lagrangian $m_q(\bar{q}_L q_R + \bar{q}_R q_L)$ are the only term mixing chiralities, its absence allows to make two *independent* rotations in flavor space. In particular, if $m_u = m_d = m_s = 0$, the symmetries are

$$U(3)_L \times U(3)_R = SU(3)_V \times SU(3)_A \times U(1)_V \times U(1)_A$$

Here subscripts V, A mean vector and axial vector: the former means that all quark types are rotated in the same direction, and the latter that there is extra γ_5 in the exponent, so that left- and right-handed components rotate in the opposite directions. The first $SU(3)_V$ symmetry is the "flavor symmetry" already mentioned, and $U(1)_V$ is multiplication of all quark fields by a common phase factor: it is related to quark ("baryon") number conservation.

So, "new" symmetries are the axial or "chiral" symmetries, $SU(3)_A$ and $U(1)_A$. Both are broken in the QCD vacuum: but in a completely different ways. The former is *broken spontaneously*, and the latter broken explicitly by the *anomaly* phenomenon, related with topology of the gauge fields.

Expansion in light quark masses is known as *chiral perturbation theory*.

---

[12]Electroweak interactions with W, Z bosons have only the left-handed part, and the interaction with scalar Higgs field has instead a structure $\bar{q}_L q_R + \bar{q}_R q_L$: so this theory does not have chiral symmetries.

[13]These values depend on the momentum scale at which they are normalized: the given one are at scale 2 GeV. For more details, see Particle Data.

1.3.1   Spontaneous Breaking of the $SU(N_f)_A$ Symmetry

The phenomenon of spontaneously broken symmetries is so well known in physics, that explaining it here in detail is not needed. Suffice it to say that, e.g., ground state energy of a superconductor is lower than that of a Fermi gas. Pairing near the Fermi surface creates a "gap" in dispersion relation of the fermions, resulting in certain energy gain.

Analogy with superconductor was used by Nambu and Jona-Lasinio (NJL) (1961), who proposed that pairing on the surface of the Dirac sea, of massless fermions, can also create a gap (i.e., a nonzero mass). Yet unlike the superconductor case, basically one-dimensional, the phenomenon does not happen at arbitrary weak attraction.

They postulated existence of 4-fermion interaction, with some coupling G_{NJL}, strong enough to make a superconductor-like gap even in fermionic vacuum. The second important parameter of the model was the momentum integration cutoff $\Lambda \sim 1\,\text{GeV}$, below which their hypothetical attractive 4-fermion interaction operates. Their magnitude was determined from empirical quark condensate and pion properties (for review, see (Klevansky 1992)).

With time there were many applications of the NJL model, for definiteness we use parameters from (Hutauruk et al. 2018) and other papers of these authors as an example of paper which consistently used NJL model for description of chiral symmetry breaking, quark constituent masses, pions and kaons as bound states, as well as nucleons, made of constituent quark and diquark. To mention one basic relation, the quark "mass gap" equation, which takes the form[14]

$$M = m + \frac{3G_{NJL}M}{\pi^2} \int_{1/\Lambda_{UV}^2}^{1/\Lambda_{IR}^2} \frac{d\tau}{\tau^2} \exp\left(-\tau M^2\right) \tag{1.8}$$

where m is current quark mass and M the constituent mass, to be obtained from this equation. Note that when $m = 0$, either (i) $M = 0$ or (ii) $M \neq 0$ can be canceled out in l.h.s. and r.h.s., remaining only in the (regulated) loop integral. For input parameters used in these works

$$G_{NJL} = 19\,\text{GeV}^{-2}, \quad \Lambda_{IR} = 0.24\,\text{GeV}, \quad \Lambda_{UV} = 0.645\,\text{GeV} \tag{1.9}$$

the constituent mass is found to be $M \approx 0.4\,\text{GeV}$, close to the half of the mass of "usual" ρ meson or $1/3$ of Δ baryon. The general importance of this phenomenon is seen from the fact that such "constituent quark mass" is in fact the main part of hadronic masses.

We get such details because we will use them to numerically compare the magnitude of NJL and pQCD effects, elucidating their *domain of applicability*

---

[14]The integral is a Feynman diagram with a loop, a propagator written in proper time representation.

in this simple way. Let us use one-gluon scattering of two quarks, exchanging a gluon with momentum transfer $F_{gluon}(k^2) \sim g^2/k^2$, and compare it using both Gaussian and exponential form factor with Λ_{UV} in it. The ratio of nonperturbative to perturbative 4-fermion effective vertices is then

$$\frac{G_{NJL}}{F_{gluon}} \exp\left[-\frac{k^2}{\Lambda_{UV}^2} \right] \sim \frac{19k^2}{g^2(\text{GeV}^2)} \exp\left[-(\frac{k}{0.645 \text{ GeV}})^2 \right] \quad (1.10)$$

Note that it remains above one up to momentum transfers of few GeV^2.

1.3.2 The Fate of $U(1)_A$ Symmetry

The NJL model was a brilliant guess in 1961, predated QCD by a decade. It was motivated by desire to explain *why pions are so light*, massless in the chiral limit: they are Nambu-Goldstone modes associated with any spontaneous symmetry breaking. In QCD the role of NJL 4-fermion interaction is played by instanton-induced interaction discovered by G.'tHooft, as we will explain below. What is important, this interaction violates the $U(1)_A$ symmetry *explicitly*.

We will come to its derivation in due time, but now let us explain the difference between two 4-fermion Lagrangians, using the so-called "mesonic" notations in the case of two flavors. Let us introduce bilinear combinations

$$\sigma \equiv (\bar{q}q), \vec{\pi} \equiv i(\bar{q}\gamma_5\vec{\tau}q), \eta == i(\bar{q}\gamma_5 q), \vec{\delta} \equiv (\bar{q}\vec{\tau}q)$$

in terms of which the 4-fermion operators in question look simpler.

The NJL Lagrangian has the structure

$$L_{NJL} \sim (\vec{\pi}^2 + \sigma^2) \quad (1.11)$$

where a vector sign over the pion indicates existence of three pion species, π^+, π^-, π^0. The bracket with sum of four bilinears squared is mapping into itself under $SU(2)_A$ rotations, so NJL Lagrangian has this symmetry.

The instanton-induced one has similar structure (remember, we are in the $N_f = 2$ case)

$$L_{t'Hooft} \sim (\vec{\pi}^2 + \sigma^2 - \vec{\delta}^2 - \eta^2) \quad (1.12)$$

The second four-squared bilinears also transform into itself by $SU(2)_A$ rotations.

Note however the crucial *minus sign* between the first pair of terms and the last ones. The $U(1)_a$ rotation mixes, say, σ with η: and this minus shows that t'Hooft Lagrangian is explicitly *not* $U(1)_A$ symmetric.

Therefore, in mesons excited by such bilinears, the η and $\vec{\delta}$, the instanton-induced interaction is not attraction but *repulsion*, making them much heavier than the former pair.[15]

The attentive reader may be quite confused by this section: indeed we first argued that the basic QCD Lagrangian with massless quarks has the $U(1)_A$ chiral symmetry, and now we say that one can derive an effective interaction which explicitly violates it. Indeed, it was rather unexpected, therefore getting the name *axial anomaly*. Not going into it here, let me just say that in fact a classical symmetry of the Lagrangian may *not* be the symmetry of a full QFT with this very Lagrangian! QFT is the path integral, which contains the action but also the integration measure, and the latter not always respects the symmetries of the former.

1.4   Introduction to Color Confinement

This phenomenon, more precisely called *color-electric* confinement, is perhaps the most complicated nonperturbative phenomenon. In this section we introduce several ways in which its presence (or absence called *deconfinement* at high temperature/densities), can be defined and studied.

In a qualitative form, its definition can be that no object with a *nonzero* color-electric charge (such as quarks or gluons) can appear in physical spectrum of states. However, to prove it in theory or in practice is rather hard, so many other definitions are used.

In pure gauge $SU(N_c)$ theories, deconfinement transition is related with breaking of certain well-defined symmetry, called center symmetry. Therefore the transition is a phase transition with a well-defined order parameter. So we will start with the corresponding explanations and lattice data in Sect. 1.4.1.

Another manifestation of confinement is formation of flux tubes between quarks and related linear confining potential $V(r) = \sigma r$. In (Wilson 1974) this idea was reformulated for lattice studies, requiring that the so-called Wilson loop defined with a contour C has expectation value decreasing as an exponent of its *area* if the area is large

$$\langle W(large\ C) \rangle \sim \exp[-\sigma_T Area(C)] \tag{1.13}$$

Discussion of this statement will be done in Sect. 1.4.2.

Another interesting definition of confinement is related with Bose-Einstein condensation (BEC) of magnetic monopoles. The introduction of this idea is given below, with much more detailed discussion of it in chapter devoted to magnetic monopoles. Let me here only mention that BEC can be detected in a number of ways, with certain specially constructed order parameters. However, those

---

[15]In $SU(3)$ flavor the former is called *eta'* in meson tables: the observation of its very large mass was the so-called $U(1)_A$ problem, first emphasized by Weinberg.

parameters are *nonlocal* ones, and it is not clear where they do or do not show up as a singularity of the free energy.

Confinement is also associated with existence of the electric flux tubes with a nonzero tension. We will discuss those in chapter devoted to QCD flux tubes. This "operational definition" is not in fact correct: it has been theoretically argued and recently observed on the lattice that the electric flux tubes can in fact exist even in the deconfined phase. Finally, the so-called Hagedorn phenomenon—apparent divergence of the partition function of hadronic matter—served historically as an (approximate) indicator of the location of the deconfinement transition on the phase diagram.

1.4.1 Polyakov Lines

The Polyakov line (Polyakov 1977) is defined as a similar integral as in W, but over a line along the Euclidean time τ. Because of its periodicity, C is a *closed loop* winding around (see Fig. 1.1). Its circumference is related to temperature by the so called Matsubara time relation $\beta = \hbar/T$

$$\hat{P} = P exp\left(i \int_C dx_4 \hat{A}_4\right), \quad P = \frac{1}{N_c} Tr\left(\hat{P}\right) \tag{1.14}$$

The hats (here and below) remind us that it is still a matrix in color space. Since $\hat{A}$ is a hermitian matrix, $\hat{P}$ is a unitary one, $\in U(N_c)$.

Pexp means path-order exponent, defined as a limit of a product of matrices describing small steps in time. On the lattice $\hat{P}$ can be seen as a product of link variables $\Pi_C(U)$, and the product should be done over the same circular contour C. The corresponding integrals in mathematics are known as "holonomies." Unlike gauge potential itself, $\hat{P}$ is *gauge invariant* due to closeness of the contour C.

The example of the simplest non-Abelian group $SU(2)$ will often be discussed. Assuming $\hat{P}$ is diagonal, it has the form $diag(e^{i\phi}, e^{-i\phi})$ where the phase is

$$\phi = (1/T)A_4^3(1/2) = \pi v$$

Fig. 1.1 The Polyakov line $P(\vec{x})$ on the lattice. The vertical direction is that of Euclidean time $x_4 = \tau$. Periodicity in it means that the fields on upper and the lower planes are identified

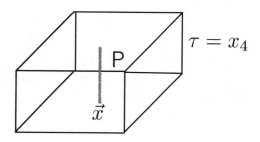

where 1/2 is from the fact that color generator is $\hat{\tau}^a/2$. The field is represented by a new parameter $2\pi T \nu = A_4^3$. In such notations the $SU(2)$ trace is simply

$$P = \frac{1}{N_c} Tr(\hat{P}) = cos(\pi \nu)$$

The physical meaning of vacuum expectation value (VEV) of P is the effective quark free energy,

$$\langle P \rangle = \exp(-F_q/T)$$

In the confining phase of pure gauge theories $\langle P \rangle = 0$ (which corresponds to infinitely heavy point quark), while deconfinement means that it is finite $\langle P \rangle \neq 0$. So, for $SU(2)$ case the so-called *trivial holonomy* corresponds to values $P = 1$, $\nu = 0$, while the confining values are $P = 0$, $\nu = 1/2$.

QCD is of course based on $SU(3)$ gauge theory and thus $\hat{P}$, $\hat{A}$ are 3×3 matrices: we will discuss those when needed. Let us give an example of the numerical simulations of lattice QCD shown in Fig. 1.2. In QCD the phase transition is smooth crossover, while in pure gauge $SU(3)$ theory without quarks (not shown), the VEV of P at $T < T_c$ is strictly zero, with a jump at T_c. Small $\langle P \rangle$ means large effective quark free energy and strong suppression of their contribution to physical processes at low temperatures.

1.4.2   Wilson Lines and Vortices

Using the language of color charges—in particular very heavy external quarks—one explains absence of colored states by the existence of color-electric flux tubes. As shown in Fig. 1.3, it has a nonzero tension—energy per length—and thus creates a linear potential between charges. In QCD with light quarks, those string can be broken if its length is sufficient for production of two heavy-light mesons.

Ken Wilson in the mid-1970s played key role in formulation of the nonperturbative definition of QCD-like theories on the lattice. He also promoted the statement about a linear potential to a more abstract mathematical form: the VEV of the Wilson line

$$\hat{W} = Tr P exp\left(i \int_C dx_\mu \hat{A}_\mu\right)$$

over some closed contour C of sufficiently large size with (the matrix valued) color gauge field. $Pexp$ means product of exponents along a given contour C. Its VEV should behave as follows

$$< W >= e^{-\sigma * Area}$$

the Area means to be of a surface enclosed by the contour C. If it is a rectangular contour $T * L$ in 0-1 plane, it indeed corresponds to the $area = T * L$ and σ is identified with the string tension. This statement is known as the so-called Wilson's confinement criterium. The main achievement of the very first numerical lattice studies, by M. Creutz in 1980, was demonstration of the validity of the area law, both in the strong and weak coupling settings.

What kind of gauge field configurations may lead to such "area law"? This question focused attention to the configurations with a nontrivial topology, known as *vortices*. Quantized vortices in liquid helium and superconductors are well known, and they are characterized by the fact that integral $\int_C dx_\mu A_\mu$ over any contour keep the same value.

The center $Z_n \in SU(N)$ is defined as the set of elements with commute with each group member: those for $n = 0...N - 1$ are

$$z_n = e^{i2\pi \frac{n}{N}}$$

Note that $(z_n)^N = 1$. In the simplest non-Abelian group with $N = 2$, there are two elements $z_0 = 1$, $z_1 = -1$ with squares equal to 1.

In the gauge theory people looked at the so-called *center vortices* for which the circulation integral around them is z_n. Let us in particular focus on $z_1 = -1$ element in the $N = 2$ case: each time such vortex pierces the Wilson line, there is a sign

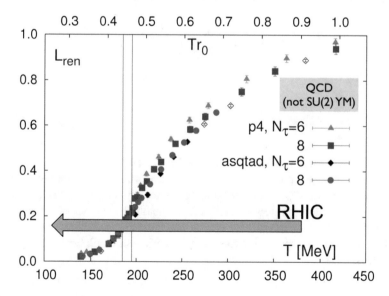

Fig. 1.2 Lattice data on the average value of the renormalized Polyakov line $L_{\text{ren}} = P$, as a function of the temperature T in QCD. Different points correspond to different lattice actions. Two vertical lines indicate location of the critical point, following from studies of the thermodynamical observables. The yellow arrow indicates the range of the temperatures studied at Relativistic Heavy Ion Collider (RHIC)

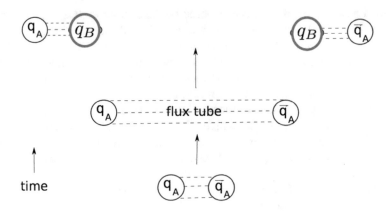

Fig. 1.3 A sketch of a heavy-heavy evolution, from small distances (bottom) to broken flux tube and formation of two heavy-light meson excitations (top)

change. Note that in four-dimensional linkage of the 2-d Wilson line with the 2-d vortex line history is a topological concept. Note also, that since one thinks about W in 0-1 plane, the 2-d vortex should be extended in the dual 2-3 plane.

Now, if there is a certain ensemble of center vortices, the area law follows. Suppose their locations is random and n of them are linked with smaller Wilson line with area A. The probability to have n piercing is

$$P(n) = C_N^n \left(\frac{A}{L^2} \right)^n \left(1 - \frac{A}{L^2} \right)^{N-n}$$

and

$$W = \sum_n (-1)^n P_n = \left(1 - \frac{2A}{L^2} \right)^N \to_{(N,L\to\infty, \rho=N/L^2=fixed)} = e^{-2\rho A}$$

The argument in this form is due to Langfeld et al. (1999).

If it is sufficiently dense, as lattice studies had shown, one obtains nearly all experimental value of the string tension σ. Removing center vortices from lattice gauge field configurations leads to zero string tension: thus the so-called dominance of center vortices claimed.

More details about the center vortices as the origin of confinement can be found in (Greensite 2003).

1.4.3 Hadronic Matter at $T < T_c$ and the Hagedorn Phenomenon

Thermodynamics is normally derived from a statistical sum over physical excited states of the system

$$Z = e^{-F(T)/T} = \sum_n \exp(-E_n/T) = \int \frac{d^3 p V}{(2\pi)^3} dM \rho(M) \exp\left(-\sqrt{p^2 + M^2}/T\right)$$

where we introduced the spectral density $\rho(M)$ of hadronic masses. Confinement in QCD-like theories is often stated as *the absence of all colored states from the physical spectrum*. All excited states are colorless hadrons: mesons $\bar{q}q$, baryons qqq, recently found $\bar{q}\bar{q}qq$ tetraquarks (with heavy quarks), etc.

The chiral symmetry is the property of the theories with massless quarks. Crudely speaking, it means that the left- and right-handed polarizations of the quark fields are independent of each other and can be gauge rotated separately. One may wander how the small quark masses in QCD Lagrangian can be seen in a hadronic framework. In fact they can: via *massless* pions, the Goldstone modes of spontaneously broken chiral symmetry.

One may also wonder if the phenomenon of the *deconfinement* can be really expressed in hadronic framework. Yes it can: via the so-called Hagedorn phenomenon. Hagedorn noticed that the spectral density of hadronic masses grows very rapidly, approximately exponentially

$$\rho(M) \sim e^{M/T_H}$$

and as a result when T approaches T_H, the statistical sum Z diverges due to proliferation of many states.

The reason why $\rho(M)$ grows so fast is very nontrivial. Already in the 1960s, hadronic phenomenology—the Regge trajectories and Veneziano scattering amplitudes—could be explained by so-called *QCD strings* or flux tubes, connecting the quarks. Important observation is that strings have much more states (configurations) than particles, illustrated in Fig. 1.4.

(Attempts to create *effective theory of the QCD strings* of course later lead to the appearance of the string theory, which with time switched to much larger space-

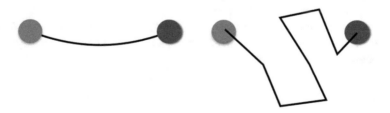

Fig. 1.4 A sketch of a meson structure at low T (left) and high T (right), with a string excitation

time dimensions and claimed a high title of "theory of everything" including gravity.
Nobody has a clue if this is or is not true.)

1.5   Particle-Monopoles, Including the Real-Time (Minkowskian) Applications

How magnetic charges may coexist with quantum mechanics[16] was explained
by (Dirac 1931), who found that it may only happen when the electric and
magnetic charges satisfy a particular relation, which makes singular lines between
monopoles—the Dirac strings—*invisible*.

t'Hooft and Polyakov discovered monopole solution in Non-Abelian gauge
theories with scalars ('t Hooft 1974; Polyakov 1974). Existence of monopoles was
used in famous model of confinement, due to (Nambu 1974; Mandelstam 1976;
't Hooft 1978a) who argued that if the monopole density is large enough for their
Bose-Einstein condensation, the resulting "dual superconductor" will expel electric
color field via dual Meissner effect, creating electric flux tubes. Monopoles were
identified on the lattice, starting from the 1980s, their properties, spatial correlations,
and paths in time $x_m(\tau)$ analyzed. It was particularly observed that these monopoles
do indeed rotate around the flux tubes, producing solenoidal "magnetic current"
needed to stabilize the flux tubes. Their paths do indeed indicate their Bose-Einstein
condensation at $T < T_c$. Elimination of monopoles from lattice configurations
kills confinement: thus there are also papers on "monopole dominance" in the
confinement problem.

Note that since magnetic monopoles are the 3-d topological objects, they are
"particles." Although lattice simulations can only work with a Euclidean (imaginary) time, nothing prevents one to use monopoles in real-time applications. Such
applications included studies of the quark and gluon scatterings on monopoles,
significantly contributing to small value of kinetic coefficient (viscosity) of quark-
gluon plasma. Recently there was identified a contribution of the monopoles to jet
quenching and kinetic phenomena to be discussed below.

1.6   Instantons and Its Constituents, the Instanton-Dyons

Finally, non-Abelian gauge theories also have some 4-d solitons with nontrivial
topology (Belavin et al. 1975), known as *instantons*. They do not explain confinement in four dimensions, as their fields fall too quickly to generate a Wilson's
area law. But they induce important effects associated with light quarks: as we will
show below, they break the chiral symmetries.

---

[16]Which was just 4 years old then!

Fig. 1.5 Short and long loops in the fermionic determinant

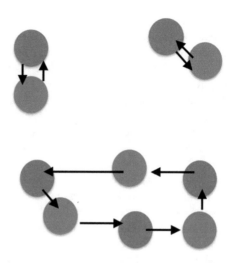

Instantons have fermionic zero modes, solutions of the Dirac equation in such fields with the zero r.h.s.

$$\gamma_\mu(i\partial_\mu + gA_\mu)\psi = 0$$

Because the contribution for a gauge field configuration to the partition function is proportional to the determinant of the Dirac operator, naively zero eigenvalues mean that instantons cannot appear in the ensemble.

This is indeed true for a single instanton; however, if there is an ensemble of them—the so-called *instanton liquid*—it is possible. The phenomenon can be described as *a collectivization* of these fermionic zero modes into the so-called *zero mode zone* (ZMZ) of quasi-zero Dirac eigenstates (Fig. 1.5).

If the instanton density is sufficiently large, ZMZ has states arbitrarily close to zero, which forms the so-called quark condensate. The ZMZ has been observed on the lattice and indeed shown to be made of linear superposition of zero modes of the individual instantons. Removing ZMZ—which constitutes only about 10^{-4} of all fermionic states—leads to effective restoration of the chiral symmetry and change in hadronic masses by large amount, typically 20–50%. This is also true for the nucleons, which mass is also that of all of us and of the whole visible matter—so it is a very important effect.

We will also discuss several other instanton-induced effects. One of them is pairing not in the $\bar{q}q$ channel, producing chiral symmetry breaking, but in the diquark qq channel, leading to color superconductivity. The celebrated Seiberg-Witten solution of $N=2$ supersymmetric gluodynamics is mostly a series of all order instanton effects, derived by the explicit calculation of all instanton amplitudes by N. Nekrasov in (2003).

Decreasing the temperature below $2T_c$ one finds a nontrivial average value of the Polyakov line $\langle P \rangle \neq 1$, indicating that an expectation value of the gauge potential is nonzero $\langle A_4 \rangle = v \neq 0$. This calls for redefining the boundary condition of A_4 at infinity, for all solitons including instantons. That lead to the 1998 discovery (by Kraan and van Baal, Lee and Li) that nonzero v splits instantons into N_c (number of colors) constituents, the self-dual *instanton-dyons*.[17] Since these objects have nonzero electric and magnetic charges and source Abelian (diagonal) massless gluons, the corresponding ensemble is an "instanton-dyon plasma," with long-range Coulomb-like forces between constituents. By tradition the $SU(2)$ self-dual ones are called M with charges $(e, m) = (+, +)$ and L with charges $(e, m) = (-, -)$, and the anti-self-dual antidyons are called $\bar{M}$, $(e, m) = (+, -)$ and $\bar{L}$, $(e, m) = (-, +)$.

Diakonov and collaborators emphasized that, unlike the (topologically protected) instantons, the dyons interact directly with the holonomy field. They suggested that since such dyon (antidyon) becomes denser at low temperature, their back reaction may overcome perturbative holonomy potential and drive it to its confining value, leading to vanishing of the mean Polyakov line, or confinement.

In order to study instanton-dyon plasma, one needs to know the dyon-antidyon interaction. This was recently achieved, and several works have been studied about the instanton-dyon plasma, both analytically in the mean field approximation and numerically, by a direct simulation.

It has indeed been confirmed that instanton-dyons in gauge theory lead to confining phase, provided their density is large enough. In QCD-like theories with light quarks, both deconfinement and chiral restoration transition happen at about the same dyon density. More recent studies focused on QCD deformations by fermion phases, which were found to modify both phase transitions drastically.

1.7   Interrelation of Various Topology Manifestations and the Generalized Phase Diagrams

The objective of this course can be defined as derivation of effective theories based on topological objects.

Various topological objects present in gauge field configurations are all interrelated. For example, intersection of two center vortices where they disappear are the monopoles: two fluxes with angle π each make one with flux 2π known as the Dirac string, ending on a magnetic monopole. Monopole paths may end at the instanton. Elimination of center vortices in lattice gauge configurations also eliminates monopoles, and elimination of the monopoles eliminates the instanton-dyons.

---

[17]They are called "instanton-monopoles" by Unsal et al. and are similar but not identical to "instanton quarks" discussed by Zhitnitsky et al.

If so, which objects should we study most? My answer is—from the top down, the instantons first—based on the following arguments. The instantons and their constituents have noticeable large action $S_{inst}/\hbar \sim O(10)$ therefore their effective theory based on semiclassical approximation can be self-consistently constructed. Furthermore, this effective theory has the form of "classical statistical mechanics," with integration over certain collective coordinates (positions of the instanton-dyons). For monopoles their effective theory would include path integral over their path, corresponding to "quantum manybody theory." For vortices one would need an analog of "quantum string theory," integrating over all of their world sheets: attempts to reformulate gauge theory as such were made, but basically abandoned.

Note, however, that even in the absence of consistent theory, one can still study effects of these objects qualitatively. Probing QFTs in various conditions produces different responses in its different versions, and comparing to phenomenology (real and lattice experiments), one may identify the best—if not unique—explanations. This is what we will do in this course a lot.

Infinitesimal probes with various quantum numbers excite corresponding elementary excitations. In QCD those are hadrons—mesons and baryons.

Yet we will not be interested that much in hadronic spectroscopy. We will focus on the study of the "vacuum structure." Generally speaking, those are revealed by a multitude of the vacuum correlation functions, with various local or nonlocal operators. Yet their detailed discussion will take us too far.

What we will discuss are vacuum expectation values (VEVs) of various fields or field combinations, known as "condensates." For example, we will discuss in detail how and under which conditions a nonzero VEV of the scalar bilinear of the quark field $\langle \bar{q}q \rangle \neq 0$ known as the quark condensate can be formed or whether it can coexist with color superconductivity in which another bilinear condensates $\langle qq \rangle \neq 0$.

Let me at the onset specify strategy we will follow. The vacuum state, with its condensates and strong coupling effects, is very complicated. To understand it gradually, it is convenient to start with high T setting, in which nonperturbative phenomena are power-suppressed. Monopoles and instantons appear as rare individual excitations. With decreasing scale T the coupling grows, and nonperturbative phenomena increase their presence, till finally they get dominant and completely reshape the statistical ensemble, creating the world we live in. Therefore we will focus at deconfinement and chiral transitions.

1.8   Which Quantum Field Theories Will We Discuss?

Many books and reviews focus on cataloging various QFTs, their properties, and interconnections in different dimensions. The main focus of this book is understanding of nonperturbative physical phenomena occurring *in our physical world*. Therefore we focus on only two ingredients of the standard model (SM):

(i) the *electroweak* one based on $SU(2)$ non-Abelian gauge theory plus quarks, leptons, and (Higgs) scalar;
(ii) Quantum chromodynamics (QCD) describing *strong* interactions, based on non-Abelian gauge theory with the $SU(3)$ color group and quarks.

The electroweak theory is in a weak coupling regime, and therefore most of its nonperturbative/topological effects are hard to address experimentally. We will only study electroweak sphalerons in conjunction with cosmological phase transition in sphaleron chapter.

Since in QCD the charge is "running" (rapidly changes) as a function of the momentum scale considered, from weak to rather strong regime, here nonperturbative phenomena are responsible for many phenomena we will discuss. In particular, the topological effects appear in its full glory, being our main focus below.

However, in order to understand these phenomena better, it would be sometimes desirable to consider not just physical QCD but also wider set of theories with and various settings of those, with certain *variable parameters*, and investigate how effects considered get modified.

The most obvious parameter is the number of colors N_c of the gauge group, and this is the first parameter of the theory which we will change whenever it is useful to do so. In some cases we will use the smallest value of it, $N_c = 2$, and in some case we would discuss the opposite limit of large $N_c \to \infty$.

Much less obvious is the parameter θ, in a CP-odd gauge theories with $\theta(\vec{E}\vec{B})$ term in the Lagrangian. We will also discuss in some cases the so-called *deformed* QCD, with added (gauge invariant) terms containing powers of the Polyakov loop: this will be needed to affect the confinement phenomenon.

Furthermore, one can consider the non-Abelian gauge theories possessing various *fermions*, different from the quarks we have in the real-world QCD.

Let us first remind how six known quark flavors are naturally divided in SM. The electroweak sector splits them in three dublets (ud), (cs), (tb).

In QCD we instead split them into three light (uds) and three heavy (cbt) ones. Putting the masses of three light to zero one finds *the chiral symmetry* $SU(3)_L \times SU(3)_R$ for left- and right-handed polarizations. Expansion in light quark masses is known as *chiral perturbation theory*.

Putting the masses of three heavy quarks to infinity $m_q \to \infty$, one finds the *heavy quark symmetry* (Shuryak 1982a; Isgur and Wise 1989), and expansion in $1/m_q$ is known as *heavy quark effective theory*.

Of course, one can add more quarks to QCD, for example, change the number of light quark flavors to arbitrary integer N_f, $f = u, d, s, \ldots$.

Another deformation or the quark fields to be considered is related with their periodicity on the Matsubara circle (in Euclidean time). The quarks, being fermions, are *antiperiodic* on this circle. If they instead assumed to be *periodic*, quarks effectively become bosons: this setting is used if one would like to preserve the supersymmetry.[18]

More generally, we will also use arbitrary periodicity phases, which also can be different for each flavor, θ_f. So these quarks would have intermediate statistics. As we will see, the deconfinement and chiral symmetry restoration phase transition are sensitive to the value of these phases, revealing the topological objects which cause them.

All of these theories can be generically called versions of *deformed QCD*, to be used in conjunction with our discussion of this or that phenomena when needed.

One should however carefully single out some of these deformations, the special cases in which there are certain *new symmetries*, absent in the real-world QCD. Those may have specific behavior which is *absent* in QCD. One example is two-color case, in which antiquarks can be related to quarks by Pauli-Gursey symmetry.

Much more important case are theories with several number of *supersymmetries* $\mathcal{N}$. They, as well as *supergravity* theories, occupy large portion of modern QFT literature, textbooks, and lecture notes in the last three decades. So, let me explain to what extent it will be discussed in this book.

As of the time of this writing, we see no evidences that supersymmetry exists in nature. We will also *not* focus on supersymmetry in technical sense, for example, we will not use superspace notations. And yet, we will turn to some supersymmetric theories which proved to be useful toy models. In many cases they provide valuable lessons, or even analytically derived expressions for some quantities, the analog of which in non-supersymmetric theories are only available from numerical simulations if at all.

To be specific, we will discuss the $\mathcal{N}=1$ super-gluodynamics in connection with the first application in chapter on the instanton-dyons: in this theory the calculation of the gluino condensate resolved some long-standing puzzle.

It would be impossible not to mention many fascinating aspects of the $\mathcal{N}=2$ (twice-supersymmetric Yang-Mills theory, two gluinoes, one scalar), also known as Seiberg-Witten theory, especially in connection with the issues of electric-magnetic duality in Chap. 2. Some minimal information about it is therefore given in Appendix. Conformal $\mathcal{N}=4$ theory (four gluinos, six scalars) leads to many more fascinating features, which we will partly cover in the next chapter and Appendix.

References

Belavin, A.A., Polyakov, A.M., Schwartz, A.S., Tyupkin, Yu.S.: Pseudoparticle solutions of the Yang-Mills equations. Phys. Lett. **B59**, 85–87 (1975). [350 (1975)]
Brodsky, S.J., de Teramond, G.F.: Hadronic spectra and light-front wavefunctions in holographic QCD. Phys. Rev. Lett. **96**, 201601 (2006)

[18]For this to be the case, one also would need either to change the color quark representation from the fundamental to *adjoint* one, in which case those will be called *gluinos* and the number of the types of those will be denoted by N_a, or introduce fundamental scalars or squarks.

Brodsky, S.J., Lepage, G.P.: Exclusive processes in quantum chromodynamics. Adv. Ser. Direct.
    High Energy Phys. **5**, 93–240 (1989)
Chernyak, V.L., Zhitnitsky, A.R.: Asymptotic behavior of exclusive processes in QCD. Phys. Rept.
    112, 173 (1984)
Dirac, P.A.M.: Quantized singularities in the electromagnetic field. Proc. Roy. Soc. Lond. **A133**,
    60–72 (1931). [278 (1931)]
Erlich, J., Katz, E., Son, D.T., Stephanov, M.A.: QCD and a holographic model of hadrons. Phys.
    Rev. Lett. **95**, 261602 (2005)
Feynman, R.P., Cline, J.M.: Feynman Lectures on the Strong Interactions (2020)
Greensite, J.: The confinement problem in lattice gauge theory. Prog. Part. Nucl. Phys. **51**, 1 (2003).
    hep-lat/0301023
Gross, D.J., Wilczek, F.: Asymptotically free gauge theories - I. Phys. Rev. **D8**, 3633–3652 (1973)
Hutauruk, P.T., Bentz, W., Cloet, I.C., Thomas, A.W.: Charge symmetry breaking effects in Pion
    and Kaon structure. Phys. Rev. C **97**(5), 055210 (2018)
Isgur, N., Wise, M.B.: Weak decays of heavy mesons in the static quark approximation. Phys.
    Lett. **B232**, 113–117 (1989)
Jarvinen, M., Kiritsis, E.: Holographic models for QCD in the veneziano limit. JHEP **03**, 002
    (2012)
Klevansky, S.: The Nambu-Jona-Lasinio model of quantum chromodynamics. Rev. Mod. Phys.
    64, 649–708 (1992)
Langfeld, K., Tennert, O., Engelhardt, M., Reinhardt, H.: Center vortices of Yang-Mills theory at
    finite temperatures. Phys. Lett. B **452**, 301 (1999)
Mandelstam, S.: Vortices and quark confinement in nonabelian gauge theories. Phys. Rept. **23**,
    245–249 (1976)
Nambu, Y.: Strings, monopoles and gauge fields. Phys. Rev. **D10**, 4262 (1974). [310 (1974)]
Nambu, Y., Jona-Lasinio, G.: Dynamical model of elementary particles based on an analogy with
    superconductivity. 1. Phys. Rev. **122**, 345–358 (1961). [127 (1961)]
Nekrasov, N.A.: Seiberg-Witten prepotential from instanton counting. Adv. Theor. Math. Phys.
    7(5), 831–864 (2003). hep-th/0206161
Politzer, H.D.: Reliable perturbative results for strong interactions? Phys. Rev. Lett. **30**, 1346–1349
    (1973). [274 (1973)]
Polyakov, A.M.: Particle spectrum in the quantum field theory. JETP Lett. **20**, 194–195 (1974)
Polyakov, A.M.: Quark confinement and topology of gauge groups. Nucl. Phys. **B120**, 429–458
    (1977)
Shuryak, E.V.: Hadrons containing a heavy quark and QCD sum rules. Nucl. Phys. **B198**, 83–101
    (1982a)
Shuryak, E.V.: Correlation functions in the QCD vacuum. Rev. Mod. Phys. **65**, 1–46 (1993)
't Hooft, G.: Magnetic monopoles in unified gauge theories. Nucl. Phys. **B79**, 276–284 (1974)
't Hooft, G.: On the phase transition towards permanent quark confinement. Nucl. Phys. **B138**,
    1–25 (1978a)
Wilson, K.G.: Confinement of quarks. Phys. Rev. **D10**, 2445–2459 (1974). [319 (1974)]
Yang, C.-N., Mills, R.L.: Conservation of isotopic spin and isotopic gauge invariance. Phys. Rev.
    96, 191–195 (1954). [150 (1954)]

Monopoles

2

> *"One would be surprised if Nature had made no use of it."*
>
> Dirac

2.1 Magnetic Monopoles in Electrodynamics

Not discussing the origin of electricity and magnetism in antiquity, let me jump to pre-Maxwellian nineteenth century, in which it was clearly stated that electric charge can be divided into positive and negative charges, while the magnets, if cut, still produce only dipoles, with the so-called north and south poles each. In other words, it was observed that it is not experimentally possible to separate magnetic charges. And, in well-known subsequent developments which lead to Maxwellian electrodynamics, magnetism is ascribed to motion of the electric charges. Yet by the end of that era some—notably J.J. Thomson and H. Poincare—were discussing possible existence of some hypothetical particles possessing a magnetic charge. Since then, this idea has a very unusual history: it tends to be dormant for two to three decades and then becomes very active and then dormant again. It did so at least 5 times, to my count.

Development of quantum mechanics brought in an issue of quantization. At that stage the greatest impetus to the whole problem has been provided by Dirac (1931) and subsequent works. He argued that the equation $\vec{\nabla}\vec{B} = 0$ can be consistent with nonzero magnetic charges if there are singular lines—the *Dirac strings*—supplying the magnetic flux from outside. And, under special Dirac quantization condition, the Dirac strings can be made invisible! Dirac stressed that this condition seems to be the only known reason explaining why all electric charges are *quantized*.

An introductory review on electromagnetic magnetic monopoles is, e.g., that by Milton (2006); for an in-depth source on monopoles in non-Abelian theories, one may consult the book by Shnir (2005b). Multiple searches for QED magnetic monopoles have produced no convincing candidate events. An argument why this

© The Author(s), under exclusive license to Springer Nature Switzerland AG 2021
E. Shuryak, *Nonperturbative Topological Phenomena in QCD and Related Theories*, Lecture Notes in Physics 977,
https://doi.org/10.1007/978-3-030-62990-8_2

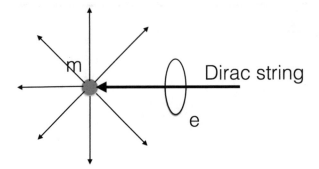

Fig. 2.1 The magnetic monopole with the Dirac string (thick line) which carries the ingoing magnetic flux, equal to the total flux carried out by Coulomb-like field (thin lines with arrows). The circle around the Dirac string is a path of an electron e moving around it

can be the case follows from the Dirac condition itself. Since the electric fine structure constant[1] $\alpha = e^2 \approx 1/137 \ll 1$ is very small, the magnetic one should be very large $g^2 \sim 137 \gg 1$: perhaps at such strong magnetic coupling, any separation of the charges is problematic.

The gauge field configuration with a Dirac string is sketched in Fig. 2.1. What one would like to obtain is a Coulomb-like magnetic field

$$\vec{B} = g \frac{\vec{r}}{r^3} = \vec{\nabla} \times \vec{A} \qquad (2.1)$$

from some vector potential configuration. Here is one which does so

$$\vec{A} = g \left(sin(\phi) \frac{1 + cos(\theta)}{rsin(\theta)}, -cos(\phi) \frac{1 + cos(\theta)}{rsin(\theta)}, 0 \right) \qquad (2.2)$$

where r, ϕ, θ are spherical coordinates. Note that for a half-line $\theta = 0$, the numerator is 2, and thus the $\vec{A}$ is singular, but for a half-line $\theta = \pi$, it is zero of higher degree, and it is in fact zero. A more general form of the *Dirac potential* is

$$\vec{A} = \frac{g}{r} \frac{[\vec{r} \times \vec{n}]}{r - (\vec{r}\vec{n})} \qquad (2.3)$$

where now $\vec{n}$ is a unit vector pointing in any direction we like.

The question however remains whether the Dirac string is or is not visible in any physical experiment. The answer to it goes back to the so-called Aharonov–Bohm

---

[1]This section is written in QED notations in which the Coulomb field is $\vec{E} = e\vec{r}/r^3$. In QCD notations, to be used elsewhere, the field and charge normalization changes, so that $\vec{E} = e\vec{r}/4\pi r^3$.

Fig. 2.2 J.J. Thomson: static
electric and magnetic charges
create a rotating field, as
indicated by the rotating
Poynting vector $\vec{S} = [\vec{E} \times \vec{B}]$

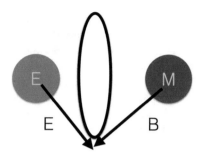

effect (Aharonov and Bohm 1959). An electron making some closed loop C around the string will pick up a phase $phase = e \int_C A_\mu dx_\mu$. For some thin solenoid,[2] this phase can take any value, and be observable, even if the electron is never allowed into the region where $\vec{B} \neq 0$. Plugging in the Dirac potential, one finds that the phase is

$$phase = 4\pi eg$$

In order for the Dirac string to be invisible, the phase needs to be $2\pi n$, and so one thus needs to enforce the following *Dirac quantization condition*[3]

$$eg = \frac{n}{2} \qquad (2.4)$$

There exists also another instructive way toward understanding the Dirac condition. In 1904, J. J. Thomson[4] observed that even *static* (non-moving) charges, electric e, and magnetic g create a rotating electromagnetic field. Indeed, two Coulomb fields from such charges meet at a generic point at some angle (see Fig. 2.2) and thus create a nonzero Poynting vector

$$\vec{S} = [\vec{E} \times \vec{B}] \neq 0$$

circling around the line connecting two charges. Thus, the field is rotating, even while the charges themselves are not moving!

---

[2]The AB effect has even been observed for quantized vortices in a superconductor, in spite of the fact that quantization means periodic wave function of the condensate. That is possible because the Cooper pairs of the condensate have electric charge 2e: so if their loop results in an "invisible phase" $2\pi n$, a single electron gets half of the phase πn which is visible for odd n.

[3]Note that in the sections below, with different normalizations of the field and couplings, the extra factor $1/4\pi$ will appear in the l.h.s.

[4]J. J. Thomson was the recipient of the 1906 Nobel Prize in Physics, for measuring the charge-to-mass ratio of the electron, effectively discovering the first elementary particle known. His other high distinction was that seven (!) of his students, including his son, also became Nobel Prize winners.

If one or all charges are allowed to move, the existence of the field angular momentum dramatically changes their trajectories, since only the *total* angular momentum—of the particles and the field combined—is conserved. We will discuss those changes later in the book: for now it is enough to say that this observation would be the key to understanding of properties of the quark–gluon plasma.

In quantum context, the angular momentum carried by the field must be quantized, as usual, to an integer times $\hbar$. The consequences of that are explained in the following exercise:

Exercise Calculate the angular momentum J, and show that its quantization just mentioned leads to the Dirac quantization condition of the charges.

2.2   The Non-Abelian Gauge Fields and t' Hooft–Polyakov Monopole

The solution, found independently by 't Hooft (1974) and Polyakov (1974), implements these ideas in the setting with non-Abelian gauge theory. What is however required for success of the program is an ingredient (which QCD-like theories do *not* possess), namely, *a scalar field in the adjoint color representation ϕ.*

Here is the Lagrangian of the so-called Georgi–Glashow (GG) model

$$L = -\frac{1}{4}\left(G_{\mu\nu}^a\right)^2 + \frac{1}{2}(D_\mu\phi)^2 - \frac{\lambda}{4}\left(\phi^2 - v^2\right)^2 \qquad (2.5)$$

a direct descendant of Ginzburg–Landau free energy of a superconductor. It differs from the electroweak sector of the standard model, the Weinberg–Salam model, exactly by the fact that ϕ has adjoint color representation. Nonzero VEV, shown as v, provides different "Higgsing" of the gauge fields: their mass is proportional to the commutator of their color generator with that of the VEV. In this model the $SU(N_c)$ color group is split by nonzero v into its $N_c - 1$ diagonal subgroups.

The simplest case (we will only discuss) is the $N_c = 2$ gauge theory. The corresponding algebra has three generators T^A, $A = 1, 2, 3$. Since we will be dealing with $N_c = 2$ QCD also, some explanation of the notations on color representations are in order.

In the $N_c = 2$ QCD, we will be dealing with quarks, which are in *fundamental* (or spinor) representation of the $SU(2)$ group. This means that quark color indices run over $a = 1, 2$ and color generators are $T^A = \tau^A/2$, where τ^A are three Pauli matrices familiar from quantum mechanics description of spin.[5]

---

[5]This is not surprising, since there is a relation between the $SU(2)$ and rotational $O(3)$ groups.

In the GG model, the scalar is in *adjoint* (or vector) representation. This means that scalar color indices run over $a = 1, 2, 3$. The color matrices are then given directly by group structure tensor, which in this group is simply $(T^A)_{bc} = \epsilon_{Abc}$.

Let us take the nonzero VEV to be along the diagonal, so $< \phi^3 > = v$. Then two gauge bosons ($W^{+,-}$) get nonzero masses, while the boson number 3 (neutral "photon") remains massless:[6] this field is called a "photon."

Since there is such a drastic difference between those components, one may like to introduce a special notations for the Abelian-projected fields (without color indices).

$$A_\mu = A_\mu^a \hat{\phi}^a, \quad \hat{\phi}^a = \frac{\phi^a}{|\phi^a|} \tag{2.6}$$

where we introduced a unit vector indicated by a hat. In order to define also the Abelian field strength, the field definition should not be just the usual Abelian expression based on A_μ because it should be supplemented by a term canceling possible derivatives of the Higgs color direction. Its definition is

$$F_{\mu\nu} = \partial_\mu A_\nu - \partial_\nu A_\mu - \frac{1}{e} \epsilon_{abc} \hat{\phi}^a \partial_\mu \hat{\phi}^b \partial_\nu \hat{\phi}^c. \tag{2.7}$$

This last term is of course zero for constant Higgs field. Furthermore, in fact it vanishes for all topologically trivial configurations of $\phi^a(x)$: it will be nonzero only for topologically nontrivial ones, and the possibility to have a gauge in which there is no Dirac string is based on this observation.

The *magnetic current* can now be defined from the definitions given above

$$k_\mu = \partial_\nu \tilde{F}_{\mu\nu} = \epsilon^{\mu\nu\rho\sigma} \epsilon_{abc} \partial_\nu \phi^a \partial_\rho \phi^b \partial_\sigma \phi^c \left(\frac{1}{2v^3 e} \right) \tag{2.8}$$

Here and in many occasions below, we will use tilde for 4-d dual field, given by application of the 4-index epsilon tensor.[7]

Unlike the usual Nether currents, this magnetic current k^μ is conserved by definition, without any underlying symmetry. The integral of its density is known in mathematics as the *Brouwer degree*. As any other topological quantity, it gives in appropriate normalization an integer, which defined topologically distinct Higgs field.

---

[6]The Georgi–Glashow model was designed to avoid existence of the Z boson, then unknown. Experiments eventually had shown that it is the Weinberg–Salam version, with fundamental representation of the scalar, which describes the weak interactions of quarks and leptons: it is now known as *The Standard Model*. Glashow was still, quite correctly, awarded the Nobel Prize.

[7]Note that magnetic field is dual to electric one, and "selfdual" fields we will be discussing later have equal electric and magnetic fields.

How it may happen is clear from an example of a "hedgehog"-like field[8]

$$\phi^a(r \to \infty) \to v \frac{r^a}{r} \qquad (2.9)$$

in which the "needles" go radially: the magnetic charge for it is

$$g = \int d^3x k_0 = \frac{4\pi}{e} \qquad (2.10)$$

Let us now look for a solution consistent with that asymptotical trend, in terms of two spherically symmetric functions

$$\phi^a = \frac{r^a}{er^2} H(ver); \quad A_n^a = \epsilon_{amn} \frac{r^a}{er^2}[1 - K(ver)]; \quad A_0^a = 0 \qquad (2.11)$$

When those are plugged back into the expression for the Hamiltonian, one finds the following expression for the monopole mass

$$E = \frac{4\pi v}{e} \int_0^\infty \frac{d\xi}{\xi^2} \left[\xi^2 \dot{K}^2 + (1/2)(\xi \dot{H} - H)^2 \right.$$
$$\left. +(1/2) \left(K^2 - 1 \right)^2 + K^2 H^2 + \frac{\lambda}{4e^2} \left(H^2 - \xi^2 \right)^2 \right] \qquad (2.12)$$

Here we rescaled the radial coordinate $\xi = evr$, and the derivative over ξ is denoted by a dot. This expression can be minimized by variational methods. The equations obtained can be viewed as corresponding to some classical motion in the K, H plane of a particle, with ξ being the time. The standard way to get equations of motion is as described in Classical Mechanics courses.

The boundary conditions at small r correspond to $H \to 0$ and $K \to 1$ as then the Higgs field is smooth in spite of the "hedgehog" direction. At large distances $H \to \xi$, and Higgs becomes of the magnitude v, while $K \to 0$.

At large distance only Abelan part of the field survives, as it is massless. The obtained soliton happens to be a Dirac monopole, with the magnetic charge

$$g = \frac{4\pi}{e} \qquad (2.13)$$

Note that it is indeed consistent with the Dirac quantization condition. In fact we prefer to write it later in a more symmetric form, using electric and magnetic "fine

---

[8]The relation between such field configuration and this cute animal—unfortunately absent in America—has appeared for the first time (to my knowledge) in the Polyakov's paper.

Fig. 2.3 Classical solutions $K(\xi)$ and $H(\xi)$ for the 't Hooft–Polyakov monopole at $\lambda = 0$ (the BPS limit) and $\lambda = 1$

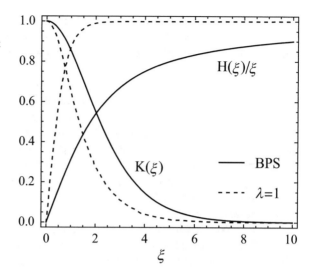

structure constants," namely,

$$\left(\frac{g^2}{4\pi}\right)\left(\frac{e^2}{4\pi}\right) = \alpha_e \alpha_m = 1 \tag{2.14}$$

which are then inverse to each other.

The solution for the EOM can generally be obtained numerically. Two of them are shown in Fig. 2.3 taken from Shnir (2005b). On general grounds, the monopole mass can be written as

$$M = \frac{4\pi v}{e} f\left(\frac{\lambda}{e^2}\right) \tag{2.15}$$

with smooth function f depending on the ration of two dimensionless couplings of the GG model. Some of its values are $f(0) = 1$ and $f(1) = 1.787$.

The special case $\lambda = 0$, in which scalar has no potential at all,[9] is the so-called Bogomolny–Prasad–Zommerfeld (BPS) limit. In this limit both profile functions are simplified and known analytically

$$K = \frac{\xi}{sinh(\xi)}, \quad H = \frac{\xi}{tanh(\xi)} - 1 \tag{2.16}$$

---

[9]At this point one invariably asks: if there is no potential and thus no minimum of it, how one can select the scalar VEV v? The answer is: all values of v are in this case possible. This means in the BPS limit all *nonequivalent* vacua with different v values may occur: there is no way to say which one is better than the other.

Exercise Using this definition of the BPS functions, calculate the Abelian (2.7) magnetic field for the BPS monopole. Observe the difference in r dependence between the color-diagonal and non-diagonal components.

Classical monopole solution has four symmetries: Three of them is a shift in the monopole position, and one is a gauge rotation $U = exp(i\alpha\tau^3)$ which leaves scalar's VEV unchanged. When quantum corrections are added to classical fields $A \rightarrow A_{cl} + a$, one finds that quadratic Lagrangian part $O(a^2)$ has four zero modes, corresponding to four directions in Hilbert space of all possible deformations a in which the action remains unchanged. Those zero modes, generated by symmetries, create significant problems in semiclassical theory: they should be taken out before calculation of quantum corrections. We will study that using the example of instantons later.

Solutions with the integer monopole number $M > 1$ obviously have $4M$ zero modes. The corresponding collective coordinate keeps the same simple meanings when one speaks about well-separated monopoles, but the situation changes when they overlap. The shapes of the solutions with $M > 1$ and the metric of the collective coordinates turned out to be a very nontrivial problem, involving high-power mathematics. For example, the $M = 2$ solution deforms from two monopoles into a doughnut, and for larger M there appear even more exotic shapes. The four-dimensional "moduli space" of relative collective coordinates called M_2, known as *Atiyah–Hitchin manifold*, has explicitly written metrics and very nontrivial geometry. Manton has further introduced a notion of "slow motion" in moduli space *along its geodesics*, in nice analogy to motion in general relativity. However, we will not have time to discuss those beautiful results. For a pedagogical introduction of them, the interested reader can consult (Shnir 2005b). Some key elements of it will be discussed later, in connection to self-dual gauge fields in connection to instantons. Two-monopole configurations, bound by fermions, will be discussed in Chap. 4.3 in connection to the so-called unusual confinement issue.

2.3   Polyakov's Confinement in Three Dimensions

This section belongs to this chapter only partly, since in it the monopole solution is not used as a particle, having paths in 4-d, but as an instanton (or pseudo-particle, as Polyakov calls it) of the three-dimensional setting. Still, it is a very famous application, from classic Polyakov's paper (Polyakov 1977),[10] touching on the confinement problem we will discuss a lot later.

---

[10] In fact this 1977 paper starts with instantons in the double well quantum-mechanical problem, which we discussed at the beginning of the book, and ends with acknowledgement to Gribov who suggested tunneling interpretation of them.

The part of this paper we will discuss here is a chapter on Georgi–Glashow (GG) model, which, after Higgsing, is renamed as "compact QED." The monopole is the solution to YM equations of motion we discussed in the monopole chapter. Now we however will not discuss particle–monopoles in $(1 + 3)$ dimensional space-time, as before, but consider this solution to be *the instanton* in Euclidean 3-d space.

Recall that generic GG model has two couplings, the gauge one e and the Higgs self-coupling called λ. The classical action for a set of monopoles is

$$S = \frac{M_W}{e^2} \epsilon \left(\frac{\lambda}{e^2} \right) \sum_a q_a^2 + \frac{\pi}{2e^2} \sum_{a \neq b} \frac{q_a q_b}{|x_a - x_b|} \tag{2.17}$$

where M_W is the mass of the non-diagonal gluons, called W by analogy to weak interaction bosons, and $\epsilon(\frac{\lambda}{e^2})$ defines the monopole mass. (Recall that for generic couplings, it can only be calculated numerically.)

The second term is the magnetic Coulomb interaction term, so the manybody problem one needs to solve is that of the 3-d Coulomb gas, with the partition function of the type

$$Z = \sum \frac{\xi^N}{N!} \int \left(\Pi_i d^3 x_i \right) e^{-\frac{\pi}{2e^2} \sum_{a \neq b} \frac{q_a q_b}{|x_a - x_b|}} \tag{2.18}$$

where fugacity $\xi \sim exp\left[-\frac{M_W}{e^2} \epsilon(\frac{\lambda}{e^2}) \right]$ is exponentially small when the monopole action is large, in weak coupling.

Polyakov uses standard mean field (or Debye) approximation, introducing a field χ coupled to charges

$$\int d\chi e^{-\frac{\pi e^2}{2} \int (\partial \chi)^2} \sum_N \sum_{q_a = \pm 1} \frac{\xi^N}{N!} \int \left(\Pi_i d^3 x_i \right) e^{i \sum q_a \chi(x_a)} \tag{2.19}$$

The physical meaning of χ is of course a gauge potential coupled to the charges. Note however that since the charges are magnetic ones, the gauge potential is "dual" to A_μ, and its gradients are not the electric but magnetic field.

Note that taking Gaussian integral in χ will return us to the original Coulomb gas. Instead we will keep χ as is, sum over two types of charges

$$\sum_{q_a = \pm 1} e^{i \sum q_a \chi(x_a)} = 2cos(\chi(x_a))$$

and exponentiate the series in monopoles.

$$Z \sim \int D\chi e^{-\frac{\pi e^2}{2} \int d^3 x \left[(\partial \chi)^2 - \frac{4\xi}{\pi e^2} cos(\chi) \right]} \tag{2.20}$$

Expanding the cosine, one finds that the (exponentially small!) coefficient in front of it is basically the mass squared of the field χ.

Now, the magnetic potential (and thus field) was the last massless field left after Higgsing: now it is also gapped. Polyakov showed that the usual criteria of confinement—like the area law of the Wilson loop—hold, and so 3-d GG theory is confining. Let me add that the mean field criterium is that there are many particles in the Debye cloud $nM^{-3} \gg 1$: it is satisfied in weak coupling because M is exponentially small.

Polyakov then goes on to the 4-d instantons, only to find that their interaction law is not Coulombic ($1/r^2$ in 4-d) but $1/r^4$ or short range, thus no Debye screening and no confinement by the same mechanism.[11]

2.4   Electric–Magnetic Duality

The term "duality" generally means that the same theory may have very different effective descriptions, depending on the dynamical regime in question. For example, QCD—a theory of colored quarks and gluons—has a dual low-energy description in terms of the so-called chiral effective Lagrangian, describing interaction of the lightest particles, the pions and other Goldstone mesons.

Electric–magnetic duality is basically a similar arrangement: depending which particles are the lightest one, effective low-energy description should use appropriate degrees of freedom. Above we discussed magnetic monopoles in a weak coupling regime $e^2/4\pi \ll 1$, in which they are heavy solitons with a large mass $M \sim v/e^2$. The previous section defined 't Hooft–Polyakov monopole as a classical solution. The reader perhaps expect that we will now develop a semiclassical theory of them, with small fluctuations around classical solutions included in 1,2, 3, etc. number of loops, like we did in the previous chapter for quantum mechanical extreme paths. And indeed, one can do so: we will in particularly return to the issue of fermions coupled to the monopoles in one-loop approximation below.

Let us however for now focus on the following question: since the coupling "runs" as a function of scale, becoming stronger as momentum scale decreases, one may ask *what happens with the monopoles when $e^2/4\pi \sim 1$ or even become large?*

Due to classic work (Seiberg and Witten 1994a), we know what happens with monopoles in the $\mathcal{N} = 2$ supersymmetric theories (pure gauge or with quark–squark multiplets. They were able to make a "quantum leap" over all such steps, to *exact* results. Not only do those include any number of loops in perturbation theory, but any number of instantons as well. We will not discuss how they figured out the answer—for this one has to read their original paper—but simply present the results.

---

[11]I remember how disappointed he was. Four years later, I came to see him and told that his instantons beautifully solve another famous QCD problem, the chiral symmetry breaking. Polyakov's answer was "but I have not invented them for that."

Fig. 2.4 The map of the moduli space according to Seiberg–Witten solution. The axes are real and imaginary parts of u (2.22)

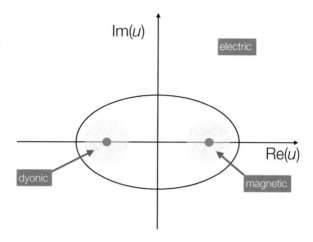

The theory in question is the simplest supersymmetric theory which *needs* to include a complex scalar field in the adjoint representation, like the Georgi–Glashow model discussed above. The $\mathcal{N} = 2$ gluodynamics or super-Yang-Mills (SYM) theory has gluons (spin 1), two real gluinoes λ, χ (spin 1/2), and a complex scalar (spin 0) which we will call a. Each of them has two degrees of freedom, thus 4 bosonic and 4 fermionic ones.

The $\mathcal{N} = 2$ QCD is a theory with additional matter supermultiplets of structure ψ_f, ϕ_f with spin 1/2 and 0, respectively. We will call N_f the number of Dirac quarks, as in QCD, or $2N_f$ Majorana ones. The **coupling renormalization** in these theories is done only via the one-loop beta function, with the one-loop coefficient of the beta function equal to[12]

$$b = 2N_c - N_f \tag{2.21}$$

while in the two-loop and higher orders, all coefficients of the beta function are zeroes. We will only consider the simplest case with two colors $N_c = 2$. Note further that if $N_f = 4$, this version of supersymmetric QCD has *zero* beta function and is therefore a scale-invariant (in fact conformal) theory: we will however not discuss it.

Let us return to the basic supersymmetric gluodynamics, $N_f = 0, N_c = 2$ or $\mathcal{N} = 2$ SYM. The only complex scalar of this theory may have a nonzero VEV, denoted by some complex number $v = \langle \phi \rangle$. Supersymmetry requires that there is no potential, and thus one has the BPS limit we discussed already. The set of all possible v fills a complex plane, which is in general called the *moduli space* of all possible vacua of the theory. A schematic plot of this space is given in Fig. 2.4.

---

[12]Note that unlike the same coefficient for non-sypersymmetric theories, it is strictly integer. The explanation for that is given in the instanton chapter where it will be explicitly derived; see the section on NSVZ beta function.

Using another variable

$$u = < tr(\phi)^2 > \qquad (2.22)$$

independent of the particular direction of the VEV.

The setting allows for monopoles, as the scalar is in the adjoint color representaion. Supersymmetry does not allow any self-interaction potential for scalars, so (in Georgi–Glashow notations) the coupling $\lambda = 0$, so that the monopoles are automatically in the BPS limit (for which we gave above the explicit classical solution).

All properties of the system are expressed as derivatives of one fundamental holomorphic function $F(u)$; in particular the effective charge and the theta angle are combined into a variable τ which is given by its second derivative

$$\tau(u) = \frac{\theta}{2\pi} + \frac{4\pi i}{e^2(u)} = \frac{\partial^2 F(u)}{\partial u^2} \qquad (2.23)$$

We will return to this function later, in connection with its instanton-based description.

The map of the moduli space is schematically given in Fig. 2.4 There are three distinct patches on the v plane:

(1) At large values of $v \rightarrow \infty$, there is a "perturbative patch," in which the coupling $e^2(v)/4\pi \ll 1$ is weak. It is dominated by electric particles—gluons, gluinoes, and higgses—with small masses $O(ev)$, which determine the beta function. Monopoles have large masses $4\pi v/e^2(v)$ there and can be treated semiclassically.
(2) A "magnetic patch" around the $Re(u) = \Lambda^2$ point, in which the coupling is infinitely strong $e^2 \rightarrow \infty$; the monopole mass goes to zero as well as the magnetic charge $g \sim 1/e \rightarrow 0$.
(3) A "dyonic patch" around the $Re(u) = -\Lambda^2$ point, in which a dyon (particle with electric and magnetic charges both being 1) gets massless.

Let us, for definiteness, follow development along the real axes, from right to left. At large $|v|$, the coupling is weak $e^2(v)/4\pi \ll 1$, and the effective "electric" theory resembles the electroweak sector of the standard model. The lightest particles are gauge bosons (W's) and their superpartners, with small masses induced by Higgsing $\sim ev$.

In the intermediate region, indicated by an oval on the plot, both the electric and magnetic couplings are $O(1)$ and thus comparable. Physics here is very complicated. For example, non-BPS-bound states appear and disappear there, both electric, magnetic, and dyonic. Positronium-like-bound state of a monopole with anti-dyon, with magnetic charge zero and electric charge one, can mix with electric states: so the classification of even the lowest excitations gets quite complicated. What is however remarkable is that, unlike in the case of non-supersymmetric

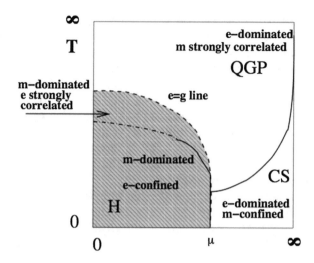

Fig. 2.5 A schematic phase diagram on a ("compactified") plane of temperature and baryonic chemical potential $T - \mu$. The (blue) shaded region shows "magnetically dominated" region $g < e$, which includes the e-confined hadronic phase as well as "postconfined" part of the QGP domain. Light region includes "electrically dominated" part of QGP and also color superconductivity (CS) region, which has electrically charged diquark condensates and therefore obviously m-confined. The dashed line called the "$e = g$ line" is the line of electric–magnetic equilibrium. The solid lines indicate true phase transitions, while the dash-dotted line is a deconfinement crossover line

theories, supersymmetry prevents any phase transitions on the oval. Exact solution tells us how gradual RG flow of the charge leads to a smooth transition.

Reaching the "magnetic" patches on the plot, one finds that the lightest degrees of freedom are now monopoles. Obviously here they cannot be treated by semiclassical approach anymore: but one can easily formulate "dual" magnetic theory of them, Abelian supersymmetric electrodynamics. The beta function in that region is indeed in agreement with such *weakly coupled magnetic theory*, and it has the opposite sign, as indeed required by the Dirac condition.

We will now terminate our discussion of the $\mathcal{N} = 2$ supersymmetric theories[13] and return to the real world where (so far?) supersymmetry is absent.

The topic will be a qualitative map of the phase diagram of QCD-like theories with finite temperature and chemical potential following the so-called magnetic scenario (Liao and Shuryak 2007) aiming at the description of quark–gluon plasma (QGP) near the phase boundary. The arguments of this paper are based on two very generic "pillars": (1) the direction of the RG flow and (2) the Dirac quantization condition. Their combination requires the "magnetic coupling" (denoted by g in this chapter) to run in the direction opposite to e, thus becoming weak in the IR.

---

[13]We will return to it few times later. In this chapter we will discuss how fermions are bound to monopoles.

In Fig. 2.5 from that paper, one finds the schematic phase diagram on a ("compactified") plane of temperature and baryonic chemical potential $T - \mu$. It resembles the upper right part of the plot Fig. 2.4 in that along its periphery the electric coupling is weak and theory is "e-dominated." Its physics is weakly coupled QGP, or wQGP for short, described by the QCD perturbation theory.

The near-circular line marked "$e = g$ line" is analogous to the oval in Fig. 2.4: here electric and magnetic quasiparticles become of comparable mass, their interactions are all strong, and no simple effective description of it is possible. We now know from heavy ion experiments that in this region QGP is a near-perfect fluid, with very small mean free path of all constituents, also known as sQGP. We will discuss its phenomenology a bit later.

The main difference between the supersymmetric world (and Fig. 2.4) with the real one (and Fig. 2.5) is that in the latter case, there are phase transitions. The blue-shaded region is the region of confinement. On the plot it is called "m-dominated" since in this phase, the color-electric field is expelled, into the flux tubes.

2.5   Lattice Monopoles in QCD-like Theories

QCD-like theories have gauge fields and fermions, but there are *no adjoint scalars*. Therefore, strictly speaking, there are no 't Hooft–Polyakov magnetic monopoles available in those theories.

Yet both the pure gauge theory and QCD certainly do possess *some* quantities which are in adjoint color representation. One option out is to use A_4 (the fourth component of the adjoint vector field) instead of a scalar. At non-zero temperature, Lorentz symmetry is broken anyway, and, as already mentioned in the Introduction, it does have a nonzero VEV (Polyakov loops). This option leads to the so-called *instanton-dyons*, which we will discuss later in a separate chapter. However, this option would not work outside the Euclidean theory: attempts to return to Minkowski time would also transform $A_4 \rightarrow i A_4$ which would ruin a would-be monopole solution.

Another option is to perform lattice simulations with the gauge field and then look for magnetic monopoles in the configurations of the ensemble. A motivation for that, e.g., following from the example of the $\mathcal{N} = 2$ supersymmetric gauge theory, is that in the infrared, the effective theory should be something resembling magnetic Abelian Higgs model. 't Hooft in particular argued that in trying to identify physical degrees of freedom from pure gauge ones, one would inevitably locate monopole-like singularities on the lattice. We already know that Dirac strings are not observable: but their endpoints are as 3-cubes in which suddenly magnetic flux appears.[14]

---

[14]In one talk I depicted monopoles as some dogs on a leash. While the leash (the Dirac string) is unphysical and thus invisible, the corresponding dog's collars (at leash end) can be detected.

Suppose we select some "composite" (not present in the Lagrangian) adjoint quantity X gauge transforming like

$$X(x)-> U(x)X(x)U^{-1}(X)$$

(Examples: a Polyakov line $\hat{P}$, or some component of the field strength like F_{12}^{ij}, or quark bilinear $\bar{\psi}^i \psi^j$ but not A_μ^{ij} as its gauge transform is different.) One can go to a gauge in which X is diagonal. This separates non-diagonal "charged" gluons from diagonal "neutral photons." We already discussed this for the Polyakov line before.

The operator X is local, and its eigenvalues depend on the point x_μ. 't Hooft argued that the locations at which these eigenvalues cross lead to singularities of this gauge fixing procedure, characteristic for the monopoles. The procedure depends on our (rather arbitrary) selection of X.

Separating the gauge field into diagonal "photonic" and non-diagonal "W" fields (with notation reminding the Georgi–Glashow model)

$$A_\mu = a_\mu + W_\mu \qquad (2.24)$$

one can further define the so-called *maximal abelian gauge* (MAG) in which the local gauge rotations are chosen in such a way as to maximize the following functional

$$F_{MAG} = \sum_{\mu,n} ReTr\left[U_\mu(n)\tau^3 U_\mu^+(n)\tau^3 \right] \qquad (2.25)$$

This gauge is widely used by lattice practitioners. Another widely used option is the so-called Polyakov gauge[15] in which the local value of the Polyakov line is used to define the abelian subgroup.

Looking at the "abelian-projected" part of the field strength (calculated from a_μ without the commutator), one indeed finds what is known as "lattice monopoles." Some 3-cubes have abelian magnetic fluxes through its surface, and those cubes make continuous paths in 4-d. The dimensionless density of monopoles in the $SU(2)$ pure gauge theory at finite temperatures is shown in Fig. 2.6 from D'Alessandro and D'Elia (2008). As expected, the dimensionless density is small at high temperature but grows rapidly toward T_c. However the best fit to these data are not the inverse power[16] but the inverse power of the log

$$\frac{\rho(T)}{T^3} \sim \frac{1}{log(T/T_c)^2}$$

---

[15] Note however that the Polyakov line is a function of 3-d coordinate, while the maximal abelian gauge is defined locally for each link in 4-d.

[16] This suggests that the effective action of a monopole is not $O(1/g^2)$ but rather $O(log(1/g^2))$, in good agreement with the Poisson duality discussion in Chap. 16.

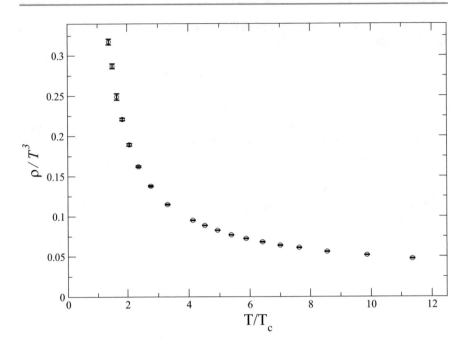

Fig. 2.6 The normalized monopole density ρ/T^3 for the $SU(2)$ pure gauge theory as a function of the temperature, in units of the critical temperature T/T_c, above the deconfinement transition

The question is whether such lattice monopoles *do or do not reflect properties of some real excitations of the system*. In fact lattice monopoles—elementary cubes of size a^3 with a magnetic flux through their surface—are nothing but the *endpoints* of the Dirac strings. Obviously the Dirac string themselves are gauge artifacts; in different gauges they can go in different directions, etc. In the continuum limit $a \to 0$, physical monopoles are expected to have a finite size, and the question is whether lattice singular monopoles do or do not correlate with the physical ones.

It was shown by Laursen and Schierholz (1988) that the lattice monopoles do correlate strongly with such gauge-invariant quantities as squared magnetic field and action. The lattice monopoles do rotate around electric flux tubes, creating the magnetic "coil" stabilizing them (Koma et al. 2003). Eliminating abelian-projected monopoles from lattice configurations does kill confinement (Suzuki et al. 2009), while keeping *only* monopoles produce nearly all string tension.

We will not follow these arguments here, but switch to another argument, by Liao and Shuryak (2008b), that lattice monopoles are real physical objects: they do display correlations expected for Coulombic magnetic plasmas. And, last but not the least, we will see that they do respect the famous Dirac condition!

In Fig. 2.7(left) from Liao and Shuryak (2008b), one sees two examples of the monopole–monopole and monopole–antimonopole correlators, as a function of distance between them, calculated from paths of the lattice monopoles by

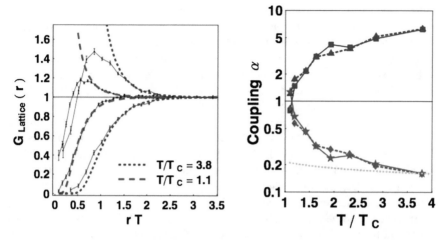

Fig. 2.7 (Left) Monopole–antimonopole (the upper two curves) and monopole–monopole (the lower two curves) correlators at $T = 1.1T_c$ (red long dashed) and $3.8T_c$ (blue dot dashed): points with error bars are lattice data, and the dashed lines are used for fits of the coupling strength. (Right) The magnetic coupling g (on Log10 scale) versus T/T_c fitted from the monopole–antimonopole (boxes with solid blue curve) and monopole–monopole (triangles with dashed blue curve) correlators. Their inverse, the corresponding to $\alpha_e = e^2/4\pi$ from the Dirac condition, are shown as stars with solid red curve and diamonds with dashed red curve, respectively, together with an asymptotic freedom (green dotted) curve

D'Alessandro and D'Elia (2008). Positive correlations for monopole–antimonopole correspond to attraction and negative ones for monopole–monopole pair to repulsion. The shape of the correlator is exactly what one expects in a Coulomb plasma of charges. The dashed lines are fits to the part of the correlators where the effect is small and can be treated by a linearized Debye theory: such fits produce values of the *effective magnetic coupling* $g^2/4\pi = \alpha_m$.

In Fig. 2.7(right) from Liao and Shuryak (2008b), the fitted couplings are plotted versus the temperature. As one can see, they indeed run *opposite* to the asymptotic freedom, becoming stronger at high T. Furthermore, its reflection (the bottom of the plot) is in qualitative agreement with the perturbative asymptotic freedom formula.

As one can also realize from these plots, by $T = T_c$, magnetic coupling decreases only to become $\alpha_m \approx 1$, not yet small. This means that the *magnetic component of sQGP is also a liquid*—the title of Liao and Shuryak (2008b). If it would be otherwise, monopoles would have large mean free paths, in contradiction to heavy ion data!

2.6   Brief Summary

The monopole story had amusingly many unpredictable twists.

- In a setting of Electrodynamics combined with Quantum Mechanics, Dirac suggested in the 1930s a beautiful possibility for magnetic monopole existence: all what is in fact needed is the Dirac condition. Electric charge quantization is apparently satisfied in Nature. And yet, no QED monopoles were ever found!
- Discovery of the 't Hooft–Polyakov solution in the 1970s Georgi–Glashow model has put them back on theorist's minds. It does play an important role in extended supersymmetry models in the 1990s.
- Electric–magnetic duality ideas were quantitatively confirmed in that setting. In fact one can follow these ideas all the way to the regime in which the gauge coupling $g \to \infty$. As shown by Seiberg and Witten (1994a), this happens at two points of the moduli plane, and there the monopole mass $M_{mono} \to 0$. In its vicinity one can find out what the *dual magnetic theory* is, which turns out to be scalar QED with the inverse coupling $1/g$.
- And yet, this solution cannot be directly used, neither in QCD nor in electroweak parts of the Standard model, as both lack adjoint scalars.
- Nevertheless, lattice practitioners do observe, since the 1990s, that there exist some localized objects with *nonzero magnetic charge* (using Gaussian surfaces around). As we will see in the next chapter, they go into Bose–Einstein condensation at $T < T_c$, explaining confinement.
- Their density qualitatively fits well with the electric–magnetic duality picture: it grows as temperature changes from high to T_c. One can formulate the QCD phase diagram in terms of electric–magnetic duality ideas; see Fig. 2.5.
- As we learned at the end of the chapter, the correlation between monopoles, derived from lattice, can be naturally explained by a picture of *monopole Coulombic plasma*. Furthermore, the Coulomb coupling constant seems to be *the inverse* of the gauge coupling g: in complete agreement with electric–magnetic duality.
- The puzzle was that the dependence of their density on temperature has apparently shown an inverse power of $log(T)$, not the inverse power of T (as needed for semiclassical objects). This puzzle will be resolved in the chapter about Poisson duality, between monopole and instanton-dyon descriptions.

References

Aharonov, Y., Bohm, D.: Significance of electromagnetic potentials in the quantum theory. Phys. Rev. **115**, 485–491 (1959). [95 (1959)]

D'Alessandro, A., D'Elia, M.: Magnetic monopoles in the high temperature phase of Yang-Mills theories. Nucl. Phys. **B799**, 241–254 (2008). 0711.1266

Dirac, P.A.M.: Quantized singularities in the electromagnetic field. Proc. R. Soc. Lond. **A133**, 60–72 (1931). [278(1931)]

Koma, Y., Koma, M., Ilgenfritz, E.-M., Suzuki, T., Polikarpov, M.I.: Duality of gauge field singularities and the structure of the flux tube in Abelian projected SU(2) gauge theory and the dual Abelian Higgs model. Phys. Rev. **D68**, 094018 (2003). hep-lat/0302006

Laursen, M.L., Schierholz, G.: Evidence for monopoles in the quantized SU(2) lattice vacuum: a study at finite temperature. Z. Phys. **C38**, 501 (1988)

Liao, J., Shuryak, E.: Strongly coupled plasma with electric and magnetic charges. Phys. Rev. **C75**, 054907 (2007). hep-ph/0611131

Liao, J., Shuryak, E.: Magnetic component of Quark-Gluon plasma is also a liquid! Phys. Rev. Lett. **101**, 162302 (2008b). 0804.0255

Milton, K.A.: Theoretical and experimental status of magnetic monopoles. Rep. Prog. Phys. **69**, 1637–1712 (2006). hep-ex/0602040

Polyakov, A.M.: Particle spectrum in the quantum field theory. JETP Lett. **20**, 194–195 (1974)

Polyakov, A.M.: Quark confinement and topology of gauge groups. Nucl. Phys. **B120**, 429–458 (1977)

Seiberg, N., Witten, E.: Electric - magnetic duality, monopole condensation, and confinement in N=2 supersymmetric Yang-Mills theory. Nucl. Phys. **B426**, 19–52 (1994a). [Erratum: Nucl. Phys. **B430**, 485 (1994)]

Shnir, Y.M.: Magnetic monopoles. In: Text and Monographs in Physics. Springer, Berlin (2005b)

Suzuki, T., Hasegawa, M., Ishiguro, K., Koma, Y., Sekido, T.: Gauge invariance of color confinement due to the dual Meissner effect caused by Abelian monopoles. Phys. Rev. **D80**, 054504 (2009). 0907.0583

't Hooft, G.: Magnetic monopoles in unified gauge theories. Nucl. Phys. **B79**, 276–284 (1974)

Monopole Ensembles

3

So far we discussed a single monopole solution: now is the time to consider interacting monopoles and statistical ensembles of them. This is a complicated subject, and we will approach it in steps, considering:

- classical systems made of increasing number of charges and monopoles;
- classical plasmas made of many charges and monopoles;
- jet quenching and charge-monopole scattering;
- quantum-mechanical charge-monopole scattering ;
- gluon-monopole scattering ;
- kinetic properties of the dual plasmas using the obtained cross sections;
- quantum Coulomb Bose gas, with Bose-Einstein condensation (BEC) ;

At this point let me make few comments on the term "plasma" and other terminology to be used. Plasma, by definition, is a matter made of particles with a long-range interactions, $1/r$ in three dimensions. It is mostly used for electromagnetic plasmas made of ionized atoms or quark-gluon plasma in which long-range forces are due to gluon exchange. Since monopoles have Coulomb-type Abelian tails, and since they come with all signs (mono and anti-mono) available, their ensemble fits this definition as well.

Furthermore, we will mostly discuss what we will call a "dual plasma", which includes *both* charges and monopoles. As we will see, their mutual interaction is not Coulomb-like, but appears due to Lorentz forces. While they do not change energy of the particle, they change direction of motion. Therefore they are not important for thermodynamics of the dual plasmas, but are central for understanding of their kinetics.

© The Author(s), under exclusive license to Springer Nature Switzerland AG 2021   45
E. Shuryak, *Nonperturbative Topological Phenomena in QCD
and Related Theories*, Lecture Notes in Physics 977,
https://doi.org/10.1007/978-3-030-62990-8_3

Classical plasmas have one key parameter

$$\Gamma = \frac{\langle V \rangle}{T} \tag{3.1}$$

characterizing the ratio of the potential and kinetic energies.

If it is small, $\Gamma \ll 1$, the potential is a perturbation, and those are called *weakly coupled* plasmas, and their properties are adequately described by the linearized Debye theory. QGP at high temperatures is an example of this kind.

If Γ is in the range between ~ 1 and about 10^3, the matter is a *strongly coupled liquid*. If Γ is even larger, it makes the so-called ionic solids. (The common table salt $NaCl$ is an example of the latter type.) QGP near T_c is a dual plasma with two gamma parameters—electric and magnetic—both in the range 1–10. It is thus an example of strongly coupled dual plasma in a liquid phase.

3.1   Classical Charge-Monopole Dynamics

Let us start with a very old (nineteenth-century) problem in classical electrodynamics: an electrically charged particle with charge e is moving in the magnetic field of a static monopole with magnetic coupling g. Classically, one does not need the vector potential; thus many subtleties are absent. The interaction is simply given by the Lorentz force

$$m\ddot{\vec{r}} = -eg \frac{[\dot{\vec{r}} \times \hat{r}]}{r^2} \tag{3.2}$$

where $\times$ indicates the vector product and $\hat{r} \equiv \vec{r}/r$ is the unit vector along the line connecting both charges. It is worth noticing that only the product of the couplings (eg) appears (Fig. 3.1). Example of a trajectory following from this equation is shown in Fig. 3.1.

Furthermore, the cross product of the magnetic field of the monopole and electric field of the charge leads to a nonzero Poynting vector, which rotates around $\vec{r}$: thus, the field itself has a nonzero angular momentum. The total conserved angular momentum for this problem has two parts

$$\vec{J} = m[\vec{r} \times \dot{\vec{r}}] + eg\hat{r}. \tag{3.3}$$

The traditional potential scattering only has the first part: therefore, in that case, the motion entirely takes place in the so-called reaction plane normal to $\vec{J}$. In the charge-monopole problem, however, the second term restricts the motion to the so-called Poincare cone: its half-opening angle $\pi/2 - \xi$ being

$$\sin(\xi) = \frac{eg}{J}. \tag{3.4}$$

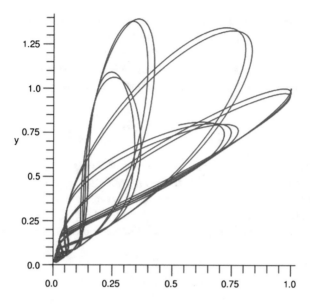

Fig. 3.1 Trajectory of a monopole in static electric Coulomb field

Only at large J (large impact parameter scattering) the angle ξ is small and thus the cone opens up, approaching the scattering plane.

Following Boulware et al. (1976), one can project the motion on the cone to a planar motion, by introducing

$$\vec{R} = \frac{1}{\cos(\xi)} [\vec{r} - \hat{J}(\vec{r} \cdot \hat{J})] \tag{3.5}$$

where the first scale factor is introduced to keep the same length for both vectors $\vec{R}^2 = \vec{r}^{\,2}$. Now, two integrals of motion are the angular momentum and the energy

$$\vec{J} = m\vec{R} \times \dot{\vec{R}} \tag{3.6}$$

$$E = \frac{m\dot{\vec{R}}^2}{2} - \frac{(eg)^2}{2mR^2} \tag{3.7}$$

and the problem seems to be reduced to the motion of a particle of mass m in an inverse-square potential. The scattering angle $\Delta\psi$ for this *planar* problem can be readily found: it is the variation of ψ as R goes from ∞ to its minimum b and back to ∞

$$\Delta\psi = \pi \left(\frac{1}{\cos\xi} - 1 \right) = \pi \left(\sqrt{1 + \left(\frac{eg}{mvb}\right)^2} - 1 \right). \tag{3.8}$$

Note that at large b (small ξ) we have $\Delta\psi \sim 1/b^2$, as expected for the inverse-square potential. Yet this is *not* the scattering angle of the original problem, because one has to project the motion back to the Poincaré cone. The true scattering angle—namely, the angle between the initial and final velocities—is $\cos\theta = -(\hat{v}_i \cdot \hat{v}_f)$. By relating velocities on the plane and on the cone, one can find it to be

$$\left(\cos\frac{\theta}{2}\right)^2 = (\cos\xi)^2 \left(\sin\frac{\pi}{2\cos\xi}\right)^2. \tag{3.9}$$

Thus for distant scattering—small ξ—one gets $\theta \approx 2\xi = 2eg/(mvb)$, which is much larger than $\Delta\psi \sim 1/b^2$. The important lesson that we learn from these formulae is that the small scattering angle is given by the opening angle of the cone, rather than the scattering angle in the planar, inverse-square effective potential. Calculating the cross section by $d\sigma = 2\pi b \, db$, one finds that, at small angles, it is

$$\frac{d\sigma}{d\Omega} = \left(\frac{2eg}{mv}\right)^2 \frac{1}{\theta^4}, \tag{3.10}$$

similar to the Rutherford scattering of two charges. The difference (apart from different charges) is also the additional second power of velocity, originating from the Lorentz force.

3.2 Monopole Motion in the Field of Several Charges

Liao (my grad student at a time) and myself (Liao and Shuryak 2007) started by investigating curios few-body motion. Suppose we take two static electric charges, e and $-e$, and put a monopole into this field. We found (see Fig. 3.2) that it can be trapped between them, bouncing from one to the other on a surface which consists of two smoothly connected Poincare cones with ends on two charges.[1] Needless to say, the same happens in a dual setting, with two static monopoles and a charge bouncing between them.[2]

Our next configuration, shown in Fig. 3.3a, is called "a grain of salt." it consists of eight charges located at the corners of 3-d cube, with alternating plus and minus charges. It was found that a monopole can get out of such "cage," but with difficulty,

---

[1] So to say, charges can play ping-pong with monopole, without even moving!

[2] Note that this last setting is very similar to famous invention by one of my teachers G.I. Budker, the *magnetic bottle*, in which the magnetic monopole is substituted by a coil with a current. This device traps an electron and is used a lot in plasma research.

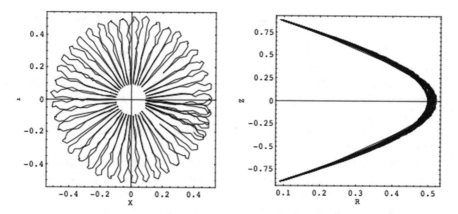

Fig. 3.2 Trajectory of monopole motion in a static electric dipole field (with charges at $\pm 1\hat{z}$) as (left panel) projected on x–y plane and (right panel) projected on R–z plane ($R = \sqrt{x^2 + y^2}$)

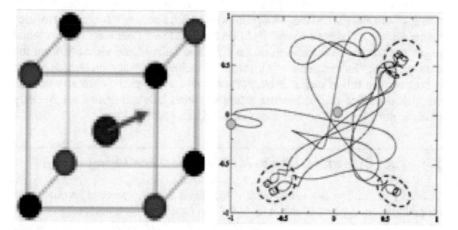

Fig. 3.3 A monopole moving in a set of eight electric charges, with alternating signs. Left picture shows the setting and right is an example of the monopole path, obtained by numerically solving classical equation of motion

suffering multiple collision with the charges at cube corners (see a typical path in Fig. 3.3b). It happens again, due to focusing mechanism due to Lorentz force just described.[3]

---

[3]I would say the monopole behaves as a proverbial drunkard who cannot go home because there are several lampposts on the street, with which he mysteriously collides all the time.

3.3   Strongly Coupled QGP as a "Dual" Plasma with Monopoles

Discovery that QGP is a "perfect liquid," with extremely small *viscosity*, had led to significant interest in strongly coupled systems in general. (It is even now referred to as strong QGP or sQGP for short in literature.) We will now show that electric-magnetic duality, emphasizing the RG flow and transition from weakly coupled electric theory in UV to mixed strongly coupled electric-magnetic sectors around T_c, can reproduce the observed unusually high collision rate (small mean free path or small viscosity).

The first one now can be characterized as a mainstream, but it does not belong to these lectures as it requires a lot of background knowledge which the advanced undergrad/beginning graduate students do not generally have. The second one, on the contrary, is rather accessible and pedagogically fruitful, and so I decided to include it here.

Classical *molecular dynamics* studies of the *"dual plasmas"* have been performed by Liao and Shuryak (2007). What this means is that we took few hundred charges, both electric and magnetic, both with positive and negative charges to keep matter neutral, and solve numerically EOM for all of them. Details are in the original paper: the output are calculation of the diffusion constant and viscosity, from the corresponding "Kubo formulae." Qualitative conclusion is that each electric charge is trapped by a cell of magnetic ones around and each magnetic charge by electric ones, as described above. Lorentz nature of forces between them does decrease particle (diffusion) and momentum (viscosity) transport by a lot!

3.4   Jet Quenching Due to Jet-Monopole Scattering

The story started by observation (Shuryak 2002) that while theory of jet quenching by radiative energy loss (Baier et al. 1996) did explained the overall magnitude of jet quenching, it failed to describe the ellipticity parameter $v_2(p)$ defined by (where ϕ is the azimuthal angle)

$$E_p \frac{dN}{d^3 p} = f(p)\big[1 + 2v_2(p)\cos(2\phi)\big] \qquad (3.11)$$

by a large factor, well beyond the experimental accuracy. This issue remained a puzzle till relatively recently: to explain it one needed the monopoles!

It was suggested by Liao and Shuryak (2009) that the puzzle can be explained if the jet quenching be strongly enhanced in the near-T_c matter. Figure 3.4 qualitatively explains the idea. The left plot shows the naive standard picture stemming from the

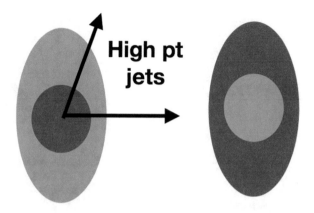

Fig. 3.4 The almond-shaped excited system created in non-central heavy ion collisions: depending on jet direction, it finds different amounts of matter. Left plot schematically shows density distribution, and the right one corresponds to $\hat{q}$ distribution

assumption that the so-called jet quenching parameter

$$\hat{q} \equiv \frac{\langle Q^2 \rangle}{length}, \tag{3.12}$$

the mean squared transverse momentum kick accumulated by a jet along certain length of propagation in matter, is simply proportional to matter density. If so, the central darker region, indicating higher density, of the almond-shaped excited system created in non-central heavy ion collisions, would dominate the jet quenching. However, this darker region is nearly azimuthally symmetric and contributes little to $v_2(p)$.

But if the assumption that quenching is proportional to the density $\hat{q} \sim n$ is wrong, one may find the way out of a puzzle. For example, if "by the eyes of the jet" quenching strength is modified, e.g., as indicated at the right picture of Fig. 3.4, then one finds large azimuthal dependence and $v_2(p)$ consistent with experimentally observed values.

But why relatively dilute periphery of the fireball can produce more kicks on the jet than its central denser part? Studies of its possible origin led to *jet-monopole scatterings*. Recall Fig. 2.6, which shows that the monopole density is large only in the vicinity of T_c, not in very hot plasma!

For detailed discussion of jet quenching, see Xu et al. (2016); , and we only illustrate the points by two plots from this paper Fig. 3.5. The upper plot shows the so-called jet quenching parameter $\hat{q}$ normalized to T^3. The parameter has dimension of the density of scatterers, to which it is supposed to be proportional. Its normalization to T^3 means that we are looking at the effective number of degrees of freedom.

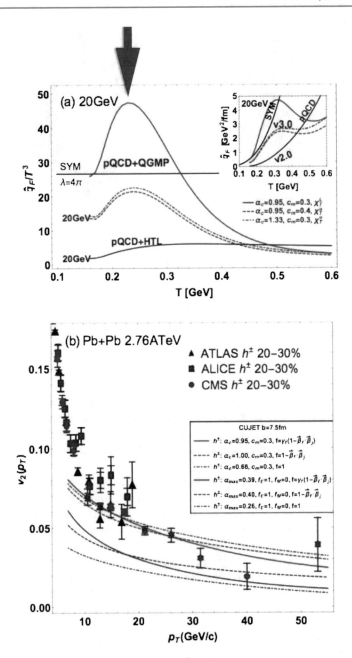

Fig. 3.5 Upper plot: jet quenching parameter $\hat{q}/T^3$ versus the temperature T. Red arrow indicates the position of a peak in the dependence. Lower plot: azimuthal asymmetry parameter $v_2 = <\cos(2\phi)>$ of jet quenching, as a function of particle transverse momentum p_T

The blue line at the bottom of it corresponds to the perturbative QGP: the effective number of degrees of freedom is constant at high T and it decreases near T_c, being basically proportional to the VEV of the Polyakov line. Red lines with a peak at certain temperature near T_c are empirical models which include scattering on the monopoles. Basically those are expected to be proportional to the normalized monopole density shown in Fig. 3.14 (right).

As shown in the lower plot of Fig. 3.5, those models are in much better agreement with the experimental data. (Red lines there correspond to those models with a peak, while several purple lines below the data correspond to the perturbative model without the monopoles.) So, in a way, we have now some experimental confirmation that a jet-monopole scattering is not only happening, but even in fact dominates certain observables.

Summary In the high-T limit the QGP is a weakly coupled plasma, amenable to perturbation theory (except the suppressed magnetic sector). As T decreases, the coupling grows. As a result of it, the density of magnetic and other topological objects grows as a power of T. By $T \approx T_c$, relevant for current experiments, one can view matter as a plasma made of both electric and magnetic particles, both with the coupling O(1) and complicated. Classical and quantum studies of such medium have been successfully done, explaining reduction of the mean free path and all forms of kinetic transport.

3.5 Quantum-Mechanical Charge-Monopole Scattering Problem

The history of *quantum monopole-charge scattering problem* is also very interesting, and not widely known, although in my opinion it clearly belongs to any good QM textbook. There are indications that Dirac, after his paper on monopoles in 1931, discussed it with Tamm, and they both tried to solve it, but did not succeed. It took a long time till it did happened, a year after non-Abelian monopoles were discovered, by two distinguished teams (Schwinger et al. 1976; Boulware et al. 1976).

The problem can be set in two ways. First, one can think of electromagnetic monopole and charge, both of size zero, so that only Coulomb fields exist. Second, we will take 't Hooft–Polyakov monopole solution in full glory (with a charged core and Coulomb+Higgs tail) and scatter small perturbation of a scalar on it.

Scattering problem results are expressed in scattering amplitudes, or scattering phase shifts, as a function of energy. Perhaps, to refresh the reader's memory of the quantum mechanics course, a couple of general formulae are in order. The wave function $\psi = e^{ikz} + f(\theta)e^{ikr}/r$ contains scattering amplitude $f(\theta)$ which defines the differential cross section of scattering

$$d\sigma = |f(\theta)|^2 2\pi \sin(\theta) d\theta$$

Here are expressions for it in partial waves

$$f(\theta) = \frac{1}{2ik} \sum_{l=0}^{\infty} (2l+1)\big(e^{2i\delta_l(k)} - 1\big) P_l(\cos(\theta))$$

expanded in Legendre polynomials. General expressions for the total cross section and the transport cross section are

$$\sigma = \int d\sigma = \frac{4\pi}{k^2} \sum_{l=0}^{\infty} (2l+1) \sin^2(\delta_l(k))$$

$$\sigma_t = \int d\sigma (1 - \cos(\theta))$$

Let us think now about parameters of the problem. If particle is pointlike, the Coulombic field of a charge has only one dimensionless parameter, the $e^2/\hbar c$ coupling. It is scale invariant and cannot provide a scale.[4] So, the scattering phase shift cannot depend on collision energy!

This observation has serious implications. Corrections to total free energy of matter and other thermal quantities contain expressions with $\int dk (d\delta_l(k)/dk) \ldots$ which are in the case considered all zero. As we will see, electric magnetic scattering will not affect the equation of state of quark-gluon plasma, but be dominant in its kinetic parameters.

Scattering depends on the couplings, the product of electric and magnetic charges. Yet according to Dirac condition, this product must be an integer

$$\frac{eg}{4\pi} = n = integer. \tag{3.13}$$

This integer is *the only input* of the problem we try to solve, provided monopole is treated as a point charge.

Let me just give the answer for the scattering phases. Indeed, they depend on the total angular momentum j and the integer n from the Dirac condition. Note that for point charges they do not depend on energy. The expression is simply proportional to t' is to be found from the r.h.s. of the equation

$$\delta_t = \frac{\pi}{2} t', \quad t'(t'+1) = t(t+1) - n^2 \tag{3.14}$$

---

[4]The usual electric Coulomb scattering does not have finite phase shifts, and it leads to logarithmic divergent expression at large r because of long-range nature of the Coulomb potential. However, in the charge-monopole problem, the force is the Lorentz force, which is *not* long range, so the phase shifts are calculable.

The first term in the r.h.s. $t(t+1)$ is the total angular momentum squared, and the last term is subtracted angular momentum of the field n^2. The t' thus has the meaning of *angular momentum of the particle*. Here comes a shocking fact: while j and n in the r.h.s. are both integers,[5] their combination t' entering the scattering phase is *not* an integer. One can easily see it by solving the quadratic equation for it: it is a square root of an integer. That is why a quite nontrivial angular distribution of charge-monopole scattering appears!

Recall that the usual angular basis used $Y_{lm}(\theta, \phi)$ at large quantum numbers $l, m \gg 1$ is in fact planar. Yet in classical charge-monopole problem, the motion happens on the *Poincare cone* rather than on a plane! So, one has to rethink the setting and select a different, more appropriate, angular functions, conical in the classical limit.

The wave function in spherical coordinates is as usual a sum of products of certain r-dependent radial functions, times the angular functions. The former basically follow from the inverse-square law potential and thus are easily solved in Bessel functions: they correspond to the auxiliary planar projection of the classical problem of the previous subsection. The nontrivial part happens to be in the unusual angular functions.

The functions that we need (e.g., in the scalar sector) must satisfy the following set of conditions

$$\left\{\begin{matrix} \vec{T}^2 \\ T_3 \\ \vec{I}^2 \\ (\hat{r} \cdot \vec{I}) \end{matrix}\right\} \phi_{ti}^{mn}(\theta, \varphi) = \left\{\begin{matrix} t(t+1) \\ m \\ i(i+1) \\ n \end{matrix}\right\} \phi_{ti}^{mn}(\theta, \varphi). \tag{3.15}$$

where $\vec{T}$ is the total angular momentum $\vec{T} = \vec{L} + \vec{I}$ and $\vec{I}$ is the isospin. The unusual condition in the above set is the last one, since the vector $\vec{I}$ must be projected to the (space-dependent) radial unit vector. The functions satisfying this requirement (Boulware et al. 1976; Schwinger et al. 1976) will be introduced below. Here it is enough to explore the large l, n limit of the D-functions involved. The result

$$D_{nl}^l \sim e^{i(l-n)\varphi} exp[-l(\theta - \theta^*)^2/2] \tag{3.16}$$

where $\cos(\theta^*) = n/l$, shows that they indeed correspond to the Poincaré cone.

Now, we move to second setting, discussing small scalar perturbations scattered on the 't Hooft–Polyakov monopole soliton, in the Georgi–Glashow framework . We introduce a total angular momentum operator $\vec{T}$, which is the sum of the orbital

---

[5]The second is due to Dirac condition.

angular momentum $\vec{L}$ and the isotopic spin $\vec{I}$:

$$\vec{T} = -i\vec{r} \times \vec{\nabla} + \vec{I} \tag{3.17}$$

with $(I^a)_{bc} = -i\epsilon_{abc}$. In terms of these operators, the wave equation for the scalar fluctuations can be written in the form

$$\left[\frac{\partial^2}{\partial r^2} - \frac{2}{r}\frac{\partial}{\partial r} - \frac{\left(\vec{T}^2 - \left(\hat{r}\cdot\vec{I}\right)^2\right)}{r^2} - \partial_0^2 \right]\vec{\chi} + \frac{2K(\xi)\left(\vec{I}\cdot\vec{T} - \left(\hat{r}\cdot\vec{I}\right)^2\right)}{r^2}\vec{\chi}$$

$$\tag{3.18}$$

$$-\frac{K(\xi)^2\left[\vec{I}^2 - \left(\hat{r}\cdot\vec{I}\right)^2\right]}{r^2}\vec{\chi} - \lambda\left[2\frac{\vec{r}\cdot\vec{\chi}}{e^2 r^4}H(\xi)^2\vec{r} + \left(\frac{H(\xi)^2}{e^2 r^2} - v^2\right)\vec{\chi}\right] = 0.$$

The term which is proportional to $K[\xi]$ induces charge-exchange reactions.

We can define a simultaneous eigenfunction $\phi_{ti}^{mn}(\hat{r})_a$ of the commuting operators $\vec{T}^2$, T_3, $\vec{I}$, and $\hat{r}\cdot\vec{I}$ (see Eq. (3.15)). This function depends only on the angular variables specified by $\hat{r}$. A solution to the equation (a specific partial wave) can be written as the product of the angular function $\phi_{ti}^{mn}(\hat{r})$ and a radial function $S_t^n(r)$

$$\chi(\vec{r})_a = \phi_{ti}^{mn}(\hat{r})_a S_t^n(r). \tag{3.19}$$

The angular function $\phi_{ti}^{mn}(\hat{r})_a$ is peculiar because the operator $\vec{I}$ is projected along $\hat{r}$. Therefore, the angular function must be rotated, from the standard Cartesian frame, to a "radial" frame. This construction can be achieved by making use of a *spatially dependent* unitary matrix which rotates $\hat{r}\cdot\vec{I}$ into I_3:

$$U(-\varphi, -\theta, \varphi) = e^{-i\varphi I_3}e^{-i\theta I_2}e^{i\varphi I_3}. \tag{3.20}$$

We therefore have

$$\hat{r}\cdot\vec{I}\,U(-\varphi, -\theta, \varphi) = U(-\varphi, -\theta, \varphi)I_3$$

$$\vec{T}\,U(-\varphi, -\theta, \varphi) = U(-\varphi, -\theta, \varphi)\vec{\mathcal{T}} \tag{3.21}$$

where

$$\vec{\mathcal{T}} = -\vec{r} \times \left(i\vec{\nabla} + e\vec{\mathcal{A}}I_3\right) + \hat{r}I_3 \tag{3.22}$$

and

$$e\vec{A} = \frac{\hat{r} \times \hat{z}}{r + z}. \tag{3.23}$$

Equation (3.15) is satisfied by the following function

$$\phi_{ti}^{mn}(\hat{r})_a = (U(-\varphi, -\theta, \varphi)\chi_i^n)_a \mathcal{D}(\hat{r}) \tag{3.24}$$

where $(\chi_i^n)_a$ is an eigenvector of I_3 in the Cartesian basis

$$I_3(\chi_i^n)_a = n(\chi_i^n)_a \tag{3.25}$$

and the function $\mathcal{D}(\hat{r})$ obeys

$$\left\{ \begin{matrix} \vec{T}^2 \\ T_3 \end{matrix} \right\} \mathcal{D}(\hat{r}) = \left\{ \begin{matrix} t(t+1) \\ m \end{matrix} \right\} \mathcal{D}(\hat{r}) \tag{3.26}$$

where I_3 in Eq. (3.22) is now replaced by its eigenvalue, n. We have

$$\mathcal{D}(\hat{r}) = \mathcal{D}_{nm}^{(t)}(-\varphi, \theta, \varphi) = \langle t, n | e^{-i\varphi T_3} e^{i\theta T_2} e^{i\varphi T_3} | t, m \rangle. \tag{3.27}$$

We can write the function $\phi_{ti}^{mn}(\hat{r})_a$ by making use of the following expansion

$$\phi_{ti}^{mn}(\hat{r})_a = \sum_{n'} (\chi_i^{n'})_a (-1)^{n-n'} \tag{3.28}$$

$$\times \sum_{l=|t-i|}^{t+i} \langle i, -n, t, n | l, 0 \rangle \mathcal{D}_{0,m-n'}^{(l)}(-\varphi, \theta, \varphi) \langle l, m - n' | i, -n', t, m \rangle$$

where

$$\mathcal{D}_{0,m}^{(l)}(\alpha, \beta, \gamma) = \sqrt{\frac{4\pi}{2l+1}} Y_l^m(\beta, \gamma). \tag{3.29}$$

These functions are normalized in the following way

$$\int_0^{2\pi} d\varphi \int_{-1}^{1} d\cos\theta\, \phi_{t_1 i}^{m_1 n_1}(\theta, \varphi)^\dagger \phi_{t_2 i}^{m_2 n_2}(\theta, \varphi) = \frac{4\pi}{2t_1 + 1} \delta_{t_1, t_2} \delta_{m_1, m_2} \delta_{n_1, n_2}. \tag{3.30}$$

Since the above function is a polynomial in x/r, y/r, z/r, the function in Eq. (3.28) is analytic everywhere. This clearly shows that there are compensating singularities in $U(-\varphi, -\theta, \varphi)$, and $\mathcal{D}(\hat{r})$. Since Eq. (3.18) in its most general form admits mixing between particles of different charges, in order to get the equation for the radial functions S_t^n, we have to write the order 1 fluctuations of the field around the classical solutions as a superposition of the functions describing a particle with definite charge n:

$$\chi(\vec{r}, x_0)_a = \sum_{n=-1,0,1} e^{i\omega x_0} \frac{S_t^n(r)}{r} \phi_{ti}^{mn}(\theta, \varphi)_a \tag{3.31}$$

where the three functions $S_t^n(r)$ correspond to the three physical fluctuations with charge $n = 0, \pm 1$. We plug the above expansion for $\vec{\chi}(\vec{r})$ into Eq. (3.18); through this procedure we obtain the following system of equations for the radial functions

$$S_t^{0''}(\xi) - \left(\frac{t(t+1)}{\xi^2} + 2\frac{K(\xi)^2}{\xi^2} - \omega^2\right) S_t^0(\xi) - \lambda \left(3\frac{H(\xi)^2}{\xi^2} - 1\right) S_t^0(\xi)$$

$$+ \frac{\sqrt{2t(t+1)}}{\xi^2} K(\xi) \left(S_t^1(\xi) + S_t^{-1}(\xi)\right) = 0 \tag{3.32a}$$

$$S_t^{1''}(\xi) - \left(\frac{t(t+1)-1}{\xi^2} + \frac{K(\xi)^2}{\xi^2} - \omega^2\right) S_t^1(\xi) - \lambda \left(\frac{H(\xi)^2}{\xi^2} - 1\right) S_t^1(\xi)$$

$$+ \frac{\sqrt{2t(t+1)}}{\xi^2} K(\xi) S_t^0(\xi) = 0 \tag{3.32b}$$

$$S_t^{-1''}(\xi) - \left(\frac{t(t+1)-1}{\xi^2} + \frac{K(\xi)^2}{\xi^2} - \omega^2\right) S_t^{-1}(\xi) - \lambda \left(\frac{H(\xi)^2}{\xi^2} - 1\right) S_t^{-1}(\xi)$$

$$+ \frac{\sqrt{2t(t+1)}}{\xi^2} K(\xi) S_t^0(\xi) = 0. \tag{3.32c}$$

where we have introduced the dimensionless variable $\xi = evr$. At the end of this section we will discuss how to fix the scale and go to physical units. As it is evident from the above system, a mixing occurs in the monopole core between different charges: the term $\propto K(\xi)$ involves a mixing between charges that differ by one unit. The above system of equations has been obtained in the most general case, for generic angular momentum t. Nevertheless, we have to keep in mind that there are some restrictions due to the following requirement:

$$\hat{r} \cdot \vec{T} = \hat{r} \cdot \vec{L} + \hat{r} \cdot \vec{I} = \hat{r} \cdot \vec{I} = n. \tag{3.33}$$

For this reason, in the case $t = 0$ only, the $n = 0$ scalar fluctuation is allowed. The equation for this special case and its solution will be discussed in the following.

For $t > 0$, the system of equations (3.35a) is difficult to solve, due to the mixing between the different radial functions. This mixing is due to the charge-exchange reactions that can occur inside the monopole core. If the monopole core is small (we have seen in Sect. 2.1 that lattice-based estimates for the monopole size give $r_m \simeq 0.15$ fm) we can neglect the charge-exchange reactions. This corresponds to considering the above system of equations (3.35a) in the limit

$$K(\xi) \to 0 \qquad H(\xi) \to \xi. \tag{3.34}$$

In this approximation, it reduces to

$$S_t^{0''}(\xi) - \left(\frac{t(t+1)}{\xi^2} - \omega^2 + 2\lambda \right) S_t^0(\xi) = 0 \tag{3.35a}$$

$$S_t^{1''}(\xi) - \left(\frac{t(t+1) - 1}{\xi^2} - \omega^2 \right) S_t^1(\xi) = 0 \tag{3.35b}$$

$$S_t^{-1''}(\xi) - \left(\frac{t(t+1) - 1}{\xi^2} - \omega^2 \right) S_t^{-1}(\xi) = 0. \tag{3.35c}$$

From the above system it is clear that we can identify the radial functions with spherical Bessel functions having index t', which is *the positive root of quadratic equation*

$$t'(t'+1) = t(t+1) - n^2. \tag{3.36}$$

This makes scattering rather unusual. Namely, in the limit of small monopole core, we have $S_t^n(r) \to j_{t'}(kr)$. The corresponding scattering phase will be $\delta_{t'} = t'\pi/2$, independent of the energy of the incoming particle.

For $t = 0$ we have only one fluctuation allowed, namely, the one having zero charge: $S_0^0(\xi)$. It obeys the following equation

$$S_0^{0''}(\xi) - \left(2\frac{K(\xi)^2}{\xi^2} \right) S_0^0(\xi) - \lambda \left(3\frac{H(\xi)^2}{\xi^2} - 1 \right) S_0^0(\xi) = -\omega^2 S_0^0(\xi). \tag{3.37}$$

In this case, there is no Coulomb potential of the form $1/\xi^2$, which is obvious since a charge-neutral particle does not feel the Lorentz force. The scattering in this case is entirely due to the monopole core. We can solve the above equation numerically, thus obtaining the scattering phase as a function of the energy of the incoming particle, by imposing the following boundary conditions

$$S_0^0(\xi) = \sin\left[\xi\sqrt{\omega^2 - 2\lambda} - \delta_0 \right] \qquad\qquad \xi \to \infty$$

$$S_0^{0'}(\xi) = \sqrt{\omega^2 - 2\lambda} \cos\left[\xi\sqrt{\omega^2 - 2\lambda} - \delta_0 \right] \qquad \xi \to \infty. \tag{3.38}$$

Figure 3.6 shows the classical solutions $H(\xi)$ and $K(\xi)$, both in the BPS limit and for $\lambda = 1$.

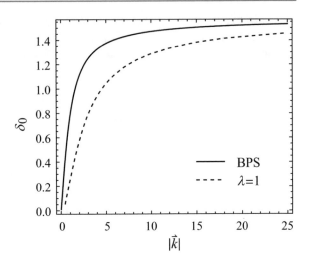

Fig. 3.6 Scattering phase δ_0 as a function of $|\vec{k}| = \sqrt{\omega^2 - 2\lambda}$. δ_0 is obtained by solving Eq. (3.37) with the boundary conditions (3.38). The continuous line corresponds to the BPS limit ($\lambda = 0$), while the dashed line corresponds to $\lambda = 1$

At this point we need to fix the scale in our problem and to estimate the scattering length in physical units. In order to do it, we have to connect the physical size of the core of the monopole, r_m, to the dimensionless units that we used in the Georgi–Glashow model. We recall that we obtained r_m through the width at half height of the "non-Abelianicity," which is defined as the square of the non-Abelian components of the gauge potential. Therefore, it is natural to fix the scale by imposing that, in the Georgi–Glashow model, r_m coincides with ξ_m, the width at half height of the function $K(\xi)^2$. By looking at Fig. 3.6, we can see that, in the BPS limit, we have $\xi_m = 1.49$, while for $\lambda = 1$ we have $\xi_m = 0.87$. Thus, $ev = 9.94\,\mathrm{fm}^{-1}$ in the BPS limit, while $ev = 5.8\,\mathrm{fm}^{-1}$ for $\lambda = 1$.

3.6   Quark and Gluon Scattering on Monopoles and Viscosity of QGP

In QGP with a monopole, one needs to find scattering amplitudes of quarks and gluons. The former problem has been solved by another distinguished team (Kazama et al. 1977), but the latter problem turned out to be especially tedious, because orbital angular momentum, spin, and color-spin of the gluon are all mixed in. It has been solved by Ratti and Shuryak (2009).

In the case of vector particles, the generalized angular momentum $\vec{J}$ is made up of three components: the orbital angular momentum $\vec{L}$, the isotopic spin $\vec{I}$, and the spin $\vec{S}$:

$$\vec{J} = \vec{L} + \vec{I} + \vec{S} = \vec{T} + \vec{S}. \tag{3.39}$$

There are generally two different ways of composing three vectors, depending on the two vectors to be added first. The monopole vector spherical harmonics are eigenfunctions of $\vec{J}^2$ and J_3. Due to the following relation

$$\hat{r} \cdot \vec{J} = \hat{r} \cdot \vec{L} + \hat{r} \cdot \vec{I} + \hat{r} \cdot \vec{S} = \hat{r} \cdot \vec{I} + \hat{r} \cdot \vec{S} = n + \sigma, \qquad (3.40)$$

the allowed values of the total angular momentum quantum number j are $|n| - 1, |n|, \ldots$ except in the case of $n = 0$, where $|n| - 1$ is absent. There are different ways of building the monopole vector spherical harmonics; for example, they can be constructed by making use of the standard Clebsch–Gordan technique of addition of momenta.

Another possibility, which we will adopt here, is to build the vector harmonics with $j \geq n$ by applying vector operators to the scalar harmonics, as we will see in the following (Weinberg 1994). By definition, these harmonics can be introduced as eigenfunctions of the operator of the radial component of the spin, $\vec{S} \cdot \hat{r}$. We will show that this is a very useful choice for the "hedgehog" configuration we are working in: in fact, there is a natural separation between radial and transverse vectors, making this choice particularly useful for studying spherically symmetric problems. The vector harmonics with the minimum allowed angular momentum $j = |n| - 1$ cannot be constructed in this way and must be treated specially. As already mentioned, there is more than one way to obtain a given value of j and thus several multiplets of harmonics with the same total angular momentum. In the following, we will classify the multiplets by the eigenvalue of $\hat{r} \cdot \vec{S} = \sigma$. In general, $\sigma = 0, \pm 1$, but it is further restricted by the requirement (3.40), which implies that $n + \sigma$ lies in the range $-j$ to j. This gives

- for $j = 0$:
  - $n = 0$ and $\sigma = 0$
  - $n = 1$ and $\sigma = -1$
  - $n = -1$ and $\sigma = 1$
- for $j = 1$ all combinations are allowed, except:
  - $n = -1$ and $\sigma = -1$
  - $n = 1$ and $\sigma = 1$

We denote the vector harmonics by $\Phi_{j,n}^{m,\sigma}(\theta, \varphi)_{ai}$. They obey the following eigenvalue equations

$$\begin{Bmatrix} \vec{J}^2 \\ J_3 \\ (\hat{r} \cdot \vec{I}) \\ (\hat{r} \cdot \vec{S}) \end{Bmatrix} \Phi_{j,n}^{m,\sigma}(\theta, \varphi)_{ai} = \begin{Bmatrix} j(j+1) \\ m \\ n \\ \sigma \end{Bmatrix} \Phi_{j,n}^{m,\sigma}(\theta, \varphi)_{ai}. \qquad (3.41)$$

In the case $j \geq 2$ all the possible combinations of n and σ are allowed.

3.7   Transport Coefficients from Binary Quantum Scattering

In the gas approximation, all transport coefficients are inversely proportional to the so-called transport cross section, normally defined as

$$\sigma_t = \int (1 - \cos\theta)d\sigma \tag{3.42}$$

where θ is the scattering angle. While the factor in brackets vanishes at small angles, the Rutherford singularity in the cross section, for any charged particle, leads to its logarithmic divergence. Since we will be comparing the gluon scattering on monopoles with that on gluons, let us first introduce those benchmarks, namely, the well-known (lowest order) QCD processes, the gg and $\bar{q}q$ scatterings:

$$\frac{d\sigma_{\bar{q}q}}{dt} = \frac{e^4}{36\pi}\left(\frac{s^4 + t^4 + u^4}{s^2 t^2 u^2} - \frac{8}{3tu}\right) \tag{3.43}$$

$$\frac{d\sigma_{gg}}{dt} = \frac{9e^4}{128\pi}\frac{(s^4 + t^4 + u^4)(s^2 + t^2 + u^2)}{s^4 t^2 u^2} \tag{3.44}$$

where (we remind) the electric coupling is related to α_s as usual: $e^2/4\pi = \alpha_s$.

While for nonidentical particles the transport cross section is simply given by the cross section weighted by momentum transport $t \sim (1 - z)$, for identical ones such as gg, one needs to introduce the additional factor $(1 + z)/2$ in order to suppress backward scattering as well. The integrated transport cross sections themselves are given by

$$\sigma_{gg}^t = \frac{3e^4}{320\pi s}\left(105\log(3) - 16 + 30\log\left(\frac{4}{\theta_{min}^2}\right)\right) \tag{3.45}$$

$$\sigma_{\bar{q}q}^t = \frac{e^4}{54\pi s}\left(4 + 7\log(3) + 3\log\left(\frac{4}{\theta_{min}^2}\right)\right) \tag{3.46}$$

where the smallest scattering angle can be related to the (electric) screening mass by $\theta_{min}^2 = 2 * M_D^2/s$. Note that the forward scattering log in the gg case has a coefficient which is roughly four times larger, as a consequence of the gluon color being roughly twice that of a fundamental quark. Note also that the gg scattering is significantly larger at large angles, as compared to the $\bar{q}q$ scattering.

Fig. 3.7 A charge scattering
on a two-dimensional array of
correlated monopoles (open
points) and antimonopoles
(closed points). The dotted
circle indicates a region of
impact parameters for which
scattering on a single
monopole is a reasonable
approximation

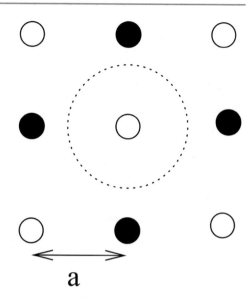

We expect that the charge-monopole scattering is Rutherford-like at small
angles: this comes from harmonics with large angular momenta (large impact
parameters). However, in matter there is a finite density of monopoles, so the
issue of the scattering should be reconsidered. A sketch of the setting, assuming
strong correlation of monopoles into a crystal-like structure, is shown in Fig. 3.7. A
"sphere of influence of one monopole" (the dotted circle) gives the maximal impact
parameter to be used.

As a result, the impact parameter is limited from above by some b_{max}, which
implies that only a finite number of partial waves should be included. The range of
partial waves to be included in the scattering amplitude can be estimated as follows

$$j_{max} = \langle p_x \rangle n_{mono}^{-1/3}/2 \sim aT \sim 1/e^2(T) \sim \log(T) \qquad (3.47)$$

Since at asymptotically high T the monopole density $n_{mono} \sim (e^2 T)^3$ is small
compared to the density of quarks and gluons $\sim T^3$, j_{max} asymptotically grows
logarithmically with T. So, only in the academic limit $T \to \infty$ one gets $j_{max} \to \infty$
and the usual free-space scattering amplitudes calculated in Boulware et al. (1976)
with all partial waves are recovered. However, in reality we have to recalculate the
scattering, retaining only several lowest partial waves from the sum. As we will see,
this dramatically changes the angular distribution, by strongly depleting scattering
at small angles and enhancing scattering backward.

The integrands of the transport cross section $(1 - \cos\theta)|f(\theta)|^2$ are shown in
Fig. 3.8 for $n = 0$, $j_{max} = 2, 4, 6$ (left panel), $n = \pm 1$, $j_{max} = 2, 4, 6$ (right panel).

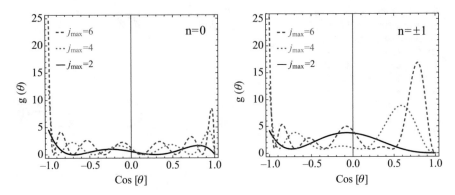

Fig. 3.8 Integrand of the transport cross section $g(\theta) = (1 - \cos\theta)|f(\theta)|^2$ with only two, four, and six lowest partial waves included, for a scalar particle with $n = 0$ (left) and $n = \pm 1$ (right). The curves can be easily recognized by higher j_{max} having more oscillations

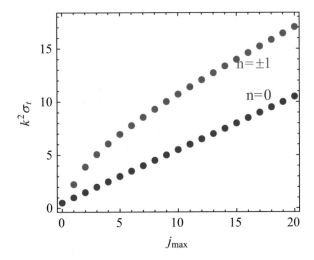

Fig. 3.9 Normalized transport cross section as a function of cutoff in maximal harmonics retained, for $n = 0, 1$

One can see how much their angular distribution is distorted. Strong oscillations of this function occur because we use a sharp cutoff for the higher harmonics, which represents diffraction of a sharp edge. This edge in reality does not exist and can be removed by any smooth edge prescription (e.g., a Gaussian weight). However, we further found that the transport cross section itself is rather insensitive to these oscillations, and thus there is no need in smoothening the scattering amplitude. The transport cross section as a function of j_{max} is shown in Fig. 3.9: it is large and smoothly rising with the cutoff.

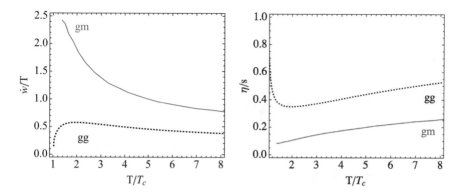

Fig. 3.10 Left panel: gluon-monopole and gluon-gluon scattering rate. Right panel: gluon-monopole and gluon-gluon viscosity over entropy ratio, η/s

Convoluting the cross sections found with the monopole density and gluon momentum distribution, we plot the scattering rates $n\sigma_t$ per gluon vs. T in Fig. 3.10.

It follows from this comparison of the gluon-monopole curve with the gluon-gluon one that the former remains the leading effect till very high T, although asymptotically it is expected to get subleading. This maximal T expected at LHC does not exceed $4T_c$, where $\eta/s \sim .2$. This value is well in the region which would ensure hydrodynamical radial and elliptic flows, although deviations from ideal hydro would be larger than at RHIC (and measurable!).

Kinetic approximate relation of these rates to viscosity/entropy ratio is

$$\frac{\eta}{s} \approx \frac{T}{5\dot{w}};\tag{3.48}$$

We plot η/s in the right panel of Fig. 3.10.

The results of this calculation are compared to lattice and experimental data in Fig. 3.11. The meaning of s/η ratio is basically the interparticle distance $\sim 1/T$ divided by the mean free path. The fact that it is about 6 means that particle in average collides with something at distance 6 times *smaller* than the distance to the next particle! This seems impossible: but one should recall that it is not just geometry, but due to the Lorentz force enhancing scattering.

The lower plot compares the scattering rates for gluon-gluon and monopole-gluon scatterings: the latter is clearly dominant in the near-T_c region. The results do indeed indicate that gluon-monopole scattering in sQGP dominates its kinetic properties and explains a small QGP viscosity observed.

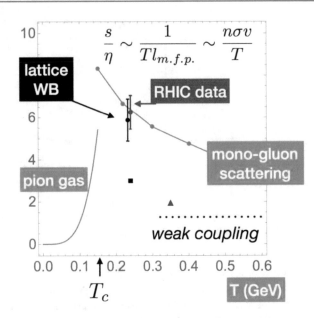

Fig. 3.11 Entropy density divided to shear viscosity s/η as a function of the temperature T (GeV). Two points with error bars at the center are for lattice and experimental data from RHIC higher harmonics flow. The points connected by line are for monopole-gluon scattering described above

3.8   Monopoles and the Flux Tubes

Let us start with two popular statements:

1. *Existence of flux tubes between two fundamental charges in QCD-like gauge theories is among the most direct manifestations of the confinement phenomenon.*
2. *Confinement is well described by the "dual superconductor" model (Mandelstam 1976; 't Hooft 1978b) , relating it to known properties of superconductors via electric-magnetic duality.*

In this section we discuss both of them, arguing that they are only *partially* correct. In fact we are going to argue that flux tubes do exist even above the deconfinement transition temperature and that there are much better analogies to confinement than superconductors. Most facts and considerations needed to understand the phenomena discussed have in fact been in literature for some time. The purpose of these comments is simply to remind them, "connecting the dots" once again, since these questions continue to be asked at the meetings. We will also point out certain aspects of the phenomena which still need to be clarified.

3.8.1 Flux Tubes on the Lattice, at Zero T, and Near T_c

Lattice gauge theory simulations have addressed the confinement issue from their beginning, and by now there are many works which studied the electric flux tubes between static charges. Most of those are done at zero/low T. This documented well the profile of the electric field and the magnetic current "coiling" around it.

The "dual superconductor" analogy leads to a comparison with Ginzburg-Landau theory (also called "the dual Higgs model"), and good agreement with it has been found. These results are well known for two decades (see, e.g., the review Bali (1998)).

However, recent lattice studies (Cea et al. 2018) exploring a near-deconfinement range of temperatures have found that a tubelike profile of the electric field persists even above the critical temperature T_c, to at least $1.5T_c$. One of the plots from this work is reproduced in Fig. 3.12. It corresponds to pure gauge $SU(3)$ theory, which has the first-order transition, seen as a jump in the field strength. Note however that at $T > T_c$ the shape of the electric field transverse profile remains about the same, while the width is *decreasing* with T, making it even more tubelike, rather than expected near-spherical Coulomb behavior. (Note also that the length of the flux tube remains constant, 0.76 fm for this plot.)

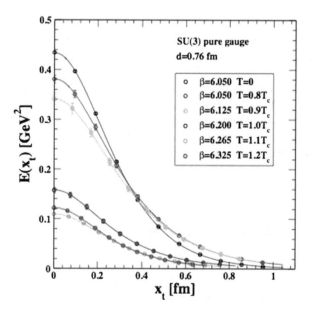

Fig. 3.12 The longitudinal electric field as a function of *transverse* coordinate is measured, for a number of temperatures, for pure gauge $SU(3)$ theory, from Cea et al. (2018)

Such behavior clearly contradicts the "dual superconductor" model: in superconductors, the flux tubes are only observed in the superconducting phase. Why do we observe flux tubes in the "normal" phase, and do we really have any contradictions with theory here?

3.8.2   Does the T_c Indeed Represent the Monopole Condensation Temperature?

Let us start by critically examining the very notion of the deconfinement transition itself, focusing on whether it is indeed the transition between the super and the normal phases.

The T_c itself is defined from thermodynamical quantities, and for pure gauge theories (we will only consider here), its definition has no ambiguities.

At this point it is worth reminding that in fact the electric-magnetic duality relates QCD not to the BCS superconductors, but rather to Bose-Einstein condensation (BEC) of bosons, the magnetic monopoles. Multiple lattice studies did confirmed that T_c does coincide with BEC of monopoles.

One such study I was involved in D'Alessandro et al. (2010) has calculated the probability of the so-called Bose (or rather Feynman's) clusters, a set of k monopoles interchanging their locations over the Matsubara time period. Its dependence on k leads to the definition of the effective chemical potential, which is shown to vanish exactly at $T = T_c$. This means that monopoles do behave as any other bosons, and they indeed undergo Bose-Einstein condensation at exactly $T = T_c$.

Earlier studies by Di Giacomo and collaborators in Pisa group over the years (see, e.g., Bonati et al. (2012)) were based on the idea to construct (highly nonlocal) order parameter for monopole BEC. It calculates the temperature dependence of the expectation value of the operator, effectively inserting/annihilating a monopole, and indeed finds a jump exactly at T_c.

In summary, it has been shown beyond a reasonable doubt that T_c does indeed separate the "super" and "normal" phases.

3.8.3   Constructing the Flux Tubes in the "Normal" Phase

Since electric-magnetic duality relates QCD not to the BCS superconductors, but rather to BEC, let us at this point emphasize a significant difference between them: the *un-condensed* bosons are also present in the system, both above and even below T_c, while the BCS Cooper pairs of superconductor exist at $T < T_c$ only.

Fig. 3.13 A sketch of a monopole traversing the electric flux tube

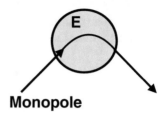

The first construction of the flux tube in the normal phase has been made in Liao and Shuryak (2008a). The first point to explain is that flux tubes as such do not require supercurrents! Indeed, flux tubes are found in various plasmas: e.g., one can even see them in solar corona in a telescope. What is needed for QCD electric flux tube formation is in fact the presence of the *dual plasma*, a medium with movable magnetic charges.

Their scattering on the electric flux tube schematically shown in Fig. 3.13 does not change the monopole energy but changes direction of its momentum, thus creating a force on the flux tube. If it is strong enough to be able to confine the electric field, a flux tube solution can be constructed.

For further Details, see the original paper (Liao and Shuryak 2008a). Let us only comment that (1) the "uncondensed" monopoles exert a *larger* force than those in the condensate, as their momenta are larger, and (2) it has in fact been *predicted* there that the highest T at which such solution may exist is about $1.5T_c$, see further discussion in Sect. 17.4.

3.9   Lattice Studies of the Bose-Einstein Condensation of Monopoles at the Deconfinement Transition

It is convenient to split lattice studies to two kinds: (1) those at $T < T_c$, addressing the Bose-condensed state, and (2) at $T > T_c$ exploring the onset of BEC by looking at finite-size clusters.

At $T < T_c$ (or even $T = 0$) the lattice studies focused on direct detection of the monopole condensate. The idea is to do an *insertion* of a monopole into the vacuum state. This method has been developed by Di Giacomo and collaborators in Pisa group over the years: see, e.g., the most self-contained articles of that series (Bonati et al. 2012).

It is based on the idea that the condensate is analogous to the "coherent state" of a harmonic oscillator, with large average number of quanta $> n <\gg 1$. Such classical-like states are superpositions of many states $|n >$ with different numbers of quanta, Created, e.g., by exponentiation of the field (coordinate) operator. Therefore they do not have fixed number of quanta.

Since the quantum mechanical momentum operator acts as a derivative over the coordinate, $\hat{p} = -i\frac{d}{dx}$ on the wave function. Its exponential $exp(i\hat{p}a)$ can be expanded and considered to be a Taylor series, corresponding to a shift of the wave function argument by a

$$exp(i\hat{p}a)|\psi(x) >= |\psi(x + a) > \qquad (3.49)$$

and thus is known as a "shift operator." Its field theory version generalizes the coordinate shift by a into a field shift

$$\phi(x) \to \phi(x) + a(x) \qquad (3.50)$$

In a gauge theory (and the $A_0 = 0$ gauge) the shift of the gauge field we will call $\vec{a}(x, t)$ and the conjugated canonical momentum is the electric field, so the appropriate operator is a shift by the monopole field

$$\mu(x) = exp\left[i\int d^3y\big(\vec{E}(y,t)\cdot\vec{A}_{mono}(y - x, t)\big)\right] \qquad (3.51)$$

(In the non-Abelian theory the color index is implied.) The object we would like to add to the vacuum, $\vec{a}(x, t)$, can be of any shape, for example, a magnetic monopole. If the state $|\psi >$ is "normal" and has a definite (zero) number of monopoles, the average of it would be zero $< \psi|\mu|\psi >= 0$, but if the state has a monopole condensate, the average would be nonzero! So the $< \mu > (T)$ is the order parameter for monopole BEC.

The details of the definition/normalization can be found in Bonati et al. (2012): we only comment that it is more convenient to measure not the average but its derivative $\bar{\rho} = \partial \log < \mu > /\partial\beta$ over $\beta = 2N_c/g^2$, the coefficient in front of the lattice action. (The bar has no particular meaning here, just as it was called like this in the original work.) The dependence on the coupling β is shown in Fig. 3.14 (from Bonati et al. (2012)): its singularity near $\beta_c = 2.2986$ known independently proves that this is indeed singular at the deconfinement critical coupling. (The particular monopole inserted is here a Yang-Wu monopole (two half-spaces glued together, no Dirac string) of the charge 4: but similar results are obtained also for other monopoles.) The expected critical scaling with the known 3-d Ising index for the order parameter $\bar{\rho} - \bar{\rho}_c \sim L^{1/\nu}$, $\nu = 0.6301$ has also been demonstrated in the same paper.

We already qualitatively discussed the "dual superconductor" paradigm ('t Hooft 1978a; Mandelstam 1976) according to which in the confinement phase at $T < T_c$ monopoles Bose-condense into a—magnetically charged—condensate. In this section we will discuss lattice evidences showing that this is indeed what happens.

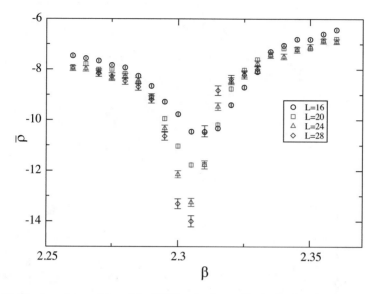

Fig. 3.14 The dependence of $\bar{\rho}$ on β for SU(2) gauge theory. The lattice time size $N_t = 4$ and the spatial sizes are shown as L in the insert for corresponding symbols. A singularity at the critical deconfinement coupling is revealed

The second approach is based on study of the so-called Feynman clusters. As $T \to T_c$ from above, the "dual superconductor" paradigm requires that the behavior of monopoles should change, revealing a "preparation" to form a Bose-Einstein condensate (BEC). The idea of identical clusters is explained in Fig. 3.15: identical bosons may have "periodic paths" in which some number k of them exchange places. Such clusters are widely known to community doing manybody path integral simulations for bosons, e.g., liquid He^4. Feynman argued that in order for statistical sum to get singular at T_c, a sum over k must diverge. In other words, one may see how the probability to observe k-clusters P_k grows as $T \to T_c$ from above.

In Fig. 3.15 (lower) from D'Alessandro et al. (2010), one see the corresponding data for the cluster density. Their dependence on k was fitted by the expression

$$\rho_k \sim \frac{exp\left(-k\mu_{eff}(T)\right)}{k^{5/2}} \tag{3.52}$$

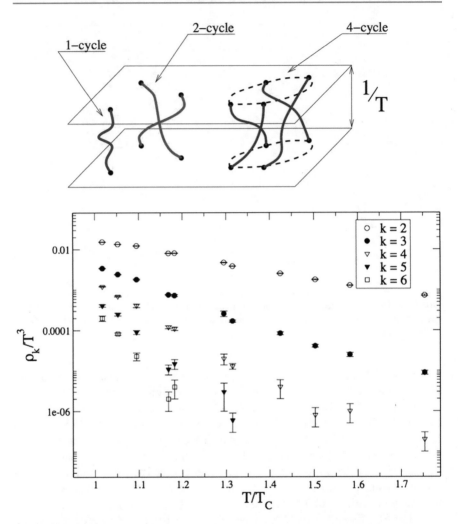

Fig. 3.15 (upper) Example of paths of seven identical particles which undergo a permutation made up of a one-cycle, a two-cycle, and a four-cycle. (lower) Normalized densities ρ_k/T^3 as a function of T/T_c

and the resulting effective chemical potential $\mu_{eff}(T)$ is plotted versus temperature at Fig. 3.16. It vanishes exactly at $T = T_c$. This means that monopoles indeed undergo Bose-Einstein condensation at exactly $T = T_c$.

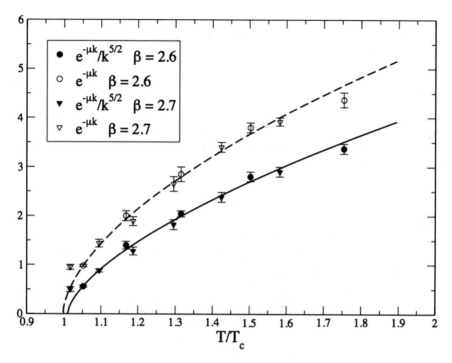

Fig. 3.16 The effective chemical potential extracted by two different fits: it should vanish at the Bose-Einstein condensation point

3.10   Quantum Coulomb Gases Studied by Path Integral Monte Carlo (PIMC)

Quantum studies of interacting particles displaying BEC are a very well-developed area of manybody physics. Its traditional applications are weakly coupled Bose gases and liquid ^4He. The former problem was intensely studied in the 1950s by C.N. Yang and collaborators and, then, by Feynman diagrams, by S.T. Belyaev. Its theory reactivated at the end of the 1990s, after experimental observation of BEC in ultracold atomic gases. Liquid helium was a frontier of experimental low-T research from the beginning of the twentieth century till the 1950s. Its theoretical treatment from the first principles becomes possible in the 1970s, with the development of numerical PIMC simulations. Here is not a place to review these works, so let me only comment that while atoms of ^4He interact with each other quite weakly, by atomic standards, in comparison to the temperature considered $T \sim 2K$ the interatomic potentials are very large. So, one may call liquid ^4He a "strongly coupled" system.

In literature one can find studies of BEC for other interactions, e.g., for a Bose gas of solid spheres, but not for the Coulomb forces. Therefore, simulations for (one- and two-component) Coulomb Bose gases have been made by ourselves (Ramamurti and Shuryak 2017) for the first time, using numerical simulation of the manybody path integral method. Unlike the previous works, it focused on temperature dependence of the density ρ_k of the Bose clusters. Their T dependence for liquid ^4He calculated in this paper looks nearly the same as the lattice data on monopoles shown in Fig. 3.15, and the procedure used to locate the BEC critical temperature accurately reproduces the value of T_c for liquid ^4He, known both from experiments and multiple previous numerical simulations.

The correlations of monopoles have been studied as well, and the coupling strength α defined by

$$V_{ij} = \alpha \frac{q_i q_j}{r_{ij}} \tag{3.53}$$

has been tuned to reproduce lattice data.

We will not discuss that and only show one nontrivial result of this paper, namely, the dependence of the critical BEC temperature T_c on α (see Fig. 3.17). Note that we found the same behavior at small values of the coupling as in the case of low-density hard spheres: the critical temperature for the BEC phase transition grows with the coupling. Yet if the coupling becomes large enough, Tc rapidly drops below

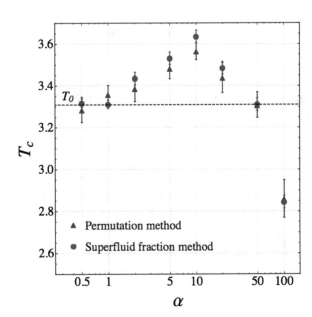

Fig. 3.17 The critical temperature for the BEC phase transition as a function of the coupling, α. The red circles are the results of the finite-size scaling superfluid fraction calculation for systems of 8, 16, and 32 particles; the blue triangles are the results of the permutation-cycle calculation for a system with 32 particles. The black dashed line denotes the Einstein ideal Bose gas critical temperature

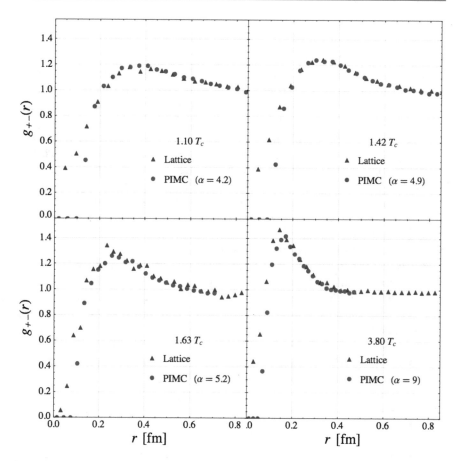

Fig. 3.18 Spatial correlations of particles in quantum Coulomb Bose gas, from PIMC simulations (red circles) compared to lattice data for monopoles

the critical temperature for an ideal Bose gas. Eventually, as the particles are "too repulsive," the BEC phenomenon becomes impossible since it becomes essentially too costly (in terms of the action) to permute them.

We have discussed at the end of the previous chapter the lattice data on the spatial monopole-monopole and monopole-antimonopole correlations. In Fig. 3.18 these data are compared with PIMC simulation for a Coulomb Bose gas (Ramamurti and Shuryak 2017). The comparison shows very good agreement, increasing the confidence that a quantum ensemble of monopoles is described by this model well. It also allowed us to fix the effective magnetic coupling rather accurate, without any reliance on the Debye fits.

3.11   Brief Summary

- Using classical equation of motion, one can play with a set of charges and monopoles, in increasing number of those. The motion is amusing and quite different from motion of charges of one kind alone. In particular, there is no familiar "scattering plane." Motion is not planar because angular momentum resigns not only in particles but also in crossed electric and magnetic fields.
- Classical "dual plasmas," made of many charges and monopoles, are also different from the conventional Coulomb systems. Large backward scattering makes even classical kinetics different, with large transport cross section.
- Another practically important setting is a high energy charge moving through "dual plasma." In heavy ion collisions, it is observed as "jet quenching." As we have shown, *charge-monopole* scattering dominates it because there is no small coupling in cross section.
- A problem of charge-monopole scattering in quantum-mechanical setting is highly nontrivial, solved only 1976, in spite of many efforts of luminaries like Dirac himself since the 1930s. Amazingly, it is not the radial equation which is difficult, but the angular part.
- It has been since carried for monopole-quark and monopole-gluon scattering, with all relevant transport cross sections calculated.
- Manybody studies of quantum Coulomb Bose gas, with its Bose-Einstein condensation (BEC), were applied to the monopole ensemble reproducing lattice data on monopole ensemble quite accurately.

References

Baier, R., Dokshitzer, Y.L., Mueller, A.H., Peigne, S., Schiff, D.: The Landau-Pomeranchuk-Migdal effect in QED. Nucl. Phys. B **478**, 577–597 (1996)

Bali, G.S.: The Mechanism of quark confinement. In: Quark confinement and the hadron spectrum III. Proceedings, 3rd International Conference, Newport News, USA, June 7–12, 1998, pp. 17–36 (1998). arXiv:hep-ph/9809351

Bonati, C., Cossu, G., D'Elia, M., Di Giacomo, A.: The disorder parameter of dual superconductivity in QCD revisited. Phys. Rev. D **85**, 065001 (2012). arXiv:1111.1541

Boulware, D.G., Brown, L.S., Cahn, R.N., Ellis, S.D., Lee, C.-K.: Scattering on magnetic charge. Phys. Rev. D **14**, 2708 (1976)

Cea, P., Cosmai, L., Cuteri, F., Papa, A.: QCD flux tubes across the deconfinement phase transition. EPJ Web Conf. **175**, 12006 (2018). arXiv:1710.01963

D'Alessandro, A., D'Elia, M., Shuryak, E.V.: Thermal monopole condensation and confinement in finite temperature Yang-Mills theories. Phys. Rev. D **81**, 094501 (2010). arXiv:1002.4161

Kaczmarek, O., Zantow, F.: Static quark anti-quark interactions at zero and finite temperature QCD. II. Quark anti-quark internal energy and entropy (2005). arXiv:hep-lat/0506019

Kazama, Y., Yang, C.N., Goldhaber, A.S.: Scattering of a Dirac particle with charge Ze by a fixed magnetic monopole. Phys. Rev. D **15**, 2287–2299 (1977)

Liao, J., Shuryak, E.: Strongly coupled plasma with electric and magnetic charges. Phys. Rev. C **75**, 054907 (2007). arXiv:hep-ph/0611131

Liao, J., Shuryak, E.: Electric flux tube in magnetic plasma. Phys. Rev. C **77**, 064905 (2008a). arXiv:0706.4465

Liao, J., Shuryak, E.: Angular dependence of jet quenching indicates its strong enhancement near the QCD phase transition. Phys. Rev. Lett. **102**, 202302 (2009). arXiv:0810.4116

Mandelstam, S.: Vortices and quark confinement in nonabelian gauge theories. Phys. Rep. **23**, 245–249 (1976)

Ramamurti, A., Shuryak, E.: Effective model of QCD magnetic monopoles from numerical study of one- and two-component coulomb quantum Bose gases. Phys. Rev. D **95**(7), 076019 (2017). arXiv:1702.07723

Ramamurti, A., Shuryak, E.: Role of QCD monopoles in jet quenching. Phys. Rev. D **97**(1), 016010 (2018). arXiv:1708.04254

Ratti, C., Shuryak, E.: The role of monopoles in a gluon plasma. Phys. Rev. D **80**, 034004 (2009). arXiv:0811.4174

Schwinger, J.S., Milton, K.A., Tsai, W.-Y., DeRaad, Jr., L.L., Clark, D.C.: Nonrelativistic Dyon-Dyon scattering. Ann. Phys. **101**, 451 (1976)

Shuryak, E.V.: The Azimuthal asymmetry at large p(t) seem to be too large for a 'jet quenching'. Phys. Rev. C **66**, 027902 (2002)

't Hooft, G.: On the phase transition towards permanent quark confinement. Nucl. Phys. B **138**, 1–25 (1978a)

't Hooft, G.: On the phase transition towards permanent quark confinement. Nucl. Phys. B **138**, 1–25 (1978b)

Weinberg, E.J.: Monopole vector spherical harmonics. Phys. Rev. D **49**, 1086–1092 (1994)

Xu, J., Liao, J., Gyulassy, M.: Bridging soft-hard transport properties of quark-gluon plasmas with CUJET3.0. J. High Energy Phys. **02**, 169 (2016). arXiv:1508.00552

Fermions Bound to Monopoles

4

4.1   Fermionic Zero Modes

Let me start this section with a small historical introduction. In 1974, when the monopole solution by 't Hooft and Polyakov has been discovered, it was not yet clear whether either Georgi-Glashow or Weinberg-Salam model is the correct description of the electroweak interactions.[1] So some discussion of electroweak monopoles happened. Afterward it shifted to the context of Grand Unification models, most of which do possess such monopoles. As noticed by Zeldovich in 1978, their cosmological production versus non-observation became an issue: Guth in 1980 famously invented cosmological inflation, in order to get rid of the undesired Grand Unification monopoles. In all of that, scattering of monopoles on ordinary matter was studied: for Grand Unification monopoles, violation of the baryon number was noticed and much discussed.

Not going into discussion of electroweak or Grand Unification models as such, let us return to the original Georgi-Glashow model, to which we will add some fermions. In fact, we already discussed the $\mathcal{N}=2$ supersymmetric models, which have the "gluinoes," fermionic partners of the gluons, or quarks, with the fundamental color representation. It is in this context that people wandered what is the spectrum of the Dirac equation, in the background field of the monopole solution.

The central observation is that there exist the so-called zero fermionic modes. There is rather extensive literature on the meaning of these states, starting from (Jackiw and Rebbi 1976). Like for other topological solitons, there are topological index theorems, relating the number of zero modes to the monopole quantum

---

[1]While some evidences for neutral weak current were known, e.g., atomic parity violation, direct observation of the Z boson only happened in 1983.

© The Author(s), under exclusive license to Springer Nature Switzerland AG 2021
E. Shuryak, *Nonperturbative Topological Phenomena in QCD and Related Theories*, Lecture Notes in Physics 977,
https://doi.org/10.1007/978-3-030-62990-8_4

number M. They require *one* zero mode for a fundamental fermion (quark) and N_c of them for the adjoint (gluino) fermion.

As argued by (Jackiw and Rebbi 1976) in their classic paper, the operator algebra involved corresponds to a pair of creation/annihilation operators, with the algebra $\{aa^+\} = 1$, requiring representation in the form of two states, the "empty" and "occupied" ones. Symmetries of the problem, such as CP conjugation, plus a requirement that these two states differ by one unit of the fermion number, let them to (at the time revolutionary) conclusion, that the baryon number of such states are $semi - integer$, namely, $\pm 1/2$.

Let us look at these two states from the viewpoint of the "Dirac sea" picture of the fermionic vacuum state. All levels with positive energy are supposed to be empty, and all negative energy states occupied. But what to do with extra two *zero energy states* sitting on the monopole? If occupied we still call it a *particle*, if not, a *hole*.

What about the *spin* of these zero mode states? Starting in the simplest case of the $N_c = 2$ theory, we will use the term "isospin" instead of the color (or "weak isospin" of electroweak sector). Thus the quarks in fundamental representation we will call isospin-1/2 fermion, and gluinoes in the adjoint representation will be called isospin-1 fermions. The monopole field is a "hedgehog," relating direction in coordinate space and in the isospin space: therefore isospin and spin are not separately conserved. Yet after some observation of the Dirac equation one may prove that so-called *grandspin*

$$\vec{K} = \vec{I} + \vec{S}$$

is in this case conserved. Following standard rules of angular momentum representations, one finds that K can have values $\vec{1}/2 + \vec{1}/2 = 0, 1$ in the case of quark, and $\vec{1} + \vec{1}/2 = 1/2, 3/2$ in the case of gluino. Explicit solution of the Dirac equation shows that in both cases zero modes correspond to the lowest values: thus the number of states is $2K + 1$. It is 1 for the quark $K = 0$ state and 2 for $K = 1/2$ gluino state. Conclusion: a quark bound to a monopole is a single state; thus it is a *scalar* spin-0 object. A gluino bound to a monopole makes two states; thus it makes a *spinor* state. Standard spin-statistics theorem then requires that in the former case one produces a boson and in the latter case a fermion.

Further discussion naturally extends to the problem of counting the number of states and their statistics when there are several species of the fermions. Explicit construction of such states has obtained another motivation in the mid-1990s, with the discussion of various supersymmetric theories. For a review see Harvey (1996). Here is the main idea: all fermions of the theory coupled to the monopole are expected to produce only "magnetically charged" multiplets which are *consistent with the underlying symmetries of the theory*.

Example: consider the $\mathcal{N}=2$ SYM, which we already mentioned in relation to Seiberg-Witten famous papers. It has two adjoint *real* gluinoes. Each can be bound to a monopole, leading to a fermion state.

Now, in order to have creation and annihilation operator algebra, with $\{\hat{a}^+\hat{a}\}$ =1 etc., the fields need to be complex. So, like in the harmonic oscillator, in which they are built from two hermitian (=real eigenvalued) operators $\hat{p}, \hat{x}$, one needs to combine two gluinoes into one complex (Dirac-like) fermion. Thus there is a *single* set of creation and annihilation operators. The maximal spin state is 1/2—two partners of a scalar monopole.

When both are bound at the same time, it is spin $\vec{1}/2 + \vec{1}/2 = 0, 1$ states. Spin zero can be combined with the unoccupied monopole, producing two scalar states. We in fact do *not* have a spin-1 state here, as its wave function is not antisymmetric as fermions should. As a result, total magnetic supermultiplet gives the "short" representation of the $\mathcal{N}=2$ supersymmetry, with two fermions and their scalar partners. So, the effective magnetic theory is the $\mathcal{N}=2$ electrodynamics.

Further study reveals, among other things, beautiful examples of theories with complete electric-magnetic $self - duality$. It has been explicitly shown for two theories, the $\mathcal{N}=4$ SYM and $\mathcal{N}=2$ SQCD with $N_f = 4$ quark flavors, that a magnetic monopole "dressed" by all available fermions of these theories, does indeed generates the "long" supermultiplet, starting from spin-1 states to $\mathcal{N}$ spin 1/2 states, and correct number of scalars. The $\mathcal{N}=4$ SYM has 4 real gluinoes, so there are *two* sets of creation and annihilation operators. Thus the maximal achievable spin is 1—one now has vector magnetic particle. There are 16 states, the "long" supermultiplet, starting from spin-1 states to $\mathcal{N}$ spin 1/2 states, and correct number of scalars (six of them).

What this means is that the effective magnetic theory and the effective electric theory happen to be in these two cases the same, up to the fact that one has electric and another magnetic—inverse to electric—coupling. At one hand Dirac condition requires that their beta functions should have the opposite sign; on the other hand they must be the same, as both theories have the same Lagrangian. The only solution possible here is the following one: both those theories are $conformal$, with their RG beta function equal to $zero$!

Let us now return to the non-supersymmetric world and ask what fermion binding to monopoles can imply for various QCD-like theories. Without adjoint fermions, only spin-zero new states can be generated, no matter how many quarks are or are not bound to the monopole. So the effective magnetic theory can only be a scalar electrodynamics.

The number of magnetic states can be rather large. Both states exist for each flavor, so in QCD with N_f quark flavors the number of magnetic states we start to consider is thus 2^{N_f}, half with integer and half with semi-integer fermion number. Of course, not all of these states can exist: since each bound state is a boson, the allowed wave functions must all be symmetric under permutations. For example, if all N_f zero modes are occupied—the states with maximal possible number of flavor indices—it should be symmetric tensor.Its baryon number is $N_f/2$, say 5 for $N_f = 10$. If all quarks have the same mass—e.g., zero—there is unbroken flavor symmetry, and thus such magnetic states should fall into its proper flavor multiplets.

Now, how would the deconfinement transition be affected by adding more and more light fermions? As argued by (Liao and Shuryak 2012), it will dramatically shift downward the transition temperature (or, more precisely, shift the transition

Fig. 4.1 Dependence of the critical lattice coupling β_c at scale T_c versus the number of fundamental quark flavors N_f in QCD-like theories. The thick blue line is the fitting curve, extended as dashed blue line beyond $N_f = 12$. The black/purple/red curves on the right are lines for vanishing beta function at 1,2,3-loop levels: they indicate the boundaries of the so-called conformal window at large N_f

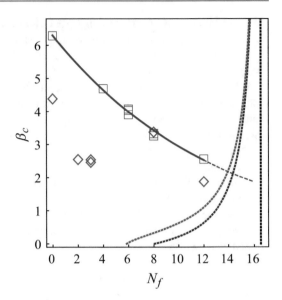

to stronger coupling). The reason for it is simple: since some number of fermions can be attached to the monopoles via zero modes, the resulting states are no longer identical particles. And only the identical ones—most likely the monopoles *without* fermions—will undergo the Bose-Einstein condensation, as soon as their density becomes critical. Shifting to IR direction, with the lower T_c or stronger coupling, decreases the monopole mass and increase the overall monopole density, compensating for "occupied" monopoles.

For details see the original paper: let me just show one plot, Fig. 4.1, which shows that the expected trend is indeed observed, by lattice studies of the theories with many quark favors, nowadays up to $N_f = 12$. Blue boxes are from Miura et al. (2012): near-coincident boxes being lattice data for the same N_f with different number of lattice cites N_τ which demonstrate lattice spacing consistency. Red diamonds are from various other lattice studies.

4.2 Chiral Symmetry Breaking by Monopoles

In chapter on instantons, we will discuss in detail how a collectivization of the 4-d fermionic instanton zero modes, resulting in breaking of the chiral symmetry by a nonzero quark condensate $\langle \bar{q}q \rangle \neq 0$. We will then show in chapter on instanton-dyons that this approach can be directly generalized to finite temperatures, since (some) of the instanton-dyons also possess the 4-d zero modes.

Above in Chap. 2, we have argued that Euclidean semiclassical theory based on instanton-dyons is *Poisson dual* to the monopole approach. If so, one should be able to derive chiral symmetry breaking using monopoles as well. This was indeed accomplished by Ramamurti and Shuryak (2018a), which we follow in this section.

One obvious difficulty of the problem is the fact that a detailed understanding of the "lattice monopoles" is lacking; they are treated as effective objects whose parameters and behavior we can observe on the lattice and parameterize, but their microscopic structure has yet to be understood. In particular, the 't Hooft-Polyakov monopole solution includes a chiral-symmetry-breaking scalar field, while we know that, in massless QCD-like theories, chiral symmetry is locally unbroken. We assume that the zero modes in question are chiral themselves, like they are in the instanton-dyon theory, and that chiral symmetry breaking can only be achieved by a spontaneous breaking of the symmetry.

The other difficulty of the problem is the important distinction between fermionic zero modes of (1) monopoles and (2) instanton-dyons. As follows from Banks-Casher relation (Banks and Casher 1980), the quark condensate is proportional to density of Dirac eigenstates at zero eigenvalue. The monopoles also have fermionic zero modes (Jackiw and Rebbi 1976), which are three-dimensional. They are, therefore, simply a bound state of a fermion and a monopole. In theories with extended supersymmetries, such objects do exist, fulfilling an important general requirement that monopoles need to come in particular supermultiplets, with fermionic spin 1/2 for $\mathcal{N} = 2$ or spins 1/2 and 1 for $\mathcal{N} = 4$. The antiperiodic boundary conditions for fermions in Matsubara time imply certain time dependence of the quark fields, and (as we will discuss in detail below) the lowest four-dimensional Dirac eigenvalues produced by quarks bound to monopoles are the values $\lambda = \pm \pi T$, not at zero.

This, however, is only true for a single monopole. In a monopole *ensemble* with nonzero density, the monopole-quark bound states are collectivized and Dirac eigenvalue spectrum is modified. The question is whether this effect can lead to a nonzero $\rho(\lambda = 0) \propto \langle \bar{q}q \rangle$ and, if so, whether it happens at the temperature at which chiral symmetry breaking is observed. As we will show below, we find affirmative answers to both these questions. The phenomenological monopole model parameters are such that a nonzero quark condensate is generated by monopoles at $T \approx T_c$.

Recognizing fermionic binding to monopoles, we now proceed to description of their dynamics in the presence of ensembles of monopoles. The basis of the description is assumed to be the set of zero modes described in the previous section. The Dirac operator is written as a matrix in this basis, so that $i - j$ element is related to "hopping" between them. The matrix elements of the "hopping matrix"

$$\mathbf{T} = \begin{pmatrix} 0 & i T_{ij} \\ i T_{ji} & 0 \end{pmatrix} \qquad (4.1)$$

where the T_{ij}s are defined as the matrix element,

$$T_{ij} \equiv \langle i | -i \not{D} | j \rangle ,$$

between the zero modes located on monopoles i and antimonopoles j. In the SU(2) case we are considering, this is equivalent to

$$
\begin{aligned}
T_{ij} &= \langle \psi_i | x | \rangle \langle x | -i\not{D} | y \rangle \langle y | \psi_j | \rangle \\
&= \int d^3 x \, \psi_{kn}^\dagger (x - x_i)(-i\not{D}) \psi_{lm}(x - x_j) \\
&= \int d^3 x \, \psi_{kn}^\dagger (x - x_i) \Big[-i(\vec{\alpha} \cdot \vec{\partial} + \vec{\alpha} \cdot \vec{\partial} - \vec{\alpha} \cdot \vec{\partial}) \delta_{nm} \\
&\quad + \frac{1}{2}(A(x - x_i) + A(x - x_j)) \sigma_{nm}^a (\vec{\alpha} \times \hat{r})_a \\
&\quad + \frac{G(\phi(x - x_i) + \phi(x - x_j))}{2} \sigma_{nm}^a \hat{r}_a \beta \Big] \psi_{lm}(x - x_j) \\
&= \int \sum_m d^3 x \, \psi_{km}^\dagger (x - x_i)[-i\vec{\alpha} \cdot \vec{\partial}]^{kl} \psi_{lm}(x - x_j) \qquad (4.2)
\end{aligned}
$$

where ψs are zero modes with origin at $x_{i,j}$, the locations of the two monopoles, n, m are the isospin/color indices, and we have used the fact that applying the Dirac operator to these wave functions gives zero.

Omitting further details, let us explain the quantization procedure adopted in that work, the *evolution matrix* U, defined as time-ordered integral of the hopping matrix in the previous section over the Matsubara periodic time $\tau \in [0, \beta]$. This matrix will then be diagonalized to find the eigenvalues for the fermion states. Because each eigenstate is still fermionic, each is required to fulfill the fermionic boundary conditions, namely, that the state must return to minus itself after one rotation around the Matsubara circle. This defines quantization condition of the Dirac eigenvalues by,

$$
\lambda_i + \omega_{i,n} = \left(n + \frac{1}{2} \right) \frac{2\pi}{\beta}
$$

where λ_is are the eigenvalues of the hopping matrix **T**. For monopoles that move in Euclidean time, we must integrate over the Matsubara circle to find the antiperiodic evolution operator, and define the fermion frequencies,

$$
U = \oint_\beta d\tau e^{iH\tau} = -1.
$$

One needs to diagonalize the resulting matrix on the right-hand side and solve to find the quantity $\lambda + \omega$. One can then compute the eigenvalues of the Dirac operator with

$$
\omega_{i,n} = \left(n + \frac{1}{2} \right) \frac{2\pi}{\beta} - \lambda_i .
$$

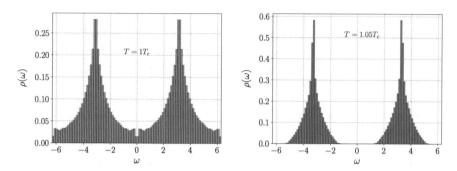

Fig. 4.2 Distributions of Dirac eigenvalues for $T/T_c =:$ (**a**) 1 and (**b**) 1.05, respectively

Considering only the $n = \pm 1$ case, so that $\omega_i = \pm \pi T - \lambda_i$, we get the distributions shown in Fig. 4.2a and b, for $T/T_c = 1$ and 1.05, respectively.

The first, and most important, thing to notice is that when $T = T_c$, the eigenvalue distribution has a finite density at $\omega = 0$ (see Fig. 4.2), which indicates the nonzero value of the chiral condensate; there is no gap in the spectrum present at $T = T_c$. (A small dip seen around zero is a consequence of finite size of the system, well known and studied on the lattice and in topological models. It should be essentially ignored in extrapolation to zero.)

Let us summarize this section as follows. The mechanism of chiral symmetry breaking based on monopoles is as follows. A single monopole (or antimonopole) generates additional quark and antiquark bound states. At high temperatures, the monopoles have large mass and the probability of hopping is therefore low. The 4-d Dirac operator eigenvalues are well localized near the fermionic Matsubara frequencies $2\pi T (n + 1/2)$. Using the condensed matter analogy, one may say that a matter is an insulator. However, as T decreases toward T_c, the amplitudes of quark "hopping" from one monopole to an antimonopole (and vice versa) grow. Eventually, at some critical density, quarks become "collectivized" and are able to travel very far from their original locations. The physics of the mechanism is similar to insulator-metal transition in condensed matter under pressure.

4.3 More on Fermions Bound to Monopoles, in the SUSY World and Perhaps Beyond

This is an advanced topic section on a very specific but fascinating topic, namely, the possible role of $multi - monopole$ states and BEC condensation.

My interest to the subject was inspired by the second Seiberg-Witten work, on $\mathcal{N}=2$ QCD (Seiberg and Witten 1994b), on the supersymmetric QCD with fundamental quarks/scalars. As the first, it deals with the simplest $N_c = 2$ color group $SU(2)$, but now with N_f flavors of quark-squark multiplets. Addition of massless quarks are restricted to only four cases: $N_f = 4$ already has zero beta function and is thus a conformal theory, as already was mentioned before.

The case of $N_f = 3$ quark/squark flavors is "normal," with the asymptotic freedom. Seiberg and Witten predicted two district singularities on the moduli space, which correspond to the following particles becoming massless:

1. a quartet of states with magnetic charge $n_m = 1$ and electric charge $n_e = 0$;
2. a singlet with $n_m = 2, n_e = 1$.

Various SUSY breaking terms would transform those singularities into two nonequivalent vacua, with two different confinement phases. The second singularity thus induces a "dual superconductor" in which not just a monopole but a *monopole pair* Bose condenses! One may wander what should this state with charges $n_m = 2, n_e = 1$ be, whether it can be identified semiclassically in weak coupling, and why does it only appear in the $N_f = 3$ case?

Studies of two- (and multi-) monopole states lead me to an influential paper by Sen (1994) just predated Seiberg-Witten works.

Recall that if there are k monopoles, there should be 4k-dimensional moduli space, which has the form $R^3 S^1 M_{4k-4}$ after the global shifts and phase is separated from the relative coordinates. Let X^α be the collective coordinates and λ^α its fermionic counterparts required by supersymmetries. For the $\mathcal{N}=4$ theory, the Lagrangian of the so-called supersymmetric quantum mechanics on the moduli space takes the form

$$L = \left(\frac{1}{2}\right) g_{\alpha\beta} \partial_0 X^\alpha \partial_0 X^\beta + \left(\frac{1}{2}\right) g_{\alpha\beta} \bar{\lambda}^\alpha i D_0 \lambda^\beta + \left(\frac{1}{12}\right) R_{\alpha\beta\gamma\delta} \bar{\lambda}^\alpha \lambda^\beta \bar{\lambda}^\gamma \lambda^\delta$$

$$(4.3)$$

where $g_{\alpha\beta}$ is the metric on the moduli spaces, the covariant derivative $D_0 \lambda^\beta = \partial_0 \lambda^\beta + \Gamma^\beta_{\alpha\gamma} \partial_0 X^\alpha \lambda^\beta$ contains the Crystoffel connection, and the last term contains the full Riemann tensor, calculated from $g_{\alpha\beta}$ in a standard way. Of course, global coordinate and phase $R^3 S^1$ part are flat and do not have such additions, which only appear for relative motion.

The metric is a nontrivial function of the coordinates X, explicitly known for the famous 4-d Atiyah–Hitchin manifold of the two monopoles. The fermionic part can in principle be rewritten in terms of creation-annihilation operators and quantized in a usual way. There are "fermion-empty" states, then one-fermion, two-fermion, etc. all the way to maximal number of fermions one can put in. It has been pointed out by Witten in the index paper of 1982, that such p-fermion states for a supersymmetric sigma model on a manyfold correspond to the p-differential form. The number of harmonic (zero energy) forms is known in mathematics as Betty number B_p and is a topological property of the manyfold itself. $\sum(-)^p B_p$ gives the Euler characteristics. Thus the number of bound multi-monopole states depends only on topology of the moduli spaces!

It turns out that in the case of the Atiyah–Hitchin two-monopole manyfold, the only nonzero Betty number is $B_2 = 1$: thus there is a *single* zero energy state, corresponding to such molecular state, with $n_m = 2$. This state was explicitly found by Sen (1994), who argued that populating gluinoes associated with the "trivial" coordinates provides exactly 16 states needed for completing the supermultiplet. Then he noticed that while the maximal p (and thus number of fermions) is the dimension of the manifold (equal to 4), one can always convolute with the epsilon and come to $4 - p$ form and get a Hodge-dual solution. Since he knew it would be a single solution, it must be a self-dual case, with $p = 4 - p$ or $p = 2$. This means that we need to find an antisymmetric function of two variables, the wave function of two gluinoes. The function itself is a bit technical and I would not give it here, but only notice that the Riemann term is crucial and at large distances between the monopoles (when it goes to zero) the 2-gluino wave function exponentially goes to zero as well.

Going to less supersymmetric theories and introducing fundamental quarks/ squarks lead to similar but more complex supersymmetric quantum mechanics on the manyfold (see Gauntlett (1994)). For $\mathcal{N}=2$ QCD, there is also a "molecular" state made of two monopoles bound together, like the Sen's state, but now by (at least) three quarks. Its existence has been shown in Gauntlett and Harvey (1996) using appropriate index theorem and properties of the Atiyah–Hitchin manyfold: but as far as I know, its wave function was not obtained, and thus the normalizability (both at large distances and near the "bolt" or hole in the manifold) may still be problematic.

What these examples of what I call the "unusual confinements" tell us? Well, in weak coupling domain, the monopoles are heavy, and monopole molecules like the ones just discussed are of course twice heavier than one monopole. The molecular state is handicapped at the start. But, when one moves along the v plane toward the stronger coupling, the binding energy of this molecular state seems to grow so much, as to make it the champion! Indeed, it gets massless and undergoes Bose-Einstein condensation *before* any of the single-monopole states! This binding is so large because of (a bit mysterious) coupling to large curvature of the two-monopole space.

Let me end by asking a (so far unanswered) question: *are there other unusual confinement cases in non-supersymmetric theories?* The natural place to look for such phenomena is the theories with the maximal N_f close to the conformal window, because they have confinement at stronger coupling, with longer RG flow.

4.4   Brief Summary

- We learned in this chapter that a massless fermion can get bound to the (t'Hooft-Polyakov) monopole, and the state has zero energy. Thus the problem appears: should we count it as a particle (positive energy states) or "hole" (negative energy states)? True answer is, *both* states exist, different by one unit of baryonic charge. This means a bound state can be either occupied or empty.

- In an ensemble of many monopoles, fermions can get attached to many of them at the same time. It creates certain collectivized fermionic states. If monopole density is larger than some critical value, these states spontaneously break the chiral symmetry.
- We also discussed the issue of fermions bound to monopoles in somewhat more general standapoint, outside QCD. In theories with unbroken supersymmetry, the set of states monopole+fermions should also form certain supermultiplets. One can indeed see how exactly this happens, in a number of examples.
- The most famous is $\mathcal{N} = 4$ SYM theory, in which all bound states of gluinoes and monopoles form also $\mathcal{N} = 4$ SYM theory, but with the inverse coupling. Beta function is the zero, as it should be itself with the minus sign.

References

Banks, T., Casher, A.: Chiral symmetry breaking in confining theories. Nucl. Phys. B **169**, 103–125 (1980)

Gauntlett, J.P.: Low-energy dynamics of N=2 supersymmetric monopoles. Nucl. Phys. B **411**, 443–460 (1994)

Gauntlett, J.P., Harvey, J.A.: S duality and the dyon spectrum in N=2 superYang-Mills theory. Nucl. Phys. B **463**, 287–314 (1996). arXiv:hep-th/9508156

Harvey, J.A.: Magnetic monopoles, duality and supersymmetry. In: Fields, strings and duality. Proceedings, Summer School, Theoretical Advanced Study Institute in Elementary Particle Physics, TASI'96, Boulder, June 2–28, 1996 (1996). arXiv:hep-th/9603086.

Jackiw, R., Rebbi, C.: Solitons with Fermion number 1/2. Phys. Rev. D **13**, 3398–3409 (1976)

Liao, J., Shuryak, E.: Effect of light fermions on the confinement transition in QCD-like theories. Phys. Rev. Lett. **109**, 152001 (2012). arXiv:1206.3989

Miura, K., Lombardo, M.P., Pallante, E.: Chiral phase transition at finite temperature and conformal dynamics in large NF QCD. Phys. Lett. B **710**, 676–682 (2012) arXiv:1110.3152.

Ramamurti, A., Shuryak, E.: Chiral symmetry breaking and monopoles in gauge theories (2018a). arXiv:1801.06922

Seiberg, N., Witten, E.: Monopoles, duality and chiral symmetry breaking in N = 2 supersymmetric QCD. Nucl. Phys. **B 431**, 484–550 (1994b). arXiv:hep-th/9408099

Sen, A.: Dyon-monopole bound states, selfdual harmonic forms on the multi-monopole moduli space, and SL(2,Z) invariance in string theory. Phys. Lett. B **329**, 217–221 (1994). arXiv:hep-th/9402032

Semiclassical Theory Based on Euclidean Path Integral

5

The quantum mechanics courses include semiclassical methods based on certain representation of the wave function, starting with the celebrated Bohr–Sommerfeld quantization condition, applied to the oscillator and hydrogen atom, and the Wentzel–Kramers–Brillouin (WKB) approximation, developed in 1926. Unfortunately, subsequent study has shown that generalization of those to systems with more than one degree of freedom, as well as to systematic order-by-order account for quantum fluctuations are difficult.

However, in this book we will only use quantum mechanical examples as pedagogical tools, while our true interest would be in applications to systems with many degrees of freedom, and eventually to QFT's, at zero and finite temperatures. Fortunately, such generalizable methods exist, based on *the Feynman path integrals*.

For pedagogical reasons, we will deviate from historical path of the development and start with a version of semiclassical theory recently developed by Escobar-Ruiz et al. (2016) for the *density matrix*, and only later will move to (somewhat technically more involved) semiclassical theory of the tunneling effects, described by quantum-mechanical *instantons*.

5.1 Euclidean Path Integrals and Thermal Density Matrix

5.1.1 Generalities

By definition the Feynman path integral gives the *density matrix* in coordinate representation, see e.g. a very pedagogical book (Feynman and Hibbs 1965)

$$\rho(x_i, x_f, t_{tot}) = \int_{x(0)=x_i}^{x(t_{tot})=x_f} Dx(t) e^{i \, S[x(t)]/\hbar} . \tag{5.1}$$

This object is a function of the initial and final coordinates, as well as the time needed for the transition between them. Here S is the classical action of the system we study, e.g. for a particle of mass m in a static potential $V(x)$ it is

$$S = \int_0^{t_{tot}} dt \left[\frac{m}{2}\left(\frac{dx}{dt}\right)^2 - V(x) \right] ,$$

Feynman had shown that the oscillating exponent of it on the path provides the correct weight of the paths integral (5.1).

For reference, the same object can also be written in forms closer to standard quantum mechanics courses. Heisenberg would write it as a matrix element of the time evolution operator, the exponential of the Hamiltonian[1]

$$\rho(x_i, x_f, t_{tot}) = \left\langle x_f | e^{i\hat{H}t_{tot}} | x_i \right\rangle \tag{5.2}$$

between states in which particle is localized at two locations considered.

Schreodinger set of stationary states $\hat{H}|n\rangle = E_n|n\rangle$ can also be used as the state basis. Because Hamiltonian is diagonal in this basis, there is a single (not double) sum over them

$$\rho(x_i, x_f, t) = \sum_n \psi_n^*(x_f)\psi_n(x_i)e^{iE_nt} \tag{5.3}$$

with $\psi_n(x) = \langle n|x\rangle$.

Oscillating weights for different states are often hard to calculate, and one may wonder if it is possible to perform analytic continuation in time to its Euclidean version with i absorbed into it. For reasons which will soon be clear, we will also define this imaginary time on a *circle* with circumference β

$$\tau = it \in [0, \beta]$$

In this way we will be able to describe *quantum + statistical* mechanics of a particle in a heat bath with temperature T related to the circle circumference

$$\beta = \frac{\hbar}{T}$$

Such periodic time is known as the Matsubara time.

Indeed the expression (5.3) will look as

$$\rho(x, x, t) = \sum_n |\psi_n(x)|^2 e^{-E_n/T} \tag{5.4}$$

---

[1] We here assumed that motion happens in time-independent potential: otherwise it would be time-ordered exponential.

combining quantum-mechanical probability to find particle at point x with the thermal weight. Taking integral over all x and using normalization of the weight functions one find the expression for thermal partition function

$$Z = \sum_n e^{-E_n/T} \tag{5.5}$$

While in this chapter we will focus on the zero temperature limit and the ground state (vacuum), it is beneficial to view quantum mechanics as the limit $T \to 0$, $\beta \to \infty$.

So, returning to the path integral, the expression we will use below would be Feynman path integral, but (1) taken over all *periodic* paths, with the same endpoints, and (2) with Euclidean or rotated time. The probability to find particle at certain point is then

$$P(x_0, t_{tot}) = \int_{x(0)=x_0}^{x(\beta)=x_0} Dx(\tau)e^{-S_E[x(\tau)]/\hbar} . \tag{5.6}$$

Note here the exponent is not oscillating, including with the minus sign and the so called Euclidean action

$$S_E = \int_0^\beta d\tau \left[\frac{m}{2} \left(\frac{dx}{d\tau} \right)^2 + V(x) \right] \tag{5.7}$$

in which the sign of the potential is reversed and the time derivative are understood to be over τ.

There are *two* basic approaches to thermodynamics, based of these expressions. While the particular formulae for the statistical sum (and other quantities) obtained by them look different, with different dependencies on the temperature and other parameters of the problem, if the sums are exact one can prove that they in fact lead to exactly *the same* results. In some applications this proof is related to the Poisson summation formula, and therefore the phenomenon is known as the *Poisson duality* of two approaches.

One approach, which one may call a *Hamiltonian* one, use the standard definition of the density matrix in terms of stationary states, the eigenstates of the Hamiltonian with definite energy $\hat{H}|n\rangle = E_n|n\rangle$

$$P(x_0, \beta) = \sum_n |\psi_n(x_0)|^2 e^{-E_n\beta} , \tag{5.8}$$

The sum over stationary states is obviously better convergent in the case of large β, or low T. In the limit $\beta \to \infty$ only the lowest—the ground state dominates

$$P(x_0, \beta \to \infty) \sim |\psi_0(x_0)|^2 \tag{5.9}$$

Another approach, which one may call a *Lagrangian* one, looks for the periodic paths with the minimal action. The simplest of such paths is obviously those for which particles do not move at all, $x(\tau) = const$! Such path would dominate in the case of small Matsubara circle, $\beta \to 0$ (or high T).[2] If one ignores the time dependence and velocity on the paths, there is no kinetic term and only the potential one in the action contributes. So,

$$P(x_0, \beta) \sim e^{-\frac{V(x_0)}{T}}, \qquad (5.10)$$

which corresponds to classical[3] thermal distribution for a particle in a potential V.

In general, the periodic paths on the circle falls into the topological classes, depending on the number of rotations—or *the winding number n_w*—the path makes. The time Fourier transform of such paths are described by a *discrete* Fourier series, with discrete *Matsubara frequencies* $2\pi n_w/\beta$. Needless to say, the general expression for the statistical sum including *all* Matsubara frequencies is still exact, valid for any T.

5.1.2 The Harmonic Oscillator

The density matrix for this example has been calculated by the path integral by Feynman himself (Feynman and Hibbs 1965), and it is impossible not to mention this result. The integrals one encounter, using the definition of the path integrals, are all Gaussian, and thus the results can be obtained exactly, without any approximations.

The harmonic oscillator is a particle with mass m moving in a one-dimensional potential

$$V = \frac{m^2\Omega^2}{2}x^2 \qquad (5.11)$$

Feynman's result for the transition amplitude from the initial point x to the final point y rotated into the Euclidean time τ has the form

$$G_{osc}(x, y, \tau) = \sqrt{\frac{m\Omega}{2\pi\hbar \sinh \Omega\tau}} exp\left[-\left(\frac{m\Omega}{2\hbar \sinh \Omega\tau}\right)\right.$$
$$\left. \times \left(\left(x^2 + y^2\right)\cosh(\Omega\tau) - 2xy\right)\right] \qquad (5.12)$$

---

[2]Note that it is opposite to the limit discussed above for the Hamiltonian approach.
[3]Note that if we would keep $\hbar \neq 1$, the one in β and in the exponent $exp(-S/\hbar)$ would cancel out, confirming the classical nature of this limit.

Although it is not very transparent yet at this point, let us note that the expression in the exponent has a simple physical meaning: it is the classical action $S[x(\tau)]/\hbar$ for the classical path, connecting the points. The pre-exponent factor includes all quantum/thermal fluctuations around this classical path. All semiclassical expressions for the amplitude we will get below will have such form, although only for the harmonic oscillator (and a couple of other related problems, like motion in magnetic field) such expressions are exact.

The diagonal element of the density matrix, or the probability to find a particle at the point x corresponds to periodic paths, as we argued above. So, setting $y = x$ and $\tau = \beta = \hbar/T$ one finds that the particle distribution of a harmonic oscillator at *any* temperature has Gaussian form

$$P(x) = \sqrt{\frac{m\Omega}{2\pi\hbar sinh(\hbar\Omega\beta)}} exp\left(-\frac{x^2}{2\langle x^2\rangle}\right) \tag{5.13}$$

The (temperature-dependent) width is given by

$$\langle x^2\rangle = \frac{1}{2m\Omega}\coth\left(\frac{\Omega}{2T}\right) \tag{5.14}$$

This expression, which we will meet a lot later in the book, has two important limits. At small $T \to 0$ the width corresponds to the quantum mechanical ground state wave function $\psi_0(x)$ of the oscillator. In the opposite limit of high $T \to \infty$ it corresponds to the *classical* thermal result $\langle x^2\rangle = \frac{T}{m\Omega^2}$. Let me rewrite this expression with coth once again, in order to elucidate its physical nature. Since for harmonic oscillator the total energy is just twice the potential energy, which is related to mean $\langle x^2\rangle$, we also have an expression for the mean energy of the oscillator at temperature T. It can be put into the familiar "physical" form

$$\langle E\rangle = \Omega\left(\frac{1}{2} + \frac{1}{e^{\Omega/T} - 1}\right) \tag{5.15}$$

Now one sees the meaning of the two terms in the bracket: they are the energies corresponding to (T-independent) zero-point quantum oscillations (familiar from the QM courses) plus the energy of the thermal excitation (familiar from the SM courses). Note that we automatically get correct Planck (or Bose) distribution from the transition amplitude in Euclidean time.

5.2 Euclidean Minimal Action (Classical) Paths: Fluctons

Before we go into technical detail, let us clarify the goals and the setting in which semiclassical approximation will be used. Imagine, for pedagogical reasons, a particle in a potential $V(x)$. Without quantum/thermal fluctuations, a classical particle would be located at its minimum x_{min} (for now, let it be the only one). Including those, one however finds certain nonzero probability $P(x)$ for a particle to be at any point. In Euclidean time path integral formalism, this probability is given by an integral over periodic paths, which start and end at x. Since the weight is $exp\big(-S[x(\tau)]\big)$, the path with the smallest action should give the largest contribution. Furthermore, we know how to find such a path: it satisfies classical (Euclidean) equation of motion, and this is what we will do below in this section. The semiclassical approximation—the *dominance* of this path—would be justified, as soon as the corresponding action is large

$$S_{cl} \equiv S[x_{cl}(\tau)] \gg \hbar \tag{5.16}$$

Such classical paths were called *fluctons* in Shuryak (1988). In order to find them note first that putting imaginary unit into time flips the sign of the kinetic energy. For the equation of motion it is the same as flipping the sign of the potential, so that its minimum becomes a maximum.

The paths should have Euclidean time period $\beta = \hbar/T$. For simplicity, let us start with "cold" QM, or vanishingly small T, $\beta \to \infty$. Little thinking of how to arrange a classical path with a very long period leads to the following solution: the particle should roll to the top of the (flipped) potential with exactly such energy as to sit there for very long time, before it will rall back to the (arbitrary) point x_0 from which the path started. The classical paths corresponding to relaxation toward the potential bottom take the form of a path *"climbing up"* from arbitrary point x_0 to the maximum, see Fig. 5.1

(I) Let us start with the harmonic oscillator, as the unavoidable first example. For simplicity, let us use units in which the particle mass $m = 1$ and the oscillator frequency $\Omega = 1$, so that our (Euclidean) Lagrangian[4] is written as

$$L_E = \frac{\dot{x}(\tau)^2}{2} + \frac{x(\tau)^2}{2} . \tag{5.17}$$

Fig. 5.1 Sketch of the flucton path climbing toward the (flipped) minimum of the potential

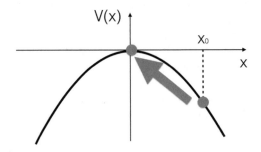

---

[4]Note again the flipped sign of the potential term: in Minkowski time the potential has sign minus.

Because of this sign, in Euclidean time τ the oscillator does not oscillate e^{it} but relaxes $e^{-\tau}$. For harmonic oscillator, the classical equations of motion (EOM) are of course not difficult to solve: but it is always easier to get solutions using energy conservation. Since we are interested in solution with zero energy $E = 0$ they correspond to $\dot{x}^2 = 2V(x)$. The boundary conditions are x_0 at $\tau = 0$ plus periodicity on a circle with circumference β. This solution is

$$x_{flucton}(\tau) = x_0 \frac{\left(e^{\beta-\tau} + e^{\tau}\right)}{e^{\beta} + 1} . \tag{5.18}$$

defined for $\tau \in [0, \beta]$. The particle moves toward $x = 0$ and reach some minimal value, at $\tau = \beta/2$, and then returns to the initial point x_0 again at $\tau = \beta$. Due to periodicity in τ, one may shift its range to $\tau \in [-\beta/2, \beta/2]$: The minimal value at $\tau = \beta/2$

$$x_{min} = \frac{x_0}{cosh(\beta/2)}$$

is exponentially small at low temperature: climbing to the potential top at $x = 0$ is nearly accomplished, if the period is large.

The solution in the zero temperature or $\beta \to \infty$ limit simplifies to $x_0 e^{-|\tau|}$. In the opposite limit of small β or high T, there is no time to move far from x_0, so in this case the particle does not move at all.

The classical action of the flucton path is

$$S_{flucton} = x_0^2 \tanh\left(\frac{\beta}{2}\right), \tag{5.19}$$

it tells us that the particle distribution

$$P(x_0) \sim \exp\left(-\frac{x_0^2}{\coth\left(\frac{\beta}{2}\right)}\right), \tag{5.20}$$

is Gaussian at any temperature. Note furthermore, that the width of the distribution

$$< x^2 > = \frac{1}{2}\coth\left(\frac{\beta}{2}\right) = \frac{1}{2} + \frac{1}{e^{\beta} - 1}, \tag{5.21}$$

can be recognized as the ground state energy plus one due to thermal excitation, which we already mentioned at the beginning of the chapter. So, we have reproduced well known results for the harmonic oscillator, see e.g. Feynman's Statistical Mechanics (Feynman 1972).

(II) Our next example is the symmetric power-like potential

$$V = \frac{g^2}{2}x^{2N}, \quad N = 1, 2, 3, \dots, \tag{5.22}$$

for which we discuss only the zero temperature $\beta = 1/T \rightarrow \infty$ case. The (Euclidean) classical equation at zero energy $\frac{\dot{x}^2}{2} = V(x)$ has the following solution

$$x_{fluct}(\tau) = \frac{x_0}{\left(1 + g(N-1)x_0^{N-1}|\tau|\right)^{N-1}} , \quad x_0 > 0, \quad (5.23)$$

with the action

$$S[x_{fluct}] = \frac{2 g x_0^{N+1}}{N+1} , \quad (5.24)$$

hence

$$P(x_0) \sim \exp\left(-\frac{2 g x_0^{N+1}}{N+1}\right) , \quad (5.25)$$

which is in a complete agreement with WKB asymptotics at $x_0 \rightarrow \infty$

(III) The third example is the anharmonic potential of the kind

$$V = \frac{1}{2}x^2 \left(1 + g x^2\right) , \quad g > 0 , \quad (5.26)$$

at zero temperature $\beta = 1/T \rightarrow \infty$. The classical flucton solution with the energy $E = 0$ is given by

$$x_{fluct}(\tau) = \frac{\sqrt{g}\, x_0}{\cosh(\tau) + \sqrt{1 + g x_0^2}\, \sinh(\tau)} , \quad (5.27)$$

which leads to the flucton action

$$S[x_0] = \frac{2}{3} \frac{\left(1 + g x_0^2\right)^{\frac{3}{2}} - 1}{g} . \quad (5.28)$$

In the limit $x_0 \rightarrow \infty$ we obtain

$$S[x_{fluct}(\tau)] = \frac{2\sqrt{g}}{3} x_0^3 + \frac{1}{\sqrt{g}} x_0 - \frac{2}{3g} + O\left(\frac{1}{x_0}\right) . \quad (5.29)$$

in complete agreement with the asymptotic expansion of the ground state wave function squared.

(IV) However, for the most detailed studies we select another example, the *quartic one-dimensional potential*, also known as the double-well problem

$$V(x) = \lambda \left(x^2 - \eta^2\right)^2 , \qquad (5.30)$$

with two degenerate minima. Tunneling between them will be subject of our subsequent studies later.

Standard steps are selecting units for η such that motion in a single well are in first approximation like in harmonic oscillator with frequency $\omega = 1$. We will also shift the coordinate by

$$x(\tau) = y(\tau) + \eta , \qquad (5.31)$$

so that the potential (5.30) takes the form

$$V = \frac{y(\tau)^2}{2} \left(1 + \sqrt{2\lambda} y(\tau)\right)^2 , \qquad (5.32)$$

corresponding to harmonic oscillator well at small y. The second minimum is shifted to negative y.

The *flucton* path for the case x_0 is outside of a maximum (see Fig. 5.2) now takes the form

$$y_{fluct}(\tau) = \frac{x_0}{e^{|\tau|}(1 + \sqrt{2\lambda}\, x_0) - \sqrt{2\lambda}\, x_0} , \qquad (5.33)$$

We remind that in zero T case, or infinite circle $\beta \to \infty$, $\tau \in (-\infty, \infty)$, and solution exponentially decreases to both infinities, see Fig. 5.1. Its generalization to finite T is straightforward.

The action of this solution is

$$S[y_{fluct}] = x_0^2(1 + \frac{2\sqrt{2\lambda} x_0}{3}) , \qquad (5.34)$$

Fig. 5.2 Sketch of the flucton path climbing toward the (flipped) double-well potential minimum

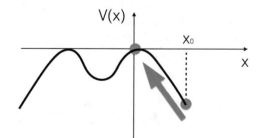

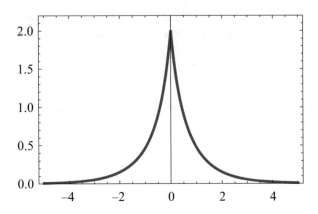

Fig. 5.3 Time dependence of the classical flucton solution $y_{fluct}(\tau)$ for $x_0 = 2, \lambda = 0.1$

and thus in the leading semiclassical approximation the probability to find the particle at x_0 takes the form

$$P(x_0) \sim \exp\left(-x_0^2 - \frac{2\sqrt{2\lambda}}{3}x_0^3\right) \tag{5.35}$$

In the weak coupling limit only the first term remains, corresponding to Gaussian ground state wave function of the harmonic oscillator. In the strong coupling limit the second term is dominant, and the distribution then corresponds to well known cubic dependence on the coordinate. These classical-order results are of course the same as one gets from a standard WKB approximation (Fig. 5.3).

5.3   Quantum/Thermal Fluctuations in One Loop

The paths close to classical ones can be written as

$$y(\tau) = y_{cl} + f(\tau) \tag{5.36}$$

Substituting it to the action, one can expand the result in powers of f, which is presumed to be small. Since classical paths are extrema of the action, the expansion always starts from the second order terms $O(f^2)$.

Taking the path integral over fluctuations around the classical path, in the Gaussian approximation, leads to the following formal expression

$$P(x_0) = \frac{exp\left(-S[x_{flucton}]\right)}{\sqrt{\mathbf{Det}\left(O_{flucton}\right)}} \times [1 + O(two\ and\ more\ loops)]\ , \tag{5.37}$$

with the "flucton operator" $O_{flucton}$ defined as (where dot means a derivative over τ and prime a derivative over coordinate)

$$Of \equiv -\ddot{f}(\tau) + V''(y_{fluct})f(\tau) , \qquad (5.38)$$

In the case of the flucton classical solution (5.33) the potential of the fluctuations we put into the form

$$V''(y_{fluct}) = 1 + W ,$$

where

$$W = \frac{6X(1 + X)e^{|\tau|}}{\left(e^{|\tau|} - X + Xe^{|\tau|}\right)^2} . \qquad (5.39)$$

The term equal to 1 is taken out, as it correspond to harmonic oscillator. The expression with the determinant is "formal" because it is divergent, and get defined *via* its (re)normalization to that of the harmonic oscillator. Note that at $X = 0$ we return to the harmonic oscillator case.

The classical path depends on 3 parameters of the problem, λ, x_0 and Ω (which we already put to 1): but in W the first two appear in one combination only

$$X \equiv x_0 \sqrt{2\lambda} . \qquad (5.40)$$

This observation will be important later.

There are several different method to calculate the determinant of the differential operator: we will use two of them subsequently.

Method 1 is based on straightforward *diagonalization* of the operator. Like for a finite matrix, this includes tedious calculation of all its eigenvalues and eigenmodes. This was also the first method we used, and for pedagogical reasons we will start with it.

Note that for $X > 0$ we discuss, $W > 0$ as well, and it exponentially decreases at large τ. This potential is repulsive, and obviously it has no bound states.[5] At large $|\tau|$ the nontrivial part of the potential disappears and solutions have a generic form

$$\psi_p(\tau) \sim \sin(p\,\tau + \delta_p) , \qquad (5.41)$$

with the so called scattering phase δ_p. Thus the eigenvalues of the operator O are, for the double well example (5.39), simply,

$$\lambda_p = 1 + p^2 , \qquad (5.42)$$

---

[5]Another classical path, the *instanton*, much more discussed in literature, is different precisely at this point: it leads to a zero and bound states, which lead to extra complications.

and the determinant **Det** O is their infinite product. Its logarithm is the sum

$$\log \textbf{Det}\, O \;=\; \sum_n \log\left(1 + p_n^2\right), \tag{5.43}$$

where the sum is taken over all states satisfying zero boundary condition on the boundary of some large box.

The nontrivial part of the problem is not in the eigenvalues themselves, but in the *counting of levels*. Standard vanishing boundary conditions at the boundary of some large box, at $\tau = L$, lead to

$$p_n L + \delta_{p_n} = \pi\, n \;, \quad n = 1, 2, \dots\,. \tag{5.44}$$

At large L and n one can replace summation to an integral, resulting in the generic expression

$$\log \textbf{Det}\, O \;=\; \sum_n \log\left(1 + p_n^2\right) \;=\; \int_0^\infty \frac{dp}{\pi}\frac{d\delta_p}{dp} \log\left(1 + p^2\right). \tag{5.45}$$

After using few different numerical methods for particular values of the parameter X, we discovered that there exist *exact* (non-normalized) analytic solution for the eigenfunctions of the form

$$\psi_p(\tau) \;\sim\; \sin\left(p\,\tau + \Delta(p, \tau)\right)\, F(p, \tau), \tag{5.46}$$

with the following two functions

$$\Delta(p, \tau) = arctan\left[\frac{-3p\,(1 + 2X)}{1 - 2p^2 + 6X + 6X^2}\right]$$
$$+\ arctan\left[\frac{N}{D}\right],$$

where

$$N \;=\; 3p\left[1 + 2X + X^2 - X^2 e^{-2\tau}\right],$$
$$D \;=\; \left(2p^2 - 1\right)\left(1 + X^2\right) - 2e^{-\tau}\left(2(1 + p^2) - e^{-\tau}(2p^2 - 1)\right) X$$
$$+\ \left(2p^2 - 1\right) e^{-2\tau} - 4e^{-\tau}\left(1 + p^2\right),$$

$$F(p, \tau) = \frac{1}{(e^\tau - X + e^\tau X)^2}$$

$$\times \left[e^{4\tau} \left(1 + 5p^2 + 4p^4 \right) + 4e^{3\tau} \left(1 + p^2 \right) \left(2 - 4p^2 + e^\tau (1 + 4p^2) \right) X \right.$$

$$+ 6e^{2\tau} \left(3 + p^2 + 4p^4 + 4e^\tau \left(1 - p^2 - 2p^4 \right) + e^{2\tau} \left(1 + 5p^2 + 4p^4 \right) \right) X^2$$

$$+ 4e^\tau \left(2 \left(1 - p^2 - 2p^4 \right) + 6e^{2\tau} \left(1 - p^2 - 2p^4 \right) \right.$$

$$+ 3e^\tau \left(3 + p^2 + 4p^4 \right) + e^{3\tau} \left(1 + 5p^2 + 4p^4 \right) \right) X^3$$

$$+ \left(1 + 5p^2 + 4p^4 + 8e^\tau \left(1 - p^2 - 2p^4 \right) + 8e^{3\tau} \left(1 - p^2 - 2p^4 \right) \right.$$

$$\left. + 6e^{2\tau} \left(3 + p^2 + 4p^4 \right) + e^{4\tau} \left(1 + 5p^2 + 4p^4 \right) \right) X^4 \right]^{1/2} .$$

It is important that at $\tau = 0$ the solution (5.46) goes to zero: according to the flucton definition *all fluctuations at this time must vanish*. At large time, where all terms with decreasing exponents in $\Delta(p, \tau)$ disappear and the remaining constant terms define the scattering phase

$$\delta_p = \arctan \left[\frac{3p(1 + 2X)}{1 - 2p^2 + 6X + 6X^2} \right] - \arctan \left[\frac{3p}{1 - 2p^2} \right]. \tag{5.47}$$

Comments:

1. the scattering phase is $O(p)$ at small p;
2. it is $O(1/p)$ at large p and, thus, there must be a maximum at some p;
3. for $X = 0$ two terms in (5.47) cancel out. This needs to be the case since in this limit the nontrivial potential W of the operator also disappears;
4. at large time the amplitude F (5.46) goes to a constant, as it should.

The arctan-function provides an angle, defined modulo the period, and thus it experiences jumps by π. Fortunately, its derivative $d\delta_p/dp$ entering the determinant (5.45) is single-valued and smooth. The momentum dependence of the integrand of this expression for $X = 4$ is shown in Fig. 5.4a. Analytic evaluation of the integral (5.45) was not successful, the results of the numerical evaluation are shown by points in Fig. 5.4b. However, the *guess* $2 \log(1 + X)$, shown by the curve in Fig. 5.4b happens to be accurate to numerical accuracy, and thus it must be correct. We will demonstrate that it is exact below.

Since the calculation above includes only a half of the time line, $\tau > 0$, and the other half is symmetric, the complete result for the log $Det\ O$ should be doubled. Substituting (5.47) to (5.45) we obtain a (surprisingly simple) exact result

$$Det\ (O) = (1 + X)^4 . \tag{5.48}$$

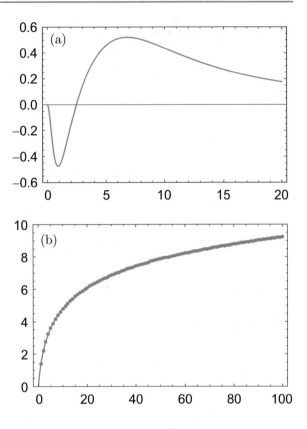

Fig. 5.4 (a) The integrand of (5.45), $\log(1 + p^2)d\delta_p/dp$, versus p, for $X = 4$. (b) The integral (5.45) vs parameter X: points are numerical evaluation, line is defined in the text

Method 2 to calculate the determinant uses the Green function of the operator.[6] It satisfies the standard equation

$$-\frac{\partial^2 G(\tau_1, \tau_2)}{\partial \tau_1^2} + V''(y_{fluct}(\tau_1))G(\tau_1, \tau_2) = \delta(\tau_1 - \tau_2) \qquad (5.49)$$

and one can find two independent solutions of the l.h.s. being zero, and then find the Green function itself. The result (for $\tau_1, \tau_2 > 0$) is

$$G(\tau_1, \tau_2) = \frac{e^{-|\tau_1 - \tau_2|}}{2\left(e^{\tau_1}(1 + X) - X\right)^2 \left(e^{\tau_2}(1 + X) - X\right)^2}$$

$$\times \left[8\, e^{\frac{1}{2}(\tau_1 + \tau_2 + 3\,|\tau_1 - \tau_2|)}\, X^3\, (1 + X) \right.$$

---

[6]For finite matrices, one also may find it easier to find the inverse matrix rather than do complete diagonalization.

$$- 8\, e^{\frac{1}{2}(3\tau_1+3\tau_2+|\tau_1-\tau_2|)}\, X\, (1+X)^3 + e^{2\,(\tau_1+\tau_2)}\, (1+X)^4$$

$$- 6\, e^{(\tau_1+\tau_2+|\tau_1-\tau_2|)}\, X^2\, (1+X)^2\, |\tau_1-\tau_2|$$

$$+ e^{(\tau_1+\tau_2+|\tau_1-\tau_2|)}\left(6\, X^4\, (\tau_1+\tau_2) + 12\, X^3\, (1+\tau_1+\tau_2)\right.$$

$$\left. + 6\, X^2\, (3+\tau_1+\tau_2) + 4\, X - 1\right) - e^{2\,|\tau_1-\tau_2|}\, X^4\,\bigg],\qquad (5.50)$$

The method relies on the following observation. When the fluctuation potential depends on some parameter, it can be varied. In the case at hand (5.39), the potential we write as

$$V_{flucton} = 1 + W(X,\tau),$$

depends on the combination (5.40). Its variation resulting in extra potential

$$\delta V_{flucton} = \frac{\partial W}{\partial X}\delta X \qquad (5.51)$$

which can be treated as perturbation: its effect can be evaluated by the following Feynman diagram

$$\frac{\partial \log \mathbf{Det}\,(O_{flucton})}{\partial X} = \int d\tau\, G(\tau,\tau)\frac{\partial V_{flucton}(\tau)}{\partial X}, \qquad (5.52)$$

containing derivative of the potential as a vertex and the "loop", the Green function returning to the same time, see Fig. 5.5. This idea relates the determinant and the Green function:[7] if the r.h.s. of it can be calculated, the derivative over X can be integrated back.

In the quartic double-well problem the "Green function loop" is

$$G(\tau,\tau) = \frac{1}{2\big(X - e^\tau(1+X)\big)^4}$$

$$\times \bigg(- X^4 + 8e^\tau X^3(1+X) - 8e^{3\tau}X(1+X)^3 + e^{4\tau}(1+X)^4$$

$$+ e^{2\tau}(-1+4X+18X^2+12X^3+12X^2(1+X)^2\tau)\bigg), \qquad (5.53)$$

---

[7]The historical origin of this idea goes back to Brown and Creamer (1978), see also Corrigan et al. (1979), for gauge theory instanton. Zarembo (1996) applied it for the monopole and Diakonov et al. (2004) for the calorons at nonzero holonomy.

Fig. 5.5 Symbolic one-loop
diagram, including variation
of the fluctuation potential
δV and the simplified
"single-loop" Green function
$G(\tau, \tau)$

and the "vertex"

$$\frac{\partial V_{flucton}(\tau)}{\partial X} = \frac{6e^\tau \big(X + e^\tau(1+X)\big)}{\big(-X + e^\tau(1+X)\big)^3} . \tag{5.54}$$

With these expressions one can evaluate the r.h.s. of the relation (5.52), and adding the same expression for negative time, one gets the result

$$\frac{\partial \log \mathbf{Det}\,(O_{flucton})}{\partial X} = \frac{4}{1+X} , \tag{5.55}$$

which exactly agrees with the result (5.48) from the direct evaluation of the determinant using the phase shift.

5.4   Two and More Loops

Unlike the WKB-like semiclassical theory, the flucton-based version of the semiclassical theory we discuss allows for systematic derivation of higher order quantum corrections, defined in terms of conventional Feynman diagrams. The vertices can be calculated from higher order terms in expansion of the action in powers of fluctuation f, of the order f^3, f^4, etc.[8]

The other ingredient of the Feynman diagram method—the Green function inverting the operator quadratic in f—should be calculated in a standard way, in the flucton background. For some examples discussed above we already have such Green functions, which has even passed a nontrivial test—producing the correct determinant. So, using only standard tools from quantum field theory, the Feynman diagrams in the flucton background, one can compute the loop correction to the density matrix (5.37) of any order. For definiteness, we will show here the results for the double-well potential. The only vertices we will need are the triple and quartic

---

[8]Later in this chapter we will apply similar approach to the tunneling and "instanton" paths. In that case the quadratic operator admits zero modes, and the Green function needs to be defined in a subspace normal to them. As shown in (Aleinikov and Shuryak 1987; Escobar-Ruiz et al. 2015), the Jacobian of the orthogonality condition generates additional diagrams, not following from the action. It is for this reason we postpone discussion of instantons in this text.

ones, which follow from the cubic and quartic potential derivatives over x taken at classical solution

$$v_3(\tau) = \frac{6\sqrt{2}\lambda\,(X + e^\tau(1 + X))}{-X + e^\tau(1 + X)}, \qquad (5.56)$$

$$v_4 = 24\lambda. \qquad (5.57)$$

The loop corrections in (5.37) are written in the form

$$[1 + O(\text{two and more loops})] = 2\sum_{n=0}^{\infty} B_n\,\lambda^n, \qquad B_0 = \frac{1}{2}, $$

where $B_n = B_n(X)$.

In 1+0 dimension of time-space we discuss (1-dimensonal quantum mechanics) there are no ultraviolet divergences. There are infrared ones, which can be cancelled out by subtraction from each diagram in the flucton background its analog for trivial or vacuum path $x(\tau) = 0$. This is done by subtracting the same expression with the "vacuum vertices"

$$v_{3,0} = 6\sqrt{2\lambda}, \qquad (5.58)$$

$$v_{4,0} = 24\lambda, \qquad (5.59)$$

and the "vacuum propagator"

$$G_0 = G(\tau_1,\ \tau_2)\,|_{x\to 0} = \frac{e^{-|\tau_1 - \tau_2|}}{2} - \frac{e^{-\tau_1 - \tau_2}}{2}. \qquad (5.60)$$

The two-loop correction B_1 we are interested in can be written as the sum of three diagrams, see Fig. 5.6, diagram a which is a one-dimensional integral and diagrams b_1 and b_2 corresponding to two-dimensional ones.

Explicitly, we have

$$a \equiv -\frac{1}{8\lambda}v_4 \int_0^\infty [G^2(\tau,\ \tau) - G_0^2(\tau,\ \tau)]d\tau = \frac{3}{560X^2(1 + X)^4}$$

$$\times \Bigg(24X - 60X^2 - 520X^3 - 1024X^4 - 832X^5 - 245X^6$$

$$+ 24(1 + X)^2(1 + 2X)(-1 + 6X(1 + X))\log(1 + X)$$

$$+ 288X^2(1 + X)^4\mathbf{PolyLog}\left[2,\ \frac{X}{1 + X}\right]\Bigg), \qquad (5.61)$$

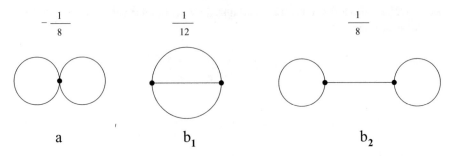

$$-\frac{1}{8} \qquad\qquad \frac{1}{12} \qquad\qquad \frac{1}{8}$$

$$a \qquad\qquad b_1 \qquad\qquad b_2$$

Fig. 5.6 Diagrams contributing to the two-loop correction $B_1 = a + b_1 + b_2$. The signs of contributions and symmetry factors are indicated

here **PolyLog**$[n, z] = \sum_{k=1}^{\infty} z^k / k^n$ is the polylogarithm function and

$$b_1 \equiv \frac{1}{12\,\lambda} \int_0^{\infty} \int_0^{\infty} [v_3(\tau_1)\, v_3\,(\tau_2) G^3(\tau_1,\ \tau_2) - v_{3,0} v_{3,0} G_0^3(\tau_1,\ \tau_2)]\, d\tau_1\, d\tau_2$$

$$= \frac{1}{280 X^2 (1+X)^4} \times \left(- 24X + 60X^2 + 520X^3 + 1024X^4 + 832X^5 \right.$$

$$+ 245X^6 + 24(1+X)^2 \left(1 - 4X - 18X^2 - 12X^3\right) \log(1+X)$$

$$\left. - 288 X^2 (1+X)^4 \mathbf{PolyLog}\left[2, \frac{X}{1+X}\right] \right),$$

$$b_2 \equiv \frac{1}{8\,\lambda} \int_0^{\infty} \int_0^{\infty} \left[v_3(\tau_1)\, v_3\,(\tau_2) G(\tau_1,\ \tau_1) G(\tau_1,\ \tau_2) G(\tau_2,\ \tau_2) \right.$$

$$\left. - v_{3,0} v_{3,0} G_0(\tau_1,\ \tau_1) G_0(\tau_1,\ \tau_2) G_0(\tau_2,\ \tau_2) \right] d\tau_1\, d\tau_2$$

$$= -\frac{1}{560 X^2 (1+X)^4} \times \left(24X - 60X^2 + 1720X^3 + 5136X^4 + 4768X^5 \right.$$

$$+ 1435X^6 + 24(1+X)^2 \left(-1 + 4X + 18X^2 + 12X^3\right) \log(1+X)$$

$$\left. + 288 X^2 (1+X)^4 \mathbf{PolyLog}\left[2, \frac{X}{1+X}\right] \right). \qquad (5.62)$$

Adding all two-loop corrections one finds an amazingly simple form,

$$B_1 \equiv a + b_1 + b_2 = -\frac{X(4+3X)}{(1+X)^2}, \qquad (5.63)$$

in which all log and **PolyLog** terms *disappear!*[9]

---

[9] One way to prove that it must be so was found by Escobar-Ruiz et al. (2017) using generalized Bloch equation for the wave function. Note however that in explicit QFT calculations of multi-loop

The combined results for the probability to find a particle at point x_0 in the quartic double-well potential at zero temperature is, with the two-loop accuracy

$$P(x_0) \sim \frac{e^{-\frac{x^2}{2\lambda} - \frac{x^3}{3\lambda}}}{(1+X)^2} \left(1 - \lambda \frac{X(4+3X)}{(1+X)^2} + O(\lambda^2) \right), \qquad (5.64)$$

where, we remind, $X = \sqrt{2\lambda}\, x_0$. Note that $X = -1$ is indeed a singularity of the potential, located in the unphysical domain.

The x_0 dependence of (5.64) is plotted in Fig. 5.7 by the thick line. The thin line is asymptotics derived in appendix A: since x_0-independent constant remained unknown we normalized it to our curve at large distances. Their comparison shows good agreement for $x_0 > 1$.

Although derived semiclassically, and thus formally valid for large flucton action only, our answer is also obviously correct at small x_0, where it merges with the answer for the harmonic oscillator.

Let us finally compare the results obtained with those one get from standard asymptotic analysis of the Schreodinger eqn. where the double-well potential in shifted coordinates we use is

$$V(y) = \frac{y^2}{2} + \sqrt{2\lambda}\, y^3 + \lambda y^4 . \qquad (5.65)$$

Note that it smoothly goes to the harmonic oscillator at $\lambda \to 0$. Introducing the phase $\phi(y) = -\log \Psi(y)$ we move to the Riccati equation,

$$\partial_y^2 \phi - (\partial_y \phi)^2 = 2E - 2V(y) , \qquad (5.66)$$

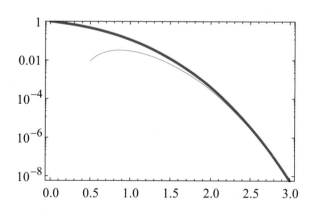

Fig. 5.7 The probability $P(x_0)$ to find particle at location x_0 for $\lambda = 0.1$. The thick line is our result (5.64), thin line is its asymptotics

---

diagrams one also finds cancellations of irrational contributions, present in individual diagrams but absent in their sum. Such cancellations, to our knowledge, remain mysterious unexplained phenomena, only seen from explicit calculations.

to which one can plug the asymptotic expansion at $|y| \to \infty$ and obtain all the coefficients (cf. Turbiner 2010)

$$\phi = \frac{1}{3}\sqrt{2}\sqrt{\lambda}|y|y^2 + \frac{1}{2}y^2 - d\log|y|^2 +$$

$$\frac{1+2E}{2\sqrt{2\lambda}}\frac{1}{|y|} - \frac{1}{8\lambda y^2} + \dots , \qquad (5.67)$$

where $d = 1/2$. The first two terms in the expansion are classical coming from classical Hamilton–Jacobi equation, log-term reflects an intrinsic property of the Laplacian: y is zero mode or kernel, this term comes from determinant, asymptotically the determinant behaves like $|y|^2$, where d is degree with which it enters to the wavefunction. Note that a constant, $O(x_0^0)$ term is absent: it can not be obtained from the Riccati equation containing derivatives only. Note also that so far the energy remains undefined: to find it one needs to solve the equation to all x. The last terms are true quantum corrections, decreasing at large distances. Intrinsically, this expansion corresponds to the ground state: it implies that the eigenphase ϕ has no logarithmic singularities at real y. Quantization for the Riccati equation implies a search for solutions growing at large y with finite number of logarithmic singularities at real finite y. For the nth excited state the first two growing terms in (5.67) remains unchanged while log-term gets integer coefficient, $(n+1)\log|y|$, see Turbiner (2010).

Multiplying by 2 (path integral is for density matrix, or wave function squared) one finds, as expected, that the first two terms coincide with the classic action of the flucton. For the determinant one needs to expand at large x_0

$$\log(1 + \sqrt{2\lambda}x_0) = \log(x_0) + \log(\sqrt{2\lambda}) + \frac{1}{\sqrt{2\lambda}x_0} + \dots , \qquad (5.68)$$

and observe that the leading term agrees with the $\log|y|$ term in the asymptotic expansion (5.67).

The two-loop correction $B_1\lambda$ found in the text (5.63) expands in inverse powers of x_0 as follows

$$-\lambda\frac{X(4+3X)}{(1+X)^2} = -3\lambda + \frac{\sqrt{2\lambda}}{x_0} + \dots , \qquad (5.69)$$

where $X = \sqrt{2\lambda}x_0$, see (5.40).

In order to compare the $1/x_0$ terms in the last two expressions one needs to substitute the ground state energy to $O(\lambda)$ accuracy

$$E = \frac{1}{2} - 2\lambda + \dots , \qquad (5.70)$$

to the $O(\frac{1}{x_0})$ term in (5.67). After that one finds agreement with both $O(\frac{1}{x_0})$ terms given in (5.68,5.69).

Similar calculations has been made for other quantum-mechanical examples. Let me just mention Sine-Gordon potential

$$V = \frac{1}{g^2}\left(1 - cos(gx)\right) \tag{5.71}$$

for which the expression for the 0-1-2 loop expansion has a simpler form

$$-\log\left(|\psi_0(x)|^2\right) = \frac{16}{g^2}\sin^2\left(\frac{gx}{4}\right) - 2\log|\cos\left(\frac{gx}{4}\right)| + \frac{g^2}{32}\tan^2\left(\frac{gx}{4}\right) + \dots \tag{5.72}$$

One can see that even for $g = 1$ the second term is few percent and the third few per mill of the classical term, for all x. So the series over the fluctuations corrections are indeed well convergent. This is in striking contrast with the WKB, in which the next-to-classical correction $1/\sqrt{p(x)}$ (where $p(x)$ is momentum) has an unphysical singularity at the turning point.

5.5   Path Integrals and the Tunneling

Tunneling through classically impenetrable barrier is perhaps the most amazing consequence of quantum mechanics. It was discovered already in 1927–1928 when it was just few months old. First came Friedrich Hund who in 1927 studied splitting of molecular states, and came to the problem of the energy levels of a double-well potential, exactly the problem we discuss in detail. Soon after that came a note (Mandelstam and Leontowitsch 1928) that (then-new) Schreodinger equation allows particle to go through classically forbidden regions.

But by far the most famous paper[10] is that by Gamow (1928), who demystified alpha-decay of heavy nuclei. The apparent puzzle was why Rutherford scattering experiments with alpha particles demonstrated existence of the repulsive barrier of the height of 10 MeV or more, while in the decay of the same nuclei alpha particles emerged with smaller energies, of only few MeV. The tunneling probability is the square of the famous Gamov factor in the amplitude, which is

$$P_{Gamov} \sim exp\left[-\frac{2\pi\left(2e^2\right)(Z-2)}{h\upsilon}\right] \tag{5.73}$$

---

[10]The paper Gurney and Condon (1928) with a similar idea (but without the analytic Gamow factor) was submitted the next *day* after the Gamow's one. My advice to the reader is obvious; never delay a paper, even for a day.

The $2(Z-2)$ in the numerator is the product of the electric charges of the alpha-particle and the remaining nucleus. The main factor in the denominator, v, is the *velocity* of the outgoing alpha particle. This expression thus has the characteristic small parameter v in denominator of the exponent, as many other non-perturbative phenomena we discuss. It explains why it takes up to billion years to decay, in spite of the fact that would-be alpha-clusters beat against the wall about 10^{22} times per second.[11] The first semiclassical WKB approximation, which we all learned in quantum mechanical courses, was developed in 1926. It was sufficient to explain tunneling in one-dimensional (or rather spherically symmetric radial) case.

Many years later, in 1950s, when one might think the subject is completely worked out, Feynman introduced his formulation of quantum mechanics in terms of path integrals. He also realized that paths integrals are easier to do in an imaginary Euclidean time $\tau = it$, defined (what we now call) a "Matsubara circle" with the circumference $\beta = \frac{\hbar}{T}$, and proceeded to paths integral formulation of the finite temperature theory.

Using Euclidean-time paths to evaluate the tunneling rate has been pioneered in 1931, by Sauter (1931) for electron pair production in constant electric field. This calculation has been confirmed in 1930s by Heisenberg and Euler via their effective ED Lagrangian, and finally in 1950s solved exactly by Schwinger. Sauter's path starts as a e^+e^- pair at the point $x = 0$. The energy of an electron is $\sqrt{p^2 + m^2} - eEx$, and it must be zero at the initial point: so $p = im$, the momentum is Euclidean. At the final points of the Euclidean path $x = \pm m/eE$ the momenta can vanish $p = 0$: from this point it will be real or Minkowskian. The rest I would leave to the reader as

Exercise Derive the Euclidean tunneling path and the corresponding action for the production of a pair of massive scalar particle in a homogeneous electric field.

Discussion of the Euclidean particle paths related to tunneling has been revived by Polyakov (1977), following a discovery of the (Euclidean) instanton solution of the gauge theory. For early pedagogical review, including the one-loop (determinant) calculation by the same tedious scattering-phase method seeABC's of instantons (Vainshtein et al. 1982). Two-loop corrections were first calculated in Aleinikov and Shuryak (1987), some technical errors in it were corrected by Wohler and myself in Wohler and Shuryak (1994). Three loop corrections have been calculated by Escobar-Ruiz et al. (2015).

As in the discussion of *fluctons* above, the main setting is the same. A usage of Euclidean time leads to successful description of motion in the "classically forbidden" region under the barrier. The minimal action—and thus classical—paths are the most probable ones, since the weight is $exp(-S[x(\tau)])$. Therefore we look for minima of the action.

---

[11]Another advice to the reader; anything may happen once somebody is persistent enough, few or even lots of failures never prove that something is impossible.

The Euclidean path method is much more powerful than WKB since it can be used in multi-dimensional problems. While our applications will be in QFTs such as pure gauge theory, let us note that they are successfully used in chemistry and even biology (e.g. describing the protein folding paths Faccioli 2011).

We will continue to use the usual toy model, a double-well potential, with the Euclidean action in slightly different notations

$$S = \int d\tau \left[m\frac{\dot{x}^2}{2} + \lambda \left(x^2 - f^2 \right)^2 \right] \tag{5.74}$$

in which the dot means the derivative over τ, not time. This potential has two minima at $\pm f$, known as the two "classical vacua".

The energy levels of this system can be derived at three levels of sophistication;

1. If one first ignores tunneling, those are given by zero point oscillations in each well, with the energy

$$E_0 = \omega/2, \qquad \omega = f\sqrt{8\lambda} \tag{5.75}$$

In order to have better contact with the gauge theory later in the chapter, we will eliminate f from all expressions substituting it by the ω just defined. The maximal hight of the barrier, for example, is then $V_{max} = \omega^4/64\lambda$, etc.

2. At small value of the only dimensionless parameter of the model $\lambda/\omega^3 \ll 1$ (the high barrier limit), one can further calculate a whole series of *perturbative corrections* which go as powers of this parameter,

$$E_0 = \frac{\omega}{2} \left[1 + \Sigma_n C_n \left(\lambda/\omega^3 \right)^n \right] \tag{5.76}$$

One may use Feynman diagrams to calculate these corrections, corresponding to fluctuations around a "lazy path" $x(\tau) = 0$.

3. Finally one may take into account the tunneling phenomenon. The left-right degeneracy of the levels is then lifted, substituted by *symmetric* and *antisymmetric* wave functions under the parity transformation $x \leftrightarrow -x$. Their energies

$$E_\pm = \frac{\omega}{2} \left[1 \mp \sqrt{\frac{2\omega^3}{\pi\lambda}} e^{-\frac{\omega^3}{12\lambda}} \right] \tag{5.77}$$

are thus separated by exponentially small gap.

Standard textbook description of tunneling goes as follows. Since the energy is conserved, one can read the Hamiltonian $H = E_{kinetic} + V(x)$ where

$$E_{kinetic} = \hat{p}^2/2m = -\frac{\partial_x^2 \psi}{\psi} \frac{1}{2m}$$

In a classically allowed region, $E_{kinetic} > 0$ the wave function resembles a wave $\psi \sim exp(ipx)$ with *real* p. In a classically *forbidden* region $E_{kinetic} < 0$, the momentum should be *imaginary* and therefore the wave function just decreases $\psi \sim exp(-|p|x)$.

The word 'tunneling' hints that one can pass the mountain (the barrier due to a repulsive potential) not by climbing and then descending from it, but by going through it, *as if* there be a path through, the tunnel. The point of this section is to show that not only *one can imagine* a path through the mountain, it is useful to find them, and even identify the *best* ones among them, the *instantons*.

If p is imaginary, one may interpret it as a motion in *imaginary time* $\tau = it$. As we already mentioned, classical equation of motion

$$m\frac{d^2x}{d\tau^2} = +\frac{dV}{dx} \tag{5.78}$$

correspond to flipping the potential upside down!

The *instantons* would be classical paths between two classical vacua, in a flipped potential $V \to -V$, see Fig. 5.8. As usual in one-dimensional problems it is easier to find it using energy conservation. Note that the path we are looking goes from one minimum to another one (see Fig. 5.8) must be at the total energy $E = 0$. The resulting solution, the *instanton*, is thus

$$x_{cl}(\tau) = f \tanh\left[\frac{\omega(\tau - \tau_0)}{2}\right] \tag{5.79}$$

The tunneling probability is given by the corresponding action $P \sim exp(-S_{cl})$, which is twice the average potential energy, so

$$S_{cl} = 2\int d\tau \lambda \left[x_{cl}^2 - f^2\right]^2 = \frac{\omega^3}{12\lambda} \tag{5.80}$$

Note that it reproduces the exponent in (5.77).

Exercise Derive the Gamov factor (5.73) using the classical under-the-barrier path in Euclidean time.

Fig. 5.8 Sketch of the instanton path, going from one (flipped) minimum of the potential to the other

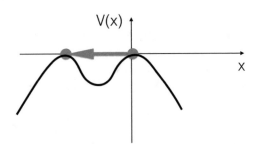

5.6   The Zero Modes and the Dilute Instanton Gas

The next step, as for the flucton, is to write the general tunneling path as small quantum deviations from the classical path

$$x(\tau) = x_{cl}(\tau) + \delta x(\tau) \tag{5.81}$$

and expand the action in powers of quantum corrections $\delta x(\tau)$. In the quadratic order

$$S = S_{classical} + (1/2) \int d\tau \delta x(\tau) \hat{O} \delta \delta x(\tau) \tag{5.82}$$

the differential operator at the instanton path takes the form

$$\hat{O} = -\frac{m}{2} \frac{d^2}{d\tau^2} + \frac{\delta^2 V}{\delta x^2}|_{x=x_{cl}} = -\frac{m}{2} \frac{d^2}{d\tau^2} + 4\lambda \left(3x_{cl}^2 - f^2\right) \tag{5.83}$$

At time distant from the mid-tunneling moment (to be called also the position of the instanton) τ_0, the last term is just constant, $8\lambda f^2$. But around τ_0 and this last term is strongly changing and it is *negative*, unlike that for the flucton.

This observation leads to significant modifications of the one-loop theory for instantons. Unlike in the flucton case, now the equation has not only the scattering states, but also two localized (bound states) solutions. The lowest eigenvalue is zero, $\epsilon_0 = 0$, and the next is positive $\epsilon_1 = \frac{3}{4}\omega^2$ but below the scattering state energy.

The wave function for the zero mode is

$$x_0(\tau) \sim \frac{1}{\cosh^2(\omega\tau/2)} \tag{5.84}$$

and one can find the normalization constant from the usual normalization condition $\int d\tau x_n^2 = 1$ to be $const = \sqrt{3\omega/8}$.

Every zero should have a simple explanation. And indeed, this one has a simple origin: it is due to time translational symmetry in the problem. Indeed, the action cannot depend on the instanton displacement in time.[12]   The zero mode can be obtained very simply, by differentiation of the instanton solution over its time τ_0

$$x_0(\tau) = S_0^{-1/2} \frac{d}{d\tau_0} x_{cl}(\tau - \tau_0) \tag{5.85}$$

Exercise Check this statement by putting this function into the fluctuation equation $\hat{O}\psi_0 = 0$.

---

[12] Note that in the case of flucton the path is "pinned" at certain point at $\tau = 0$, so there was no time shift symmetry and no zero mode.

Now, if the operator in question has a zero eigenvalue, its determinant is zero too. Since for Gaussian integral over fluctuations the (sqrt) of the determinant appears in denominator, the tunneling probability we discuss is in fact infinite! Indeed, we encounter here (the simplest case of) the so called 'valleys' in the functional space: one of the integrals is actually non-Gaussian; zero eigenvalue means that nothing prevents large amplitude of fluctuations in the corresponding direction in the Hilbert space.

The solution to this problem is however quite simple; one may not take this integral at all! All we have to do is to rewrite the integral over dC_0 as the integral over the collective coordinate τ_0. Consider a modification of C_0 by dC_0. The path changes by $dx = x_{(\tau)} dC_0$ (remember the definition $x(\tau) = \Sigma_n c_n x_n(\tau)$). At the same time we have another definition of the zero mode from which it follows

$$dx = \frac{dx_{cl}}{d\tau_0} d\tau_0 = -\sqrt{S_0} x_0(\tau) d\tau_0 \qquad (5.86)$$

Equalizing two expressions for dx, we have

$$dC_0 = \sqrt{S_0} d\tau_0 \qquad (5.87)$$

Returning to our functional integral over the quantum fluctuation, we now have the following form for it

$$\int D\delta x(\tau) e^{-S} = e^{-S_{cl}} \prod_{n>0} \sqrt{\frac{2\pi}{\epsilon_n}} \sqrt{S_0} \int d\tau_0 \qquad (5.88)$$

The product here is the determinant *without* the zero mode, often called $det'(\hat{O})$. It is finite and its calculation will be done below. But before we engaged in it, let us first explain what to do with the divergent integral over the τ_0. Suppose the whole path integrals is taken over some finite time, from 0 to τ_{max}; then the contribution of tunneling—described by the instanton-type path– grows linearly with τ_{max}. One may say that the finite quantity is the *transition probability per unite time*, but it is not a satisfactory solution in the long run. If τ_{max} is very large, it may overcome the smallness of the exponent. If so, the amplitude is no longer small and one has to think about ensemble of *many* instantons.

Suppose one has a path with n instantons (anti-instantons), placed at $\tau_1 < \ldots < \tau_n < \tau_0$. If they are all separated sufficiently far from each other, (as one says, the instanton gas is *dilute*) the action is the sum of actions. Furthermore, determinants become factorized, and the expression for the transition amplitude in this case reads as

$$G(f, -f, \tau_0) = \left[\sqrt{\omega/\pi} exp^{-\omega\tau_0/2}\right]\left[\sqrt{6S_0/\pi} e^{-S_0}\right]^n \int_0^{\tau_0} \omega d\tau_n \ldots \int_0^{\tau_2} \omega d\tau_1 \qquad (5.89)$$

where integrals are done over ordered positions. The factor which repeatedly appears here

$$d = \sqrt{\frac{6S_0}{\pi}} e^{-S_0} \omega \tag{5.90}$$

we will call the *instanton density* (per unit time). One can relax the nesting condition on the tunneling moments by integrate over all interval, but then dividing the result by the n factorial. It looks then like the exponential series, but since n is actually odd one gets the following final expression for the Green function

$$G(f, -f, \tau_0) = \left[\sqrt{\omega/\pi} \exp^{-\omega \tau_0/2} \right] \sinh \left[\sqrt{\frac{6S_0}{\pi}} e^{-S_0} \omega \tau_0 \right] \tag{5.91}$$

Now, it may be used for any time. If it is very large, one has another asymptotic of the sinh, the exponential one, and notice that the total dependence on τ_0 is now again exponential, with the *corrected ground state energy*

$$E_0 = \frac{\omega}{2} - \sqrt{\frac{6S_0}{\pi}} e^{-S_0} \omega \tag{5.92}$$

Note that it is precisely what one also gets for level shift from the Schroedinger equation in the semiclassical approximation.

Even if the instanton gas is dilute, it is still interesting to ask what happens if two instantons are close, $\tau_k - \tau_{k-1} \sim 1/\omega$. Then they certainly do interact, because the total action is actually *less* than $2S_0$. Numerical studies indeed show that there is strong positive correlation between the instanton positions at smaller time intervals. We will return to interacting instantons later.

Finally, let me introduce the last issue of this section, related with the so called correlation functions of the coordinates. Those characterize properties of the ground state, and in particular show the crucial role of the dilute instanton gas. The simplest observable we can think of is just the particle coordinate, so let us define the correlator of the coordinates as

$$K(\tau_1 - \tau_2) = < x(\tau_1) x(\tau_2) > \tag{5.93}$$

where averaging is supposed to be done by appropriately weighted quantum paths. We imagine now the length of all paths τ_0 to be infinite, so the correlator depends only on the time *difference*.

Recalling the "old fashioned" quantum mechanical expressions, one can use matrix elements of the coordinate operator, and write this functions as a sum over stationary states which can be excited by the operator of the coordinate

$$K(\tau) = \Sigma_n |<0|x|n>|^2 \exp(-E_n \tau) \tag{5.94}$$

This function is not only *positive* but it is a *monotonously decreasing* one. Although the sum runs over all states, only *parity odd* ones can be excited by the coordinate operator. For example, symmetric ground state is absent in this sum, because $< 0|x|0 >= 0$.

Thus, if one knows the correlation function at large times, one knows E_1, the lowest parity odd state as well. This is analogous to what we will do in Chap. 6 for calculation of the lowest excitations in QCD, mesons and baryons.

Now, let us try to understand how the correlation function looks like, if we are not specifically interested in large times. Clearly, the problem has two time scales; (1) perturbative scale $\tau_{pert} = 1/\omega$ and (2) tunneling scale $\tau_{tunneling}$, the inverse tunneling rate.

If one imagines the particle is sitting in the same well all the time, the correlator is $K(\tau) \approx f^2$. However, the tunneling kills the correlation, and eventually $K(\tau) \to 0$. One can write approximately the paths as sequence of steps or kinks

$$x(\tau) = f \prod_i sign(\tau - \tau_i) \tag{5.95}$$

and calculate the correlation function

$$K = f^2 \frac{\Sigma_n \int \Pi_n d\tau_n d^n sign(\tau - \tau_i)sign(-\tau_i)}{\Sigma_n \int \Pi_n d\tau_n d^n} \tag{5.96}$$

where d=$\sqrt{6S_0/\pi}e^{-S_0}\omega$. Its large-time limit is

$$K(\tau) \sim exp(-2\tau d) \tag{5.97}$$

Comparing it to the expression above, one gets the *gap*, the energy splitting between the vacuum and the first excited state, $E_1 - E_0 = 2d$. Intuitively 2 appears because of instanton plus the anti-instantons.

Summarizing this discussion; if the classical action is large $S_0 \gg 1$ (actually, much larger than the Plank constant), the ensemble of paths is an exponentially dilute gas of instantons and anti-instantons. All such paths together lead to 'exponentiation' of tunneling correction, and corresponding to the negative shift of the ground state energy. They also randomize the sign of the coordinate and create the gap between the vacuum and the excited states. As we will see later in the chapter, exactly the same thing happens in gauge theories.

Exercise Derive the expression (5.97) for the correlator of coordinates in dilute instanton gas approximation.

5.7   Quantum Fluctuations Around the Instanton Path

The Feynman rules, as usual, are based on *vertices*, calculated from the derivatives of the potential over δx in various powers (large than 2), and *propagators*(Green functions), inverting the quadratic form $\hat{O}$. There are two complications:

- this inversion should be made *in the "primed" subspace excluding the zero mode.*
- Jacobian related with collective coordinate leads to new type of Feynman diagrams, not following from the action.

Starting with the one-loop calculation, one needs to calculate the "primed" determinant of the operator (5.83) over all non-zero modes. Fortunately, since the quadratic form $\hat{O}$ in this case corresponds to exactly solvable equation, one may proceed with the most direct method of its calculation—the direct diagonalization. One can determine the scattering phase $\delta(k)$ and summing log of all eigenvalues with the appropriate level counting. We would not reproduce this tedious calculation here, the interested reader can find it in Vainshtein et al. (1982).

The second method—using a loop diagram and the Green function—does not work in this case, although it provides a very nontrivial test for the Green function: discussion of this can be found in Appendix in Escobar-Ruiz et al. (2016). We will skip calculation of the instanton determinant, and proceed directly to the Green function and higher order corrections, following Wohler and Shuryak (1994); Escobar-Ruiz et al. (2015). We will of course not give full details of these calculations here, and only present the results, with some comments.

To next order in $1/S_0$, the tunneling amplitude can be written as

$$\langle -f | e^{-H\tau} | f \rangle = |\psi_0(f)|^2 \left(1 + \frac{2A}{S_0} + \dots \right)$$
$$\times e^{\left[-\frac{\omega\tau}{2}\left(1 + \frac{B}{S_0} + \dots \right) \right]} 2d \left(1 + \frac{C}{S_0} + \dots \right) \tau, \qquad (5.98)$$

where corrections, A and B are those to the wave function at the minimum and to the energy due to anharmonicity of the oscillations. These two corrections are unrelated to tunneling and we can get rid of them by dividing the amplitude by $\langle f | \exp(-H\tau) | f \rangle$, in which they appear in the same way. We are interested in the coefficient C, the next order correction to the tunneling amplitude (instanton density d) and eventually to the level splitting.

Step one is the calculation of the propagator (the Green function) of the differential operator $\hat{O}$ we defined above. It however has the following complication: unlike in the case of the flucton path, the instanton operator has a zero mode. Thus, strictly speaking, it *cannot be inverted!*

Yet the inversion is still uniquely defined in a Hilbert subspace *orthogonal to the zero mode.* Note that delta function in the r.h.s. of the Green function equation can be written as follows[13]

$$\delta\left(\tau - \tau'\right) = \sum_{\lambda \neq 0} \psi_\lambda^*(\tau)\psi_\lambda\left(\tau'\right) + \sum_{\lambda=0} \psi_\lambda^*(\tau)\psi_\lambda\left(\tau'\right) \tag{5.99}$$

which expresses completeness of the set of all eigenvalues, zero and nonzero.

In a subspace orthogonal to zero modes one can define a new Green function $G_\perp(\tau, \tau')$ satisfying a modified equation with a new r.h.s., the first term in (5.99)

$$\hat{O}G_\perp(\tau, \tau') = \delta(\tau - \tau') - \psi_0(\tau)\psi_0(\tau') \tag{5.100}$$

Although we will from now on drop the subscript $\perp$ on the Green function, this is the one which is to be used below. Its analytic form is

$$G(x, y) = G_0(x, y)\left[2 - xy + \frac{1}{4}|x - y|(11 - 3xy) + (x - y)^2\right] \tag{5.101}$$

$$+ \frac{3}{8}\left(1 - x^2\right)\left(1 - y^2\right)\left[\log\left(G_0(x, y)\right) - \frac{11}{3}\right]$$

where $x = \tanh(\tau/2)$, $y = \tanh(\tau'/2)$ and G_0 being the perturbative Green function near trivial vacuum in this notations is

$$G_0(x, y) = \frac{1 - |x - y| - xy}{1 + |x - y| - xy}$$

Exercise Check that it is indeed orthogonal to the zero mode, in respect to both of its variables.

There are four diagrams at the two loop order, see Fig. 5.9. The first three diagrams are of the same form as for the flucton path. Subtracting the contributions of the same diagrams around the trivial path $x(\tau) = 0$, one gets rid of the effects far from the tunneling event: this is so-to-say infrared renormalization of the diagrams. Their contributions are

$$a = -3\int_{-\infty}^{\infty} dt \left(G^2(t, t) - G_0^2(t, t)\right) = -\frac{97}{1680}$$

$$b_1 = 3\int_{-\infty}^{\infty}\int_{-\infty}^{\infty} dt\,dt' \left(\tanh(t/2)\tanh\left(t'/2\right)G^3\left(t, t'\right) - G_0^3\left(t, t'\right)\right) = -\frac{53}{1260}$$

---

[13] We write the sum over zero modes for generality, assuming there may be several of them. In the problem we study now there is only one zero mode, corresponding to time-shift symmetry of the instanton path.

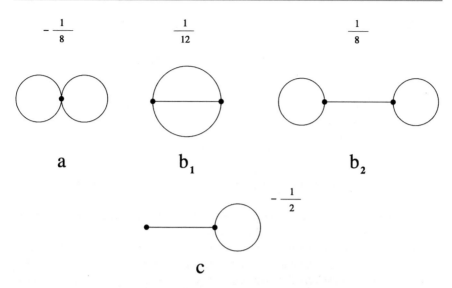

Fig. 5.9 Two-loop diagrams contributing to the instanton amplitude. The signs of contributions and symmetry factors are indicated. The only difference with the flucton case is the appearance of the new diagram c

$$b_2 = \frac{9}{2} \int_{-\infty}^{\infty} \int_{-\infty}^{\infty} dt dt' \left(\tanh(t/2) \tanh\left(t'/2\right) G\left(t, t\right) G\left(t, t'\right) G\left(t', t'\right) \right.$$

$$\left. - G_0(t, t) G_0\left(t, t'\right) G_0\left(t', t'\right) \right) = -\frac{39}{560}.$$

However, unlike for the flucton path, now there appear a *new series of diagrams* introduced in Aleinikov and Shuryak (1987) and not following from classical action. Indeed, all fluctuations under consideration should be orthogonal to the zero mode

$$\int d\tau \delta x(\tau) x_0(\tau) = 0 \tag{5.102}$$

Inserting into the functional integral the delta function with this condition *a la* Faddeev and Popov, one finds the Jacobian[14] which generates tadpole graphs proportional to a new vertex $\ddot{x}_{cl}$ which needs to appear only once. These diagrams have no counterpart in the expansion around the trivial vacuum, and thus they need

---

[14] Using identity $1 = f'(u_{root}) \int du \delta[f(u)]$ for arbitrary function with a singe root $f(u_{root}) = 0$, for function $f(\tau_0)$ being the orthogonality condition above, one finds the Jacobian to be $\partial f / \partial \tau_0 = \int d\tau (x_{cl}(\tau) - x(\tau) dx_0/d\tau_0)$.

no subtraction. Its two-loop contribution is only one diagram contributing

$$c = -9\beta \int_{-\infty}^{\infty} \int_{-\infty}^{\infty} dt\, dt' \, \frac{\tanh(t/2)}{\cosh^2(t/2)} \tanh(t'/2) G(t, t') G(t', t') = -\frac{49}{60}.$$

(5.103)

The sum of the four diagrams is

$$C = a + b_1 + b_2 + c = -71/72$$

is in agreement the result obtained from Schreodinger eqn, see e.g. review (Zinn-Justin and Jentschura 2004).

Three loop corrections (Escobar-Ruiz et al. 2015) come from the diagrams shown in Fig. 5.10, their numerical contributions are given in Table 5.1. As one can see, unfortunately not all diagrams we were able to calculate analytically. Those which we could contain irrational parts, related with Riemann zeta function. Yet the final result for three loop coefficient agrees well with the answer from Schreodinger equation we mentioned before in (5.111). This answer is *rational*, so all terms with special functions must somehow cancel.

The reason it does happen must be the existence of certain *resurgent relation* between perturbative and instanton series we discussed above. The first of such relation has been proposed in Zinn-Justin and Jentschura (2004).

5.8   Transseries and Resurgence

In the previous sections we have shown how one can calculate corrections due to quantum/thermal fluctuations, order by order, to the density matrix (fluctons) and the vacuum energy (instantons). We did not pay much attention to similar perturbative series, on top of the trivial vacuum path $x_{cl}(\tau) = 0$, except to use it to subtract infrared divergencies.

Now we shift our focus and in this section discuss more general theoretical questions. We start with the unavoidable issue, namely the fact that the coefficients of all such series are increasing with their order, so that all of them are asymptotic—strictly speaking divergent—series.

"Divergent series are the invention of the devil, and it is shameful to base on them any demonstration whatsoever" wrote N.H. Abel in 1828. But modern physicists tend to be "non-Abelian", and they use the perturbation theory widely, its divergent series notwithstanding.

The main idea why the perturbative series *must* diverge was suggested by Dyson (1952). In short, the argument went as follows. In QED one makes expansion in e^2, order by order. Let us think a bit what would happen if one analytically continue QED into negative $e^2 < 0$ domain. Then protons and electrons would no longer attract each other, so atoms would dissolve. On the other hand, electrons would attract each other and congregate in large number, and so would positrons and

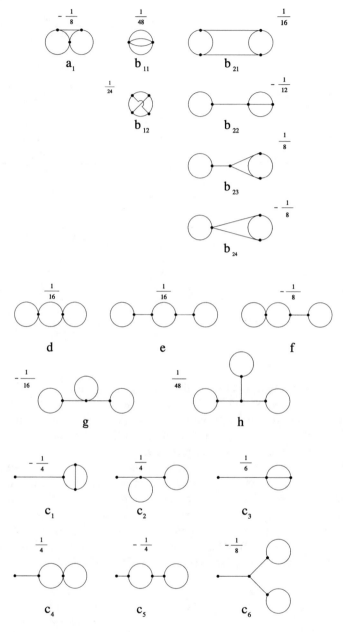

Fig. 5.10 Diagrams contributing to the three loop corrections to the instanton amplitude. The signs of contributions and symmetry factors are indicated

Table 5.1 Contribution of diagrams in Fig. (5.10) for the three-loop corrections B_2 (left) and A_2 (right). We write $B_2 = (B_{2loop} + I_{2D} + I_{3D} + I_{4D})$ where I_{2D}, I_{3D}, I_{4D} denote the sum of two-dimensional, three-dimensional and four-dimensional integrals, respectively. Similarly, $A_2 = I_{2D} + I_{3D} + I_{4D}$. The term $B_{2loop} = 39589/259200 \approx 0.152735$ (see text)

Feynman diagram	Instanton B_2	Vacuum A_2
a_1	−0.06495185	$\frac{5}{192}$
b_{12}	0.02568743	$-\frac{1}{64}$
b_{21}	0.04964284	$-\frac{11}{384}$
b_{22}	−0.13232566	$\frac{1}{24}$
b_{23}	0.28073249	$-\frac{1}{8}$
b_{24}	−0.12711935	$\frac{1}{24}$
e	0.39502676	$-\frac{9}{64}$
f	−0.35244758	$\frac{3}{32}$
g	−0.39640691	$\frac{3}{32}$
h	0.31424977	$-\frac{3}{32}$
c_1	−0.3268200	−
c_2	0.63329511	−
c_3	0.12657122	−
c_4	0.29747446	−
c_5	−0.77100484	−
c_6	−0.80821157	−
I_{2D}	0.0963	$-\frac{7}{384}$
I_{3D}	−0.0158	$\frac{19}{64}$
I_{4D}	−0.8408	$-\frac{155}{384}$

protons. A bit of thought shows that the binding energy in this case can be made arbitrarily large; so there would be a complete collapse of the theory, as it lacks any stable ground state. Now, if the positive and negative vicinity of the $e^2 = 0$ point are that different, there must be singularity at this point. Therefore any power expansion around it obviously cannot be nice (convergent).

We already mentioned in the introduction a distinction between perturbative and non-perturbative phenomena. In the toy models we studied those can be thought of as effects proportional to *powers* and *inverse exponents* of the coupling g^2. The more definite mathematical formulation of those is given by the so called *transseries*. French mathematician J. Ecalle in 1980s defined those as triple series ("standing on three whales") including not only powers of the parameter (=coupling) as in perturbation theory, but also exponential and logarithmic functions

$$f\left(g^2\right) = \sum_{p=0}^{\infty} \sum_{k=0}^{\infty} \sum_{l=1}^{k-1} c_{pkl} g^{2p} \left(exp(-\frac{c}{g^2})\right)^k \left(ln(\pm\frac{1}{g^2})\right)^l \qquad (5.104)$$

His argument was based on symmetries: such set of functions is *closed*, under manipulations including the so called Borel transform plus analytic continuations. Examples studied in mathematics usually were limited to some integrals with parameters, defining special functions. QM path integrals are infinite-dimensional, but simpler than those in QFT's, and, as we will see, in this case one can indeed build the transseries explicitly and demonstrate that indeed those two functions do appear. Furthermore, there are certain relations between the coefficients, known as *resurgence*: in some cases those can be derived, but its very existence in the QFT context remains a mystery.

Inter-relations of perturbative and non-perturbative effects are very deep. One reason of why they *must* be connected can be demonstrated using Borel transform as follows. Consider the following example: for $x > 0$ the integral of the r.h.s. generates the factorially divergent series in the l.h.s., clear from expanding the r.h.s. in x in geometric series

$$\sum_{n=0}^{\infty}(-1)^n n! x^n = \int dt\, e^{-t} \frac{1}{1+xt} \qquad (5.105)$$

Borel suggested doing the integral over t instead: since for $x > 0$ the pole is at $t = -1/x < 0$ is outside the integration region, the result is a good "Borel improved" result.[15] But for $x < 0$ the series do not have the alternated sign

$$\sum_{n=0}^{\infty} n! x^n = \int dt\, e^{-t} \frac{1}{1-xt} \qquad (5.106)$$

the Borel pole is at $t = 1/x$, right on the integration line. Shifting the integration cone up or down results in ambiguous imaginary part

$$\pm \frac{i\pi}{x} e^{-1/x}$$

The quantity of interest—say the ground state energy of our QM system—cannot have imaginary part, and yet its perturbative series, via Borel re-summation, seem to get it!

These puzzling questions are potentially resolved by a generalization of the perturbative series to the *transseries*. The idea is that the exponential terms in them are such that *all ambiguous/unphysical effects*—like unwanted imaginary parts— *must safely cancel out* to all orders. Furthermore, one expects that correct transseries define a *unique* function of the parameters.

Since we studied in detail tunneling in the double-well potential, let me use this particular example to elucidate the trasseries issue. A standard reference for results obtained using Schreodinger eqn is a summary by Zinn-Justin and Jentschura (2004). Their definition includes particle with mass $m = 1$ in the double well

---

[15] Note that Sum operation in Mathematica now have an option to try Borel-based and other regularizations, try to do this sum with *Regularization* → "*Borel*" prescription added.

potential

$$V(q, g) = \frac{q^2}{2}(1 - \sqrt{g}q)^2 \tag{5.107}$$

The quantity of interest will be the "vacuum energy" (that of the ground state). The beginning of its *perturbative expansion* in powers of g reads

$$E_0^{pert} = \frac{1}{2} - g - \frac{9}{2}g^2 - \frac{89}{2}g^3 - \frac{5013}{8}g^4 \ldots \tag{5.108}$$

Note that the coefficients grow rapidly (in fact factorially) and that all signs are the same (minus): thus we indeed recognize an example of "bad" Borel nonsummable series. Several hundreds of those terms has been generated by some recursive relation: they confirm this conclusion.

It is important to note that the perturbation theory has no knowledge of the existence of the second well: thus the two lowest levels E_0 and E_1 are degenerate. The *nonperturbative effects* are in this case representing by splitting of those levels due to tunneling effects. The first contribution is given by non-analytic exponent in g with a particular coefficient

$$E_0 = E_0^{pert} - \frac{2}{\sqrt{\pi g}}exp\left(-\frac{1}{6g}\right)(instanton\ series) \tag{5.109}$$

times another series in g called the "instanton series".

Full transseries for E_0 have the form of multi-instantons times new singular function, the log, appearing starting from the two-instanton term

$$E_0 = E_0^{pert} + \sum_{n=1} \left(\frac{2}{g}\right)^n \left(-\frac{e^{-\frac{1}{6g}}}{\sqrt{\pi g}}\right)^n \sum_{k=0}^{n-1}\left(ln(-\frac{2}{g})\right)^k E_{0,nkl}g^l \tag{5.110}$$

We will calculate below several terms in the expansion, so let me now give for reference their values

$$E_{0,100} = 1, \quad E_{0,101} = -\frac{71}{12}, \quad E_{0,102} = -\frac{6299}{288} \tag{5.111}$$

The question of *resurgence* is whether there are any relations between the perturbative series and those with different instanton number n. In Zinn-Justin and Jentschura (2004) the series generating functions $A(E, g)$, $B(E, g)$ were shown to be related by some exact relation, emanating from the condition that the ground state wave function should be symmetric and thus its derivative must be zero at the middle of the potential. Dunne and Unsal (2014) had found even more direct expression of the series, in terms of one function. We will not however discuss these developments here, following the principle that only material generalizable to QFT's is inside the scope of these lectures.

The instanton series, by themselves, also literally make no sense: negative argument of the log leads to imaginary part, which is physically meaningless for the ground state energy: there can not be any decays as we discuss the lowest (ground) state of the system. Properly defined transseries however make all those unphysical imaginary parts to cancel among themselves, producing the correct real answer. If one would be able to show that some well-defined transseries appear, at least for some QFT's, this would be a dramatic shift toward strict mathematical formulation of what these QFT's are. Yet for now no such examples are known, and all this remains just a theorist's dream.

5.9 Complexification and Lefschetz Thimbles

5.9.1 Elementary Examples Explaining the Phenomenon

Naive approach may suggest simple analytic continuation, from positive to negative x, and the pioneering papers (Dyson, Bogomolny, Zenn-Justin et al.) argued so. But it is not really justified and does not generally lead to the correct results: see e.g. the following

Exercise Two functions are defined by the following integrals

$$Z_1 = \int_{-\infty}^{\infty} dx \, e^{-\frac{1}{2\lambda} \sinh^2(\sqrt{\lambda}x)}, \quad Z_2 = \int_{-\infty}^{\infty} dx \, e^{-\frac{1}{2\lambda} \sin^2(\sqrt{\lambda}x)}$$

Expand them in powers of λ and show that one leads to Borel summable and another to non-summable series. Naively they are related by analytic continuation $Z_1(-\lambda) = Z_2(\lambda)$ but this is not true. Expressing both integrals in terms of Bessel functions, derive a correct relation between them which includes an imaginary part:

$$Z_1(e^{\pm i\pi}\lambda) = Z_2(\lambda) \mp i e^{\frac{-1}{2\lambda}} Z_1(\lambda)$$

Examples like that show that naive analytic continuations in parameters often lead to wrong answers. In general, these are consequences of the so called Stokes phenomenon. The general theory to be discussed elucidate how and why the very geometry of the integration contours can be abruptly changed, bringing in or out new saddle points and thus new asymptotic representation of the function.

In general the idea to complexify the integration variable of some integrals and change the integration contour is very old, used in particularly in "saddle point" method. In quantum mechanics and QFT's applications, with path integrals, it was tried also in cases with complex action (real time path integrals, Euclidean theories with finite chemical potential or theta angle, etc) in form of the so called "complex Langevin quantization". Many people (including myself) experimented with it in 1970s and found that it worked for some integrals and failed for the others. Recent wave of using complexification are based on more solid ground related with Lefschetz–Picard theory.

It has been introduced in QM setting by Witten in Witten (2010), which we now follow. Suppose we study a one-dimensional integral

$$I(a) = \int_{-\infty}^{\infty} dx e^{S}, \qquad S(x) = a * x^2 - x^4 \tag{5.112}$$

where the function in exponent S is chosen to resemble actions with typically have in QFT's, and the "mass" a is in general some complex parameter.

Let us promote x to a complex variable $z = x + iy$ and replace real axes of integration by some contour Γ in the complex plane. Since this particular function has no singularities, any integrals over closed contours Γ must vanish: and thus only the open contours (with different endpoints) are of interest.

What those contour can be? At large $|z|$ the quartic term is dominant, and the integral is well defined only provided the integration contour C ends up at some $Re(z^4) > 0$ lines, or $z_n \sim e^{i2\pi \frac{n}{4}}$ where $n = 0, 1, 2, 3$, see Fig. 5.11. So Γ should approach one of those four lines for the integral to be well defined.

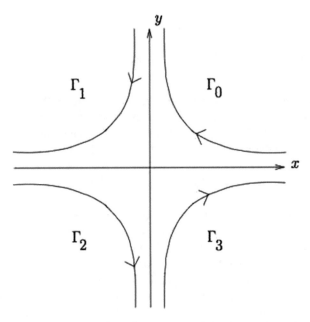

Fig. 5.11 Four basic contours on which integral is well defined: since their sum is zero, only three are independent

There are 3 extrema of this action, solutions to

$$0 = S'(z_m) = 2*a*z_m - 4z_m^3 \tag{5.113}$$

which are at the locations

$$z_0 = 0, z_1 = \sqrt{\frac{a}{2}}, z_2 = -\sqrt{\frac{a}{2}}$$

The simplest to discuss is the first one at the origin: action near it can be approximated by the first term. For $a > 0$, it is clear that the real axis has a minimum at it, while along the imaginary axis it is the maximum. For complex a this direction rotates accordingly.

So, rotating the integral to the imaginary axes makes sense. The other two also have directions on which the integral has a maximum. All of this is well known, as "saddle point" method to do the integrals.

Further improvements start with the question whether there are lines along which ReS are increasing (or decreasing) *monotonously*. Those can be followed by "gradient flow" equation

$$\dot{z} = \frac{\partial \bar{S}}{\partial z} \tag{5.114}$$

where dot stands for the derivative over some "computer time". The asymptotics of the lines approaches four "good" directions, along which the integral is convergent.

Exercise Solve this equation and find these 6 lines, originating from all three extrema in both directions, for $a = i$.

These lines are called *Lefschetz thimbles*: on them the real part of the action grows monotonously, from its values at the extreme points to infinity. The important statement of the Lefschetz theory is that the imaginary part of the action, $Im(S)$, *remains constant on these lines*. Instead of proving it, let me suggest to check it for the example at hand.

Exercise Find the thimble lines for the example at hand using this theorem.

Important consequence of this theorem is that since any integral over some contour C can be rewritten as a superposition of integrals over the thimbles

$$I = \sum_i c_i \int_{\Gamma^i} dz e^{S(z)}$$

with some coefficients c_i, basically ± 1 or 0. Each of them has fixed $Im(S)$ which can be taken out of the remaining integrals, which are therefore all real and well convergent.

Further consequence is that Lefschetz thimbles can only cross each other if their phases match,

$$Im(S_i) = Im(S_j), \quad (i \neq j) \tag{5.115}$$

Those can be easily found at the extremal points. Rapid changes of the integral—the "phase transitions"—can be caused by crossing and change of the thimble geometry and thus c_i. Thus matching of the phases is a very useful tool.

If the integral has more than one dimension, the geometry of thimbles gets more complicated. And yet there are successful practical applications of that, e.g. for models similar to finite density QCD in which Euclidean path integral has complex factors $e^{i\mu/T}$ (Alexandru et al. 2016).

Generic contributions of the thimbles have distinct complex phases, and this prevent their crossing. However at some values of parameters the phases can be the same. Often this happens when phases are multiple of π with some integer n, and the corresponding contributions are real. Those cases obviously split into two groups: the even n allows for addition of such contributions, but odd n lead to subtractions, sometimes to outright cancellation of them! A prototype model (from Behtash et al. 2015) nicely illustrate this last point.

Consider the integral

$$I(k, \lambda) = \int_\Gamma dw e^{2\lambda sinh(w) + kw} \tag{5.116}$$

with complex parameters k, λ. The three extrema of it and thimbles are shown in Fig. 5.12, and the contribution of the leading 1 and 3 thimbles have the same modulus of the exponential factor and form the following combination

$$I \sim (1 + e^{2\pi i(k+1/2)}) e^{-S_1}$$

which, for integer k, vanishes because of cancellation between thimbles. It was then shown by these authors that in quasi-exactly solvable and supersymmetric cases when no instanton contribution was present a similar cancellation—between the instantons and some *complex saddles*—take place!

5.9.2 Quasi-Exactly Solvable Models and the Necessity of Complex Saddles

Let me start with the definition: quasi-exactly solvable (QES) quantum mechanical problems are those which allow *some group of states* to be solved explicitly, with their wave functions and energies exactly known. (They are different from exactly

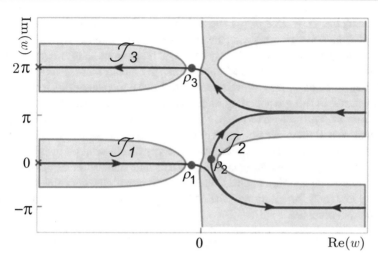

Fig. 5.12 The blue areas show "good regions" in which the integrand falls sufficiently rapidly at infinity to guarantee convergence. The red dots give the locations of three saddle points, and the blue contours show the "Lefschetz thimble" paths

solvable models in which all states can be found.) References to the original works can be found in a review (Turbiner 2016), in which also the underlying algebraic structure of QES problems is described in detail. We will not go into general discussion, focusing below on just one example .

For general orientation, let me give the simplest example of a familiar general statement: all' problems with supersymmetry, which remains unbroken, have the ground state energy *exactly equal to zero*. That means, that not only perturbative series needs to be canceled term by term (which is often possible to see), but also all nonperturbative effects should somehow get cancelled. How exactly that happens needs to be understood.

We will return to QES example a bit later, but we will first address a problem which do not even belong to this class, and yet it provides a very instructive puzzle. Its proposed resolution is from Kozcaz et al. (2016) and Behtash et al. (2016) which we here follow. The problem is known as *tilted double well potential* (TDW), which is the double well problem with added linear term $p \cdot x$, with some parameter p.

Of course, at small p one can proceed perturbatively. The TDW is related to supersymmetric quantum mechanics with the action

$$S = \int dt \left(\frac{\dot{x}^2}{2} + \frac{(W')^2}{2} + \bar{\psi}\psi + pW''\bar{\psi}\psi \right) \tag{5.117}$$

where $W(x)$ is known as superpotential and ψ is time-dependent Grassmanian field. We will use

$$W = \frac{x^3}{3} - x, \quad W' = x^2 - 1, \quad W'' = 2x \tag{5.118}$$

reproducing the TDW. For $p = 1$ the action is supersymmetric, and perturbatively the ground state energy vanishes. However, following a famous Witten's argument, supersymmetry is broken dynamically and the actual ground state energy is *nonzero* (and of course *positive*, since the Hamiltonian is a square).

For non-tilted case, $p = 0$, we developed above a nice theory of instantons and antiinstantons, in full glory and up to quantum corrections with one, two and three loop diagrams. The non-perturbative correction to ground state energy is *negative*, But when $p \neq 0$ there are no classical solutions going from one maximum to the other, since their hight is now different and energy on classical paths is conserved.

The following authors Kozcaz et al. (2016) and Behtash et al. (2016) argued that such classical solution is obtained if one complexify the coordinate, $x \to z = x + i \cdot y$ and look for solution of the *holomorphic Newton's equation* for inverted complexified potential

$$\frac{d^2z}{dt^2} = +\frac{\partial V}{\partial z} \tag{5.119}$$

Since the solution is going to have finite action, it must start from the highest point of the (inverted) potential we call x_{max}: but where can it go? Some thinking leads to the solution: at $p \neq 0$ the lowest maximum splits into a pair of two turning points, and our quartic potential can be re-written in a form convenient for motion with the maximal energy $E(x_{max})$ as

$$V = E(x_{max}) + (z - x_{max})^2(z - z_1)(z - z_2) \tag{5.120}$$

where $z_1^* = z_2$. At the turning point the velocity is zero, and so recoiling back (or going into an entirely new direction) is possible: so the paths we need to construct should go between x_{max} and z_1.

An example of such path, called *complexified bion* is found by Behtash et al. (2016). The action of this path is complex

$$S_{cb} = \frac{8}{3g} + p \cdot log(\frac{16}{pg}) + \ldots \pm ip\pi \tag{5.121}$$

but for p integer e^{-S} is real. Furthemore, for $p = 1$ one finds the sign opposite from what is normally expected from a semiclassical expression (in the Euclidean time). A number of numerical evidences that this is the right solution are given in Kozcaz et al. (2016), but those include using Dunne–Unsal relation instead of a direct evaluation of the determinant.

Now we turn to the second example from Kozcaz et al. (2016): its Hamiltonian is

$$H = \frac{g}{2}p^2 + \frac{1}{2g}(W'(x))^2 + \frac{p}{2}W''(x) \tag{5.122}$$

with $W = -\omega cos(x)$.

Note that one can put $\omega = 1$, recovering all answers by dimensional arguments later. The first term in Hamiltonian $\sim sin^2(x)$ has half natural period, while the last $\sim cos(x)$ has a natural period and the coefficient containing an extra parameter p. So at a generic value of p every second max/min is different from the first: that is why it is called Double Sine Gordon (DSG) problem.

The interesting thing about this particular problem is that it is an example of QES problems. Let us start by following its usual logics, by proposing an Ansatz for the wave functions

$$\psi(x) = u(x)e^{-\frac{W(x)}{g}} \tag{5.123}$$

Schreodinger equation for $u(x)$ is

$$\left[-\frac{g}{2}\frac{d^2}{dx^2} + sin(x)\frac{d}{dx} - \frac{(p-1)}{2}cos(x)\right]u = Eu \tag{5.124}$$

The operator in brackets can be represented in terms of generators of the SU(2) algebra

$$J_+ = e^{ix}\left(j - i\frac{d}{dx}\right), \quad J_- = e^{ix}\left(j + i\frac{d}{dx}\right), \quad J_3 = i\frac{d}{dx} \tag{5.125}$$

The operator in bracket for our problem can be written in terms of those operators

$$[] = (g/2)J_3^2 - (1/2)(J_+ + J_-) \tag{5.126}$$

provided $j = (p-1)/2$. The representations of SU(2) require j to be integer or semi-integer: thus the trick works only when p is integer. The number of states $2j + 1 = p$ can be separated from the rest, and using $p \times p$ matrix representation of the operators

$$J_\pm = \sqrt{(j \mp m)(j \mp m + 1)} \tag{5.127}$$

one can get exact Hamiltonian. Its diagonalization should produce the energies of the p states in question. In the paper Kozcaz et al. (2016) that is all done for $p = 1, 2, 3, 4$.

Exercise check that all the commutators are indeed those for SU(2) algebra. Also check that the Casimir operator is just number, $(1/2)J_+J_- + (1/2)J_-J_+ + J_3^2 = j(j + 1)$. For $j = 1/2, p = 2$ calculate the hamiltonian in the matrix form and find the energy eigenvalues. Expand them in powers of g and see the convergent series without $exp(-1/g^2)$ terms.

Summarizing this discussion: if p is some integer, the problem at hand belongs to the QES set. Indeed, there are exactly p levels whose energy and wave functions can be exactly calculated.

Since the energy has no instanton-like terms, one needs to explain a puzzle *how this may be possible*, since physically the tunneling between different minima of the potential cannot possibly disappear. The direct indication of that is that energies of all levels *other than the chosen p* do have the characteristic tunneling corrections!

The resolution of the puzzle proposed in Kozcaz et al. (2016) is based on the interference of the contributions of the two tunneling paths. One, is real tunneling between $x = 0$ and $x = 2\pi$ minima: its contribution is straightforward to calculate. Also clear is that it shifts the ground energy downward.

New element is the "complex bion" solution, with the action having imaginary part, as in the case we discussed at the beginning of this section. The solution

$$z_{cb} = 2\pi \pm 4\left[arctan(e^{-\omega_{cb}(t-t_0)}) + arctan(e^{-\omega_{cb}(t+t_0)})\right] \qquad (5.128)$$

can be obtained from the analytic continuation from the real bounce by $p \to pe^{i\theta}$ and taking $\theta = \pi$. Here $\omega_{cb} = \sqrt{1 + pg/8}$ and complex $t_0 \approx \frac{1}{2\omega_{cb}}log(-\frac{32}{pg})$ is such that its real part is related to the duration of the bounce.

The exponent of its action $exp(-S_{CB}) = exp(-ReS_{CB} - i\pi p)$ for odd p gets additional minus sign, meaning the energy is modified upward. It is this contribution which has *the potential*[16] *to cancel* that of the real tunneling. It is argued indirectly, from asymptotic of the perturbative series, that such cancellation happens for p levels, but not others. Why it is the case is still not clear.

Let us end with a wider question. Two examples discussed in this section show a *necessity* to include some complex extrema of the path integral, beyond the obvious real tunneling paths. Most likely the complex ones proposed so far in the papers mentioned do indeed fit the bill, do the right job, explaining these particular puzzles. Yet the general question—which of all possible complex classical paths (or solitons for QFTs) one should include, and which one should not—remains unanswered. The Picard-Lefshetz theory provide some light on this issue, so to say "in principle", since the original integration contour can be deformed only to some particular combination of their thimbles. Yet to use it outside some single (or few-)dimensional integrals is at the moment well beyond our abilities.

5.10   Brief Summary

- Feynman path integral formulation has opened completely new avenues in quantum mechanics, statistical mechanics, and QFTs. One of them is novel semiclassical approximation, based on certain solutions of *classical* equations of motion. In QM/SM setting the density matrix to find a system at the point x,

---

[16]So far it was shown that both real parts of the action are the same. For two amplitudes to cancel each other *exactly* one need to be sure that the real and complex paths generate exactly the same perturbative corrections: to see this is beyond current technical means.

$P(x)$, (x is a nofiguration in QFTs) is given by "fluctons". In SM setting it is periodic in euclidean time, with the period related to temperature $\hbar / T$

- for a number of systems—quartic potential well, the double-well potential or sin-Gordon potential—fluctons can be readily found.
- Novel semiclassical theory includes small fluctuations around flucton paths, following systematic Feynman rules, order by order. Quite elaborate examples of those are given, with the semiclassical series in complete agreement with series obtainable in QM/SM by other means.
- Tunneling through quantum-mechanical barriers are described by *instanton* paths. Their semiclassical theory is similar to that of fluctons. It is made more complicated by existence of *zero modes* (related to classical symmetries of the solution) which makes derivation of Green functions (propagators) more complicated.
- Perturbative and instanton contributions (e.g. to the vacuum energy) are specific examples of *transseries*. Those believed to give a unique representation of a function, as divergencies of perturbative and instanton series do cancel each other in all orders. The relations between perturbative and instanton series is called *resurgence*: we discuss some examples of those. In general, we know next to nothing about resurgence relations in QFTs, or whether those even exist at all.
- Semiclassical theory discussed are generic part of theory of complexified integrals, many-dimensional or infinite-dimensional ones. Extrema ((classical solutions) are connected by the so called "Lefschetz thimbles". Attempts to applied such theory to numerical lattice gauge theory are under way.

References

Aleinikov, A.A., Shuryak, E.V.: Instantons in quantum mechanics, two-loop effects. Yad. Fiz. **46**, 122–129 (1987)

Alexandru, A., Basar, G., Bedaque, P.: Monte Carlo algorithm for simulating fermions on Lefschetz thimbles. Phys. Rev. **D93**(1), 014504 (2016). 1510.03258

Behtash, A., Sulejmanpasic, T., Schafer, T., Unsal, M.: Hidden topological angles and Lefschetz thimbles. Phys. Rev. Lett. **115**(4), 041601 (2015). 1502.06624

Behtash, A., Dunne, G.V., Schafer, T., Sulejmanpasic, T., Unsal, M.: Complexified path integrals, exact saddles and supersymmetry. Phys. Rev. Lett. **116**(1), 011601 (2016). 1510.00978

Brown, L.S., Creamer, D.B.: Vacuum polarization about instantons. Phys. Rev. **D18**, 3695 (1978)

Corrigan, E., Goddard, P., Osborn, H., Templeton, S.: Zeta function regularization and multi - instanton determinants. Nucl. Phys. **B159**, 469–496 (1979)

Diakonov, D., Gromov, N., Petrov, V., Slizovskiy, S.: Quantum weights of dyons and of instantons with nontrivial holonomy. Phys. Rev. **D70**, 036003 (2004). hep-th/0404042

Dunne, G.V., Unsal, M.: Generating nonperturbative physics from perturbation theory. Phys. Rev. **D89**(4), 041701 (2014). 1306.4405

Dyson, F.J.: Divergence of perturbation theory in quantum electrodynamics. Phys. Rev. **85**, 631–632 (1952)

Escobar-Ruiz, M.A., Shuryak, E., Turbiner, A.V.: Three-loop correction to the instanton density. I. The quartic double well potential. Phys. Rev. **D92**(2), 025046 (2015). [Erratum: Phys. Rev.D92,no.8,089902(2015)]

Escobar-Ruiz, M.A., Shuryak, E., Turbiner, A.V.: Quantum and thermal fluctuations in quantum mechanics and field theories from a new version of semiclassical theory. Phys. Rev. **D93**(10), 105039 (2016). 1601.03964

Escobar-Ruiz, M.A., Shuryak, E., Turbiner, A.V.: Fluctuations in quantum mechanics and field theories from a new version of semiclassical theory. II. Phys. Rev. **D96**(4), 045005 (2017). 1705.06159

Faccioli, P.: Investigating biological matter with theoretical nuclear physics methods. J. Phys. Conf. Ser. **336**, 012030 (2011). 1108.5074

Feynman, R.: Statistical Mechanics: A Set of Lectures. W.A.Benjamin, Reading (1972)

Feynman, R., Hibbs, H.: Quantum Mechanics and Path Integrals. Mcgraw-Hill, New York (1965)

Gamow, G.: Zur Quantentheorie des Atomkernes. Zeitschrift fur Physik **51**, 204 (1928)

Gurney, R., Condon, E.: Quantum mechanics and radioactive disintegration. Nature **122**, 439 (1928)

Kozcaz, C., Sulejmanpasic, T., Tanizaki, Y., Unsal, M.: Cheshire cat resurgence, self-resurgence and quasi-exact solvable systems (2016). 1609.06198

Mandelstam, L., Leontowitsch, M.: Zur Theorie der Schroedingerschen Gleichung. Zeitschrift fur Physik **47**(1–2), 131–136 (1928)

Polyakov, A.M.: Quark confinement and topology of gauge groups. Nucl. Phys. **B120**, 429–458 (1977)

Sauter, F.: Uber das Verhalten eines Elektrons im homogenen elektrischen Feld nach der relativistischen Theorie Diracs. Z. Phys. **69**, 742–764 (1931)

Shuryak, E.V.: Toward the quantitative theory of the 'Instanton Liquid' 4. Tunneling in the double well potential. Nucl. Phys. **B302**, 621–644 (1988)

Turbiner, A.V.: Double well potential: perturbation theory, tunneling, WKB (beyond instantons). Int. J. Mod. Phys. **A25**(02n03), 647–658 (2010). 0907.4485

Turbiner, A.V.: One-dimensional quasi-exactly solvable Schroedinger equations. Phys. Rep. **642**, 1–71 (2016). 1603.02992

Vainshtein, A.I., Zakharov, V.I., Novikov, V.A., Shifman, M.A.: ABC's of instantons. Sov. Phys. Usp. **25**, 195 (1982). [201(1981)]

Witten, E.: A new look at the path integral of quantum mechanics (2010). 1009.6032

Wohler, C.F., Shuryak, E.V.: Two loop correction to the instanton density for the double well potential. Phys. Lett. **B333**, 467–470 (1994). hep-ph/9402287

Zarembo, K.: Monopole determinant in Yang-Mills theory at finite temperature. Nucl. Phys. **463**, 73–98 (1996). hep-th/9510031

Zinn-Justin, J., Jentschura, U.D.: Multi-instantons and exact results II: specific cases, higher-order effects, and numerical calculations. Ann. Phys. **313**, 269–325 (2004). quant-ph/0501137

Gauge Field Topology and Instantons

6

6.1 Chern–Simons Number and Topologically Nontrivial Gauges

Topological invariants are a traditional field in mathematics, and we will need those in a form discovered by Chern and Simons (1974). Generally, they exist in a different form in odd and even dimensional spaces and are related in a curious way.

We will start with $d = 3$ topology, physically relevant for a gauge theory defined in four dimensions.[1] The so called 3-d Chern–Simons number density is defined as the 4-th component of the following *topological current*

$$K_\mu = \frac{1}{16\pi^2}\epsilon^{\mu\alpha\beta\gamma}\left(A_\alpha^a\partial_\beta A_\gamma^a + \frac{1}{3}\epsilon^{abc}A_\alpha^a A_\beta^b A_\gamma^c\right) \tag{6.1}$$

$$N_{CS} = \int d^3x\, K_4$$

Let us select $t_1 = -\infty, t_2 = \infty$, and think of the gauge field at such times being "pure gauge," with zero field strength:

$$A_i = U^+(\vec{x})i\partial_i U(\vec{x}) \tag{6.2}$$

---

[1]Interesting gauge theories in three dimensions can be defined using N_{CS} as the Lagrangian: such construction was introduced by Witten in 1988 and is called the *topological field theory*. While it has applications in physics, e.g., in quantum Hall effect, we will not discuss it.

© The Author(s), under exclusive license to Springer Nature Switzerland AG 2021
E. Shuryak, *Nonperturbative Topological Phenomena in QCD
and Related Theories*, Lecture Notes in Physics 977,
https://doi.org/10.1007/978-3-030-62990-8_6

Substituting it to N_{CS}, one finds the following expression

$$N_{CS} = \frac{1}{24\pi^2} \int d^3 x \epsilon^{ijk} (U^+ \partial_i U)(U^+ \partial_j U)(U^+ \partial_k U) \qquad (6.3)$$

Now, $U(\vec{x})$ is a map from a 3-d space to the gauge group. If it is $SU(2)$, with three generators and three Euler angles, the group is basically the 3-sphere. The expression above is in fact nothing else but the topological invariant of such map, known as the *winding number*: it is the integer number of times the map covers the group.

Exercise Consider a "hedgehog" form for

$$U = \exp[i \frac{(\vec{r} \vec{\tau})}{r} P(r)]$$

with τ being Pauli matices, $P(0) = 0$ and $P(\infty) = \pi n$ with integer n. Note that only with such $P(r)$ the map of the point $r = \infty$ is smooth on the group. Substitute it into the previous expression and show that the result is equal to n.

What we learned is that pure gauge fields can be split into some topologically distinct classes, and, because of the relation (6.8), if before and after of certain gauge field configurations the pure gauge unphysical fields change this class, there must be some 4-d topological charge in between. Usually we do not think much about pure gauge fields, considering them to be unphysical and basically completely irrelevant. Now we learned that the so-called large gauge transformations, which change the winding number, cannot be irrelevant because their change is related to the 4-d topological charge Q_T which is expressed in terms of gauge fields and is clearly gauge invariant and physical.

Suppose one would like to start with a classical vacuum in a trivial gauge, with $N_{CS}(-\infty) = 0$, and interpolate it somehow with time-dependent intermediate field to the one with $N_{CS}(\infty) = 1$. This relation tells us that in doing so one necessarily has to go through intermediate fields with a nonzero field strength and thus energy: topologically distinct vacua must be separated by some kind of a *physical barrier*. Since there is no reason transition from 0 to 1 is different from n to $n+1$, we come to the conclusion that the 3-d gauge field configurations have a periodic potential as a function of N_{CS}. The optimal path, leading from one minimum to another, is known as the *sphaleron path*. In Chap. 11 we will derive explicit set of configurations along which N_{CS} changes from 0 to 1. Their energy E_{stat} will be calculated as well, explaining the shape and in particularly the *height* of the barrier separating different topological sectors. Let me here give the answer

$$E = 3\pi^2 (1 - \kappa^2)^2 / (g^2 \rho) \qquad (6.4)$$

$$N_{CS} = \text{sign}(\kappa)(1 - |\kappa|)^2 (2 + |\kappa|)/4$$

Fig. 6.1 The potential energy E (in units of $1/g^2\rho$) versus the Chern–Simons number $\tilde{N}_{CS}$, for the "sphaleron path" solution to be derived in Sphaleron chapter

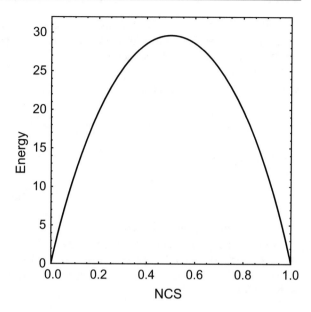

in parametric form. The corresponding plot $E(N_{CS})$, Fig. 6.1, shows the energy of the "sphaleron path" configurations between two subsequent values of Chern–Simons number, 0 and 1. Here ρ is arbitrary r.m.s. size of the field distribution: it appears because classical Yang–Mills is scale invariant and the energy is simply inversely proportional to the size.

Since $N_{CS} \in [integers]$ takes all integer values, from $-\infty$ to ∞, this potential repeats itself infinitely many times. This tells us that, as a function of this topological coordinate N_{CS}, the gauge theory resembles an *infinitely long crystal*. Therefore the states in it can be written as plain waves

$$\langle \theta | N_{CS} \rangle = \sum e^{i\theta N_{CS}} \tag{6.5}$$

with quasimomentum $\theta \in [-\pi, \pi]$. We will return to this in the next section.

6.2 Tunneling in Gauge Theories and the BPST Instanton

So, we already know that there is infinite set of pure gauge fields—therefore, with zero field strength and zero energy, called *classical vacua*—classified by the integer N_{CS}. We also know that there are also field configurations with non-integer N_{CS}: but those do have nonzero energy and therefore form a kind of a barrier separating classical vacua.

This barrier separating adjusted classical vacua, e.g., $N_{CS} = 0, 1$, turns out to be penetrable for quantum tunneling. Furthermore, as we will see, the tunneling rate can be found even without knowledge of the detailed shape of the potential.

This is possible because of the following key relation, between the topological current and the topological charge in 4-d

$$\partial_\mu K_\mu = \frac{1}{32\pi^2} G^a_{\mu\nu} \tilde{G}^a_{\mu\nu} \qquad (6.6)$$

The integral over the r.h.s.

$$Q_T = \frac{1}{32\pi^2} \int d^4x \, G^a_{\mu\nu} \tilde{G}^a_{\mu\nu} \qquad (6.7)$$

is the so-called 4-d topological charge.

Let us think about the consequence of this relation. Assuming one deals with some gauge field which decays well at spatial infinity, in a spirit of Gauss theorem, let us consider two time surfaces and integrate this relation in four-volume V_4 between them

$$Q_T(V_4) = \int_{t_1}^{t_2} dt \, \frac{\partial N_{CS}}{\partial t} = N_{CS}(t_2) - N_{CS}(t_1) \qquad (6.8)$$

It means that the topological charge in a volume between two time surfaces is equal to the *difference* of N_{CS} defined at those time moments.

(This is completely analogous to what one learns studying static electrodynamics: if there is a difference between electric field fluxes through two planes, you know how much charge is enclosed in between.)

In this section we are going to look for a tunneling path in gauge theory, which connects topologically different classical vacua, found in the famous work by Belavin, Polyakov, Schwartz, and Tyupkin and thus known as the BPST instanton (Belavin et al. 1975).

To find classical solution corresponding to tunneling, BPST used the following four-dimensional spherical ansatz depending on *radial* trial function f

$$g A^a_\mu = \eta_{a\mu\nu} \partial_\nu F(y), \quad F(y) = 2 \int_0^{\xi(y)} d\xi' f(\xi') \qquad (6.9)$$

with $\xi = ln(x^2/\rho^2)$ and η the 't Hooft symbol[2] defined by

$$\eta_{a\mu\nu} = \begin{cases} \epsilon_{a\mu\nu} & \mu, \nu = 1, 2, 3, \\ \delta_{a\mu} & \nu = 4, \\ -\delta_{a\nu} & \mu = 4. \end{cases} \qquad (6.10)$$

---

[2]More on its meaning, related with angular momentum in 4d, and properties one can find in Appendix A4.

We also define $\bar{\eta}_{a\mu\nu}$ by changing the sign of the last two equations. Further properties of $\eta_{a\mu\nu}$ are summarized in Appendix section "Angular Momentum in Four Dimensions and t'Hooft η Symbol". Upon substitution of the gauge fields in the gauge Lagrangian $(G_{\mu\nu})^2$, one finds that the effective Lagrangian has the form

$$L_{eff} = \int d\xi \left[\frac{\dot{f}^2}{2} + 2f^2(1-f)^2 \right] \tag{6.11}$$

corresponding to the motion of a particle in a double-well potential. We need Euclidean solution, the same as for the quantum mechanical instanton, connecting the *maxima* of the flipped potential. The corresponding field is

$$A_\mu^a(x) = \frac{2}{g} \frac{\eta_{a\mu\nu} x_\nu}{x^2 + \rho^2} \tag{6.12}$$

Here ρ is an arbitrary parameter characterizing the size of the instanton. Its appearance is dictated by the scale invariance of classical Yang–Mills equations.

The ansatz itself perhaps needs some explanation. The 't Hooft symbol projects to self-dual fields. The reason we selected it is related with the following identity

$$S = \frac{1}{4g^2} \int d^4x \, G_{\mu\nu}^a G_{\mu\nu}^a = \frac{1}{4g^2} \int d^4x \left[\pm G_{\mu\nu}^a \tilde{G}_{\mu\nu}^a + \frac{1}{2} \left(G_{\mu\nu}^a \mp \tilde{G}_{\mu\nu}^a \right)^2 \right], \tag{6.13}$$

where $\tilde{G}_{\mu\nu} = 1/2\epsilon_{\mu\nu\rho\sigma} G_{\rho\sigma}$ is the dual field strength tensor (the field tensor in which the roles of electric and magnetic fields are interchanged). Since the first term is a topological invariant (see below) and the last term is always positive, it is clear that the action is minimal if the field is (anti) *self-dual*[3]

$$G_{\mu\nu}^a = \pm \tilde{G}_{\mu\nu}^a, \tag{6.14}$$

In a simpler language, it means that Euclidean electric and magnetic fields are the same.[4] The action density is given by

$$(G_{\mu\nu}^a)^2 = \frac{192\rho^4}{(x^2 + \rho^2)^4}, \tag{6.15}$$

---

[3]This condition written in Euclidean notations; in Minkowski space extra i appears in the electric field.

[4]In the BPST paper, the self-duality condition (first-order differential equation) was solved, rather than (second-order) EOM for the quartic oscillator.

and one can see that it is indeed spherically symmetric and very well localized; at large distances, it is $\sim x^{-8}$. The action depends on scale only via the running coupling

$$S = \frac{8\pi^2}{g^2(\rho)} \tag{6.16}$$

which we will discuss more in the next section.

Note that while the gauge potential is long range, $A_\mu \sim 1/x$, in the field strength, the gradient and the commutator terms cancel each other. So physical effects are not long range: it suggests that the tail of the potential is gauge artifact.

Using invariance of the Yang–Mills equations under coordinate inversion implies that the singularity of the potential can be shifted from infinity to the origin by means of a (singular) gauge transformation $U = i\hat{x}_\mu \tau^+$. The gauge potential in singular gauge is given by

$$A_\mu^a(x) = \frac{2}{g}\frac{x_\nu}{x^2}\frac{\bar{\eta}_{a\mu\nu}\rho^2}{x^2+\rho^2}. \tag{6.17}$$

This singularity at the origin is unphysical, pure gauge, like the one for regular gauge at infinity. While there is only one infinity, each instanton has its own center, and so the singular gauge is better suited to make a superposition of many instantons.

What about the multi-instanton solutions? It is a long story, and the solution known as ADHM construction (for Atiyah et al. 1978) in principle solved it. What "in principle" means is that if one can solve all equations, the number of the parameters in the solution is equal to the number of zero modes, $4N_c$. Let me just mention that when the solution is found, one also gets Green functions, Dirac zero modes, etc. for free, automatically. It all started from one brilliant idea. If the field is pure gauge, $A_\mu = \Omega^+\partial_\mu\Omega$, the unitary gauge matrix can be eliminated from long derivatives in a standard way. The unitary gauge matrix can be written as $\Omega = \exp(i\vec{n}\vec{\tau})$ with real $\vec{n}$. If one takes *complex* $\vec{n}$ instead, Ω is not unitary and the field is therefore not a pure gauge. And yet, in many equations one can still get rid of this Ω as if it would be a gauge matrix. For example, the quark Green function is still $S(x, y) = \Omega(x)S_0(x, y)\Omega(x)^+$.

6.2.1 The Theta-Vacua

The fact that the action for the instanton is finite means that the barrier separating valleys in the topological landscape, with different N_{CS}, is penetrable. Since the potential as a function of N_{CS} is periodic, the complete set of states ψ_θ, characterized by a phase $\theta \in [0, 2\pi]$, is the so called theta-vacua, with the theta parameter—"quasimomentum"—defined by the periodicity condition

$$\psi_\theta(x + n) = e^{i\theta n}\psi_\theta(x) \tag{6.18}$$

Let us see how this band arises from tunneling events. If instantons are sufficiently dilute, then the amplitude to go from one topological vacuum $|i\rangle$ to another $|j\rangle$ is given by

$$\langle j| \exp(-H\tau)|i\rangle = \sum_{N_+} \sum_{N_-} \frac{\delta_{N_+ - N_- - j + i}}{N_+! N_-!} \left(K\tau e^{-S}\right)^{N_+ + N_-}, \tag{6.19}$$

where K is the preexponential factor in the tunneling amplitude and $N_{\pm}$ are the numbers of instantons and antiinstantons. Using the identity

$$\delta_{ab} = \frac{1}{2\pi} \int_0^{2\pi} d\theta \, e^{i\theta(a-b)} \tag{6.20}$$

the sum over instantons and anti-instantons can rewritten as

$$\langle j| \exp(-H\tau)|i\rangle = \frac{1}{2\pi} \int_0^{2\pi} d\theta \, e^{i\theta(i-j)} \exp\left[2K\tau \cos(\theta) \exp(-S)\right]. \tag{6.21}$$

This result shows that the quantum energy eigenstates are the theta-vacua $|\theta\rangle = \sum_n e^{in\theta}|n\rangle$. Their energy is

$$E(\theta) = -2K \cos(\theta) \exp(-S) \tag{6.22}$$

As usual, the width of the zone is on the order of the tunneling rate. The lowest state corresponds to $\theta = 0$ and has *negative* energy. This is as it should be, since the tunneling lowers the ground state energy.

At nonzero $\theta \neq 0$ the vacuum is not T or CP invariant: indeed it has "an arrow of time." The instanton amplitude has complex phase $e^{i\theta}$, and antiinstanton gets the conjugate phase $e^{-i\theta}$. In a world with nonzero θ, there exists the so-called Witten effect: electric and magnetic fields get admixed. For example, a magnetic monopole obtains some electric charge as well. Neutrons, together with their usual magnetic moment, obtain also an electric dipole, etc.

Experiments[5] show that CP symmetry is satisfied in strong interactions, so $|\theta| < 10^{-10}$. So we do live in the bottom (the lowest state) of the θ zone.[6]

It is obvious that all effects—e.g., the vacuum energy—are periodic in θ with period 2π. An interesting fact is that two branches of the vacuum meet at $\theta = \pi$, crossing as shown schematically in Fig. 6.2. The world with $\theta = \pi$ is T and CP even, as instantons and antiinstantons get the same phase factor -1. The two branches lead however to the double-degenerate vacua, and selecting one of them

---

[5]Specifically, the hunt for a nonzero electric dipole moment of the neutron.

[6]Why? The value of θ cannot be changed within the QCD. Hypothetical new particles, called *axions*, were suggested to relax any theta-vacuum to the bottom of the zone. Multiple searches for axions were made, so far without success.

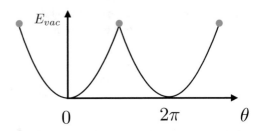

Fig. 6.2 Vacuum energy versus theta, schematically

breaks T and CP symmetry *spontaneously*. If one arranges domains of both types of vacuum, they are separated with the 1-d topological object, the *domain wall*. The excitations living of this wall have been studied and provided interesting window into the QCD-like theories compactified to three dimensions.

Presence of light quarks affects theta-vacua dramatically. It is enough to say that if *any* quark flavor be truly massless, the fermionic determinant of all gauge field configurations with global $Q \neq 0$ would vanish. This would mean that the whole theta-zone would collapse into a single vacuum, as theta-dependency would be erased. In the real world, however, this is not the case.

6.2.2   The One-Loop Correction to the Instanton: The Bosonic Determinant

The next natural step is the one-loop calculation of the preexponent in the tunneling amplitude. In gauge theory, this is a rather tedious calculation which was done in the classic paper by 't Hooft (1976). Basically, the procedure is completely analogous to what we did in the context of quantum mechanics. The field is expanded around the classical solution, $A_\mu = A_\mu^{cl} + \delta A_\mu$. In QCD, we have to make a gauge choice. In this case, it is most convenient to work in a background field gauge $D_\mu(A_\nu^{cl})\delta A_\mu = 0$.

We have to calculate the one-loop determinants for gauge fields, ghosts, and possible matter fields (which we will deal with later). The determinants are divergent both in the ultraviolet, like any other one-loop graph, and in the infrared, due to the presence of zero modes. As we will see below, the two are actually related. In fact, the QCD beta function is only *partly* determined by the zero modes, while in certain supersymmetric theories, the beta function is completely determined by zero modes, as we will discuss later.

First one has to deal with the $4N_c$ zero modes of the system. (For two groups, one has to deal with, in practice, electroweak $SU(2)$ and QCD $SU(3)$; let us enumerate them explicitly. In both cases, there are four coordinates plus one size ρ. For $SU(2)$, there are three Euler angle for rotations, either in space or in color space—which does not matter as those are directly related: thus $5 + 3 = 8$. For $SU(3)$ group, one simply does imbedding of the two-color solution into some subgroup of $SU(3)$. Out of $N_c^2 - 1 = 8$ rotation angles, one can see that only *one* does not affect the $SU(2)$ instanton, so there are seven angles, and $5 + 7 = 12$.)

The integral over the zero mode is traded for an integral over the corresponding collective variable. For each zero mode, we get one factor of the Jacobian $\sqrt{S_0}$. The group integration is compact, so it just gives a factor, but we have to keep the integral over size and position. As a result, we get a differential tunneling rate

$$dn_I \sim \left(\frac{8\pi^2}{g^2}\right)^{2N_c} \exp\left(-\frac{8\pi^2}{g^2}\right) \rho^{-5} d\rho d^4 z, \tag{6.23}$$

where the power of ρ can be determined from dimensional considerations.[7]

The ultraviolet divergence is regulated using the Pauli–Villars scheme, the only known method to perform instanton calculations (the final result can be converted into any other scheme using a perturbative calculation). This means that the determinant det O of the differential operator O is divided by $\det(O + M^2)$, where M is the regulator mass. Since we have to extract $4N_c$ zero modes from det O, this gives a factor M^{4N_c} in the numerator of the tunneling probability.

In addition to that, there will be a logarithmic dependence on M coming from the ultraviolet divergence. To one-loop order, it is just the logarithmic part of the polarization operator. For any classical field A_μ^{cl}, the result can be written as a contribution to the effective action (Brown and Creamer 1978)

$$\delta S_{NZM} = \frac{2}{3}\frac{g^2}{8\pi^2}\log(M\rho)S(A^{cl}) \tag{6.24}$$

In the background field of an instanton, the classical action cancels the prefactor $\frac{g^2}{8\pi^2}$, and $\exp(-\delta S_{NZM}) \sim (M\rho)^{-2/3}$. Now, we can collect all terms in the exponent of the tunneling rate

$$dn_I \sim \exp\left(-\frac{8\pi^2}{g^2} + 4N_c\log(M\rho) - \frac{N_c}{3}\log(M\rho)\right)\rho^{-5}d\rho dz_\mu \tag{6.25}$$

$$\equiv \exp\left(-\frac{8\pi^2}{g^2(\rho)}\right)\rho^{-5}d\rho dz_\mu,$$

where we have recovered the running coupling constant $(8\pi^2)/g^2(\rho) = (8\pi^2)/g^2 - (11N_c/3)\log(M\rho)$. Thus, the infrared and ultraviolet divergent terms combine to give the coefficient of the one-loop beta function, $b = 11N_c/3$. The bare charge and the regulator mass M can be combined into to a running coupling constant. At two-loop order, the renormalization group requires the miracle to happen once again, and the nonzero mode determinant can be combined with the bare charge to give

---

[7]Note that we for the first time meet here 5-d Anti-de Sitter space with the 5-th coordinate being a scale; it is the same as in famous Maldacena duality.

the two-loop beta function in the exponent and the one-loop running coupling in the preexponent.

The remaining constant from the determinant of the nonzero modes was calculated in ('t Hooft 1976). The result is

$$dn_I = \frac{0.466\exp(-1.679N_c)}{(N_c-1)!(N_c-2)!}\left(\frac{8\pi^2}{g^2}\right)^{2N_c}\exp\left(-\frac{8\pi^2}{g^2(\rho)}\right)\frac{d^4z\,d\rho}{\rho^5}. \tag{6.26}$$

The tunneling rate dn_A for antiinstantons is of course identical. Using the one-loop beta function, the result can also be written as

$$\frac{dn_I}{d^4z} \sim \frac{d\rho}{\rho^5}(\rho\Lambda)^b \tag{6.27}$$

and because of the large coefficient $b = (11N_c/3) = 11$, the exponent overcomes the Jacobian and small-size instantons are strongly suppressed. On the other hand, there appears to be a divergence at large ρ, although the perturbative beta function is not applicable in this regime.

6.2.3 Propagators in the Instanton Background

In the chapter on semiclassics, we discussed a systematic semiclassical loop expansion, allowing to calculate the tunneling probability and vacuum energy order by order, using Feynman diagrams. The same method may of course be used for QFTs and the gauge theory in particular. Unfortunately, the gauge theory instanton amplitudes have not yet been calculated even to the two-loop order.[8] In this section, we will discuss technical difficulties related with such calculation, as well as some ideas how to go around those.

The central objects we would need for Feynman diagrams are of course the propagators of the relevant fields, the quarks and gluons and perhaps ghosts, in the instanton background. As always, the fields are written as classical plus quantum ones

$$A^\mu = A^\mu_{cl} + a^\mu$$

the action expanding in powers of the latter. The propagators (or Green functions) are the inverse of the corresponding differential operator defined by the quadratic form $O(a^2)$. The technical problems mentioned are related to the fact that inversion can only be performed in a part of the functional space "normal" to zero modes.

---

[8]For a comparison, some QCD quantities like vector current polarization operator is now calculated to five loops.

But let us follow the problem methodically. The first step is the calculation of a propagator for *scalar* particles, both in fundamental and adjoint color representations. They satisfy the equation

$$-D_\mu^2 G(x, y) = \delta(x - y) \tag{6.28}$$

with the covariant derivative containing the background field of the instanton

$$iD_\mu = i\partial_\mu + T_a A_\mu^a \tag{6.29}$$

and so, symbolically the scalar propagator is

$$\Delta(x, y) = \left\langle x \left| \frac{1}{-D^2} \right| y \right\rangle.$$

There are no zero modes and the explicit solution was found by Brown et al. (1978). It is instructive to explain why they were able to do so. The instanton potential in the singular gauge can be written as

$$A_\mu^a(x) = -\bar{\eta}_{\mu\nu}^a \partial_\nu ln \left[1 + \frac{\rho^2}{(x - z)^2} \right]. \tag{6.30}$$

If the quantity in the square bracket is some general function $log[\Pi(x)]$ and the field is supposed to be self-dual, the condition on Π turns out to be the Laplacian

$$\partial_\mu^2 \Pi = 0 \tag{6.31}$$

in which the usual derivative, *not the covariant one*, appears. Thus, one can use it to generate a multi-instanton solution of the form[9]

$$\Pi = 1 + \sum_i \frac{\rho_i^2}{(x - z_i)^2}. \tag{6.32}$$

The crucial observation is that the instanton potential has the form $\Omega(x - z)^{-1} i\partial_\mu \Omega(x - z)$ which looks like pure gauge, except that Ω is *not* an unitary matrix (and therefore the field strength is *not zero*). This feature allows to factor the D_μ^2 operator and therefore also factor out its inverse. Using ansatz

$$\Delta(x, y) = \Pi(x)^{-1/2} \frac{F(x, y)}{4\pi^2(x - y)^2} \Pi(y)^{-1/2} \tag{6.33}$$

---

[9]Counting parameters one would find that this cannot be the most general multi-instanton solution.

with $\Pi(x)$ defined above, one can finally find the form for the function[10]

$$F(x, y) = 1 + \sum \rho_i^2 \frac{\left(\tau_\mu (x - z_i)_\mu\right) \left(\tau_\mu^+ (y - z_i)_\mu\right)}{(x - z_i)^2 \ (y - z_i)^2}. \tag{6.34}$$

Note that it defines the scalar propagator not just for one instanton but for multi-instanton solution of the form considered. Note also that at large distances $x, y \gg \rho$ all the correction factors are small compared to one in them, and the propagator reduces to free (no background) scalar propagator

$$\Delta \rightarrow \frac{1}{4\pi^2 (x - y)^2}.$$

Exercise Check that it satisfies the equation

The next step is getting the isospinor spinor (quark) propagator or more precisely its part normal to the zero mode. This was achieved by using the fact that zero mode resigns for one chirality only. This leads to the form of the propagator

$$S_{nz} = \left(\gamma_\mu D_\mu\right) \frac{1}{-D^2} \left(\frac{1 + \gamma_5}{2}\right) + \frac{1}{-D^2} \left(\gamma_\mu D_\mu\right) \left(\frac{1 - \gamma_5}{2}\right) \tag{6.35}$$

and since the inverse of D^2 we know already, the scalar propagator, the task is accomplished just by differentiation of it, which in the first term is directed to the right, and in the second to the left.

At this point, the reader perhaps expects the gauge propagator to be obtained by similar tricks. This is correct: the symbolic expression for the vector field propagator can indeed be written as

$$D_{\mu\nu}^{nz} \sim D_\mu \left(\frac{1}{D^2}\right) \left(\frac{1}{D^2}\right) D_\nu. \tag{6.36}$$

The long derivatives at the left and right of the expression act on the middle of it. The product of two inversions of D^2 should be understood as the *convolution* of two

---

[10]We use here and elsewhere the 4-d Pauli matrix extension $\tau_\mu = (\vec{\tau}, i)$.

scalar propagators, which includes the integration over some intermediate point[11] z_μ, namely,

$$\sum_z \left\langle x \left| \left(\frac{1}{D^2} \right) \right| z \right\rangle \left\langle z \left| \left(\frac{1}{D^2} \right) \right| y \right\rangle$$

$$= \int \frac{d^4 z}{4\pi^2} \frac{(1/2) tr \left[\tau_a (\tau_x^+ \tau_z + \rho^2)(\tau_z^+ \tau_y + \rho^2) \tau_b (\tau_y^+ \tau_z + \rho^2)(\tau_z^+ \tau_x + \rho^2) \right]}{(x^2 + \rho^2)(y^2 + \rho^2)(z^2 + \rho^2)^2 (x-z)^2 (y-z)^2}$$

(6.37)

where we used shorthand notations, $\tau_z = (\tau_\mu z_\mu)$, etc. The trace is a certain polynomial in components of x, y, z vectors. So, if the integral over z can somehow be calculated, the propagator is obtained by differentiation. The problem remains that only some analytic limits of the convolution integral is analytically known, not the complete integral.

The equation for the nonzero mode vector Green function has the explicit form

$$-\left(D^2 \delta_{\mu\lambda} + 2 G_{\mu\lambda} \right) D_{\lambda\nu}^{nz}(x, y) = P_\perp(x, y) = \delta_{\mu\nu} \delta(x - y) - \sum_{i=(\text{zero modes})} \phi_\mu^i(x) \phi_\nu^i(y)$$

(6.38)

in which the r.h.s. is the projector to all *nonzero* modes. It is a complete delta function minus projector to *all* zero modes.

The SU(2) instanton has eight of them: four translations, one scale transformation, and three Euler angles of rotation. All modes can be obtained by differentiation over corresponding collective coordinates.

Already from the equation itself one can see a coming problem. Let us look at it at large distances from the instanton. The operator in the l.h.s. becomes the ordinary Laplacian ∂^2, and the r.h.s. has the large-distance tails of the modes. Let us take one of them, the scale one, as example.

$$\phi_{scale} \sim \frac{\partial A_\mu}{\partial \rho^2} \sim \frac{x}{(x^2 + \rho^2)^2} \sim \frac{1}{x^3}$$

Since $\partial^2 D^{nz}(x, y) \sim x^{-3}$, the tail of the vector Green function must be $D^{nz} \sim 1/x$. If so, the Feynman diagrams involving 4-d integrals of it can have bad infrared divergences.

It has then been pointed out by Levine and Yaffe (1979) that this difficulty can be eliminated by a redefinition of "orthogonality": in Hilbert space, it can be defined with some weight function, which can be chosen to decay with distance appropriately. Their "improved" propagator was shown to have no $D^{nz} \sim 1/x$ tail.

---

[11]Not to be confused with the instanton center we also called z above. Currently, we have a single instanton with the center at the origin.

And yet, neither these authors nor anybody else used their improved propagator for four decades.

One of the methods for evaluation of the determinant is its relation to the Green function. The reader may be reminded that the method was based on differentiation of the classical solution over some parameter and relating the results to a one-loop Feynman diagram including the propagator, which can be calculated if the latter is known.

A comment about ADHM construction above sheds some light on how exact propagators in the field of instanton (or many instantons) were calculated—see the actual derivations in the original paper (Brown et al. 1978). Now, in the case of the instanton, there is indeed such parameter—the size ρ—and the method can be applied. In fact that method was used by Brown and Creamer (1978) for this purpose for the first time. One may expect that following this route one can cut off many difficulties of the problem, resolved by 't Hooft by brute force diagonalization.

Brown and Creamer were able to show that all UV divergences occur as expected, leading to the correct renormalized charge. But, attempting to calculate the finite part, Brown and Creamer unexpectedly found infrared divergences stemming from projector normal to zero modes.

Although the setting of next-order calculations is in principle quite analogous to those we have discussed in quantum mechanical context in chapter semiclassical, and even the diagrams are the same, in all the years passed since 1976 the two-loop correction to 't Hooft formula has not yet been calculated.

6.2.4 The Exact NSVZ Beta Function for Supersymmetric Theories

At first glance, instanton amplitudes seem to violate supersymmetry: the number of zero modes for gauge fields and fermions does not match, while scalars have no zero modes at all. We will however not discuss translation of the problem to a superspace and other related issues here, sticking to the standard notation. The remarkable fact is that the determination of the tunneling amplitude in SUSY gauge theory is actually *simpler* than in QCD. Furthermore, one can determine the complete perturbative beta function from a generic calculation of the tunneling amplitude!

The tunneling amplitude in question is given by

$$n(\rho) \sim \exp\left(-\frac{2\pi}{\alpha(M)}\right) M_0^{n_g - n_f/2} \left(\frac{2\pi}{\alpha(M)}\right)^{n_g/2} d^4x \frac{d\rho}{\rho^5} \rho^k \prod_f (fermions)_f,$$

$$(6.39)$$

where all factors can be understood from the 't Hooft calculation discussed above. There are $n_g = 4N_c$ bosonic zero modes that have to be removed from the determinant and give one power of the regulator mass M each. Similarly, each of the n_f fermionic zero modes gives a factor $M^{1/2}$. Introducing collective coordinates for the bosonic zero modes gives a Jacobian $\sqrt{S_0}$ for every zero mode. Finally, the last factor in the integral is related to fermionic collective coordinates and zero modes (to be discussed a bit later), and ρ^k is the power of ρ needed to give the correct dimension.

Here is the key observation: *supersymmetry ensures that spectra of eigenmodes for bosonic and fermionic fluctuations around the instanton are related.* As a result, one can show that the nonzero mode contributions in bosonic and fermionic determinants exactly cancel: therefore, there is no need to calculate them!

More precisely, the subset of SUSY transformations which does not rotate the instanton field itself mixes fermionic and bosonic modes nonzero modes but annihilates zero modes. This is why all nonzero modes cancel but zero modes can be unmatched. Note another consequence of the cancellation: *the power of M in the tunneling amplitude is an integer.*

Renormalizability demands that the tunneling amplitude is independent of the regulator mass. This means that the explicit M dependence of the tunneling amplitude and the M dependence of the bare coupling have to cancel. As in QCD, this allows us to determine the one-loop coefficient of the beta function $b = (4 - N)N_c - N_f$. Again note that b is an integer, a result that would appear very mysterious if we did not know about instanton zero modes.

In supersymmetric theories, one can even go one step further and determine the beta function to all loops (Novikov et al. 1986). For that purpose, let us write down the renormalized instanton measure

$$
n(\rho) \sim \exp\left(-\frac{2\pi}{\alpha(M)}\right) M_0^{n_g - n_f/2} \left(\frac{2\pi}{\alpha(M)}\right)^{n_g/2} Z_g^{n_g} \left(\prod_f Z_f^{-1/2}\right)
$$

$$
\times \, d^4x \frac{d\rho}{\rho^5} \rho^k \prod_f (fermions)_f, \tag{6.40}
$$

where we have introduced the field renormalization factors $Z_{g,f}$ for the bosonic/fermionic fields. Again, non-renormalization theorems ensure that the tunneling amplitude is not renormalized at higher orders (the cancellation between the nonzero mode determinants persists beyond one loop). For gluons the field renormalization (by definition) is the same as the charge renormalization $Z_g = \alpha_R/\alpha_0$. Furthermore, supersymmetry implies that the field renormalization is the same for gluinos and gluons. This means that the only new quantity in (6.40) is the anomalous dimension of the quark fields, $\gamma_\psi = d \log Z_f / d \log M$.

The renormalizability demands that all physical quantities—such as the amplitude under consideration—are independent of M. All powers of M we found should thus be compensated by the M−dependence of the charge. Indeed the cutoff of the integrals at M implies that the original charge was in fact the "bare one," $g(M)$. This condition gives the charge dependence on the scale, which can be reformulated as the so-called *NSVZ beta function* (Novikov et al. 1986) which, in the single supersymmetry case, $\mathcal{N} = 1$, reads

$$\beta(g) = -\frac{g^3}{16\pi^2} \frac{3N_c - N_f + N_f \gamma_\psi(g)}{1 - N_c g^2/8\pi^2}. \tag{6.41}$$

The anomalous dimension of the quarks γ_ψ has to be calculated perturbatively. To leading order, it is given by

$$\gamma_\psi(g) = -\frac{g^2}{8\pi^2} \frac{N_c^2 - 1}{N_c} + O(g^4). \tag{6.42}$$

As far as I know, the result (6.41) was checked by explicit calculations up to three loops.[12]

In theories without quarks, the NSVZ result determines the beta function completely. For $\mathcal{N}$-extended supersymmetric gluodynamics, we have

$$\beta(g) = -\frac{g^3}{16\pi^2} \frac{N_c(\mathcal{N} - 4)}{1 + (\mathcal{N} - 2)N_c g^2/(8\pi^2)}. \tag{6.43}$$

Let us recognize several interesting special cases:

(i) For $\mathcal{N} = 4$, the beta function vanishes and the theory is conformal. The reason for that we already discussed in the chapter on monopoles, where it was shown that this theory is electric–magnetic *self-dual*.

(ii) The case $\mathcal{N} = 2$ shows another curious phenomenon: the nontrivial part of the denominator vanishes, so that *the one-loop result for the beta function becomes exact*. This theory is the one partially solved by Seiberg and Witten: we will follow the charge running in it in the next section.

(iii) The next interesting case is the $\mathcal{N} = 1$ SUSY QCD, where we add N_f matter fields (quarks ψ and squarks ϕ) in the fundamental representation. Let us first look at the NSVZ beta function. The beta function vanishes at $g_*^2/(8\pi^2) = [N_c(3N_c - N_f)]/[N_f(N_c^2 - 1)]$, where we have used the one-loop anomalous dimension. This is reliable if g_* is small, which we can ensure by choosing $N_c \to \infty$ and N_f in the conformal window $3N_c/2 < N_f < 3N_c$. Seiberg showed that the conformal point exists for all N_f in the conformal window (even if N_c is not large) and clarified the structure of the theory at the conformal point. This is a phenomenon which also exists in QCD-like theories with many fermions.

---

[12]Note that the beta function is scheme dependent beyond two loops, so in order to make a comparison with high-order perturbative calculations, one has to translate from the Pauli–Villars scheme to a more standard perturbative scheme, e.g., $\overline{MS}$.

Another observation is that for $\mathcal{N} = 1$ the beta function blows up at $g_*^2 = 8\pi^2/N_c$, so the renormalization group trajectory cannot be extended beyond this point. The meaning of it remains mysterious (to me).

6.2.5 Instanton-Induced Contribution to the Renormalized Charge

Specific running of the coupling in non-Abelian gauge theories—the *asymptotic freedom*—gave birth to QCD, and so it is not surprising that higher-order corrections to the original celebrated one-loop result had attracted a lot of attention.

But one should not necessarily think only about the UV divergences in perturbative diagrams while considering the renormalized coupling constant. Let me give two examples: one very simple and one very complex. My first example is Feynman diagrams in quantum-mechanical 1+0 dimensional path integrals, say with $g^2 x^4$ interaction term. There are many loop diagrams renormalizing $g^2 x^4$ operators, but there are no divergences and UV logarithms.

The second example—conceptually simple but technically challenging—is the case of 4-d *superconformal field theories*, SCFTs. We mentioned two of them before: $\mathcal{N} = 4$ SYM and also $\mathcal{N} = 2, N_c = 2, N_f = 4$ supersymmetric QCD. SCFTs have zero beta functions and thus—unlike QCD—no Λ parameter. There is a bare coupling g_0 in the Lagrangian, which is however different from the "true" coupling g in certain exactly known observables. The relation between them can be expanded in the instanton series: for details see Alday et al. (2010), Marshakov et al. (2009).

Let us however return to QCD-like theories with the running coupling. A charge is defined, by general OPE rules, as a coefficient $1/g^2(\mu)$ in front of the operator $G_{\mu\nu}^2$ in the action, where general rules define the field G as the "soft" one, containing only Fourier harmonics with $p < \mu$. All "hard" phenomena, with $p > \mu$, are supposed to be delegated to the coefficient. All nonperturbative phenomena—and in particular instantons with sufficiently small size $\rho\mu < 1$—need to be included in the running charge. The difference between the perturbative and nonperturbative series is that the former are series in $1/\log(\mu)$ while the latter in (inverse) powers of μ.

Let us discuss the running coupling in a number of theories, following the paper by Randall et al. (1999). The best known case is the $\mathcal{N} = 2$ SYM theory, in which one knows both the exact analytic expression for the charge dependence on the scalar VEV a from Seiberg–Witten elliptic curve (see Appendix E2, in which we used variable $u = a^2$.) and also its perturbative–nonperturbative series which start as follows

$$\frac{8\pi^2}{g^2(a)} = 2\log(\frac{2a^2}{\Lambda^2}) - 6\left(\frac{\Lambda}{a}\right)^4 + \cdots \tag{6.44}$$

where the dots do not include high loop logs—they vanish in this theory, as shown in the previous section—but higher powers of the instanton terms $\left(\frac{\Lambda}{a}\right)^{4k}$. The number 4 in power, the same as the coefficient of $\log(a)$, is nothing but the one-loop coefficient of the beta function b: see NSVZ result discussed above. All terms have been explicitly calculated by Nekrasov (2003), and they confirm the expansion of the Seiberg–Witten elliptic curve.

This expression of course should only be used when the second term is much smaller than the first, but one can still make a tempting guess: since they are of the opposite sign, perhaps at some scale, the r.h.s. vanishes, which means that the coupling gets infinite! According to Seiberg–Witten, it is indeed the case, but the singularity is at a place slightly misplaced compared to what one would get from those two terms.

In Fig. 6.3 from Randall et al. (1999), the exact answer (solid thick line) is compared with the one-instanton expression (the thin dashed line). The QCD curves correspond to the OPE definition

$$\frac{8\pi^2}{g^2(a)} = b \log\left(\frac{a}{\Lambda}\right) - \frac{4\pi^2}{N_c^2 - 1} \int_0^{1/a} dn(\rho)\rho^4 \left(\frac{8\pi^2}{g^2(\rho)}\right)^2 \tag{6.45}$$

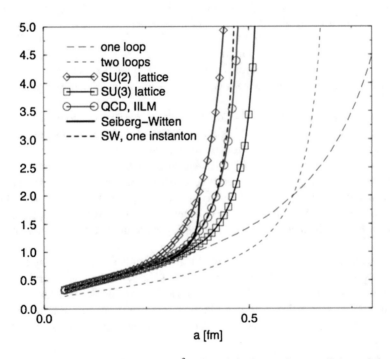

Fig. 6.3 The effective charge $bg_{eff}(\mu)/8\pi^2$, where b is the one-loop coefficient of the beta function) versus normalization scale $a[fm]$ (in units of its value at which the one-loop charge blows up). The thick solid line corresponds to exact solution [20] for the $N = 2$ SYM, while the thick dashed line shows the one-instanton correction. Lines with symbols (as indicated on figure) stand for $N = 0$ QCD-like theories, SU(2) and SU(3) pure gauge ones and QCD itself. Thin long-dashed and short-dashed lines are one- and two-loop results

including the instanton density with the size ρ extracted from either lattice simulations or models—both to be discussed below. The bottom line is that in all cases one finds a very similar behavior: at certain scale the instanton-induced power term is rapidly switched in and increases the coupling. This explains why the transition from weak to strong coupling happens rather abruptly.

6.3   Single Instanton Effects

6.3.1   Quarkonium Potential and Scattering Amplitudes

The simplest effects we will describe utilize relatively strong gluonic fields of the instantons. Perhaps the simplest effect is instanton contribution to the effective gluon mass,[13] obtained by averaging of the gluon propagator. As shown by Musakhanov and Egamberdiev (2018), with the original parameters of the instanton liquid model, it is $M_g^{inst} \approx 360\,\text{MeV}$.

The first example would be instanton contribution to the quarkonium potentials: it is one of the oldest ideas, suggested by Callan et al. (1978). Quarkonium is substituted by a color dipole, a pair of Wilson lines. In Fig. 6.4, we show a perturbative and instanton-induced settings.

Spin-independent (or central) potential is represented as

$$V = \int d\rho \frac{1}{\rho^2} \frac{dn(\rho)}{d\rho} W(x/\rho) \qquad (6.46)$$

Fig. 6.4 The one-gluon exchange and instanton-induced settings for the quarkonium potential

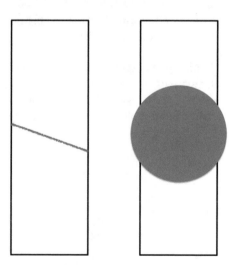

---

[13]The effective quark mass is related to chiral symmetry breaking and is thus multi-instanton effect, to be discussed in next chapters.

where density $dn(\rho)$ includes both instantons and antiinstantons, and the last factor is the convolution of two Wilson lines, each done exactly

$$W(x/\rho) = \frac{1}{3\rho^3} \int d^3r \, tr \left[1 - W(\vec{x} - \vec{r})W^+(-\vec{r})\right]$$

$$W(\vec{r}) = \cos\left(\frac{\pi r}{\sqrt{r^2 + \rho^2}}\right) + \frac{\vec{r}\vec{\tau}}{r}\sin\left(\frac{\pi r}{\sqrt{r^2 + \rho^2}}\right).$$

Since these authors were also interested in magnetic effects as well—spin–spin and spin–orbit ones—they make the lines a bit tilted relative to the time direction, so they looked as follows

$$P\exp\left(i\int A_\mu dx_\mu\right) = \exp\left(i\int d\tau \frac{-(\vec{\tau}\vec{x}) + (\vec{\tau}\vec{v} \times \vec{x})}{\tau^2 + \rho^2 + (\vec{x} + \vec{v}\tau)^2}\right) \tag{6.47}$$

and then expanded in velocity to the needed order.

One can also calculate other relativistic corrections, namely, the spin–spin, spin–orbit, and tensor potentials, defined by

$$V = V_C(r) + V_{SS}(r)(\vec{S}_Q\vec{S}_{\bar{Q}}) + V_{LS}(r)(\vec{S}\vec{L}) + V_T(r)\left[3(\vec{S}_Q\vec{n})(\vec{S}_{\bar{Q}}\vec{n}) - (\vec{S}_Q\vec{S}_{\bar{Q}})\right]. \tag{6.48}$$

As shown by Yakhshiev et al. (2018), Musakhanov (2018), the instanton liquid model Shuryak (1982) with the original parameters, $\rho = 1/3\,fm, n = 1\,fm^{-3}$, describes well spectrum of known charmonium states, including $L = 0, 1, 2$ states. The corresponding potentials are shown in Fig. 6.5 for two sets of the instanton parameters.

Similarly, Shuryak and Zahed (2000) have generalized this calculation to instanton-induced static dipole–dipole potential or the interaction *between* two quarkonia. Before describing it, let me remind the situation in QED: according to famous Casmir–Polder paper, the interaction of two distant dipole is

$$V(r) = -\frac{\alpha_1\alpha_2}{r^7} \tag{6.49}$$

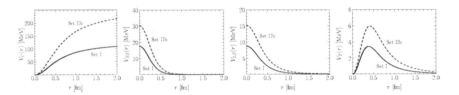

Fig. 6.5 The instanton-induced heavy quark potentials. Solid (dashed) curves are for $\rho = 1/3\,fm, R = 1\,fm$ ($\rho = 0.36\,fm, R = 0.89\,fm$) instanton parameters

Fig. 6.6 The setting of the
dipole–dipole potential
calculation: two-gluon
exchange and
instanton-induced

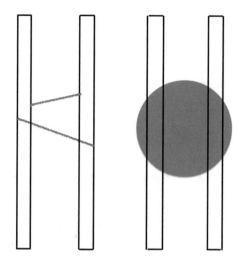

where α_i are the so-called polarizabilities. Note that it is not a square of the dipole field $\sim 1/r^6$: the difference comes from the time delay.

Let me briefly reproduce this result in Euclidean setting we use. For simplicity, let us only consider the case when dipoles are small $d \ll R, \rho$. In this case, the dipole approximation is justified, both gluons (photons) are emitted at close time, and the correlator to consider is

$$\langle \vec{E}^2(\tau_1)\vec{E}^2(\tau_2)\rangle \sim \frac{1}{\left[R^2 + (\tau_1 - \tau_2)^2\right]^4}. \tag{6.50}$$

Its integral over relative time leads to Casimir–Polder answer just mentioned (Fig. 6.6).

The correlator of two fields squared in the instanton background is calculated using its expression (6.15). The averaging over the instanton location can be carried out analytically: we then get the so-called correlation function (to be discussed systematically in the next chapter).

It is important to point out that while in the quarkonium potential the instanton-induced effect leads to relatively small energy shifts ($\sim 70\,\mathrm{MeV}$) compared to the perturbative one-gluon exchange, it is in fact dominant in the dipole–dipole case for $R > \rho$; see Fig. 6.7 . This is a manifestation of a general trend: the higher is the order of the effect in perturbation theory, the more important instanton-induced effects become. Indeed, because the instanton field is $O(1/g)$, the coupling to a quark is cancelled out and extra exchanges go "for free," without any additional penalty. The only small factor in the problem is then the instanton amplitude itself.

Introduction of the nonzero angle θ between the Wilson lines promotes the calculation into that of the scattering amplitude. A continuity between static potential and the low-energy scattering amplitude $\theta \rightarrow 0$ for two very heavy dipoles is then apparent. The untraced and tilted Wilson line in the one-instanton background reads

$$\mathbf{W}(\theta, b) = \cos\alpha - i\tau \cdot \hat{n} \sin\alpha \tag{6.51}$$

where

$$n^a = \mathbf{R}^{ab}\, \eta^b_{\mu\nu}\, \dot{x}_\mu (z - b)_\nu = \mathbf{R}^{ab}\, \mathbf{n}^b \tag{6.52}$$

$$\langle [gG^a_{\mu\nu}(x)]^2 [gG^a_{\mu\nu}(0)]^2 \rangle = \frac{384g^4}{\pi^4 x^8} + (n_0\rho_0^4)\Pi_{inst}(x/\rho)/\rho^8,$$

$$\Pi_{inst}(y) = \frac{12288\,\pi^2}{y^6\,(y^2+4)^5}\left(y^8 + 28\,y^6 - 94\,y^4 - 160\,y^2 - 120 \right.$$

$$\left. + \frac{240}{y\,\sqrt{y^2+4}}(y^6 + 2\,y^4 + 3\,y^2 + 2)\, \text{arcsinh}(y/2) \right).$$

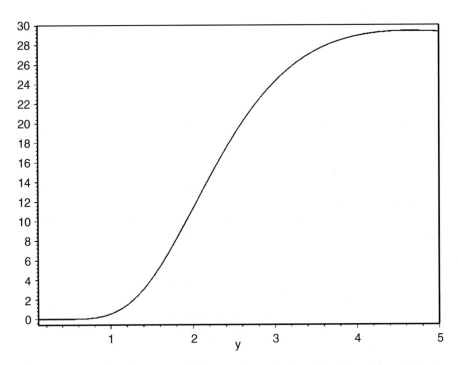

Fig. 6.7 The correlator of two gauge fields squared, perturbative and instanton-induced. The plot is the ration of the former to the latter, as a function of distance (in units of the instanton size ρ

and $\alpha = \pi \gamma / \sqrt{\gamma^2 + \rho^2}$ with

$$\gamma^2 = n \cdot n = \mathbf{n} \cdot \mathbf{n} = (z_4 \sin\theta - z_3 \cos\theta)^2 + (b - z_\perp)^2 .$$

In case of high-energy collisions, eikonalized expressions for the scattering amplitude in terms of a correlator of two Wilson lines (quarks) or Wilson loops (dipoles) are well known, developed by Nachtmann (1997), relating the scattering amplitude to the correlator of pairs of Wilson lines.

One can use these expressions, in Euclidean space-time with the angle θ between the Wilson lines and later analytically continue the result into the Minkowski world by the substitution

$$\theta \to iy \qquad\qquad (6.53)$$

where y is the Minkowski rapidity difference between the colliding objects. It has been checked in (Meggiolaroab 1998; Shuryak and Zahed 2000) and elsewhere that it works correctly for perturbative amplitudes. We used it for quark–quark and dipole–dipole scattering amplitudes as well (Nowak et al. 2001; Shuryak and Zahed 2004): the setting for the latter case is shown in Fig. 6.8.

Strictly speaking, a mutual scattering of two small dipoles corresponds to *double deep-inelastic scattering*. For example, future lepton collider can be used as a collider of two virtual photons $\gamma^*\gamma^*$. The quark–antiquark pair produced by the photons has small sizes $d \sim 1/Q$ provided each photon is highly virtual $Q \gg \Lambda_{QCD}$. But in practice, the proton ensemble is often represented as a set of dipoles,

Fig. 6.8 The setting for calculation of the instanton-induced dipole–dipole scattering amplitude. The angle between two Wilson lines is θ

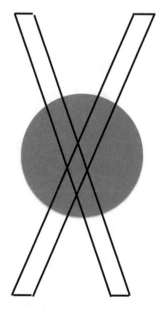

and even the proton itself sometimes is treated as a color dipole made of quark and diquark. For the details about instanton-induced effect in scattering amplitudes, see the papers mentioned.

6.4 Fermionic Transitions During Changes of Gauge Topology

6.4.1 The Fermionic Zero Mode of the Instanton

The so-called *index theorems* connect topology of the background gauge field with the *number of zero modes* of the Dirac operator defined on them. Without going into details, let me just say that if the gauge field has the topological charge Q, then

(i) the fundamental (isoscalar) fermions (e.g., quarks) should have Q zero modes;
(ii) the adjoint (isovector) fermions (e.g., gluinoes of supersymmetric gauge theories) should have QN_c zero modes.

The meaning of these facts is very profound, with important consequences for instanton-induced effects which we will be discussing.

Let us first look at the explicit form of the zero mode originally discovered by 't Hooft (1976). It is a solution of the Dirac equation $i\not{D}\psi_0(x) = 0$ in the instanton background. For an instanton in the singular gauge, the zero mode wave function takes the form

$$\psi_0(x) = \frac{\rho}{\pi} \frac{1}{(x^2 + \rho^2)^{3/2}} \frac{\gamma \cdot x}{\sqrt{x^2}} \frac{1 + \gamma_5}{2} \phi \tag{6.54}$$

where $\phi^{\alpha m} = \epsilon^{\alpha m}/\sqrt{2}$ is a constant spinor[14] in which the $SU(2)$ color index α is coupled to the spin index $m = 1, 2$. Let us briefly digress in order to show that (6.54) is indeed a solution of the Dirac equation. First observe that[15]

$$(i\not{D})^2 = \left(-D^2 + \frac{1}{2}\sigma_{\mu\nu}G_{\mu\nu}\right). \tag{6.55}$$

We can now use the fact that $\sigma_{\mu\nu}G_{\mu\nu}^{(\pm)} = \mp\gamma_5\sigma_{\mu\nu}G_{\mu\nu}^{(\pm)}$ for (anti) self-dual fields $G_{\mu\nu}^{(\pm)}$. In the case of a self-dual gauge potential, the Dirac equation $i\not{D}\psi = 0$ then

---

[14]It can therefore be called "spinor hedgehog." The norm is such that this mode is normalized to $\int d^4x\, \bar{\psi}_0\psi_0 = 1$.

[15]We use Euclidean Dirac matrices that satisfy $\{\gamma_\mu, \gamma_\nu\} = 2\delta_{\mu\nu}$. We also will use the following combinations of gamma matrices $\sigma_{\mu\nu} = i/2[\gamma_\mu, \gamma_\nu]$ and $\gamma_5 = \gamma_1\gamma_2\gamma_3\gamma_4$.

implies ($\psi = \chi_L + \chi_R$)

$$\left(-D^2 + \frac{1}{2}\sigma_{\mu\nu}G^{(+)}_{\mu\nu}\right)\chi_L = 0, \qquad -D^2\chi_R = 0, \tag{6.56}$$

and vice versa ($+ \leftrightarrow -$, $L \leftrightarrow R$) for anti-self-dual fields. Since $-D^2$ is a positive operator, χ_R has to vanish and the zero mode in the background field of an instanton has to be left-handed, while it is right-handed in the case of an antiinstanton. This result is not an accident. Indeed, there is a mathematical theorem (the Atiyah–Singer index theorem) that requires that $Q = n_L - n_R$ for every species of chiral fermions. A general analysis of the solutions of (6.56) was given in 't Hooft (1976). For (multi) instanton gauge potentials of the form $A^a_\mu = \bar{\eta}^a_{\mu\nu}\partial_\nu \log \Pi(x)$, the solution is of the form

$$\chi^m_\alpha = \sqrt{\Pi(x)}\partial_\mu\left(\frac{\Phi(x)}{\Pi(x)}\right)(\tau^{(+)}_\mu)^{\alpha m}. \tag{6.57}$$

The Dirac equation requires $\Phi(x)$ to be a harmonic function, $\Box\Phi(x) = 0$. Using this result, it is straightforward to verify (6.54).

It is also important that there is no chirality partner for zero modes: the "pairing" theorem for λ and $-\lambda$ modes holds for nonzero modes only. So what was wrong with the proof? Of course the assumption that $\gamma_5\psi_\lambda$ can be used as *another* eigenvector: it would not work for purely chiral solutions. So, what does the existence of the zero mode mean for the tunneling rate? At the first glance, zero mode is a problem since it vanishes the fermionic determinant in the partition function. Indeed, the determinant is of the operator $i\slashed{D} + im$, and since the former term gives zero on a zero mode one has to conclude that for massless fermions the tunneling probability vanishes. Not necessarily, argued 't Hooft, since the mass term can be supplemented by external scalar current. What it all means is that there is no tunneling *unless* a $\bar{q}_R q_L$ pair for each massless flavor is produced.

Still the whole process looks very mysterious. The final "demystification" of the anomaly is as follows: one can follow the tunneling configurations adiabatically, and for each value of time we are looking for static energy levels of the Dirac particle and ignoring all time derivatives. One then finds that the levels move in such a way that all left-handed states make one step down, to the next level, and all right-handed ones make one step up. A hint that this is the case can be explained as follows: in the adiabatic approximation (slow change in time), the time-dependent solution is

$$\psi(t, x) = \psi_{static}(t, x)\exp\left[-\int_0^t dt'\epsilon_{static}(t')\right]. \tag{6.58}$$

If energy is positive for large t and negative for $t \to \infty$, the corresponding time-dependent wave function is four-dimensionally normalizable. The explicit 't Hooft zero mode is such that four-dimensional normalizable solution. Thus, if only one such solution exists, it means that only *one* state has passed the zero energy mark.

So, when tunneling is finished, the spectrum is of course the same, but it is the *level occupation* which is different!

6.4.2   Electroweak Instantons Violate Baryon and Lepton Numbers

Briefly about the electroweak instantons, very little changes in terms of formulae, while the numbers involved are drastically different. The Higgs VEV sets a scale and therefore instanton size also gets fixed ('t Hooft 1976). The charge at electroweak scale is small, and therefore the probability of tunneling is now extremely small

$$P \sim \exp\left(-\frac{16\pi^2}{g_w^2}\right) \sim 10^{-169}, \tag{6.59}$$

so it seems out of question that one can observe any manifestations of it.

However, it is still instructive to discuss what should happen if one observes the consequences of electroweak instanton. Unlike QCD, weak gauge fields are only coupled to left-handed fermions. So there is no cancellation of the anomaly (by right-handed fermion loop) for vector current: it is *not* therefore conserved, as well as the axial current.

Let us think what it implies: unlike chirality in QCD, we now have nonconservation of the *baryon B* and *lepton L* numbers! More specifically, an instanton must generate production of nine quarks (three colors times three generations) and three leptons or

$$\Delta B = \Delta L = 3. \tag{6.60}$$

The difference $B - L$ is thus conserved.

While electroweak instantons (tunneling) are unobservable, the corresponding *sphaleron transitions* at finite temperature electroweak plasma are much more probable. They have quark and lepton zero modes and generate similar process, to be discussed in the sphaleron chapter.

6.4.3   Instanton-Induced ('t Hooft) Effective Lagrangian

Let us introduce quark sources $j_f(x)$ via auxiliary terms in Lagrangian $\int d^4x(j_f^+(x)\psi_f(x) + cc)$ and calculate a $2N_f$-quark Green's function

$$G(x_1\ldots x_f y_1 \ldots y_f) = \left\langle \prod_f \bar{j}_f(x_f) j_f(y_f) \right\rangle$$

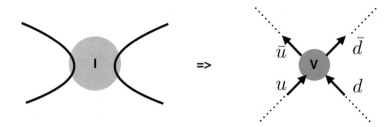

Fig. 6.9 The $2N_f$-leg Green function (left, for $N_f = 2$) at large distances can be seen as a local $2N_f$ operator V with free propagators (shown by the dotted lines). The free propagators can be amputated, leaving local multi-fermion vertex

containing a quark and an antiquark of each flavor once. Contracting all the quark fields, the Green's function is given by the tunneling amplitude multiplied by N_f fermion propagators. It is shown schematically in Fig. 6.9

It would be important to introduce also quark mass m (same for all) as an IR regulator. Every propagator has a zero mode contribution with one power of the fermion mass in the denominator:

$$S(x, y) = \frac{\psi_0(x)\psi_0^+(y)}{im} + \sum_{\lambda \neq 0} \frac{\psi_\lambda(x)\psi_\lambda^+(y)}{\lambda + im} \tag{6.61}$$

where I have written the zero mode contribution separately. Note that if both points x, y are far from the instanton center (relative to ρ), one can use asymptotic expression for ψ_0 which at large arguments behaves as constant spinor times $1/x^3$. Since this behavior is nothing else but just free propagator for a massless fermion, one sees that in this limit the first term can be interpreted as two free propagators, from x to z and from y to z, times some constant vertex. The procedure we have described is in fact standard "amputation of external legs" of the Green functions, used when one would like to derive the effective vertex or Lagrangian.

Let us now look at the dependence on the light quark masses. Suppose there are N_f light quark flavors and all masses are the same. The instanton amplitude is proportional to m^{N_f} (or, more generally, to $\prod_f m_f$) due to the fermionic determinant in the weight. But contributions of the zero modes in the propagators give us $1/m^{N_f}$! As a result, the zero mode contribution to the Green's function is finite in the chiral limit (Fig. 6.10)[16] $m \to 0$.

The result can be written in terms of a new effective Lagrangian ('t Hooft 1976). It is a local $2N_f$-fermion vertex, where the quarks are emitted or absorbed in the (x-independent) spinor states obtained after the "amputation of the free motion" from

---

[16]Note that Green's functions involving more than $2N_f$ legs are not singular as $m \to 0$. The Pauli principle always ensures that no more than $2N_f$ quarks can propagate in zero mode states.

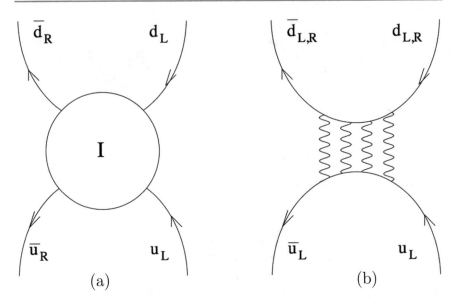

Fig. 6.10 The instanton-induced 't Hooft vertex (**a**) for two flavor QCD versus the ordinary gluon exchange diagrams (**b**). Note a very different chiral structure of the two: the latter does not violate any chiral symmetry because chirality is conserved along each line

the zero mode wave functions

$$\chi = \lim_{x \to \infty} \frac{\psi_0(x)}{S_0(x_c, x)}$$

as shown in the right of Fig. 6.9.

One may wish to simplify this general fairly complicated vertex. First, if instantons are uncorrelated, one can average over their orientation in color space. For $SU(3)$ color and $N_f = 1$, the result is

$$\mathcal{L}_{N_f=1} = \int d\rho \, n_0(\rho) \left(m\rho - \frac{4}{3}\pi^2\rho^3 \bar{q}_R q_L \right), \tag{6.62}$$

where $n_0(\rho)$ is the tunneling rate without fermions. Note that the zero mode contribution acts like a mass term. This is quite natural, because for $N_f = 1$, there is only one chiral $U(1)_A$ symmetry, which is anomalous. Unlike the case $N_f > 0$, the anomaly can therefore generate a fermion mass term.

For $N_f = 2$, the result is

$$\mathcal{L}_{N_f=2} = \int d\rho \, n_0(\rho) \left[\prod_f \left(m\rho - \frac{4}{3}\pi^2 \rho^3 \bar{q}_{f,R} q_{f,L} \right) + \frac{3}{32} \left(\frac{4}{3}\pi^2 \rho^3 \right)^2 \right.$$

$$\left. \times \left(\bar{u}_R \lambda^a u_L \bar{d}_R \lambda^a d_L - \frac{3}{4} \bar{u}_R \sigma_{\mu\nu} \lambda^a u_L \bar{d}_R \sigma_{\mu\nu} \lambda^a d_L \right) \right] + (L \leftrightarrow R) \tag{6.63}$$

where the λ^a are color Gell–Mann matrices. One can easily check that the interaction is $SU(2) \times SU(2)$ invariant, but $U(1)_A$ is broken. This means that the 't Hooft Lagrangian provides another derivation of the $U(1)_A$ anomaly. Furthermore, we will argue below that the importance of this interaction goes much beyond the anomaly and that it explains the physics of chiral symmetry breaking and the spectrum of light hadrons.

Any multi-fermion operator can be identically rewritten using the so-called Fierz identities, grouping fermions into various possible pairs. For four-fermion operators, there are three pairings possible (fermion 1 paired with any of three remaining). Therefore, $N_f = 2$ 't Hooft Lagrangian can be written in three different (but identical) forms. One or the other may be more or less convenient for particular problem, but various forms in literature may create some confusion. Following (Rapp et al. 2000) we give all three of them here. Here t^a are color generators (Gell–Mann matrices for $N_c = 3$), $\tau^- = (\vec{\tau}, i)$, and $\vec{\tau}$ is an isospin matrix, and momentum-dependent form factor $F(k)$ (which is also a matrix in the Dirac indices) is related to zero mode solution.

$$L_1 = \frac{G}{4(N_c^2 - 1)} \left[\frac{2N_c - 1}{2N_c} \left(\bar{q} F^+ \tau_\alpha^- F q \right)^2 + \frac{2N_c - 1}{2N_c} \left(\bar{q} F^+ \gamma_5 \tau_\alpha^- F q \right)^2 \right.$$

$$\left. + \frac{1}{4N_c} \left(\bar{q} F^+ \sigma_{\mu\nu} \tau_\alpha^- F q \right)^2 \right] \tag{6.64}$$

$$L_2 = \frac{G}{8N_c^2} \left[\left(\bar{q} F^+ \tau_\alpha^- F q \right)^2 + \left(\bar{q} F^+ \gamma_5 \tau_\alpha^- F q \right)^2 \right.$$

$$+ \frac{N_c - 2}{2(N_c^2 - 1)} \left(\left(\bar{q} F^+ \tau_\alpha^- t^a F q \right)^2 + \left(\bar{q} F^+ \gamma_5 \tau_\alpha^- t^a F q \right)^2 \right)$$

$$\left. - \frac{N_c}{4(N_c^2 - 1)} \left(\bar{q} F^+ \sigma_{\mu\nu} \tau_\alpha^- t^a F q \right)^2 \right] \tag{6.65}$$

$$L_3 = \frac{G}{8N_c^2} \left[-\frac{1}{N_c - 1} \left(q^T F^T C \tau_2 t_A^a F q \right) \left(\bar{q} F^+ \tau_2 t_A^a C F^* \bar{q}^T \right) \right.$$

$$- \frac{1}{N_c - 1} \left(q^T F^T C \tau_2 \gamma_5 t_A^a F q \right) \left(\bar{q} F^+ \tau_2 \gamma_5 t_A^a C F^* \bar{q}^T \right)$$

$$\left. + \frac{1}{2(N_c + 1)} \left(q^T F^T C \tau_2 \sigma_{\mu\nu} t_A^a F q \right) \left(\bar{q} F^+ \tau_2 \sigma_{\mu\nu} t_A^a C F^* \bar{q}^T \right) \right] \tag{6.66}$$

The beginning of L_1, L_2 is often called the "mesonic form," since they are squares of colorless $(\bar{q} \ldots q)$ "meson currents." These mesons include four different species, scalars, and pseudoscalars with isospin zero σ, η' and one π, δ. In the first order in L_i, the masses of these four mesons get shifted, up or down according to the signs of the terms: $\pi\sigma$ down, δ, η' up.

The Lagrangian L_3 is called the "diquark form," because the brackets include (qq) and $(\bar{q}\bar{q})$ pairs. Here T means transposed, whereas C is charge conjugation which in reality is the antisymmetric isospin matrix τ_2. Color generators t_A^a in it are only those which are antisymmetric in indices; this is what subscript A indicates. Like mesons, the first two terms have a sign opposite to the last one: it implies that to first order the scalar and vector diquarks, get mass corrections of the opposite sign. (The spin 0 is going down, spin 1 up.)

The effective Lagrangian is a kind of nonlocal vertex, in which either quarks scatter on each other or a pair of quarks is produced from initial quark. Note that this interaction explicitly violates $U_a(1)$ chiral symmetry. Instanton-induced production of the axial charge is of course just a specific case of a "sphaleron process": whether the motion in the topological landscape is by real or virtual fields does not matter; general relation between N_{CS} and Q_5 holds in any process.

In the real-world QCD, one has not two but *three* light quark flavors, u, d, s. Therefore, 't Hooft Lagrangian is in fact six-quark operator. However, at $T < T_c$ chiral symmetry is spontaneously broken, and therefore one (or two) pair of quarks can be substituted by their condensate: this leads to four- (or two-) quark effective operators. Both are extremely important for understanding of hadronic physics. The two-quark operator leads to nonzero "constituent quark mass." The resulting four-fermion quark operators, shown in Fig. 6.11, leads to many effects we will discuss below.

A comment about violation of flavor $SU(3)$ symmetry by strange quark mass, which is much larger than those of light u, d quarks

$$m_s \sim 120\,\text{MeV} \gg m_{u,d} \sim few\,\text{MeV}.$$

If one ignores the latter, one finds that operator of the type $(\bar{u}u)(\bar{d}d)$ has two contributions (diagrams (b) and (c) in the last figure), while operators of the type $(\bar{u}u)(\bar{s}s)$ and $(\bar{d}d)(\bar{s}s)$ have only contribution of the diagram (d). All of them are comparable, so it turns out that the $(\bar{u}u)(\bar{d}d)$ interaction is about twice larger than others. So, in effects caused by these operators, the flavor $SU(3)$ symmetry is violated at level $O(1)$.

Finally, in order to complete the effects of light quarks on the tunneling, we need to include the effects of nonzero modes. One effect is that the coefficient in the beta function is changed to $b = 11N_c/3 - 2N_f/3$. In addition to that, there is an overall constant that was calculated in ('t Hooft 1976; Carlitz and Creamer 1979)

$$n(\rho) \sim (1.34m\rho)^{N_f}\left(1 + N_f(m\rho)^2\log(m\rho) + \ldots\right), \tag{6.67}$$

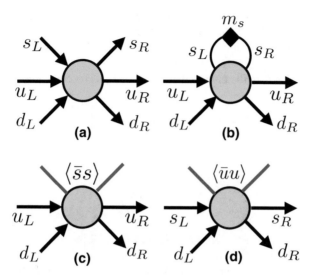

Fig. 6.11 Schematic form of the six-quark 't Hooft effective Lagrangian is shown in fig (**a**). If quarks are massive, one can make a loop shown in (**b**), reducing it to four-fermion operator. Note a black rhomb indicating the mass insertion into a propagator. We only show it for s quark, hinting that for u, d their masses are too small to make such diagram really relevant. In (**c, d**) we show other types of effective four-fermion vertices, appearing because some quark pairs can be absorbed by a nonzero quark condensates (red lines)

where we have also specified the next-order correction in the quark mass. Note that at two-loop order, one not only gets the *two-loop beta function* in the running coupling but also the *one-loop anomalous dimensions* of the quark masses.

6.4.4 Instanton-Induced Quark Anomalous Chromomagnetic Moment

This derivation follows (Kochelev 1998) and relies on the color-spin term in the 't Hooft effective Lagrangian including external gluon field, written for a quark of type q as

$$\Delta L = \int d\rho \frac{n(\rho)}{m_q^*} \frac{\pi^4 \rho^4}{g} (i\bar{q}\sigma_{\mu\nu} \frac{\lambda^a}{2} q) G_{\mu\nu}^a \qquad (6.68)$$

where, in the mean field approximation, the instanton density is written as

$$n(\rho) = n_{gluodynamics}(\rho) \prod_q (m_q^* \rho), \quad m_q^* = m_q - 2\pi^2 \rho^2 \langle \bar{q}q \rangle \qquad (6.69)$$

and contains the product of effective quark masses of all light quark flavors $q = u, d, s$. Note that in the previous formula the mass m_q^* of the quark under consideration is in denominator and thus drops out: only masses of *other* flavors appear.

Comparing this with the definition of quark anomalous chromomagnetic moment

$$\Delta L = -i\mu_q \frac{g}{2m_q^*} \left(\bar{q}\sigma_{\mu\nu} \frac{\lambda^a}{2} q \right) G_{\mu\nu}^a, \qquad (6.70)$$

one obtains its value

$$\mu_q = -\frac{\pi^3 n_{inst} \rho^4}{2\alpha_s(\rho)}. \qquad (6.71)$$

The numerator contains the instanton diluteness combination $n_{inst}\rho^4 \sim 10^{-2}$, but α_s is in denominator as it should be, due to nonperturbative nature of the instanton field. The absolute magnitude for light quarks is $\mu_{u,d} \approx -0.2$, and it is used in effective quark models of hadronic spectra.

6.4.5   Instanton-Induced Diquark–Quark Configurations in the Nucleon

This subsection just introduces this important subject: more detailed discussion of it will be continued in Chap. 10 devoted to hadronic wave functions on the light front.

In Fig. 6.12, we show the simplest valence-quark nucleon configuration (a), together with the lowest-order instanton-induced effects. The diagram (b) illustrates the *ud* diquark correlation, appearing in the first order in 't Hooft Lagrangian. Since the diquark has spin zero, the *d* quark in it does not contribute to the total spin of the nucleon. This conclusion is supported by lattice studies.

The attention to the last diagram (c) comes from the paper (Dorokhov and Kochelev 1993), where it was noted that "sea quarks" produced by instantons,

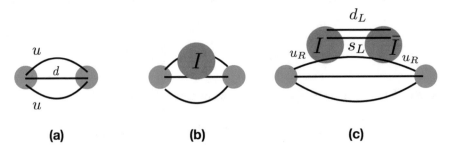

Fig. 6.12 Quark configurations of the nucleon, including the lowest-order instanton-induced ones

and resulting in the five-quark configuration, are highly polarized both in spin and isospin. Indeed, the valence u quark can only produce d, s ones (flavor polarization). Furthermore, if this quark happens to be right-handed, the sea quark pair would be left-handed (and vice versa). In that paper, this configuration was proposed as an explanation of observed deviations from Ellis-Jaffe and Gottfried sum rules, related to the famous "spin crisis" of the nucleon.

Note that the instanton-induced production of sea quarks is very different from the usual one-gluon vertex creating $\bar{q}q$ pairs, which is obviously flavor and chirality-blind. Thus the usual pQCD evolution of structure functions, while dominant at very small x, *cannot* start from simple valence quark distributions and needs asymmetric phenomenological input.

6.4.6   Instanton-Induced Decays of η_c and Scalar/Pseudoscalar Glueballs

Let me here provide one example noticed by Bjorken (2000): decays of η_c have three large three-body modes, about 5% each of the total width:

$$\eta_c \to KK\pi; \quad \pi\pi\eta; \quad \pi\pi\eta'.$$

Note that there is no $\pi\pi\pi$ decay mode or many other decay modes one may think of: that is because 't Hooft vertex *must* have all light quark flavors including the $\bar{s}s$; see Fig. 6.13. More generally, in fact the average multiplicity of J/ψ, η_c decays is significantly larger than 3, so large probability of these three-body decays is a phenomenon by itself. Bjorken pointed out that the vertex seems to be $\bar{u}u\bar{d}d\bar{s}s$ and suggested that these decays proceed via 't Hooft vertex.

The actual calculations were done by Zetocha and Schafer (2003), and it included the following two and three meson decay channels of the lowest charmonium state

$$\eta_c \to \pi\pi, KK, \eta\eta, KK\pi, \eta\pi\pi, \eta'\pi\pi \tag{6.72}$$

using the three-flavor Lagrangian shown in Fig. 6.14.

Fig. 6.13 The instanton-induced decay of the pseudoscalar η_c

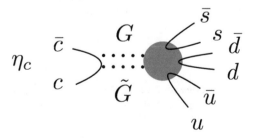

$$\mathcal{L}_{I+A} = \int dz \int d_0(\rho) \frac{d\rho}{\rho^5} \frac{1}{N_c^2-1} \left(\frac{\pi^3 \rho^4}{\alpha_s}\right) G\tilde{G}\left(\frac{1}{4}\right)\left(\frac{4}{3}\pi^2\rho^3\right)^3 \left\{ [(\bar{u}\gamma^5 u)(\bar{d}d)(\bar{s}s) + (\bar{u}u)(\bar{d}\gamma^5 d)(\bar{s}s) + (\bar{u}u)(\bar{d}d)(\bar{s}\gamma^5 s) \right.$$

$$+ (\bar{u}\gamma^5 u)(\bar{d}\gamma^5 d)(\bar{s}\gamma^5 s)] + \frac{3}{8}\left[(\bar{u}t^a\gamma^5 u)(\bar{d}t^a d)(\bar{s}s) + (\bar{u}t^a u)(\bar{d}t^a\gamma^5 d)(\bar{s}s) + (\bar{u}t^a u)(\bar{d}t^a d)(\bar{s}\gamma^5 s) + (\bar{u}t^a\gamma^5 u) \right.$$

$$\times(\bar{d}t^a\gamma^5 d)(\bar{s}\gamma^5 s) - \frac{3}{4}[(\bar{u}t^a\sigma_{\mu\nu}\gamma^5 u)(\bar{d}t^a\sigma_{\mu\nu}d)(\bar{s}s) + (\bar{u}t^a\sigma_{\mu\nu}u)(\bar{d}t^a\sigma_{\mu\nu}\gamma^5 d)(\bar{s}s) + (\bar{u}t^a\sigma_{\mu\nu}u)(\bar{d}t^a\sigma_{\mu\nu}d)(\bar{s}\gamma^5 s)$$

$$+ (\bar{u}t^a\sigma_{\mu\nu}\gamma^5 u)(\bar{d}t^a\sigma_{\mu\nu}\gamma^5 d)(\bar{s}\gamma^5 s)] - \frac{9}{20}d^{abc}[(\bar{u}t^a\sigma_{\mu\nu}\gamma^5 u)(\bar{d}t^b\sigma_{\mu\nu}d)(\bar{s}t^c s) + (\bar{u}t^a\sigma_{\mu\nu}u)(\bar{d}t^b\sigma_{\mu\nu}\gamma^5 d)(\bar{s}t^c s)$$

$$+ (\bar{u}t^a\sigma_{\mu\nu}u)(\bar{d}t^b\sigma_{\mu\nu}d)(\bar{s}t^c\gamma^5 s) + (\bar{u}t^a\sigma_{\mu\nu}\gamma^5 u)(\bar{d}t^b\sigma_{\mu\nu}\gamma^5 d)(\bar{s}t^c\gamma^5 s)] + (2 \; cyclic \; permutations \; u\leftrightarrow d\leftrightarrow s) \Big]$$

$$- \frac{9}{40}d^{abc}[(\bar{u}t^a\gamma^5 u)(\bar{d}t^b d)(\bar{s}t^c s) + (\bar{u}t^a u)(\bar{d}t^b\gamma^5 d)(\bar{s}t^c s) + (\bar{u}t^a u)(\bar{d}t^b d)(\bar{s}\gamma^5 t^c s) + (\bar{u}t^a\gamma^5 u)(\bar{d}t^b d)(\bar{s}t^c\gamma^5 s)]$$

$$- \frac{9}{32}if^{abc}[(\bar{u}t^a\sigma_{\mu\nu}\gamma^5 u)(\bar{d}t^b\sigma_{\nu\gamma}d)(\bar{s}t^c\sigma_{\gamma\mu}s) + (\bar{u}t^a\sigma_{\mu\nu}u)(\bar{d}t^b\sigma_{\nu\gamma}\gamma^5 d)(\bar{s}t^c\sigma_{\gamma\mu}s) + (\bar{u}t^a\sigma_{\mu\nu}u)(\bar{d}t^b\sigma_{\nu\gamma}d)$$

$$\times(\bar{s}t^c\sigma_{\gamma\mu}\gamma^5 s) + (\bar{u}t^a\sigma_{\mu\nu}\gamma^5 u)(\bar{d}t^b\sigma_{\nu\gamma}\gamma^5 d)(\bar{s}t^c\sigma_{\gamma\mu}\gamma^5 s)] \Big\}.$$

Fig. 6.14 The form of the $N_f = 3$ effective 't Hooft Lagrangian

Their results contain rather high power of the instanton radius and therefore strongly depend on its value. So the authors used the inverted logic, evaluating from each experimentally known decay rate the corresponding value of the mean instanton size $\bar{\rho}$. The results reasonably well reproduced the ratios between the channels and even the absolute width. Furthermore, these calculations provide about the most accurate evaluation of the average instanton size available, in the range of $\bar{\rho} = 0.29–0.30\,\text{fm}$.

6.4.7   Instanton-Induced Spin Polarization in Heavy Ion Collisions

Heavy ion collisions create new form of matter—quark–gluon plasma (QGP)—which expands and cools according to relativistic hydrodynamics. This field is rich and will be discussed in a separate lecture volume. However one particular phenomenon fits here so well that I decided to include it. The point is *noncentral* collisions lead to large orbital momentum and consequently to fluid undergoing rotational motion, with a nonzero *vorticity*

$$\left[\frac{\partial}{\partial \vec{x}} \times \vec{v}\right] = \vec{\omega} \neq 0. \tag{6.73}$$

If one goes to rotational frame, vorticity acts as magnetic field, and in thermal equilibrium the spin states get polarized

$$P \sim \exp\left[(\vec{\omega} \cdot \vec{S})/T\right]. \tag{6.74}$$

The observation of Λ hyperon polarization has been made by STAR collaboration at RHIC (Adamczyk et al. 2017).

This story may look boring, unless one asks the question of *how* exactly a spin can get equilibrated by rotating medium. Investigation of this by Kapusta et al. (2019) started with spin-flip reactions induced by thermal flow fluctuations. With disappointment it was found that this mechanism can only equilibrate spins at times longer than $>10^3$ fm/c, obviously not available in practice in the collisions. Then it was shown that perturbative amplitude of strange quark spin flip—proportional to strange quark mass m_s—also leads to unreasonably long equilibration time.

Therefore (Kapusta et al. 2019), in desperation, eventually turn their attention to nonperturbative QCD. They used four-quark operators obtained from six-quark 't Hooft operator and "instanton liquid model" parameters, which do indeed produce reasonable time for spin equilibration.

Note, however, that reduction of six- to four-quark operator involves nonzero quark condensates and therefore is only possible not in the QGP phase but only at $T < T_c$, in hadronic phase. This implies that spin polarization only occurs late in the collision.

In principle, spin polarization can be observed not only for Λ hyperon but for many other hadron species, like vector mesons or $S = 3/2$ baryons like Δ^{++}. In a comment (Shuryak 2016), I suggested the latter to be a good way to tell apart the effect of vorticity from that of magnetic field. Eventually, one should be able to compare polarization of light-to-strange quarks and compare it to predictions from 't Hooft operator. At the time of this writing there appeared measurements of spin polarization for strange vector mesons K^*, ϕ: but there is no understanding of how are they related with local vorticity and/or magnetic field yet .

6.5 Brief Summary

- We start this long chapter from defining N_{CS} Chern–Simons number (in 3-d) and showing that gauge choices can be classified by it. We then learned that at its integer values the minimal energy (pure gauge) configurations have energy zero.
- Since physics should not depend on gauge choice, it is periodic as a function of n_{CS}. By analogy to quantum mechanical problem with one-dimensional periodic potential, states are numerated by "quasimomentum" θ angle. There is a "superselection" rule not allowing to change θ inside QCD. Physical vacuum seems to correspond to $\theta = 0$ and QCD does not violate CP invariance. (Axions?)
- States with non-integer N_{CS} have minimal energy nonzero. Tunneling through such barrier is described by classical *instanton* path in Euclidean time, derived in details.
- Quantum measure, also called "instanton density," is then derived in one-loop approximation. Also propagators describing quantum fluctuations around classical path are derived. Those can be used to derive "scattering on instantons."
- One-loop fluctuations around instantons in supersymmetric theories lead to a derivation of general NSVZ beta function. Among other things, it predicted that

$\mathcal{N} = 4$ theory has zero beta function and is therefore a conformal theory without any scale.

- One particular instanton effect is their contribution to beta function. We show that this contribution known analytically in Seiberg–Witten $\mathcal{N} = 2$ theory and QCD (derived from phenomenology and lattice) is very similar.

- Instanton contributions to energies of heavy-quark mesons can be calculated via correlators of Wilson lines. The same can be done for high-energy collisions in eikonal approximation.

- Instantons have (4-d) fermionic zero modes. Their physical meaning is dumping a quark to negative energy Dirac sea and pulling another one from it, with opposite chirality. As a result, effective 't Hooft Lagrangian can be derived, which explicitly breaks $U(1)_A$ chiral symmetry. In electroweak theory, quarks and leptons with right-hand chirality do not interact with the gauge fields: thus left-handed fermions can appear or disappear. Baryon and lepton charges are no longer conserved.

- 't Hooft Lagrangian leads to a wide variety of processes, e.g., specific three-meson decay channels of η_c meson.

References

Adamczyk, L., et al.: Global Λ hyperon polarization in nuclear collisions: evidence for the most vortical fluid. Nature **548**, 62–65 (2017)

Alday, L.F., Gaiotto, D., Tachikawa, Y.: Liouville correlation functions from four-dimensional gauge theories. Lett. Math. Phys. **91**, 167–197 (2010) . 0906.3219

Atiyah, M.F., Hitchin, N.J., Drinfeld, V.G., Manin, Yu. I.: Construction of instantons. Phys. Lett. **A65**, 185–187 (1978) [,133(1978)]

Belavin, A.A., Polyakov, A.M., Schwartz, A.S., Tyupkin, Yu. S.: Pseudoparticle solutions of the Yang–Mills equations. Phys. Lett. **B59**, 85–87 (1975) [,350(1975)]

Bjorken, J.D.: Intersections 2000: What's new in hadron physics. AIP Conf. Proc. **549**(1), 211–229 (2000). hep-ph/0008048

Brown, L.S., Carlitz, R.D., Creamer, D.B., Lee, C.-K.: Propagation functions in pseudoparticle fields. Phys. Rev. **D17**, 1583 (1978) [,168(1977)]

Brown, L.S., Creamer, D.B.: Vacuum polarization about instantons. Phys. Rev. **D18**, 3695 (1978)

Callan, Jr., C.G., Dashen, R.F., Gross, D.J., Wilczek, F., Zee, A.: The effect of instantons on the heavy quark potential. Phys. Rev. **D18**, 4684 (1978)

Carlitz, R.D., Creamer, D.B.: Light quarks and instantons. Ann. Phys. **118**, 429 (1979)

Chern, S.-S., Simons, J.: Characteristic forms and geometric invariants. Ann. Math. **99**, 48–69 (1974)

Dorokhov, A.E., Kochelev, N.I.: Instanton induced asymmetric quark configurations in the nucleon and parton sum rules. Phys. Lett. **B304**, 167–175 (1993)

Kapusta, J.I., Rrapaj, E., Rudaz, S.: Is hyperon polarization in relativistic heavy ion collisions connected to axial U(1) symmetry breaking at high temperature? (2019)

Kochelev, N.I.: Anomalous quark chromomagnetic moment induced by instantons. Phys. Lett. **B426**, 149–153 (1998)

Levine, H., Yaffe, L.G.: Higher order instanton effects. Phys. Rev. **D19**, 1225 (1979)

Marshakov, A., Mironov, A., Morozov, A.: Zamolodchikov asymptotic formula and instanton expansion in $N = 2$ SUSY $N_f = 2N_c$ QCD. JHEP **11**, 048 (2009). 0909.3338.

Meggiolaroab, E.: High-energy quark-quark scattering and the Eikonal approximation. Nucl. Phys. Proc. Suppl. **64**, 191–196 (1998)

Musakhanov, M.: Gluons, heavy and light quarks in the QCD vacuum. In: Proceedings, Sixth International Conference on New Frontiers in Physics (ICNFP 2017): Crete, Greece, August 17–29, 2017 (2018)

Musakhanov, M., Egamberdiev, O.: Dynamical gluon mass in the instanton vacuum model. Phys. Lett. **B779**, 206–209 (2018)

Nachtmann, O.: High-energy collisions and nonperturbative QCD. Lect. Notes Phys. **479**, 49–138 (1997). [Lect. Notes Phys.496,1(1997)]

Nekrasov, N.A.: Seiberg–Witten prepotential from instanton counting. Adv. Theor. Math. Phys. **7**(5):831–864 (2003). hep-th/0206161

Novikov, V.A., Shifman, M.A., Vainshtein, A.I., Zakharov, V.I.: Beta function in supersymmetric gauge theories: instantons versus traditional approach. Phys. Lett. **166B**, 329–333 (1986). [Yad. Fiz.43,459(1986)].

Nowak, M.A., Shuryak, E.V., Zahed, I.: Instanton induced inelastic collisions in QCD. Phys. Rev. **D64**, 034008 (2001). hep-ph/0012232

Randall, L., Rattazzi, R., Shuryak, E.V.: Implication of exact SUSY gauge couplings for QCD. Phys. Rev. **D59**, 035005 (1999). hep-ph/9803258

Rapp, R., Schafer, T., Shuryak, E.V., Velkovsky, M.: High density QCD and instantons. Ann. Phys. **280**, 35–99 (2000)

Shuryak, E.: Comment on measurement of the rotaion frequency and the magnetic field at the freezeout of heavy ion collisions (2016)

Shuryak, E.V.: The role of instantons in quantum chromodynamics. 1. Physical vacuum. Nucl. Phys. **B203**, 93 (1982)

Shuryak, E.V., Zahed, I.: Instanton induced effects in QCD high-energy scattering. Phys. Rev. **D62**, 085014 (2000). hep-ph/0005152

Shuryak, E. V., Zahed, I.: Understanding the nonperturbative deep inelastic scattering: instanton induced inelastic dipole-dipole cross-section. Phys. Rev. **D69**, 014011 (2004). hep-ph/0307103

't Hooft, G.: Computation of the quantum effects due to a four-dimensional pseudoparticle. Phys. Rev. **D14**, 3432–3450 (1976). [,70(1976)]

Yakhshiev, U., Kim, H.-C., Hiyama, E.: Instanton effects on charmonium states (2018)

Zetocha, V., Schafer, T.: Instanton contribution to scalar charmonium and glueball decays. Phys. Rev. **D67**, 114003 (2003). hep-ph/0212125.

Topology on the Lattice

7

7.1 Global Topology: The Topological Susceptibility and the Interaction Measure

"Global topology," which we will discuss in this first section, is the total topological charge of the whole lattice: in the subsequent sections, we will review more local observables aimed at revealing the topological substructure of the gauge theory vacuums.

The overall fluctuations of the topological charge can be deduced from lattice measurements of the so-called topological susceptibility, defined by

$$\chi_{top} = \frac{< Q^2 >}{V_4} \tag{7.1}$$

where V_4 is the 4-volume in which the topological charge Q is measured. We will not go into technical details about the topological charge definitions on the lattice, and just notice that with proper definition, Q is always integer-valued. Lattice configurations provide histograms of the probabilities $P(Q)$ used in the averaging implied in the definition. If it would be Gaussian, the only parameter would be χ_{top}, related to its width.

The non-Gaussian distributions possess higher moments. The so-called *interaction parameter* is defined by the ratio of the following moments of the Q distribution, via

$$b_2 = -\frac{< Q^4 > -3 < Q^2 >^2}{12 < Q^2 >^2} \tag{7.2}$$

If one thinks that topological charge is contained in an ensemble of instantons, χ_{top} provides the combined density of instantons and anti-instantons. Below we will focus on its dependence on the temperature T, but one should also study its dependence on other parameters such as the quark masses m_f and the number of colors N_c.

Dilute instanton gas approximation (DIGA) provides some predictions. We will discuss the total density a bit later, and now notice that DIGA predicts the theta dependence of the vacuum energy to be

$$F(\theta, T) - F(0, T) = \chi_{top}(T) \left(1 - cos(\theta)\right) \tag{7.3}$$

Expanding this to powers of theta, one finds prediction for higher-order coefficients; in particular the value of b_2 is defined by

$$b_2^{DIGA} = -\frac{1}{12}$$

The interaction between instantons will influence this parameter, and so there were devoted measurements on the lattice.

Of course, it would be nice to directly simulate the ensemble at nonzero θ: but it is not possible because of the complex weight. One method used to go around this is to do simulations for imaginary $\theta = i\tilde{\theta}$ (the chemical potential conjugated to the topological charge) and then extrapolate back from imaginary to real one.

Significant efforts have been invested in studies of the dependence of these parameters on the number of colors N_c. In particular, χ_{top}/σ^2, b_2 for three, four, and six colors (Bonati et al. 2016c). The results provide a robust evidence of the large-N_c behavior predicted by standard large-N scaling arguments, i.e., $b_{2n} = O(N_c^{-2n})$. In particular,

$$b_2 = \frac{\bar{b}_2}{N_c^2} + O\left(1/N_c^4\right), \quad \bar{b}_2 = -0.23(3). \tag{7.4}$$

In QCD with light quarks, one can study the dependence of these parameters on the quark masses. Recall that gauge fields with nonzero Q have fermionic zero modes: so if $m \to 0$, the determinant is zero, and thus those configurations are impossible. So, in the massless limit, $\chi \to 0$ as well. We have noticed before that to vanish χ_{top}, in fact it is enough that a single-quark flavor be massless! For $m_u = m_d$ and ignoring strangeness, the chiral perturbation theory gives

$$\chi = m_\pi^2 f_\pi^2/4 \approx (75. \, \text{MeV})^4 \tag{7.5}$$

which does indeed vanish in the massless limit since the pion mass vanishes as $m_\pi^2 \sim m_q f_\pi$. All of that is reproduced by lattice measurements.[1]

---

[1] Note, however, that DIGA would predict much stronger vanishing, since the instanton amplitude is proportional to the *product* of all quark masses $\prod_f q_f$. It is not the case in the QCD vacuum because of chiral symmetry breaking, but becomes true at high T.

Topological susceptibility in QCD at finite T has been studied in Bonati et al. (2016a), from which we took the following Fig. 7.1. Let us start with the second plot, displaying the interaction parameter b_2. Note that at low T, it is very different from the DIGA prediction $-1/12$, but at high T, it does go to an agreement with the dilute gas prediction.

The first plot shows that $\chi(T)$ in the QGP phase does have a power dependence on T, but the fitted power $D_2 \approx 3$. On the other hand, DIGA predicts

$$\chi^{DIGA}(T) \sim T^4 \left(\frac{\Lambda}{T}\right)^b \left(\frac{m}{T}\right)^{N_f} \tag{7.6}$$

with $N_f = 3, b = 11N_c/3 - 2N_f/3 \approx 9$, the power of the temperature thus should be $D_2 = 8$ (shown on the plot by dash-dotted curve): thus there is a very serious disagreement. We will return to its discussion in the next chapters.

On the other hand, another lattice group (Borsanyi et al. 2016) had measured $\chi(T)$ up to much higher $T \approx 2300$ MeV, see Fig. 7.2. In this range $\chi(T)$ drops by about 10 orders of magnitude! This group finally does see at high T the dependence corresponding to DIGA: their fitted power of T is 8.16, close to the power from the beta function $(11N_c/3) - (2N_f/3) = 8.666$ (for 4 quark flavors, u, d, s, c).

Since their method basically included only configurations with total charge $Q = \pm 1$, which means they kept a single instanton in a box, no doubt the result agrees with a single-instanton amplitude.

What would be important to test is the coefficient, given by the semiclassical calculation. To my knowledge the coefficient does not agree with expectations, being about an order of magnitude *larger* than the prediction. In view of high power of Λ_{QCD} involved and the fact that it is only one-loop prediction available, it is perhaps not yet a problem, but the issue clearly needs more work.

7.2   "Lattice Cooling" and Instantons

Lattice field configurations are not of course classical: they do include the full extent of quantum fluctuations of all fields. But the tools used to reveal the underlying topology are basically the same, "cooling" and "gradient flow," reducing fluctuations and restoring smooth classical fields.

The simplest method to look for topological object on the lattice is "cooling," an iterative procedure reducing local values of the gauge action. Indeed, the topological solitons represent local extrema (minima) of the action, and thus those should be approached under cooling. The first works (Teper 1986; Ilgenfritz et al. 1986) applying cooling had indeed found topological lumps, with their action being equal (up to a sign) to the topological charge, $S = |Q|$.

Further works have, however, found certain technical issues, which made hard to make really quantitative results from cooling. One is that the results obtained depend on the number of cooling steps: the reason is that close instanton-anti-instanton pairs

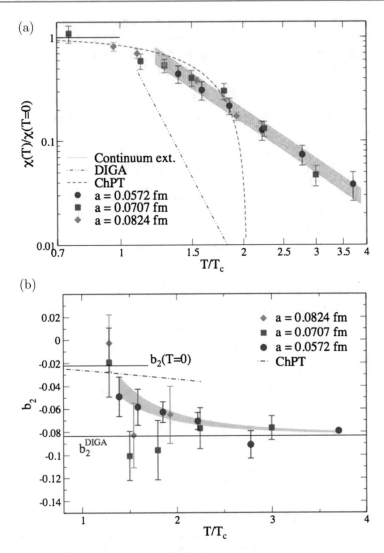

Fig. 7.1 (A) Ratio of the topological susceptibilities $\chi(T)/\chi(T=0)$, evaluated at fixed lattice spacing. The horizontal solid line describes zero T-dependence, while the dashed line is the prediction from finite-temperature ChPT and the dashed-dotted line shows the slope predicted by DIGA computation. The band corresponds to the continuum extrapolation using the function $\chi(a,T)/\chi(a,T=0) = D_0(1+D_1 a^2)(T/Tc)^{D_2}$; only data corresponding to $T > 1.2T_c$ have been used in the fit. (B) b_2 evaluated at three lattice spacing. The horizontal solid lines correspond to the value of b_2 at T = 0 predicted by ChPT (which is about -0.022) and to the instanton gas expected value $b_{DIGA} = -1/12$. The dotted-dashed line is the prediction of finite-temperature ChPT, while the two light blue bands is the result of a fit to the smallest lattice spacing data

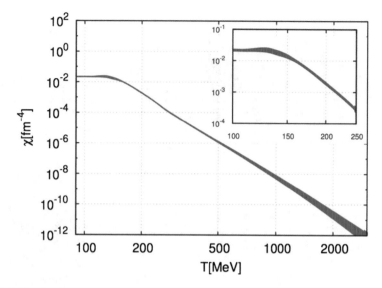

Fig. 7.2 The topological susceptibility versus temperature, for QCD with physical quark masses. The insert shows the dependence around the deconfinement phase transition

move toward each other and eventually annihilate. Trying to locate and include such annihilation events made the cooling studies more an art than a regular method.

Of similar nature is the question of "topological defects," instanton-like objects with a size comparable to lattice spacing a. In order to define the topological charge, lattice-discretized fields should be made continuous by some extrapolation procedures, which, unfortunately, by definition are not unique.

The other issue is the fact that only in the continuum limit the instanton action becomes scale-independent. On the lattice there is no scale invariance, and depending on particular choice of the lattice action, one can force instantons to either grow or shrink, as the cooling proceeds. This issue has been addressed using the *improved* lattice actions (de Forcrand et al. 1997). Without going into detail, let me just say that with tuning it, one can tune the $O(a^2)$ term in the instanton action to prevent changes of the instanton size during cooling. This allowed to measure the instanton density with the size distribution[2] $dn(\rho)/d^4xd\rho$, The average size was found to be $\langle \rho \rangle \approx 0.43\,fm$, somewhat larger than in the instanton liquid model. The distribution have a peak at $\rho_{max} \approx 0.4\,fm$. Distribution of separation between instantons have a clear peak at $0.9\,fm$, close to what the instanton liquid model value. In Fig. 7.3 we show two more examples of the vacuum topological structure

---

[2]The measurements started from some threshold value of $\rho_{min} = 2.3a$, but semiclassical theory tells us that only very negligible number of instantons can be below this threshold.

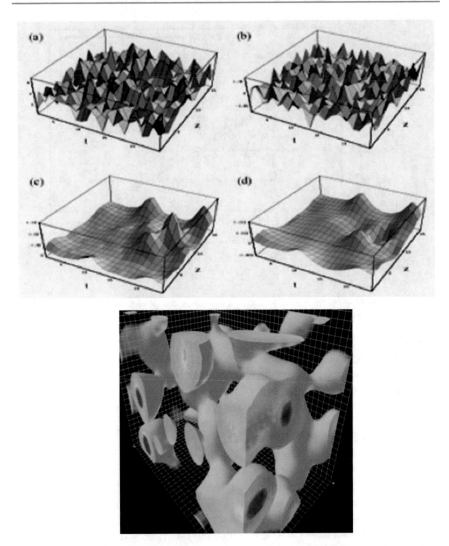

Fig. 7.3 The topological structure revealed by "cooling" of lattice gauge configurations. Four upper plots are from the MIT group by Negele et al.; they show the distributions of the action $\sim GG$ and topological charge $\sim G\tilde{G}$ (left and right). The upper plots are before and the lower ones after cooling. The lower third picture of the topological charge is from the Adelaide group (Leinweber et al.)

Later works, included extrapolation back to "no cooling" time, allowed careful determination of the density and size distributions, in pure gauge $SU(2)$, $SU(3)$ and even physical QCD with quarks (Hasenfratz 2000). The average size was found to be $\langle \rho \rangle \approx 0.30 \pm 0.01 \, fm$, a bit smaller than in the ILM. The mean distance was found instead to be $0.61 \pm 0.02 \, fm$.

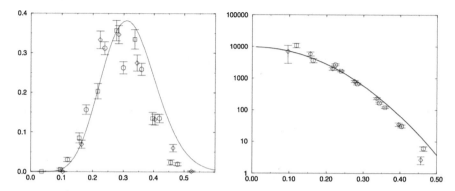

Fig. 7.4 (left) The instanton size ρ [fm] distribution $dn/d\rho d^4z$. (right) The combination $\rho^{-6}dn/d\rho d^4z$, in which the main one-loop behavior drops out for $N_c = 3$, $N_f = 0$. The points are from the lattice work for this theory, with $\beta = 5.85$ (diamonds), 6.0 (squares), and 6.1 (circles). Their comparison should demonstrate that results are lattice-independent. The line corresponds to the proposed expression; see text

Lattice data on the instanton size distribution are shown in Fig. 7.4 (the figure is taken from Shuryak (1999), the lattice data from A. Hasenfratz et al.). The left plot shows the size distribution itself. Recall that the semiclassical theory predicted it to be $dn/d\rho \sim \rho^{b-5}$ at small sizes, with $b = 11Nc/3 = 11$ for pure gauge $N_c = 3$ theory. The right plot—in which this power is taken out—is constant at small ρ, which agrees with the semiclassical prediction.

The other feature is a peak at $\rho \approx 0.3\,fm$—the value first proposed phenomenologically in Shuryak (1982b), long before these lattice data. The reason for the peak is a suppression at large sizes.

Trying to understand its origin, one may factor out all known effects. The right plot shows that after this is done, a rather simple suppression factor $\sim exp(-const * \rho^2)$ describes it well, for about three decades. What can the physical origin of this suppression be?

There is no clear answer to that question. One option is that it is due to mutual repulsion between instantons and anti-instantons: we will return to it in section on the *instanton liquid*.

Another one, proposed in Shuryak (1999) already mentioned, and "Poisson dual" to the first one, is that the coefficient is proportional to the dual magnetic condensate, that of Bose-condensed monopoles. It has been further argued there that it can be related to the string tension σ, so that the suppression factor should be

$$\frac{dn}{d\rho} = \frac{dn}{d\rho}\bigg|_{semiclassical} \cdot e^{-2\pi\sigma\rho^2} \tag{7.7}$$

If this idea is correct, this suppression factor should be missing at $T > T_c$, in which the dual magnetic condensate is absent. But, on the other hand, here quantum/thermal fluctuations generate at high T a similar factor (Pisarski and Yaffe 1980)

$$\frac{dn}{d\rho} = \frac{dn}{d\rho}\Big|_{T=0} \cdot e^{-\left(\frac{2N_c+N_f}{3}\right)(\pi\rho T)^2} \tag{7.8}$$

related to scattering of quarks and gluons of QGP on the instanton (Shuryak and Velkovsky 1994). Empirically, the suppression factor at any temperature looks Gaussian in ρ, interpolating between those limiting cases.

Another example of lattice study focusing on instanton contribution to certain Green functions is Athenodorou et al. (2018), in full quantum vacuum and with cooling. The original motivation has been extraction of the gluon coupling $\alpha_s(k)$, so the observable on which this work was focused is the following ratio of three-point to two-point Green function (in configurations transformed to Landau gauge)

$$\alpha_{MOM}(k) = \frac{k^6}{4\pi} \frac{\langle G^{(3)}\left(k^2\right)\rangle^2}{\langle G^{(2)}\left(k^2\right)\rangle^3} \tag{7.9}$$

In Fig. 7.5 the results are plotted versus the momentum scale k. At the lower curve (corresponding to uncooled quantum vacuum with gluons) at large $k > 1\,\text{GeV}$,

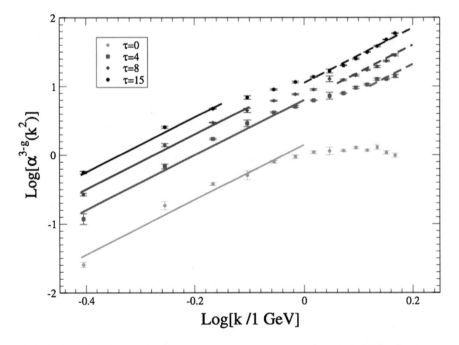

Fig. 7.5 The ratio of Green functions $\alpha_{MOM}(k)$ versus momentum scale k. Different colors of points correspond to different cooling time. The lines correspond to the instanton ensembles with fitted densities

the effective coupling starts running downward, as asymptotic freedom requires. However, in infrared, at low $k \to 0$, one finds certain positive power. Its slope was found to be exactly what one would obtain from an instanton ensemble. Furthermore, after cooling for different time τ (shown by three other curves), it was seen that the same power also persists at high $k > 1\,\text{GeV}$ as well (see dashed curves at the right). This corresponds well to the expectation that cooling eliminates perturbative gluons (the plain waves) but, for some time, preserve instantons.

The authors notice that this power corresponds well to that calculated from instantons. The ratio defined above (7.9) calculated for en ensemble of instantons reads

$$\alpha_{MOM}(k) = \frac{k^4}{18\pi n} \frac{\langle \rho^9 (I(k\rho))^3 \rangle^2}{\langle \rho^6 (I(k\rho))^2 \rangle^3} \qquad (7.10)$$

where the function containing the instanton size ρ is

$$I(s) = \frac{8\pi^2}{s} \int_0^\infty dz\, z\, J_2(sz)\phi(z)$$

a certain Bessel transform of the shape function ϕ. (To compare, the BPST instanton has $\phi = 1/(1+z^2)$.)

The gradient flow cooling allowed to identify instantons quite well. As shown in Fig. 7.6, the topological charge density at the center depends on the fitted size in a way quite close to that of BPST instantons.

While with the cooling time the instanton sizes grow, one should remember to extrapolate to *zero cooling time* $\tau \to 0$ to recover how they were in the original quantum vacuum. Therefore, this density is not what one can calculate from the right plot of Fig. 7.6, but significantly larger: their extrapolated total density is $\sim 10\,fm^{-4}$, an order of magnitude larger than in the original ILM.

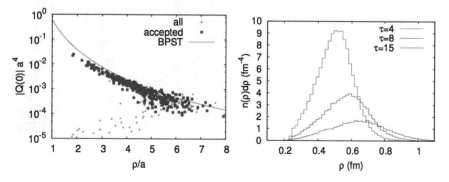

Fig. 7.6 (left): The topological charge at the origin vs the fitted radius, compared to the BPST profile shown by the line. (right): evolution of the instanton size distribution with the cooling time τ

Extrapolating these results back to *zero cooling*, the authors used these data for an estimate of the vacuum instanton density. Their conclusion is that it is much higher than the value suggested by the instanton liquid model, by about an order of magnitude. If so, it erases the diluteness parameter of that ensemble, making the "instanton liquid" really dense.

7.3 A "Constrained Cooling": Preserving the Polyakov Line Value

From the Introduction chapter, we emphasized that the VEV of the Polyakov line does play a very important role in non-perturbative phenomena. It is used as the most practical confinement measure, and as we discussed in the Introduction, it influences the instantons, effectively splitting them into the instanton-dyons (we will discuss later).

So it is rather natural, following Langfeld and Ilgenfritz (2011), to work out the so-called *constrained cooling*, with the local values of the Polyakov line preserved. For technical definition how it is achieved, one need to see the original paper: let me show only two plots in Fig. 7.7. The left figure compares the behavior of the topological susceptibility χ_{top} on the amount of the cooling steps. While the standard method shows a global topology disappearing, the constrained cooling does preserve it.

Let me briefly summarize the main findings of Langfeld and Ilgenfritz (2011) as follows: while the global topological charge Q is quantized to integer values, the actions and the topological charges of local topological clusters do not. So they cannot really be instantons!

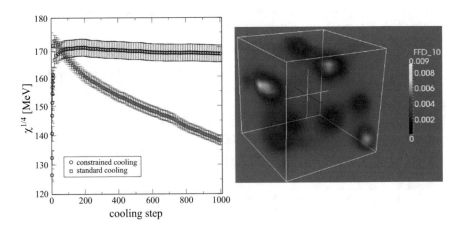

Fig. 7.7 The left plot shows the topological susceptibility χ_{top} as a function of the number of cooling sweep N_{cool} for standard (red points decreasing with N_{cool}) and constrained (black points). The right plot shows the distribution of the topological charge in configurations obtained by the constrained cooling

On the other hand, they satisfy another major requirements: the fields in these clusters are *locally self-dual*. The picture in the right side of Fig. 7.7 shows also that they are well localized and thus their fields are quite strong. Those features still suggest that the objects are topological and that their semiclassical treatment looks promising.

7.4   Brief Summary

- Gauge instantons describe tunneling between topologically different classical vacua.
- Gauge instanton amplitude is known to one loop.
- Fermionic zero modes lead to nontrivial interaction Lagrangian between light quarks, violating $U(1)_A$ symmetry.
- Instanton gauge field contributes to heavy quark potential and high energy scattering amplitudes
- Lattice configurations, when "cooled", reveal topological clusters. They are (anti) self-dual but not necessarily have integer topological charge.

References

Athenodorou, A., Boucaud, P., De Soto, F., Rodriguez-Quintero, J., Zafeiropoulos, S.: Instanton liquid properties from lattice QCD. JHEP **02**, 140 (2018). 1801.10155

Bonati, C., D'Elia, M., Mariti, M., Martinelli, G., Mesiti, M., Negro, F., Sanfilippo, F., Villadoro, G.: Axion phenomenology and θ-dependence from $N_f = 2 + 1$ lattice QCD. JHEP **03**, 155 (2016a). 1512.06746

Bonati, C., D'Elia, M., Rossi, P., Vicari, E. θ dependence of 4D $SU(N)$ gauge theories in the large-N limit. Phys. Rev. **D94**(8), 085017 (2016c). 1607.06360

Borsanyi, S. et al.: Calculation of the axion mass based on high-temperature lattice quantum chromodynamics. Nature **539**(7627), 69–71 (2016). 1606.07494

de Forcrand, P., Garcia Perez, M., Stamatescu, I.-O.: Topology of the SU(2) vacuum: A Lattice study using improved cooling. Nucl. Phys. **B499**, 409–449 (1997). hep-lat/9701012

Hasenfratz, A.: Spatial correlation of the topological charge in pure SU(3) gauge theory and in QCD. Phys. Lett. **B476**, 188–192 (2000). hep-lat/9912053

Ilgenfritz, E.-M., Laursen, M.L., Schierholz, G., Muller-Preussker, M., chiller, H.: First evidence for the existence of instantons in the quantized SU(2) lattice vacuum. Nucl. Phys. **B268**, 693 (1986)

Langfeld, K., Ilgenfritz, E.-M.: Confinement from semiclassical gluon fields in SU(2) gauge theory. Nucl. Phys. **B848**, 33–61 (2011). 1012.1214

Pisarski, R.D., Yaffe, L.G.: The density of instantons at finite temperature. Phys. Lett. **97B**, 110–112 (1980)

Shuryak, E.V.: The role of instantons in quantum chromodynamics. 1. Physical vacuum. Nucl. Phys. **B203**, 93 (1982b)

Shuryak, E.V.: Probing the boundary of the nonperturbative QCD by small size instantons **18**, 33 (1999). hep-ph/9909458

Shuryak, E.V., Velkovsky, M.: The Instanton density at finite temperatures. Phys. Rev. **D50**, 3323–3327 (1994)

Teper, M.: The topological susceptibility in SU(2) lattice gauge theory: an exploratory study. Phys. Lett. **B171**, 86–94 (1986)

Instanton Ensembles

8

8.1 Qualitative Introduction to the Instanton Ensembles

In the previous chapter, we have discussed the semiclassical theory of a single instanton, together with some applications of it. However, since the translational zero modes are substituted by an integral over collective coordinates—the instanton location in 4-d—the tunneling amplitude for an instanton is proportional to the four-volume V_4 considered. This means that if V_4 is large enough to overcome the exponential tunneling suppression, naive probability of tunneling exceeds 1, and re-summing (unitarization) of the probability is necessary. As was shown in the previous chapter, simple exponentiation of the amplitude leads to "ideal instanton gas." The instanton density is directly related to the *nonperturbative downward shift in vacuum energy density*.

The so-called "diluteness parameter"[1]

$$\kappa = n_{I+\bar{I}}\pi^2\rho^4 \sim \frac{1}{10} \qquad (8.1)$$

describes how far instantons are from each other in units of their size. When it is large enough, this ideal gas approximation does not hold, and one naturally needs to take into account instanton interactions.

This section is about different regimes in which the instanton ensemble may exist. Before we describe them in detail, let us just enumerate them (in historical order):

(1) A gas of individual instantons
(2) A gas of pairs, the *instanton-anti-instanton molecules*

---

[1]Recall that volume of a sphere in 4-d is $\pi^2\rho^4/2$. We ignore the difference between $< \rho >^4$ and $< \rho^4 >$.

© The Author(s), under exclusive license to Springer Nature Switzerland AG 2021
E. Shuryak, *Nonperturbative Topological Phenomena in QCD and Related Theories*, Lecture Notes in Physics 977,
https://doi.org/10.1007/978-3-030-62990-8_8

(3) The *"instanton liquid"*
(4) The *"instanton polymers,"* producing quasibound Cooper quark pairs, of color superconducting phases

The first two are "dilute" ensembles, as their density can be arbitrarily small, so that the interaction between constituents can be safely neglected: we will discuss those in the next subsection.

The last two are not only "dense," in the sense that the interaction needs to be accounted for, but is in fact "dense enough" for certain qualitative changes in the system to take place. The "instanton liquid" is a phase in which the chiral symmetry is spontaneously broken, so that there is a nonzero quark condensate $< \bar{q}q > \neq 0$. The color supercondor phase has nonzero condensate of diquarks $< qq > \neq 0$ (it may or may not also include $< \bar{q}q > \neq 0$, depending on details of the setting).

The value of the condensate influences back the instanton density: these parameters are all found from minimization of the free energy. In a mean field approximation, the free energy is approximated by certain analytic expressions: its derivatives over parameters, set to zero, are known as the "gap equations"—their solution may or may not be done analytically.

8.2   The Dilute Gas of Individual Instantons

At high T, the large-size instantons are screened out; only those with sizes

$$\rho T < 1$$

remain present.[2] That is why the dimensionless diluteness of the ensemble should depend on T as large inverse power

$$n\rho^4 \sim \left(\frac{\Lambda_{QCD}}{T}\right)^b \tag{8.2}$$

For instantons, with action $S = 8\pi^2/g^2$, one has $b = (11/3)N_c - (2/3)N_f \approx 9$, from one-loop beta function.

Furthermore, in QCD with light quarks, the dilute instanton gas is made far more diluted by the existence of fermionic zero modes. In the chiral limit $m_q \to 0$, any object with nonzero topological charge $Q \neq 0$, taken by itself, should have Q fermionic zero modes. Recall that in the QCD partition function, the fermionic determinant is in the numerator: so zero eigenvalue makes it to vanish. Thus, in the chiral limits, there simply cannot exist any dilute instanton gas.

---

[2]A question at this point usually comes, if the size should be instead limited by the electric Debye mass, $M_D \sim gT$. However, recall that nonperturbative instanton field is large $O(1/g)$.

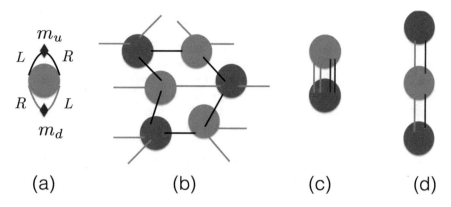

Fig. 8.1 Schematic representation of four instanton ensembles for $N_f = 2$; see text. Red and blue circles correspond to instantons and anti-instantons and black and green lines to u and d quarks, respectively. Black diamonds are mass insertions, flipping chirality from left (L) to right (R)

If one take into account small but nonzero quark masses, the dilute instanton gas is possible, but with rather small density. We will call it "the 't Hooft regime": as we already discussed in the previous chapter, in this case the instanton amplitude is additionally—on top of the tunneling exponent—suppressed by product of all light quark masses. This factor is numerically about

$$\prod_{f=u,d,s} (m_f \rho) \approx \left(\frac{2.5 \, \mathrm{MeV}}{600 \, \mathrm{MeV}}\right) \left(\frac{5. \, \mathrm{MeV}}{600 \, \mathrm{MeV}}\right) \left(\frac{100 \, \mathrm{MeV}}{600 \, \mathrm{MeV}}\right) \approx 6 \cdot 10^{-6} \qquad (8.3)$$

where we put the mean instanton size ρ for dimension. (Its numerical value will be extensively discussed below in this chapter.) If all instanton effect in QCD would be that small, perhaps there would not be a need to discuss them. Fortunately, this regime only occurs at high temperatures $T > T_c$: first indications for that have been recently demonstrated on the lattice.

Recall that these mass insertions appear in the single-instanton amplitude because all quarks emitted via zero modes *need to be reabsorbed back*, Fig. 8.1a. Zero modes for a quark and an antiquark have opposite chiralities, and the price for flipping it is the mass insertion. But such *chirality flips are not needed if a quark emitted by an instanton is absorbed by an anti-instanton!*

In other ensembles, a quark emitted by an instanton can be absorbed by anti-instantons. Therefore these other phases can exist even in the chiral limit $m_f \to 0$.

If the density is high and a quark emitted by an instanton can, with comparable probability, be absorbed by multiple anti-instantons (see Fig. 8.1b), this ensemble is called *the instanton liquid* (Shuryak 1982b). The fermionic determinant is a sum of closed loops, so if one follows one particular (say u) quark, it will always come back. In the instanton liquid ensemble, a typical length of the loop is very long, scaling with the volume V_4, becoming infinite in the $V_4 \to \infty$ limit, and thus $SU(N_f)$

chiral symmetry gets *spontaneously broken*. We will have a chapter devoted to it soon.

If the density is low, the simplest loop is a travel to the nearest anti-instanton and back (see Fig. 8.1c). This ensemble is called a "molecular phase" made of instanton-anti-instanton pairs (Ilgenfritz and Shuryak 1994). Transition from the option (ii), the instanton liquid, to (iii), $\bar{I}I$ pairs, as a function of T is the instanton-based explanation for the chiral symmetry restoration transition.[3]

The last ensemble (iv) produces condensates of *quark Cooper pairs* in cold but dense quark matter: thus it is called (the instanton-induced) color superconductor in which there is a nonzero VEV $< qq > \neq 0$. Depending on the parameters, it may exist in several interesting phases; unfortunately, we will not have time for its discussion in these lectures.

8.3   The "Instanton Liquid Model" (ILM)

Complementing the theory built from first principles, one may also look at the phenomena under consideration from empirical point of view, searching for hints and combining those into a simple approximate model.

It is precisely what happened in the early 1980s. By that time, the QCD sum rules (Shifman et al. 1979) have been widely used, and they provided some understanding of the behavior of the QCD correlation functions. Combining partonic description at small distances with hadronic description at large ones, one learned what happens in between. Furthermore, using Wilsonian Operator Product Expansion (OPE), one was able to qualitatively understand the correlator phenomenology in terms of VEVs of few "vacuum condensates," the gluonic $< G_{\mu\nu}^2 >$ and quark $< \bar{q}q >$ ones. (Some of that will be discussed in chapter on QCD correlation functions.) Furthermore, as it was pointed out by the authors of the method themselves, not "all hadrons are alike": for spin-zero channels, the OPE-based theory apparently failed to predict the magnitude of the nonperturbative effects, even qualitatively.

In order to explain available "phenomenology of the vacuum," a qualitative model was proposed in my work (Shuryak 1982a), the so-called "instanton liquid" model (ILM) of the QCD vacuum.

There were two parameters—the (mean) instanton size ρ_0 and the instanton density n_0—to be determined phenomenologically. The instanton size distribution was assumed to be just

$$\frac{dn}{d\rho} = n_0 \delta(\rho - \rho_0) \tag{8.4}$$

---

[3] Its physics is rather similar to the so-called Berezinskii-Kosterlitz-Thouless transition with the two-dimensional vortices, for which the 2016 Nobel Prize was awarded.

An idea what instanton density in the QCD vacuum can be was taken from the empirical value of the gluon condensate:

$$n_0 < \frac{< \left(g G_{\mu\nu}^a\right)^2 >}{32\pi^2} \sim (1\,fm)^{-4} \tag{8.5}$$

(where the density means both pseudoparticles together, I and $\bar{I}$).

To evaluate the typical size, one more observable was used, the quark condensate $< \bar{q}q >$, in a version of mean field estimate. The result was

$$\rho_0 \sim \frac{1}{3} fm = \frac{1}{600\,\mathrm{MeV}} \tag{8.6}$$

It was soon found that this model can reproduce well also other properties of chiral symmetry breaking, such as the pion decay constant f_π. I then proceeded forward calculating many different correlation functions (Shuryak 1983), including those where the OPE sum rules failed, and found that the model reproduces phenomenology in those channels as well.

While this model historically originated from studies of the hadronic phenomenology and related correlation functions, most of its elements have been later confirmed directly from gauge field configurations generated in lattice numerical simulations. Therefore, we will not follow the historical path, but discuss first certain lattice results. Yet, to have certain picture in mind, let me still summarize here some important qualitative features of this model:

1. The **diluteness parameter** is small

$$n_0 \rho_0^4 = (\rho/R)^4 \sim (1/3)^4 \tag{8.7}$$

where R is the typical distance between the pseudoparticles. So only few percent of the space-time is occupied by strong field. The factorization hypothesis is thus violated, by the inverse diluteness.

2. **Semiclassical formulae are applicable.** The action of the typical instanton is large enough

$$S_0 = 8\pi^2/g(\rho)^2 \sim 10 \gg 1 \tag{8.8}$$

Quantum corrections go as $1/S_0$ and are presumably small enough.

3. **The interaction does not destroy instantons.** Estimated by the dipole formula, interaction was found to be typically

$$|\delta S_{int}| \sim (2-3) \ll S_0 \tag{8.9}$$

4. **It is a liquid, not a gas.** The interaction is not small in the statistical mechanics
of instantons; on the contrary, correlations are strong

$$exp|\delta S_{int}| \sim 20 \gg 1 \qquad (8.10)$$

8.4   Statistical Mechanics of the Instanton Ensembles

Early attempts to relate instantons with practical applications to QCD were summarized in an important paper by Callan et al. (1979). Incorporating the dipole-like
forces between instantons and anti-instantons they have tried to create a self-
consistent theory of interacting instantons.[4]   After (Witten 1979) pointed out
difficulties in approaching the large N_c limit using instantons,[5] this Princeton group
abandoned this direction and switched to this limit of QCD.

In 1981 a simple model (Shuryak 1982a) with the fixed instanton size and
density was proposed. Since the instantons in it were considered uncorrelated,
it was called *Random* Instanton Liquid Model (RILM for short). While it was
phenomenologically quite successful, some consistent many-body theory of the
instanton ensemble was of course needed.

The first step was a simplified hard-core model by Ilgenfritz and Muller-
Preussker (1981). The second was the variational approach by Diakonov and Petrov
(1984). For a "sum ansatz" configurations—the gauge potential equal to a sum of
those for individual instantons and anti-instantons, in a singular gauge—classical
interaction was calculated as well as the mean field approximation (MFA) of the
statistical mechanics. The interaction was quite repulsive, leading to quite dilute
equilibrium ensemble. Inclusion of light quarks has followed Diakonov and Petrov
(1986), which lead to the first calculation of the quark condensate, also in the mean
field approximation. The accuracy of the mean field approach remained, however,
unclear, as the quark-induced interactions between instantons are very strong and
expected to produce strong correlations between them.

More direct and accurate methods have to be developed to treat the statistical
ensemble, which was done in a series of papers started with Shuryak (1989):
this approach was called the *Interacting* Instanton Liquid Model (IILM). It uses
the combined fermionic determinant for all instantons, which is is equivalent to
including the diagrams containing 't Hooft effective Lagrangian to *all orders*.[6]

---

[4]But soon this groups switched to so the called merons—half-instantons—using which they have
attempted to explain confinement. This idea did not work but in some way predated the instanton-
dyons to be discussed below.

[5]It took many years till the issue was clarified by instanton ensemble simulations; see discussion
to follow later.

[6]Well, strictly speaking, for all orders which included all instantons+anti-instantons in a box,
typically $N \sim$ few hundreds. Note that the number of diagrams thus included is of the order of
$N!$ and is astronomically large.

First, the experimentally known correlation functions were reproduced by the model at *small distances* at a quantitative level. Then, for many mesonic channels (Shuryak and Verbaarschot 1993a,b), significant numerical efforts were made, allowing to calculate the relevant correlation functions till larger distances (about 1.5 fm), where they decay by few decades. As a result, the predictive power of the model has been explored in substantial depth. Many of the coupling constants and even hadronic masses were calculated, with good agreement with experiment and lattice.

Subsequent calculations of baryonic correlators (Schafer et al. 1994) have revealed further surprising facts. In the instanton vacuum, the nucleon was shown to be made of a "constituent quark" plus a deeply bound *diquark*, with a mass nearly the same as that of constituent quarks. On the other hand, decuplet baryons (like Δ^{++}) had shown no such diquarks, remaining weakly bound set of three constituent quarks. To my knowledge, it was the first dynamical explanation of deeply bound scalar diquarks. While being a direct consequence of 't Hooft Lagrangian, this phenomenon has been missed for a long time. It also leads to a realization that diquarks can become Cooper pairs in dense quark matter (see Schafer and Shuryak (2001b) for a review on "color superconductivity").

Further elaboration of such analysis for vector and axial correlation functions has been made (Schafer and Shuryak 2001a): RILM happens to reproduce rather accurately the ALEPH data on τ decay for both vector and axial correlators. A comparison between the correlators calculated in RILM and on the lattice (Chu et al. 1994) has also found good agreement, including the baryonic channels unreachable by usual phenomenology. Studies of hadronic "wave functions" and even glueball correlation functions have followed, again with results very close to what lattice measurements had produced: for a review, see Schafer and Shuryak (1998).

8.4.1 Instanton Ensemble in the Mean Field Approximation (MFA)

The main assumption of the mean field approach is that a particle interact with its many neighbors via their common "mean" field, rather than developing specific correlations with few of them. More technically, the multibody distributions are then approximated by ansatz, made of a product of independent one-body distributions.[7]

Whether this assumption is or is not applicable needs of course be studied on a case-by-case basis.

---

[7] In this respect, this assumption is similar to *Boltzmann hypothesis* behind his kinetic equation. While true for gases, it does not hold for stronger correlated systems such as liquids. As we will see below, it is not a good approximation for instanton liquid as well.

In this section we follow Diakonov and Petrov (1984). Let us start with the partition function for a system of instantons in pure gauge theory

$$Z = \frac{1}{N_+!N_-!} \prod_i^{N_++N_-} \int [d\Omega_i \, n(\rho_i)] \exp(-S_{int}). \tag{8.11}$$

Here, $N_\pm$ are the numbers of instantons and anti-instantons, $\Omega_i = (z_i, \rho_i, U_i)$ are the collective coordinates of the instanton i, $n(\rho)$ is the semiclassical instanton distribution function (6.26), and S_{int} is the bosonic-instanton interaction.

So, MFA assumes that the partition function is evaluated via a *product* of single-instanton distributions $\mu(\rho)$

$$Z_1 = \frac{1}{N_+!N_-!} \prod_i^{N_++N_-} \int d\Omega_i \, \mu(\rho_i) = \frac{1}{N_+!N_-!}(V\mu_0)^{N_++N_-} \tag{8.12}$$

where $\mu_0 = \int d\rho \, \mu(\rho)$. The distribution $\mu(\rho)$ is then determined from the variational principle or maximization of the partition function $\delta \log Z_1/\delta\mu = 0$. In quantum mechanics, a variational wave functions always provide an upper bound on the true ground state energy. The analogous statement in statistical mechanics is known as Feynman's variational principle. Using convexity

$$Z = Z_1 \langle \exp(-(S - S_1)) \rangle \geq Z_1 \exp(-\langle S - S_1 \rangle), \tag{8.13}$$

where S_1 is the variational action, one can see that the variational vacuum energy is always higher than the true one.

In our case, the single-instanton action is given by $S_1 = \log(\mu(\rho))$, while $\langle S \rangle$ is the average action calculated from the variational distribution (8.12). Since the variational ansatz does not include any correlations, only the average interaction enters. For the sum ansatz

$$\langle S_{int} \rangle = \frac{8\pi^2}{g^2}\gamma^2\rho_1^2\rho_2^2, \qquad \gamma^2 = \frac{27}{4}\frac{N_c}{N_c^2-1}\pi^2 \tag{8.14}$$

the same for both IA and II pairs. Note that (8.14) is of the same form as the hard core mentioned above, with a different dimensionless parameter γ^2. Applying the variational principle, one finds (Diakonov and Petrov 1984)

$$\mu(\rho) = n(\rho) \exp\left[-\frac{\beta\gamma^2}{V}N\overline{\rho^2}\rho^2\right], \tag{8.15}$$

where $\beta = \beta(\overline{\rho})$ is the average instanton action and $\overline{\rho^2}$ is the average size. We observe that the single-instanton distribution is cut off at large sizes by the average instanton repulsion. Note also that the cutoff has a Gaussian dependence on ρ, the same as seen in lattice studies we mentioned above.

The average size $\overline{\rho^2}$ is determined by the self-consistency condition $\overline{\rho^2} = \mu_0^{-1} \int d\rho \mu(\rho)\rho^2$. The result is

$$\overline{\rho^2} = \left(\frac{\nu V}{\beta \gamma^2 N}\right)^{1/2}, \qquad \nu = \frac{b-4}{2}, \tag{8.16}$$

which determines the dimensionless diluteness of the ensemble, $\rho^4(N/V) = \nu/(\beta\gamma^2)$. Using the pure gauge beta function $b = 11$, $\gamma^2 \simeq 25$ from above and $\beta \simeq 15$, we get the diluteness $\rho^4(N/V) = 0.01$, even more dilute than phenomenology requires. The instanton density can be fixed from the second self-consistency requirement, $(N/V) = 2\mu_0$ (the factor 2 comes from instantons and anti-instantons). One gets

$$\frac{N}{V} = \Lambda_{PV}^4 \left[C_{N_c}\beta^{2N_c}\Gamma(\nu)\left(\beta\nu\gamma^2\right)^{-\nu/2}\right]^{\frac{2}{2+\nu}}, \tag{8.17}$$

$$\chi_{top} \simeq \frac{N}{V} = (0.65\Lambda_{PV})^4, \qquad (\overline{\rho^2})^{1/2} = 0.47\Lambda_{PV}^{-1} \simeq \frac{1}{3}R, \qquad \beta = S_0 \simeq 15, \tag{8.18}$$

It may be consistent with the phenomenological values if $\Lambda_{PV} \simeq 300\,\mathrm{MeV}$. It is instructive to calculate the free energy as a function of the instanton density. Using $F = -1/V \cdot \log Z$, one has

$$F = \frac{N}{V}\left\{\log\left(\frac{N}{2V\mu_0}\right) - \left(1 + \frac{\nu}{2}\right)\right\}. \tag{8.19}$$

The instanton density is determined by minimizing the free energy, $\partial F/(\partial(N/V)) = 0$. The vacuum energy density is given by the value of the free energy at the minimum, $\epsilon = F_0$. We find $N/V = 2\mu_0$ as above and

$$\epsilon = -\frac{b}{4}\left(\frac{N}{V}\right) \tag{8.20}$$

Estimating the value of the gluon condensate in a dilute instanton gas, $\langle g^2 G^2 \rangle = 32\pi^2(N/V)$, we see that (8.20) is consistent with the trace anomaly.

The second derivative of the free energy with respect to the instanton density, the compressibility of the instanton liquid, is given by

$$\left.\frac{\partial^2 F}{\partial(N/V)^2}\right|_{n_0} = \frac{4}{b}\left(\frac{N}{V}\right), \tag{8.21}$$

where n_0 is the equilibrium density. This observable is also determined by a low-energy theorem based on broken scale invariance

$$\int d^4x \; \langle g^2 G^2(0) g^2 G^2(x) \rangle = \left(32\pi^2\right) \frac{4}{b} \langle g^2 G^2 \rangle. \tag{8.22}$$

Here, the left hand side is given by an integral over the field strength correlator, suitably regularized and with the constant disconnected term $\langle g^2 G^2 \rangle^2$ subtracted. For a dilute system of instantons, the low-energy theorem gives

$$\langle N^2 \rangle - \langle N \rangle^2 = \frac{4}{b} \langle N \rangle. \tag{8.23}$$

Here, $\langle N \rangle$ is the average number of instantons in a volume V. The result (8.23) shows that density fluctuations in the instanton liquid are not Poissonian. Using the general relation between fluctuations and the compressibility gives the result (8.21). This means that the form of the free energy near the minimum is determined by the renormalization properties of the theory. Therefore, the functional form (8.19) is more general than the mean field approximation used to derive it.

How reliable are these results?

First of all, it cannot be better than the underlying interaction, obtained from the particular set of gauge field configurations. Later studies had shown that the sum ansatz used indeed is *not* a good representation of instanton-anti-instanton valley (which is not surprising since it was chosen without any justification other than simplicity). The interaction based on streamline configurations found by Verbaarschot (1991) should be used instead: but the results are also not satisfactory, because the ensemble contains too many close pairs and too many large instantons. Stabilization of the density is reached by an exclusion of configurations with small action—presumably already included in the perturbation theory—with is a repulsive core weaker than in the sum ansatz, but still present.

Another issue is the accuracy of the MFA itself. As the density decreases, the binary correlations are building up, which are ignored in the MFA. Its accuracy can be checked by doing statistical simulations of the full partition function.

8.4.2 Diquarks and Color Superconductivity

The second most attractive channel is the interaction of two quarks is the scalar S=I=0 channel. It also follows from 't Hooft effective Lagrangian and suppressed by a factor $1/(N_c-1)$ relative to the most attractive scalar $\bar{q}q$ channel. It was pointed out in two simultaneous papers[8] (Alford et al. 1998; Rapp et al. 1998) in 1997: the same interaction leads to a very robust Cooper pairing in high-density QCD. In few

---

[8]They were submitted to the archive on the same day.

years, this field has boomed and has now a bibliography of about 500 papers, but we will not discuss it here.

The issue of diquarks may require some further discussion. Since there exists only one zero mode per instanton, only diquarks with *different* flavors can be formed: ud, us, ds.[9]

One well-known argument for a nucleon-like hadrons being made of quark-diquark pair comes from Regge trajectories. As we will discuss in chapter on the flux tubes, the slope of the nucleon Regge trajectory is the same as for mesons. This means there are two color objects connected by a *single* flux tube.

Another interesting consequence is for a nucleon spin structure. If it is made of $u+(ud)$ spinless diquark (and never $d + (uu)$), then there is no way d quark contributes anything to the nucleon spin. Recent lattice works did indeed confirm this. d contributes half of what u does into the total momentum, but (within decreasing errors) nothing to spin.

8.4.3 Instantons for Larger Number of Colors

Recall that the (one-loop) instanton action is given by $S_0 = (8\pi^2)/g^2 = -b\log(\rho\Lambda)$ where $b = (11N_c)/3$ is the first coefficient of the beta function in pure gauge QCD. In the 't Hooft limit $N_c \to \infty$ with

$$\lambda_{tHooft} \equiv g^2 N_c = const$$

we expect $S_0 = O(N_c)$ and $\rho = O(1)$. It leads to argument by Witten (1979) that instantons do not survive the large-N_c limit.

Our discussion of finite-T theory will show rather simple realization of that idea: instantons will be indeed split into N_c instanton-dyons. The action per dyon is thus finite in the 't Hooft limit.

However, it was also shown that the IILM in fact also has a reasonable large N_c limit, although it is reached in a nontrivial way. We will discuss that now following a detailed study by Schafer (2002), who managed to show that few known paradoxes of the dilute gas approximation do disappear in the interacting instanton liquid. In fact, a self-consistent picture of the ensemble emerges, which well agrees with the preexisting theoretical expectations, including the Witten's conjectures about the topological susceptibility and the η' mass. Another remarkable feature of this regime is that the difference with what we know about instanton ensemble in QCD is not really drastically changed, even in the large N_c limit.

In brief, main features of this regime are as follows: The density of instantons is predicted to grow as N_c, whereas the typical instanton size remains finite. The

---

[9] Also Fermi statistics requirement needs to be satisfied: with total spin zero, the spin wave function is antisymmetric; color part is antisymmetric as well—so flavor should also be antisymmetric. A product of three minuses is minus, as Fermi statistics requires.

effective diluteness (accounting for the fact that instantons not overlapping in color do not interact) remains constant. Interactions between instanton are important and suppress fluctuations of the topological charge. As a result, the $U(1)_A$ anomaly is effectively restored even though the number of instantons increases. Using mean field approximation and then numerical IILM simulations, one finds that this scenario does not require fine-tuning but arises naturally if the instanton ensemble is stabilized by a classical repulsive core. Although the total instanton density is large, the instanton liquid remains effectively dilute because instantons are not strongly overlapping in color space.

Since the *total instanton density* is related to the nonperturbative gluon condensate

$$\frac{N}{V} = \frac{1}{32\pi^2} \langle g^2 G^a_{\mu\nu} G^a_{\mu\nu} \rangle, \tag{8.24}$$

The N_c counting suggests that $\langle g^2 G^2 \rangle = O(N_c)$, and we are led to the conclusion that $(N/V) = O(N_c)$. This is also consistent with the expected scaling of the vacuum energy. Using Eq. (8.24) and the trace anomaly relation

$$\langle T_{\mu\mu} \rangle = -\frac{b}{32\pi^2} \langle g^2 G^a_{\mu\nu} G^a_{\mu\nu} \rangle, \tag{8.25}$$

the vacuum energy density is given by

$$\epsilon = -\frac{b}{4} \left(\frac{N}{V} \right). \tag{8.26}$$

Using $(N/V) = O(N_c)$, we find that the vacuum energy scales as $\epsilon = O(N_c^2)$ which agrees with our expectations for a system with N_c^2 gluonic degrees of freedom.

Note that $(N/V) = O(N_c)$ implies that the effective diluteness of instantons remains constant in the large N_c limit. Indeed, in spite of large density, most instantons do not see each other: the number of mutually commuting $SU(2)$ subgroups of $SU(N_c)$ scales as N_c.

If instantons are distributed randomly, then fluctuations in the number of instantons and anti-instantons are expected to be Poissonian. This leads to the predictions

$$\langle N^2 \rangle - \langle N \rangle^2 = \langle N \rangle, \tag{8.27}$$

$$\langle Q^2 \rangle = \langle N \rangle, \tag{8.28}$$

where $N = N_I + N_A$ is the total number of instantons and $Q = N_I - N_A$ is the topological charge. Equation (8.28) implies that

$$\chi_{top} = \frac{\langle Q^2 \rangle}{V} = \left(\frac{N}{V} \right). \tag{8.29}$$

Using $(N/V) = O(N_c)$, we observe that $\chi_{top} = O(N_c)$ which is *in contradiction to Witten's conjecture* $\chi_{top} = O(1)$. However, as we shall see, the interactions between instantons suppress the fluctuations and invalidate Eqs. (8.27) and (8.28).

We now include the fermion-related dynamics and ask how the chiral condensate scales with N_c, using first analytic MFA.[10]   After averaging over the color orientation of the instanton, the effective Lagrangian is

$$\mathcal{L} = \int n(\rho) d\rho \frac{2(2\pi\rho)^4 \rho^2}{4(N_c^2 - 1)} \epsilon_{f_1 f_2} \epsilon_{g_1 g_2} \left(\frac{2N_c - 1}{2N_c} (\bar{\psi}_{L,f_1} \psi_{R,g_1})(\bar{\psi}_{L,f_2} \psi_{R,g_2}) \right. \tag{8.30}$$

$$\left. - \frac{1}{8N_c} (\bar{\psi}_{L,f_1} \sigma_{\mu\nu} \psi_{R,g_1})(\bar{\psi}_{L,f_2} \sigma_{\mu\nu} \psi_{R,g_2}) + (L \leftrightarrow R) \right).$$

We observe from it that the explicit N_c dependence is given by $1/N_c^2$. This is again related to the fact that instantons are $SU(2)$ objects. Quarks can only interact via instanton zero modes if they overlap with the color wave function of the instanton. As a result, the probability that two quarks with arbitrary color propagating in the background field of an instanton interact is $O(1/N_c^2)$.

The MFA gap equation for the spontaneously generated constituent quark mass is

$$M = GN_c \int \frac{d^4k}{(2\pi)^4} \frac{M}{M^2 + k^2}, \tag{8.31}$$

where M is the constituent mass and G is the effective coupling constant in Eq. (8.30). Factor N_c comes from doing the trace over the quark propagator. The coupling constant G scales as $1/N_c$ because the density of instantons is $O(N_c)$ and the effective Lagrangian contains an explicit factor $1/N_c^2$. We conclude that the coefficient in the gap equation is $O(1)$ and that the dynamically generated quark mass is $O(1)$ also. This also implies that the quark condensate, which involves an extra sum over color, is $O(N_c)$.

The results in the mean field approximation are summarized in the Fig. 8.2 which shows that for $N_c > 4$ the average instanton size is essentially constant, while the instanton density grows linearly with N_c. This can also be verified by expanding

---

[10] For definiteness, we will consider the case $N_f = 2$, but the conclusions are of course independent of the number of flavors.

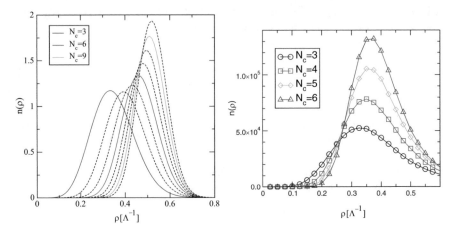

Fig. 8.2 Instanton size distribution $n(\rho)$ for different numbers of colors: (left) $N_c = 3, \ldots, 10$ calculated using mean field approximation, (right) from numerical simulations with $N = 128$ instantons

$\log(N/V)$ in powers of N_c and $\log(N_c)$: one observes that independent of the details of the interaction, the instanton density scales at most as a power, not an exponential, in N_c.

Another way to see why the instanton density scales as the number of colors is as follows: the size distribution is regularized by the interaction between instantons. This means that there has to be a balance between the average single-instanton action and the average interaction between instantons. If the average instanton action satisfies $S_0 = O(N_c)$, we expect that $\langle S_{int}^{tot} \rangle = O(N_c)$ also. Using $\langle S_{int}^{tot} \rangle = (N/V)\langle S_{int} \rangle$ and the fact that the average interaction between any two instantons satisfies $\langle S_{int} \rangle = O(1)$, we expect that the density grows as N_c.

Figure 8.2b shows the instanton size distribution for different numbers of colors. We observe that the number of small instantons is strongly suppressed as $N_c \to \infty$ but the average size stabilizes at a finite value $\bar{\rho} < \Lambda^{-1}$. We also note that there is the critical size ρ^* for which the number of instantons does not change as $N_c \to \infty$. The value of ρ^* is easy to determine analytically. We write $n(\rho) = \exp(N_c F(\rho))$ with $F(\rho) = a \log(\rho) + b\rho^2 + c$ where the coefficients $a, b, and c$ are independent of N_c in the large N_c limit. The critical value of ρ is given by the zero of $F(\rho)$. We find $\rho^* = 0.49\Lambda^{-1}$. The existence of a critical instanton size for which $n(\rho^*)$ is independent of N_c was discussed by Teper (1980); Neuberger (1980); Shuryak (1995) and indeed observed on the lattice (Lucini and Teper 2001). Fluctuations in the net instanton number are related to the second derivative of the free energy with respect to N (8.23). This result is in agreement with a low-energy theorem (8.22) based on broken scale invariance, based solely on the renormalization group equations. The left hand side is given by an integral over the field strength correlator, suitably regularized and with the constant term $\langle G^2 \rangle^2$ subtracted. For a dilute system of instantons Eq. (8.22) reduces to Eq. (8.23). The result (8.23) shows that fluctuations of the instanton ensemble are suppressed by $1/N_c$. This is in agreement with general arguments showing that fluctuations are suppressed in the large N_c limit. We also note that the result (8.23) clearly shows that even if instantons are semiclassical, interactions between instantons are crucial in the large N_c limit.

Fluctuation in the topological charge can be studied by adding a θ-term to the partition function. The mean square is just the average instanton number

$$\langle Q^2 \rangle = \langle N \rangle, \tag{8.32}$$

which is identical to the result in the random instanton liquid and not in agreement with Witten's hypothesis $\chi_{top} = O(1)$. However, Diakonov et al. noticed that Eq. (8.32) is a consequence of the fact that in the sum ansatz, the average interaction between instantons of the same charge is identical to the average interaction between instantons of opposite charge. In general there is no reason for this to be the case, and more sophisticated instanton interactions do not have this feature. If r denotes the ratio of the average interaction between instantons of opposite charge and instanton of the same charge, $r = \langle S_{IA} \rangle / \langle S_{II} \rangle$, then

$$\langle Q^2 \rangle = \frac{4}{b - r(b-4)} \langle N \rangle. \tag{8.33}$$

This result shows that for any value of $r \neq 1$, fluctuations in the topological charge are suppressed as $N_c \to \infty$. We also note that $\chi_{top} = O(1)$, in agreement with Witten's hypothesis.

In Schafer (2002), the instanton size distribution, the topological susceptibility, and the spectrum of the Dirac operator for different numbers of colors have been determined in IILM numerically. The instanton size distribution obtained shows that small instantons are strongly suppressed as the number of colors increases. We observe a clear fixed point in the size distribution at $\rho^* \Lambda \simeq 0.27$. The simulations were carried out in the total topological charge $Q_{top} = 0$ sector of the theory. One can nevertheless determine the topological susceptibility by measuring the average Q_{top}^2 in a sub-volume $V_3 \times l_4$ of the Euclidean box $V_3 \times L_4$. The topological susceptibilities for $N_c = 3$ agree well with the expectation based on Poissonian statistics, $\chi_{top} \simeq \langle N/V \rangle$. For $N_c > 3$, however, fluctuations are significantly suppressed, and the topological susceptibility increases more slowly than the density of instantons, consistent with a scenario in which χ_{top} remains finite as $N_c \to \infty$.

The chiral condensate for $m_q = 0.1\Lambda$ and topological susceptibility are shown in Fig. 8.3. We clearly see that $\langle \bar{q}q \rangle$ is linear in N_c while χ_{top} approach a constant.

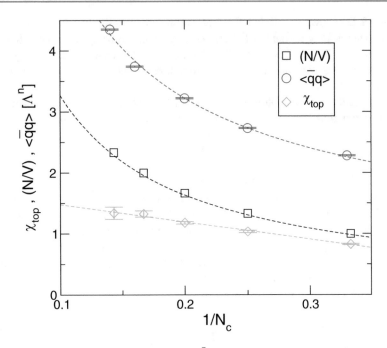

Fig. 8.3 Dependence of the chiral condensate $\langle \bar{\psi}\psi \rangle$ and the topological susceptibility χ_{top} on the number of colors. The instanton density (N/V) was assumed to scale as $(N/V) \sim N_c$. The dashed lines show fits of the form $a_1 N_c + a_2$ (for $\langle \bar{\psi}\psi \rangle$ and N/V) and $a_2 + a_3/N_c$ (for χ_{top})

8.5   Brief Summary

- This section starts from discussion of possible phases of instanton ensembles, such as a gas of single instantons (strongly suppressed by product of quark masses), a gas of $I\bar{I}$ molecules, and (last but not least) the *instanton liquid*. Its main feature is collectivization of zero modes of individual instantons into "zero-mode zone" (ZMZ). Since its density of states is finite at zero eigenvalue, this phase spontaneously breaks $SU(N_f)$ chiral symmetry.

- One more logical possibility is "crystallization" of the ensemble, with breaking of translations. Note that this would include time (no different from space in Euclidean formulation), so this would be an example of the "time crystal." In it energy is changed to quasi-energy (like momentum in periodic potential), which is *not conserved!* Fortunately, we never found a regime in which such phase actually occurs.

- The original "Instanton Liquid Model" (ILM) was constructed in 1981 using empirical values of quark and gluon condensates as input. It suggested a relatively small instanton sizes, as compared to interparticle distance $\rho \ll R$. While not an ideal gas, with interaction action not small and producing significant correlations, it is small compared to total action per instanton

$$|\delta S_{int}| \sim (2-3) \ll S_0 \sim 12$$

- The first statistical approach used versions of the mean field approximation, in which interaction with neighbors is substituted by its average.
- Yet the correlations turned out to be significant (therefore "liquid"), and therefore its statistical mechanics was addressed numerically in the 1990s.
- One form of correlation appears when a quark and antiquark (of different flavors) interact with an instanton: this creates mesons, including Nambu-Goldstone pions.
- Another form is when two quarks (again, of different flavors) meet in the instanton: this leads to relatively deeply bound *diquarks*. This explains why a nucleon is, crudely, a quark-diquark pair. It also leads to the so-called *color superconductivity* at high baryon density, in which pairing occurs at the surface of the Fermi sphere.

References

Alford, M.G., Rajagopal, K., Wilczek, F.: QCD at finite baryon density: Nucleon droplets and color superconductivity. Phys. Lett. **B422**, 247–256 (1998). hep-ph/9711395

Callan, Jr., C.G., Dashen, R.F., Gross, D.J.: A theory of hadronic structure. Phys. Rev. **D19**, 1826 (1979)

Chu, M.C., Grandy, J.M., Huang, S., Negele, J.W.: Evidence for the role of instantons in hadron structure from lattice QCD. Phys. Rev. **D49**, 6039–6050 (1994). hep-lat/9312071

Diakonov, D., Petrov, V. Yu.: Instanton based vacuum from Feynman variational principle. Nucl. Phys. **B245**, 259–292 (1984)

Diakonov, D., Petrov, V. Yu.: A theory of light quarks in the instanton vacuum. Nucl. Phys. **B272**, 457–489 (1986)

Ilgenfritz, E.-M., Muller-Preussker, M.: Interacting instantons, $1/N$ expansion and the gluon condensate. Phys. Lett. **99B**, 128 (1981). [87(1980)]

Ilgenfritz, E.-M., Shuryak, E.V.: Quark induced correlations between instantons drive the chiral phase transition. Phys. Lett. **B325**, 263–266 (1994). hep-ph/9401285

Lucini, B., Teper, M.: SU(N) gauge theories in four-dimensions: exploring the approach to N = infinity. JHEP **06**, 050 (2001)

Neuberger, H.: Instantons as a bridgehead at N = infinity. Phys. Lett. **94B**, 199–202 (1980)

Rapp, R., Schafer, T., Shuryak, E.V., Velkovsky, M.: Diquark Bose condensates in high density matter and instantons. Phys. Rev. Lett. **81**, 53–56 (1998). hep-ph/9711396

Schafer, T.: Instantons in QCD with many colors. Phys. Rev. **D66**, 076009 (2002). hep-ph/0206062

Schafer, T., Shuryak, E.V.: Instantons in QCD. Rev. Mod. Phys. **70**, 323–426 (1998). hep-ph/9610451

Schafer, T., Shuryak, E.V.: Implications of the ALEPH tau lepton decay data for perturbative and nonperturbative QCD. Phys. Rev. Lett. **86**, 3973–3976 (2001a)

Schafer, T., Shuryak, E.V.: Phases of QCD at high baryon density. Lect. Notes Phys. **578**, 203–217 (2001b). [203(2000)]

Schafer, T., Shuryak, E.V., Verbaarschot, J.J.M.: Baryonic correlators in the random instanton vacuum. Nucl. Phys. **B412**, 143–168 (1994). hep-ph/9306220

Shifman, M.A., Vainshtein, A.I., Zakharov, V.I.: QCD and resonance physics: theoretical foundations. Nucl. Phys. **B147**, 385–447 (1979)

Shuryak, E.V.: Hadrons containing a heavy quark and QCD sum rules. Nucl. Phys. **B198**, 83–101 (1982a)

Shuryak, E.V.: The role of instantons in quantum chromodynamics 1. Physical Vacuum. Nucl. Phys. **B203**, 93 (1982b)

Shuryak, E.V.: Pseudoscalar mesons and instantons. Nucl. Phys. **B214**, 237–252 (1983)

Shuryak, E.V.: Instantons in QCD 1. Properties of the 'Instanton Liquid'. Nucl. Phys. **B319**, 521–540 (1989)

Shuryak, E.V.: Instanton size distribution: repulsion or the infrared fixed point? Phys. Rev. **D52**, 5370–5373 (1995). hep-ph/9503467

Shuryak, E.V., Verbaarschot, J.J.M.: Mesonic correlation functions in the random instanton vacuum. Nucl. Phys. **B410**, 55–89 (1993a). hep-ph/9302239

Shuryak, E.V., Verbaarschot, J.J.M.: Quark propagation in the random instanton vacuum. Nucl. Phys. **B410**, 37–54 (1993b). hep-ph/9302238

Teper, M.J.: Instantons and the $1/N$ expansion. Z. Phys. **C5**, 233 (1980)

Verbaarschot, J.J.M.: Streamlines and conformal invariance in Yang-Mills theories. Nucl. Phys. **B362**, 33–53 (1991). [Erratum: Nucl. Phys. **B386**, 236 (1992)]

Witten, E.: Instantons, the quark model, and the 1/n expansion. Nucl. Phys. **B149**, 285–320 (1979)

QCD Correlation Functions and Topology

9

9.1 Generalities

9.1.1 Definitions and an Overall Picture

Before we embark on technical discussion of the correlation function, let us first summarize in the non-technical terms some pictures of the vacuum and hadronic structure, which result from what we learned about instantons.

In particularly, we know that the QCD-like theories with light quarks have spontaneously broken $SU(N_f)$ chiral symmetry. In terms of hadronic spectroscopy, this phenomenon manifests itself in two ways:

(1) quarks obtain "constituent quark masses" $M \sim 300 - 400 \, \text{MeV}$,
(2) there exist multiplets of (near)massless Goldstone modes—the pions.

We have further learned that chiral symmetry breaking is induced by collectivization of the quark zero modes, associated with nontrivial topology of the instantons. In Fig. 9.1 we sketch a picture of that: the pions exist because light quark undergo tunneling[1] events (the instantons) in pairs. In terms of the path integral, such correlated quark paths are interpreted as existence of some attraction between $\bar{u}$ and d, strong enough to cancel twice the constituent quark mass $2M$ and make it (near)zero.

Another manifestation of the correlated tunneling is in the ud di-quark pair. In two-color QCD its strength is exactly the same as in the pion channel, and thus this

---

[1]Once in a talk, T.D. Lee was explaining 't Hooft interaction by saying that it is like cars go in parallel lanes in the tunnel, like from Queens to Manhattan. I happen to be there, and said that it is more like the tunnel between England and France: whoever was left-handed become right-handed, and vice versal.

© The Author(s), under exclusive license to Springer Nature Switzerland AG 2021
E. Shuryak, *Nonperturbative Topological Phenomena in QCD and Related Theories*, Lecture Notes in Physics 977,
https://doi.org/10.1007/978-3-030-62990-8_9

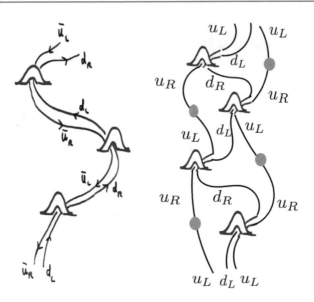

Fig. 9.1 The pion (left) and the proton (right), depicted as a sequence of tunneling events. The blue circle indicate mass insertions. Note ud diquarks inside the proton flip chirality after tunneling

diquark (and anti-diquark) are also (near)massless Goldstone modes, 5 in total for $N_f = 2$. In the three-color world we live in, the ud di-quark is not color neutral and thus is not a hadron: but it exists as a correlated diquark[2] inside the nucleons!

Note that, because of topological index theorem, quark zero modes have specific chiralities: therefore 't Hooft interaction has very specific chiral structure. For example, the picture above cannot hold for vector or tensor mesons. So, the aim of this section is to elucidate—using phenomenological or lattice correlation functions—the role of the topology in the vacuum and hadronic structure.

In fact, one could have done it even simpler, without quarks and their zero modes, in pure gauge theories. Indeed, the topological solitons themselves—being selfdual or antiselfdual—are made of so-to-say "chirally polarized gluonic fields. Scalar and pseudoscalar glueballs are strongly affected by tunneling events, while other ones—e.g. tensor ones—are not.

Correlation functions are the main tools used in studies of structure of the QCD vacuum. They can be obtained in several ways. First, they can in many cases be deduced phenomenologically, using vast set of data accumulated in hadronic physics. Second, they can be directly calculated *ab initio* using quantum field theory methods, such as lattice gauge theory, or semiclassical methods. Significant amount

---

[2]Pauli principle works for zero modes: so the third quark is prohibited from tunneling together with ud pair: it needs to "drive around" rather than take a tunnel, and so its constituent mass is not reduced.

of work has also been done in order to understand their small-distance behavior, based on the Operator Product Expansion (OPE). The large distance limit can also be understood using effective hadronic approaches or various quark models of hadronic structure. In this section we focus on available *phenomenological informa-tion* about the correlation functions, emphasizing the most important observations, which are then compared with *predictions of various theoretical approaches*; lattice numerical simulations, the operator product expansion and interacting instantons approximation. As a "common denominator" for our discussion we have chosen *the point-to-point correlation functions in coordinate representations*.

We will discuss two types of the operators, mesonic and baryonic ones

$$O_{mes}(x) = \bar{\psi}_i \delta_{ij} \psi_j, \quad O_{bar}(x) = \psi_i \psi_j \psi_k \epsilon^{ijk} \tag{9.1}$$

(Here the color indices are explicitly shown, but they will be suppressed below). As all color indices are properly contracted and all quark fields are taken at the same point x, these operators are manifestly gauge invariant.

The correlation function is the vacuum expectation value (VEV) of the product of two (or more) of them at different points

$$K(x - y) = \langle 0|O(x)O(y)|0\rangle \tag{9.2}$$

Since the vacuum is homogeneous, it depends on the relative distance. The distance is assumed to be *space-like*: we prefer to deal with *virtual* (instead of real) propagation of quarks or hadrons from one point to another, so one deals with decaying (instead of oscillating) functions.[3] One can look at a pair of points separate by Euclidean distance in terms of two distinct limits. Either they can be two different points in space, taken at the same time moment, or be two events at the same spatial point separated by non-zero interval of the *imaginary (Euclidean) time*; $ix_0 - iy_0 = \tau$. Below we will use both, depending on which is more convenient at the moment. Due to Lorentz invariance (rotational O(4) in Euclid) the answers are of course the same.

9.1.2 Small Distances: Perturbative Normalization of the Correlators

At small distances $|x|\Lambda_{QCD} \ll 1$ (remember, the other argument of the correlator we take at the origin $y=0$) the "asymptotic freedom" of QCD tells us, that (up

---

[3] At this point I was inevitably asked an old question; how can be any correlation between the fields *outside of the light cone*? It was essentially answered by Feynman long ago, who had to defend his propagator and $i\epsilon$ prescription following from Euclidean definitions. In the path integral the particles can propagate along *any* path going from x to y. The correlations outside the light cone *do not contradict causality* because one cannot use it for the signal transfer.

to small and calculable radiative corrections) quarks and gluons propagate freely. Therefore[4] $K(x) \approx K_{free}(x)$, where free quark correlator in mesonic (baryonic) case is essentially the square (cube) of the free massless quark propagator,

$$S_{free}(x) = <\bar{q}(x)q(0)> = (\gamma_\mu x_\mu)/2\pi^2 x^4 \qquad (9.3)$$

From the dimension of the free correlators they can only be for the mesonic (baryonic) channels $K_{free}(x) \sim x^{-6}$ ($\sim x^{-9}$). Of course, QCD does have a dimensional parameter Λ_{QCD}, which shows up in physical (non-free) correlators $K(x)$ and cause deviations from $K_{free}(x)$. However, in the perturbation theory it only comes in via the radiative corrections. Therefore, at small $x\Lambda \ll 1$, those produce corrections to our estimates above containing powers of $\alpha_s(x) \sim 1/log(1/x\Lambda)$. If quarks are allowed to propagate to larger distances, they start interacting with the non-perturbative vacuum fields. If corrections are not too large, one can take these effects into account using the operator product expansion (OPE) formalism. At intermediate distances description of the correlation functions becomes in general very complicated, and one may only evaluate them either using lattice numerical simulations or some "vacuum models" (e.g., the instanton ensembles we discuss now).

At *large distances* the behavior of the correlation functions is given using the time evolution of an operator $O(t) = e^{iHt}O(0)e^{-iHt}$ where H is a Hamiltonian, and then insert a complete set of *physical intermediate states* between the two operators. In Minkowski time this means

$$K(t) = \Sigma_n |\langle 0|O(0)|n\rangle|^2 e^{-itE_n}. \qquad (9.4)$$

Now one can analytically continue the correlation function into the Euclidean domain $\tau = it$, and get a sum over decreasing exponents.

Physically, application of such relation in QCD means that one considers propagation of physical excitations, or *hadrons* between two Euclidean points, so $K(x) \sim exp(-mx)$ at large x, where m is the mass of the lightest particle with the corresponding quantum numbers. Note, that this is essentially the idea of Yukawa, who had related the range of the nuclear forces to the (then hypothetical) meson mass.

After we have recollected these general facts, let us try to explain *why the correlation functions are so important in non-perturbative QCD and hadronic physics*. The answer is; it is the most effective way to study the *inter-quark effective interaction*. Models of hadronic structure—various bags, skyrmions etc.—resemble the state of the nuclear physics in its early days, when only limited information about the nuclear forces were known from properties of the simplest the bound

---

[4]Strictly speaking, small x includes the zero distance $x = 0$, and thus the reader should be aware that some correlators may have *local* terms, proportional to $\delta(x)$ and/or its derivatives. We will mostly ignore such terms, not showing them in the plots etc, unless the integral over x is done.

states (e.g. the deuteron). Those are sufficient to reproduce qualitative features of the interaction. But it is the extensive studies of the NN scattering phases in all relevant channels at all relevant energies which had eventually revealed the details of nuclear forces, with their complicated spin-isospin structure.

Quite similarly, applications of quark models are mostly are averaged over few lowest states, and the precise dependence of the inter-quark interaction on distance and momenta remains unknown. A confining potential with few additions (like spin forces) fit the spectrum. Since, due to confinement, the qq or the $\bar{q}q$ scattering is experimentally impossible, a set of correlation functions $K(x)$ per channel substitute for the the phase shifts in nuclear physics. Roughly speaking, the correlator tells us about *virtual $\bar{q}q$ or qq scattering*, using instead of physical hadrons the *wave packets of a variable size*.

9.1.3 Dispersion Relations and Sum Rules

If one makes a Fourier transform of $K(x)$, the resulting function $K_{mom}(q^2)$ depends on the momentum transfer q flowing from one operator to another. For clarity we use the following notations, introducing momentum squared with negative sign $Q^2 = -q^2$, so for *virtual* space-like momenta $q^2 < 0$ we are interested in (like in scattering experiments) $Q^2 > 0$.

Due to causality, it satisfies standard dispersion relations

$$K_{mom}(q^2) = \int \frac{ds}{\pi} \frac{Im K_{mom}(s)}{(s - q^2)} \tag{9.5}$$

where the r.h.s. contains the so called *physical spectral density* $Im K_{mom}(s)$. It describes the squared matrix elements of the operator in question between the vacuum and all hadronic states with the invariant mass $s^{1/2}$, and certainly is non-zero only for *positive s*. Note, that because we will only consider *negative q^2*, or the semi-plane without singularities, we never come across a vanishing denominator and therefore ignore $i\epsilon$ which is usually put in denominator. This simplifications are possible because our discussion is restricted to *virtual* processes (although in the r.h.s. we do use information coming from the *real* experiments of annihilation type).[5]

The dispersion relation can be a basis of the so called *sum rules*. Their general idea is as follows; suppose one knows the l.h.s. $K_{mom}(q^2)$ in some region of the argument; it means that some integral in the r.h.s. of the physical spectral density is known. It can be used to relate a set of physical parameters. Unfortunately, the so called *finite energy sum rules* using directly momentum space are not very

---

[5]In principle, virtual processes contain all the information, but of course in practice it is much more difficult to go in the opposite direction, and reproduce the physical spectral density from the correlators considered.

productive; most of the dispersion integrals are divergent, leading to usable sum
rules only after some subtractions, which introduce extra parameters and undermine
their prediction power.

For example, we have mentioned above that at small x mesonic correlators are
just given by $K_{free} \sim 1/x^6$, the free quark propagator squared. Its Fourier transform
is $K_{mom}(q^2) \sim q^2 \log q^2$ and the imaginary part of the log gives for the spectral
density in the r.h.s. $Im K_{mom}(s) \sim s$. Therefore, putting it into dispersion relation
given above one finds an ultraviolet divergent integrals.[6] A simple way to go around
this is to consider the second derivative over Q^2 of both sides of (9.5); then one finds
a convergent dispersion relation. However, while going back to the original function
$K_{mom}(s)$ from its derivative, one has to fix the integration constant.[7]

We will use the coordinate space. By applying Fourier transformation to (9.5)
one obtains the following nearly self-explanatory form (Shuryak 1993);

$$K(x) = \int \frac{ds}{\pi} Im K_{mom}(s) D(s^{1/2}, x) \tag{9.6}$$

the former function describe the amplitude of production of all intermediate states
of mass $\sqrt{s}$, while the latter function

$$D(m, x) = (m/4\pi^2 x) K_1(mx) \tag{9.7}$$

is nothing else but the Euclidean propagation amplitude of this states to the distance
x. The difference between this expression and Borel sum rules is not really very
significant; at large x the propagator goes as $exp(-mx)$, therefore one also has an
exponential cutoff, only $exp(-\sqrt{s}x)$ substitutes the $exp(-s/m^2)$. The space-time
one can be calculated numerically on the lattice or in instanton liquid, and also
analytic formulae are simpler to derive.

For completeness, let me also mention one more type of the correlation function,
the one traditionally used in LGT. This is the so called *plane-to-plane* correlation
function, obtained from K(x) by an integration over the three-dimensional plane;

$$K_{plane-to-plane}(\tau) =< \int d^3 x O(x, \tau) O(0, 0) > \tag{9.8}$$

In other terms, spatial integration makes the momentum of intermediate states to be
zero, so dispersion relation are done in energy only. This function, respectively, can
be related to physical spectral density by

$$K_{plane-to-plane}(\tau) = \int \frac{dm}{\pi} Im K_{mom}(m) exp(-\tau m) \tag{9.9}$$

---

[6]This signal that in the last argument something is missing. In this particular example it is clear
what is it; the constant under the log is lost.

[7]Note that polynomials in s generate local terms in $K(x)$ we ignore, assuming that x is never zero.

9.1.4 Flavor and Chirality Flow: Combinations of Correlators

Let us now follow the quark flavor flow in the correlators. The *flavored* and *unflavored* currents have different types of quark diagrams; as shown in Fig. 9.2 the former ones have only the one-loop contributions (a), while the latter have the two-loop diagrams (b) as well. For example, we will discuss the flavored ($I = 1$) channels like π^+, ρ^+ etc which have operators of the type $\bar{u}\Gamma d$ (where Γ is the appropriate Dirac matrix). Since two quark lines are in this case of different flavor, $\bar{u}$ and d respectively, obviously one cannot have a double-loop. On the other hand, considering similar unflavored $I = 0$ channels like $\eta, \omega \dots$ one has the $\bar{u}u, \bar{d}d, \bar{s}s$ terms which can be "looped". Thus, if one would be interested in a *difference* between say ρ and ω channels, that would be entirely due to the two-loop diagram (b). Furthermore, one can be interested in the so called *non − diagonal* correlators, for example with a different flavors such as $< \bar{u}(x)\Gamma u(x)\bar{d}(0)\Gamma d(0) >$. Again, one is restricted to the two-loop diagrams only.

The main focus of lattice work deals with the one-loop diagrams and therefore with the flavored channels; the reason is technical to which we would not go into.

It is often instructive to use specific combination of correlation functions, which focus on the phenomena we would like to study. In particular, one would like to understand how breaking of the $SU(N_f)$ and $U(1)_A$ manifest themselves in the correlation functions. Let me give two examples of such choices. The first example is the following linear combinations $\Pi_{V-A} = \Pi_V - \Pi_A$ of the $I = 1$ vector and axial correlation functions. All effects of the interaction in which quark chirality is preserved throughout the loop cancels out, because two γ_5 produced $(\pm)^2 = 1$. Furthermore, Π_{V-A} is non-zero only due to the effects of chiral symmetry breaking. This can be most clearly expressed if one writes the two currents in terms of left and right handed currents, as $\bar{q}_L\gamma_\mu q_L \pm \bar{q}_R\gamma_\mu q_R$; then this combination includes the chirality flip *twice*

$$\Pi_{\mu\nu}^{V-A} = 4 < \left(\bar{q}_L\gamma_\mu(x)q_L\right)\left(\bar{q}_R\gamma_\mu q_R(y)\right) > \tag{9.10}$$

Therefore in the chiral limit $m_f \to 0$ the charged component of it (one loop diagram) is obviously zero to any order of the perturbation theory.

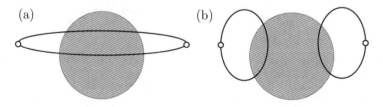

Fig. 9.2 Two diagrams for mesonic correlators in a some gluonic background (indicated by the dashed circle). The flavored currents have only the single loop diagram (**a**) while the unflavored currents have also the two-loop contributions of the type (**b**)

The second example is a similar difference, but between the $I = 1$ scalar (called a_0 or δ) and the $I = 1$ pseudoscalar (the pion π).

$$R^{NS}(\tau); = \frac{A_{flip}^{NS}(\tau)}{A_{non-flip}^{NS}(\tau)} = \frac{\Pi_\pi(\tau) - \Pi_\delta(\tau)}{\Pi_\pi(\tau) + \Pi_\delta(\tau)}, \qquad (9.11)$$

where $\Pi_\pi(\tau)$ and $\Pi_\delta(\tau)$ are pseudo-scalar and scalar NS two-point correlators related to the currents $J_\pi(\tau); = \bar{u}(\tau) i\gamma_5 d(\tau)$ and $J_\delta(\tau); = \bar{u}(\tau) d(\tau)$. If the propagation is chosen along the (Euclidean) time direction, $A_{flip(non-flip)}^{NS}(\tau)$ represents the probability amplitude for a $|q\bar{q}>$ pair with iso-spin 1 to be found after a time interval τ in a state in which the chirality of the quark and anti-quark is interchanged (not interchanged) Notice that the ratio $R^{NS}(\tau)$ must vanish as $\tau \to 0$ (no chirality flips), and must approach 1 as $\tau \to \infty$ (infinitely many chirality flips). Again, this amplitude receives no leading perturbative contributions. The difference with the example 1 is that this ratio is proportional to the chirality structure $\bar{R}L\bar{R}L + (R \leftrightarrow L)$, same as for the 't Hooft vertex, while in the example 1 it was $\bar{L}L\bar{R}R$.

Unlike the former combination, the latter one is growing surprisingly rapidly already at small distances, $\sim.3 - .6$ fm, see Fig. 9.3a from the lattice study (Faccioli and DeGrand 2003). The reason is that this correlator admits a direct instanton-induced 't Hooft vertex, while the V-A does not.

Two instanton liquid ensembles used are: the random one RILM which has no fermionic determinant, and the interacting IILM, in which it is included. Note that

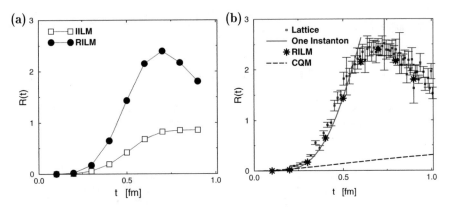

Fig. 9.3 The chirality-flip ratio, $R^{NS}(\tau)$, in lattice and in two phenomenological models. (**a**) Circles are RILM (quenched) results, squares are IILM (unquenched) results. (**b**) Squares are lattice points of previous DeGrand simulation. Stars are RILM points obtained numerically from an ensemble of 100 instantons of $1/3$ fm size in a 5.3×2.65^3 fm^4 box. The solid line is the contribution of a single-instanton, calculated analytically. The dashed curve was obtained from two free "constituent" quarks with a mass of 400 MeV. Such a curve qualitatively resembles the prediction of a model in which chiral symmetry is broken through a vector coupling (like in present DSE approaches)

in RILM the chirality flip ratio rises rapidly and exceed 1, while the unquenched IILM follows the unitarity requirement, $R^{NS}(\tau) \leq 1$. The result of quenched lattice calculation shown in Fig. 9.3b follow very accurately the same "overshooting" of 1 as the RILM results. (The unquenched lattice ones are not yet available.) At the other hand, the naive constituent quark models, which only include chirality flips due to constituent quark mass, show very slow rise shown by the lowest curve in the Fig. 9.3b.

9.1.5   General Inequalities Between the One-Quark-Loop Correlators

For correlators given by one-quark-loop diagrams there exist some important general inequalities between them. In order to derive those following Weingarten (1983), one uses the following relation for the propagator in backward direction

$$S(x, y) = -\gamma_5 S^{+}(y, x)\gamma_5 \qquad (9.12)$$

Second, one can decompose it into a sum over all 16 Dirac matrices $S = \Sigma a_i \Gamma_i$ where $\Gamma_i = 1, \gamma_5, \gamma_\mu, i\gamma_5\gamma_\mu, i\gamma_\mu\gamma_\nu$ where the last term is anti-symmetric only and $(\mu \neq \nu)$. The third step; write all one-loop correlators of the type $\Pi = Tr(S(x, y)\Gamma_i S(y, x)\Gamma_i)$, perform the traces, and write them explicitly in terms of the coefficients.

Exercise Prove the Weingarten expression for the inverse propagator. Derive the expressions for all the diagonal correlators with $\Gamma_i = 1, \gamma_5, \gamma_\mu, i\gamma_5\gamma_\mu, i\gamma_\mu\gamma_\nu$ in terms of the propagator decomposition of the same type. Use Weingarten expression for the inverse propagator.

The resulting expression for the $I = 1$ pseudoscalar (pion) correlator contains a simple sum of all coefficients squared;

$$\Pi_{PS} \sim |a_1|^2 + |a_5|^2 + |a_\mu|^2 + |a_{\mu5}|^2 + |a_{\mu\nu}|^2) \qquad (9.13)$$

while others have some negative signs, e.g. the scalar one is instead

$$\Pi_S \sim -|a_1|^2 - |a_5|^2 + |a_\mu|^2 + |a_{\mu5}|^2 - |a_{\mu\nu}|^2 \qquad (9.14)$$

As a result, the *Weingarten inequality* follows;

$$\Pi_{PS}(x) > \Pi_S(x) \qquad (9.15)$$

the pseudoscalar correlator should exceed the scalar one at all distances. The nontrivial consequence is that the masses of the lowest states should also then have the inequality, $m_{PS} < m_S$. This of course is satisfied in real world, as the physical pion is indeed lighter than any scalars. Note however that we did not said a word about chiral symmetry breaking and Goldstone theorem here; the result is more general. Note also that at large x the scalar correlator must be much much smaller than the pseudoscalar one, since the different lowest masses are in the exponent. It means that there must be a very delicate cancellation between different components of the quark propagator in all channels except the pseudoscalar one (in which all terms appear as squares with positive coefficients).

More information is provided by similar relations for vector (ρ) and axial (A_1) channels;

$$\Pi_V \sim (2|a_1|^2 - 2|a_5|^2 + |a_\mu|^2 - |a_{\mu 5}|^2) \tag{9.16}$$

$$\Pi_A \sim (-2|a_1|^2 + 2|a_5|^2 + |a_\mu|^2 - |a_{\mu 5}|^2) \tag{9.17}$$

and the Verbaarschot inequalities follow;

$$\Pi_{PS}/\Pi_{PS}^{free} > (1/2)(\Pi_V/\Pi_V^{free} + \Pi_A/\Pi_A^{free}) \tag{9.18}$$

$$\Pi_{PS}/\Pi_{PS}^{free} > (1/4)(\Pi_V/\Pi_V^{free} - \Pi_A/\Pi_A^{free}) \tag{9.19}$$

More information about other general inequalities can be found in the review (Nussinov and Lampert 2002). Note that these inequalities are identities, to be satisfied for *any configuration* of the gauge fields, not just for their ensembles.

9.2   Vector and Axial Correlators

We start the discussion of the correlation functions with the vector and axial currents. These currents really exist in nature, as the *electromagnetic* ones coupled to photons and (the parts of) the *weak* current coupled to W, Z bosons. Therefore, we know a lot about such correlators. In fact, quite complete spectral densities have been experimentally measured, in e^+e^- annihilation into hadrons and in weak decays of heavy lepton τ. The currents and their correlation functions will be denoted by the name of the lightest meson in the corresponding channel, in particular

$$j_\mu^{\rho 0} = \frac{1}{\sqrt{2}}\left[\bar{u}\gamma_\mu u - \bar{d}\gamma_\mu d\right] \qquad j_\mu^{\rho -} = \bar{u}\gamma_\mu d \tag{9.20}$$

$$j_\mu^{\omega} = \frac{1}{\sqrt{2}}\left[\bar{u}\gamma_\mu u + \bar{d}\gamma_\mu d\right] \qquad j_\mu^{\phi} = \bar{s}\gamma_\mu s \tag{9.21}$$

(see more definitions in the Table 9.1).

Table 9.1 Definition of various currents and hadronic matrix elements referred to in this work

Channel	Current	Matrix element	Experimental value
π	$j_\pi^a = \bar{q}\gamma_5 \tau^a q$	$\langle 0|j_\pi^a|\pi^b\rangle = \delta^{ab}\lambda_\pi$	$\lambda_\pi \simeq (480\,\text{MeV})^3$
	$j_{\mu 5}^a = \bar{q}\gamma_\mu\gamma_5 \frac{\tau^a}{2} q$	$\langle 0|j_{\mu 5}^a|\pi^b\rangle = \delta^{ab}q_\mu f_\pi$	$f_\pi = 93\,\text{MeV}$
δ	$j_\delta^a = \bar{q}\tau^a q$	$\langle 0|j_\delta^a|\delta^b\rangle = \delta^{ab}\lambda_\delta$	
σ	$j_\sigma = \bar{q}q$	$\langle 0|j_\sigma|\sigma\rangle = \lambda_\sigma$	
η_{ns}	$j_{\eta_{ns}} = \bar{q}\gamma_5 q$	$\langle 0|j_{\eta_{ns}}|\eta_{ns}\rangle = \lambda_{\eta_{ns}}$	
ρ	$j_\mu^a = \bar{q}\gamma_\mu \frac{\tau^a}{2} q$	$\langle 0|j_\mu^a|\rho^b\rangle = \delta^{ab}\epsilon_\mu \frac{m_\rho^2}{g_\rho}$	$g_\rho = 5.3$
a_1	$j_{\mu 5}^a = \bar{q}\gamma_\mu\gamma_5 \frac{\tau^a}{2} q$	$\langle 0|j_{\mu 5}^a|a_1^b\rangle = \delta^{ab}\epsilon_\mu \frac{m_{a_1}^2}{g_{a_1}}$	$g_{a_1} = 9.1$
N	$\eta_1 = \epsilon^{abc}(u^a C\gamma_\mu u^b)\gamma_5\gamma_\mu d^c$	$\langle 0|\eta_1|N(p,s)\rangle = \lambda_1^N u(p,s)$	
N	$\eta_2 = \epsilon^{abc}(u^a C\sigma_{\mu\nu} u^b)\gamma_5\sigma_{\mu\nu} d^c$	$\langle 0|\eta_2|N(p,s)\rangle = \lambda_2^N u(p,s)$	
Δ	$\eta_\mu = \epsilon^{abc}(u^a C\gamma_\mu u^b)u^c$	$\langle 0|\eta_\mu|N(p,s)\rangle = \lambda^\Delta u_\mu(p,s)$	

Let us remind that the electromagnetic current is the following combination

$$j_\mu^{em} = (2/3)\bar{u}\gamma_\mu u - (1/3)\bar{d}\gamma_\mu d + \ldots = (1/2^{1/2})j_\mu^\rho - (1/2^{1/2}3)j_\mu^\omega + \ldots \quad (9.22)$$

The correlation functions are defined as

$$\Pi_{\mu\nu}(x) = \langle 0|j_\mu(x)j_\nu(0)|0\rangle \quad (9.23)$$

and the Fourier transform (in Minkowski space-time) is traditionally written as

$$i\int d^4x e^{iqx}\Pi_{\mu\nu}(x) = \Pi(q^2)(q_\mu q_\nu - q^2 g_{\mu\nu}) \quad (9.24)$$

The r.h.s. is explicitly "transverse" (it vanishes if multiplied by momentum q), because all vector currents are conserved.

The dispersion relations for the scalar functions $\Pi(q^2)$ has the usual form

$$\Pi(Q^2 = -q^2) = \int \frac{ds}{\pi} \frac{Im\Pi(s)}{(s + Q^2)} \quad (9.25)$$

where the physical spectral density $Im\Pi_i(s)$ is directly related with the cross section of e^+e^- annihilation into hadrons. As this quantity is dimensionless, it is proportional to the normalized cross section

$$R_i(s) = \frac{\sigma_{e^+e^-\to i}(s)}{\sigma_{e^+e^-\to\mu^+\mu^-}(s)} \quad (9.26)$$

where the denominator includes the cross section of the muon pair production[8] $\sigma_{e^+e^- \to \mu^+\mu^-} = (4\pi\alpha^2/3s)$ and α is the fine structure constant. If the current considered has only one type of quarks (like e.g. ϕ one) one gets

$$Im\,\Pi_s(s) = \frac{R_s(s)}{12\pi e_s^2} = \frac{R_s(s)}{12\pi(1/9)} \tag{9.27}$$

where e_s is the s-quark electric charge. Generalization to ρ, ω channels is straight-forward; instead of the charge there appear the corresponding coefficients in the expression for the electromagnetic current (9.22);

$$Im\,\Pi_\rho(s) = \frac{1}{6\pi} R_\rho(s) \qquad Im\,\Pi_\omega(s) = \frac{3}{2\pi} R_\omega(s) \tag{9.28}$$

(The reader may wonder how the experimental selection of the channels is actually made. It is clear enough for heavy flavors (c and b); if the final state has a pair of such quarks, there are much more chances that they were directly produced in the electromagnetic current than that these are produced by strong "final state interaction". Below we use this idea for the strange quark as well, although some corrections should, in principle, be applied in this case. It is also possible to separate light quark ρ, ω channels; they have a different isospin $I = 1, 0$, which is conserved by any strong final state interaction. As it is well known, C-parity plus isotopic invariance leads to the so called G-parity conservation, and pions have *negative G-parity*. Therefore, strong interactions do not mix states with even and odd number of pions. The currents j_ρ, j_ω have fixed G-parity as well, and therefore pionic states created by them can have only even or odd number of pions, respectively.)

Let us start the simple and well known predictions of QCD; all the ratios $R_i(s)$ have very simple limit at high energies s. It is conjugate to the small-distance limit $x \to 0$ in which $\Pi \to \Pi_{free}$ because quarks and anti-quarks propagate there as free particles. For currents containing only one quark flavor q the only difference with muon is a different electric charge and a color factor;

$$lim_{s\to\infty} R_q(s) = e_q^2 N_c \tag{9.29}$$

which for ϕ case give $lim_{s\to\infty} R_\phi(s) = 1/3$. For ρ and ω cases one may use the following decomposition of the electromagnetic current;

$$lim_{s\to\infty} R_\rho(s) = 3/2; \qquad lim_{s\to\infty} R_\omega(s) = 1/6 \tag{9.30}$$

As we will show shortly, these relations are well satisfied experimentally (being historically one of the first and simplest justification for QCD).

---

[8]The muon mass is neglected in this expression.

Coming back to coordinate representation of the dispersion relation one obtains;

$$\Pi_{i,\mu\nu}(x) = (\partial^2 g_{\mu\nu} - \partial_\mu \partial_\nu) \frac{1}{12\pi^2} \int_0^\infty ds\, R_i(s) D(s^{1/2}, x) \qquad (9.31)$$

were, we remind, $D(m, x)$ is just the propagator of a scalar mass-m particle to distance x. Convoluting indices and using the equation $-\partial^2 D(m, x) = m^2 D(m, x) +$ contact term (which we disregard), one finally obtains the following for the dispersion relation

$$\Pi_{i,\mu\mu}(x) = \frac{1}{4\pi^2} \int_0^\infty ds\, s\, R_i(s) D(s^{1/2}, x) \qquad (9.32)$$

Since the r.h.s. is experimentally available, this equation would serve as our "experimental definition" of the l.h.s., the vector correlation functions in Euclidean space-time.

As we are interested in quark *interactions*, it is convenient to plot all correlators normalized to free motion of a massless quarks $\Pi_{\mu\mu}(x)/\Pi_{\mu\mu}^{free}(x)$ where $\Pi_{\mu\mu}^{free}(x)$ corresponds to perturbative loop diagram without interaction.

Figure 9.4 from Particle Data Group shows a sample of experimental data on e^+e^- annihilation into hadrons. One can see that this function consists of two quite different parts; (i) at $\sqrt{s} < 1.1$ GeV the prominent ρ, ω, ϕ-meson resonances (which all have very distinguishable decay channels, 2π, 3π, and $\bar{K}K$, respectively, measured but not shown in the plot); and (ii) "primed" resonances (of which only ρ' is indicated on the plot, decaying mainly into the 4 pion channel, etc.

The distinction between the lowest resonances and the primed one can be seen from a striking fact, that the contribution of the latter's are in good agreement with the horizontal curves (with and without perturbative corrections). Adding the next

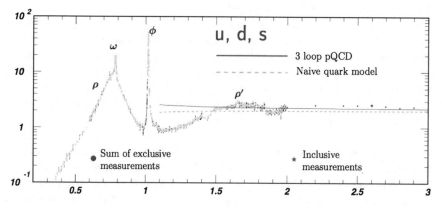

Fig. 9.4 The ratio of $R = \sigma(e^+e^- \to hadrons)/\sigma(e^+e^- \to \mu^+\mu^-)$ versus the total invariant mass of the hadronic system $\sqrt{s}$ in GeV

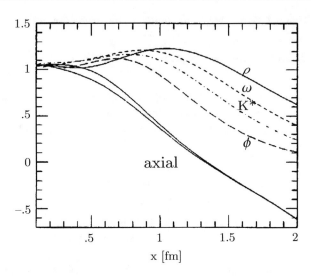

Fig. 9.5   All vector correlators together, plus the a_1 axial correlator for comparison

ones (ρ'' seen in the 6 pion channel) etc. creates a rather smooth non-resonance "continuum", corresponding to perturbative quark propagation. As seen from this plot, this happens at energies $\sqrt{s} > 1.5\,\text{GeV}$. It is not shown on this plot, but still true, that this in fact happen not just in the sum, but in each ρ, ω, ϕ channel individually.

Using parametrizations of the data in each channel, and the dispersion relation (9.32), one can calculate the Euclidean correlation function. The resulting curve is shown in Fig. 9.5. Note that, starting with rather complicated spectral densities, containing high peaks and low dips, one arrives at a very smooth function of the distance. Clearly, the way back, from the Euclidean space to the physical spectral density would be next to impossible task!

Quite striking observation made in (Shuryak 1993) (which is specific to vector currents only, as we will see later in this section) is that the resonance and continuum contributions complement each other very accurately. As a result the ratio $\Pi(x)/\Pi_{free}(x)$ remains close to one *up to the distances as large as 1.5 fm!*, while each functions falls by orders of magnitude. This "fine tuning" was called in (Shuryak 1993) a *superduality*; let me explain why this is indeed a remarkable (and so far unexplained) fact.

At small distances it is very natural to expect the so called "hadron-parton duality" between the sum over hadronic states and the pQCD quark-based description; basically it is a simple consequence of the "asymptotic freedom" up to about $x < 1/\text{GeV} = .2\,\text{fm}$. In the interval $x = .2 - 1.5\,\text{fm}$ the correlator itself drops by more than 4 orders of magnitude, while its ratio to the free loop (free quark propagation) remains close to 1 within 10–15%! What this remarkable phenomenon means is that in this channel all kind of interactions—perturbative, instanton and confinement-related ones—cancel out in wide range of distances.

Omitting the details of how other correlators has been extracted from phenomenology, let us show in Fig. 9.5 how all 4 vector correlators and also the axial one look like, as a function of Euclidean distance and normalized at small distances as explained above. Comparing them one can see, that in spite of completely different widths of the resonances and different decay states (even or odd pion numbers), all four vector correlators look *remarkably similar*. The main difference between them is the strange quark mass, which systematically suppress the correlators at larger distance. This demonstrates deep consistency between 4 independent sets of data which is rather impressive.

To complete our discussion of the vector channels, let us comment on the difference between the ρ^0 and ω correlators: what induces it? Smallness is in particular due to the following facts: (1) The rho-omega mass difference is only 12 MeV; (2) The omega-phi mixing angle is only 1–3°. Those were the basis of the so called "Zweig rule", forbidding the flavor-changing transitions.

Let us look at the transition correlator itself. The former current $\rho \sim \bar{u}u - \bar{d}d$ while the latter is $\omega \sim \bar{u}u + \bar{d}d$. Thus the difference between them is the *flavor-changing* correlator

$$K^{\omega-\rho}(x) = 2(\Pi_{\omega,\mu\mu} - \Pi_{\rho,\mu\mu}) = <\bar{u}\gamma_\mu u(x)\bar{d}\gamma_\mu d(0)> \qquad (9.33)$$

which can only come from the two-loop diagram. The reason for such strong suppression of is that in vector channels there are no direct instanton contribution in the first order in 't Hooft interaction, and the effects of the second order tend to cancel.

9.3 The Pseudoscalar Correlators

Now we move to the pseudoscalar $SU(3)$ octet π, K, η channels, and the $SU(3)$ singlet η' where we see a completely different picture. Their definitions are

$$j_\pi = \frac{i}{\sqrt{2}}\left(\bar{u}\gamma^5 u - \bar{d}\gamma^5 d\right) \qquad j_K = i\bar{u}\gamma^5 s \qquad (9.34)$$

$$j_\eta = \frac{i}{\sqrt{6}}\left(\bar{u}\gamma^5 u + \bar{d}\gamma^5 d - 2\bar{s}\gamma^5 s\right) \qquad j_{\eta'} = \frac{i}{\sqrt{3}}\left(\bar{u}\gamma^5 u + \bar{d}\gamma^5 d + \bar{s}\gamma^5 s\right)$$

Again, omitting the details of the phenomenological inputs, the resulting π, K, η, η' pseudoscalar correlators are shown in Fig. 9.6. Note huge difference compared to the vector correlators considered above; instead of changes within 10–20% in the interval of distances $x \sim 1$ fm considered, for pseudoscalars the ratio K/K_{free} has changed by up to two orders of magnitude!

Fig. 9.6 The correlation functions for the pseudoscalar nonet. Note that in contrast to the preceding figures it is now shown with the logarithmic scale

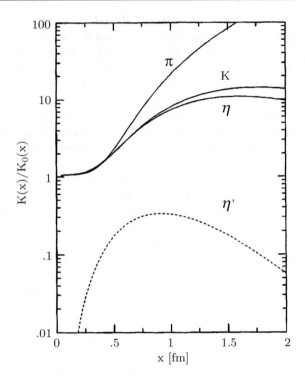

One reason for that behavior is of course small masses of the pseudoscalar mesons. In terms of the $\bar{q}q$ interaction, this implies a very strong attraction. Note also, that up to distances of the order of .5 fm there is *no marked difference between the three curves*, which means that all effects proportional to the strange quark mass are irrelevant in this region.

What is really surprising, is that the *asymptotic freedom is violated at very small distances*, about 1/5 fm, and that this happens due to the contributions of the lowest mesons themselves. This fact, noticed first by Novikov et al. (1981), shows that the pseudoscalar channels differ substantially from vector and axial ones, already at very small distances.

The SU(3) singlet meson η' can be associated with the important operators. One is the singlet axial current mentioned above and another is the gluonic pseudoscalar $G\tilde{G}$.

The relation between matrix elements of these 3 operators is given by sandwiching (between vacuum and the η' state) of the famous Adler-Bell-Jackiw anomaly relation we discussed in Chap. 1;

$$\partial_\mu j_{\eta'}^\mu = \sqrt{3}\left(2im_s\bar{s}\gamma_5 s + \frac{3g^2}{16\pi^2}G\tilde{G}\right) \tag{9.35}$$

where the contributions proportional to the light quark masses are ignored.

Omitting details, we comment that the latter matrix elements can be extracted from charmonium decay, with the result[9]

$$\langle 0|G\tilde{G}|\eta \rangle \approx .9\,\text{GeV}^3, \quad \langle 0|G\tilde{G}|\eta' \rangle \approx 2.2\,\text{GeV}^3, \tag{9.36}$$

$$\langle 0|G\tilde{G}|\eta(1440) \rangle \approx 2.9\,\text{GeV}^3$$

To complete discussion of this section let us now do some preliminary estimates of the contribution of three η states to pseudoscalar gluonic correlator. As in all other channels, at small distances the correlator is dominated by perturbative "asymptotically free" gluonic contribution, which is equal to

$$K(x) = \langle 0|G\tilde{G}(x)G\tilde{G}(0)|0\rangle, \qquad K_{free}(x) = \frac{48(N_c^2 - 1)}{\pi^4 x^8} \tag{9.37}$$

It is thus instructive to ask a simple question, at what x the contribution of $\eta, \eta', \eta(1440)$ together to $K(x)$ become equal to K_{free}. The answer is, at $x \approx .2\,\text{fm}$: a very short distance compared to $\sim 1\,\text{fm}$ for all vector resonances. And, keep in mind, that we still have not seen the contribution of "the *true pseudoscalar glueball*" yet, expected to dominate the η's.

The main conclusion one can draw from this discussion is that the boundary between pQCD and non-perturbative physics are very much channel dependent. They go to larger momenta (smaller distances) in pseudoscalar channels, and are even higher (smaller distances) in the spin-zero gluonic channels. All of this point to the topological solitons and their zero modes, the subject of the next subsection.

9.4   The First Order in the 't Hooft Effective Vertex

We have seen in Sect. 6.4.1 that the zero modes of the Dirac operator in the instanton field play a special role in the chiral limit $m_q \to 0$. For flavored currents the single-loop diagram in which only the zero mode parts of both propagators is used leads to expression of the type

$$K(x - y) = \frac{n}{m_u m_d} \int d^4 z \,\bar{\psi}_0(x - z)\Gamma\psi_0(x - z)\bar{\psi}_0(y - z)\Gamma\psi_0(y - z) \tag{9.38}$$

where for the time being $N_f = 2$ and we ignore the strange quark entirely, n is the instanton density and the matrix Γ is the gamma matrix in the vertex.

---

[9]One may wander why they grow, rather decrease, with the mass: the reason is the pseudoscalar glueball state is expected to be in the range $M_{0^- glueball} = 2 - 3\,\text{GeV}$, and the closer $\eta's$ are to it, the larger is its admixture.

Now recall that for the instanton background field the quark zero mode is right-handed, while the antiquark one is left-handed only. (It is flipped $L \leftrightarrow R$ for the anti-instanton.) This leads to the following general conclusions;

- Such contribution in the vector or the axial channels, when $\Gamma = \gamma_\mu, \gamma_\mu \gamma_5$, is *zero* since these currents are nonzero only if both spinors have the same chirality.
- For scalar and pseudoscalar[10] channels $\Gamma = 1, i\gamma_5$, it is non-zero and have the *opposite sign*.
- For flavored current one can see that the pseudoscalar π gets positive relative correction while the scalar δ or a_0 gets a negative one.
- Analogous calculation for flavor-singlet scalar σ (or f_0) and pseudoscalar we would still call η' gets split as well, with the former getting positive and the latter negative contribution.
- The absolute magnitude of these corrections for all 4 cases considered is the same.

So, all the signs are in perfect agreement with phenomenology, which (as discussed in the previous chapter) does indeed suggest light π, σ and heavy a_0, η'.

Now, as we are satisfied that signs are correct, what about the absolute magnitude of these corrections? The product of small quark masses appear in the denominator, which however can be tamed in a single-instanton background; its density n also has the product of these masses due to the fermion determinant.

However, a consistent evaluation of any effect cannot proceed without account for broken chiral symmetry in the QCD vacuum and condensates. In the single instanton approximation (SIA) discussed in Chap. 4 in place of the bare quark masses one should substitute properly defined effective masses. In the first paper on the subject (Shuryak 1983) it was done more crudely, in the MFA, for π, K, η, η' correlators.

Let us return to a single instanton background and proceed to the vector channel. We have shown above that the zero mode term does not contribute in this case, so the correlator is actually finite in the chiral limit without $m_u * m_d$ in the fermionic determinant. The calculation itself was done by Andrei and Gross (1978), and this paper created at the time significant controversy.

Non-vanishing contributions come from the non-zero mode propagator, and from the interference between the zero mode part and the mass correction. The latter term survives even in the chiral limit, because the factor m in the mass correction is canceled by the $1/m$ from the zero mode.

$$\Pi_\rho^{AG}(x, y) = \text{Tr}\left[\gamma_\mu S^{nz}(x, y)\gamma_\mu S^{nz}(y, x)\right]$$
$$+ 2\text{Tr}\left[\gamma_\mu \psi_0(x)\psi_0^\dagger(y)\gamma_\mu \Delta(y, x)\right] \qquad (9.39)$$

---

[10]Note that we put i into the current, in order to keep the zeroth order free quark loop the same in both cases.

After averaging over the instanton coordinates, the result is[11]

$$\Pi_\rho^{SIA}(x) = \Pi_\rho^0 + \int d\rho \, n(\rho) \frac{12}{\pi^2} \frac{\rho^4}{x^2} \frac{\partial}{\partial(x^2)} \left\{ \frac{\xi}{x^2} \log \frac{1+\xi}{1-\xi} \right\} \qquad (9.40)$$

where $\xi^2 = x^2/(x^2 + 4\rho^2)$.

The reason we discuss this result is its relations to the OPE. Expanding (9.40), we get

$$\Pi_\rho^{SIA}(x) = \Pi_\rho^0(x) \left(1 + \frac{\pi^2 x^4}{6} \int d\rho n(\rho) \right). \qquad (9.41)$$

This agrees exactly with the OPE expression, provided we use the average values of the operators in the dilute gas approximation

$$\langle g^2 G^2 \rangle = 32\pi^2 \int d\rho \, n(\rho), \qquad m\langle \bar{q}q \rangle = -\int d\rho \, n(\rho). \qquad (9.42)$$

Note, that the value of $m\langle \bar{q}q \rangle$ is "anomalously" large in the dilute gas limit. This means that the contribution from dimension 4 operators is attractive, in contradiction with the OPE prediction based on the canonical values of the condensates.

9.5   Correlators in the Instanton Ensemble

In this section we generalize the results of the last section to the more general case of an ensemble consisting of many pseudo-particles. The quark propagator in an arbitrary gauge field can always be expanded as

$$S = S_0 + S_0 A S_0 + S_0 A S_0 A S_0 + \ldots, \qquad (9.43)$$

where the individual terms have an obvious interpretation as arising from multiple gluon exchanges with the background field. If the gauge field is a sum of instanton contributions, $A_\mu = \sum_I A_{I\mu}$, then (9.43) becomes

$$S = S_0 + \sum_I S_0 A_I S_0 + \sum_{I,J} S_0 A_I S_0 A_J S_0 + \ldots \qquad (9.44)$$

$$= S_0 + \sum_I (S_I - S_0) + \sum_{I \neq J} (S_I - S_0) S_0^{-1} (S_J - S_0) \qquad (9.45)$$

$$+ \sum_{I \neq J, J \neq K} (S_I - S_0) S_0^{-1} (S_J - S_0) S_0^{-1} (S_K - S_0) + \ldots.$$

---

[11]There is a mistake by an overall factor 3/2 in the original work, originated from color traces. In other words, the result is correct in SU(2).

Here, $I, J, K, \ldots$ refers to both instantons and anti-instantons. In the second line, we have re-summed the contributions corresponding to an individual instanton. S_I refers to the sum of zero and non-zero mode components. At large distance from the center of the instanton, S_I approaches the free propagator S_0. Thus Eq. (9.45) has a nice physical interpretation; Quarks propagate by jumping from one instanton to the other. If $|x - z_I| \ll \rho_I$, $|y - z_I| \ll \rho_I$ for all I, the free propagator dominates. At large distance, terms involving more and more instantons become important.

In the QCD ground state, chiral symmetry is broken. The presence of a condensate implies that quarks can propagate over large distances. Therefore, we cannot expect that truncating the series (9.45) will provide a useful approximation to the propagator at low momenta. Furthermore, we know that spontaneous symmetry breaking is related to small eigenvalues of the Dirac operator. A good approximation to the propagator is obtained by assuming that $(S_I - S_0)$ is dominated by fermion zero modes

$$(S_I - S_0)(x, y) \simeq \frac{\psi_I(x)\psi_I^\dagger(y)}{im}. \tag{9.46}$$

In this case, the expansion (9.45) becomes

$$S(x, y) \simeq S_0(x, y) + \sum_I \frac{\psi_I(x)\psi_I^\dagger(y)}{im}$$

$$+ \sum_{I \neq J} \frac{\psi_I(x)}{im} \left(\int d^4r \, \psi_I^\dagger(r)(-i\slashed{\partial} - im)\psi_J(r) \right) \frac{\psi_J^\dagger(y)}{im} + \ldots, \tag{9.47}$$

which contains the overlap integrals T_{IJ}. This expansion can easily be summed to give

$$S(x, y) \simeq S_0(x, y) + \sum_{I,J} \psi_I(x) \frac{1}{T_{IJ} + imD_{IJ} - im\delta_{IJ}} \psi_J^\dagger(y). \tag{9.48}$$

Here, $D_{IJ} = \int d^4r \, \psi_I^\dagger(r)\psi_J(r) - \delta_{IJ}$ arises from the restriction $I \neq J$ in the expansion (9.45). The quantity mD_{IJ} is small in both the chiral expansion and in the packing fraction of the instanton liquid and will be neglected in what follows. Comparing the re-summed propagator (9.48) with the single instanton propagator (9.46) shows the importance of chiral symmetry breaking. While (9.46) is proportional to $1/m$, the diagonal part of the full propagator is proportional to $(T^{-1})_{II} = 1/m^*$.

The result (9.48) can also be derived by inverting the Dirac operator in the basis spanned by the zero modes of the individual instantons

$$S(x, y) \simeq S_0(x, y) + \sum_{I,J} |I\rangle\langle I| \frac{1}{i\not{D} + im} |J\rangle\langle J|. \tag{9.49}$$

The equivalence of (9.48) and (9.49) is easily seen using the fact that in the sum ansatz, the derivative in the overlap matrix element T_{IJ} can be replaced by a covariant derivative.

The propagator (9.48) can be calculated either numerically or using the mean field approximation. We will discuss the mean field propagator in the following section. For our numerical calculations, we have improved the zero mode propagator by adding the contributions from non-zero modes to first order in the expansion (9.45). The result is

$$S(x, y) = S_0(x, y) + S^{ZMZ}(x, y) + \sum_I (S_I^{NZM}(x, y) - S_0(x, y)). \tag{9.50}$$

How accurate is this propagator? We have seen that the propagator agrees with the general OPE result at short distance. We also know that it accounts for chiral symmetry breaking and spontaneous mass generation at large distances. In addition to that, we have performed a number of checks on the correlation functions that are sensitive to the degree to which (9.50) satisfies the equations of motion, for example by testing whether the vector correlator is transverse (the vector current is conserved).

9.5.1 Mesonic Correlators

In the following we will therefore discuss results from numerical calculations of hadronic correlators in the instanton liquid. These calculations go beyond the RPA in two ways; (1) the propagator includes genuine multi-instanton effects and non-zero mode contributions; (2) the ensemble is determined using the full (fermionic and bosonic) weight function, so it includes correlations among instantons. In addition to that, we will also consider baryonic correlators and three point functions that are difficult to handle in the RPA.

We will discuss correlation function in three different ensembles, the random ensemble (RILM), the quenched (QILM) and fully interacting (IILM) instanton ensembles. In the random model, the underlying ensemble is the same as in the mean field approximation, only the propagator is more sophisticated. In the quenched approximation, the ensemble includes correlations due to the bosonic action, while the fully interacting ensemble also includes correlations induced by the fermion determinant. In order to check the dependence of the results on the instanton interaction, we study correlation functions in two different unquenched ensembles,

Table 9.2 Bulk parameters
of different instanton
ensembles

		Streamline	Quenched	Ratio ansatz	RILM
	n	$0.174\,\Lambda^4$	$0.303\,\Lambda^4$	$0.659\,\Lambda^4$	$1.0\,\text{fm}^4$
	$\bar{\rho}$	$0.64\,\Lambda^{-1}$	$0.58\,\Lambda^{-1}$	$0.66\,\Lambda^{-1}$	$0.33\,\text{fm}$
		$(0.42\,\text{fm})$	$(0.43\,\text{fm})$	$(0.59\,\text{fm})$	
	$\bar{\rho}^4 n$	0.029	0.034	0.125	0.012
	$\langle \bar{q}q \rangle$	$0.359\,\Lambda^3$	$0.825\,\Lambda^3$	$0.882\,\Lambda^3$	$(264\,\text{MeV})^3$
		$(219\,\text{MeV})^3$	$(253\,\text{MeV})^3$	$(213\,\text{MeV})^3$	
	Λ	$306\,\text{MeV}$	$270\,\text{MeV}$	$222\,\text{MeV}$	–

one based on the streamline interaction (with a short-range core) and one based on the ratio ansatz interaction. The bulk parameters of these ensembles are compared in Table 9.2.

Correlation functions in the different instanton ensembles were calculated in Shuryak and Verbaarschot (1993a); Schafer et al. (1994); Schafer and Shuryak (1996) to which we refer the reader for more details. The results are shown in Fig. 9.7 and summarized in Table 9.3. The pion correlation functions in the different ensembles are qualitatively very similar. The differences are mostly due to different values of the quark condensate (and the physical quark mass) in the different ensembles. Using the Gell-Mann, Oaks, Renner relation, one can extrapolate the pion mass to the physical value of the quark masses, see Table 9.3. The results are consistent with the experimental value in the streamline ensemble (both quenched and unquenched), but clearly too small in the ratio ansatz ensemble. This is a reflection of the fact that the ratio ansatz ensemble is not sufficiently dilute.

In Fig. 9.7 we also show the results in the ρ channel. The ρ meson correlator is not affected by instanton zero modes to first order in the instanton density. The results in the different ensembles are fairly similar to each other and all fall somewhat short of the phenomenological result at intermediate distances $x \simeq 1$ fm. We have determined the ρ meson mass and coupling constant from a fit, the results are given in Table 9.3. The ρ meson mass is somewhat too heavy in the random and quenched ensembles, but in good agreement with the experimental value $m_\rho = 770\,\text{MeV}$ in the unquenched ensemble.

Since there are no interactions in the ρ meson channel to first order in the instanton density, it is important to study whether the instanton liquid provides any significant binding. In the instanton model, there is no confinement, and m_ρ is close to the two (constituent) quark threshold. In QCD, the ρ meson is also not a true bound state, but a resonance in the 2π continuum. In order to determine whether the continuum contribution in the instanton liquid is predominantly from 2-π or 2-quark states would require the determination of the corresponding three point functions, which has not been done yet. Instead, we have compared the full correlation function with the non-interacting (mean field) correlator, where we use the average (constituent quark) propagator determined in the same ensemble, see Fig. 9.7). This comparison provides a measure of the strength of interaction. We observe that there is an attractive interaction generated in the interacting liquid due

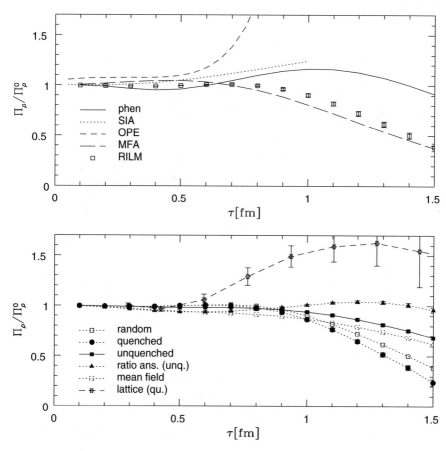

Fig. 9.7 Rho meson correlation functions. The dashed squares show the non-interacting part of the rho meson correlator in the interacting ensemble

to correlated instanton-anti-instanton pairs. This is consistent with the fact that the interaction is considerably smaller in the random ensemble. In the random model, the strength of the interaction grows as the ensemble becomes more dense. However, the interaction in the full ensemble is significantly larger than in the random model at the same diluteness. Therefore, most of the interaction is due to dynamically generated pairs.

We have already discussed ALEPH τ-decay data and have shown the data compared to OPE and the calculation in the random instanton liquid model (RILM) by Schafer and Shuryak (2001a). Another figure from this work, Fig. 9.8, shown here shows larger-x part of the correlator studied. As one can see, RILM works for the whole $V - A$ curve, and, with 10% radiative correction α_s/π, it works very well for $V + A$ as well. There is no fit of any parameter here, and in fact the calculation preceded the experiment by few years.

Table 9.3 Meson
parameters in the different
instanton ensembles. All
quantities are given in units of
GeV. The current quark mass
is $m_u = m_d = 0.1\Lambda$. Except
for the pion mass, no attempt
has been made to extrapolate
the parameters to physical
values of the quark mass

	Unquenched	Quenched	RILM	Ratio ansatz (unqu.)
m_π	0.265	0.268	0.284	0.128
m_π (extr.)	0.117	0.126	0.155	0.067
λ_π	0.214	0.268	0.369	0.156
f_π	0.071	0.091	0.091	0.183
m_ρ	0.795	0.951	1.000	0.654
g_ρ	6.491	6.006	6.130	5.827
m_{a_1}	1.265	1.479	1.353	1.624
g_{a_1}	7.582	6.908	7.816	6.668
m_σ	0.579	0.631	0.865	0.450
m_δ	2.049	3.353	4.032	1.110
$m_{\eta_{ns}}$	1.570	3.195	3.683	0.520

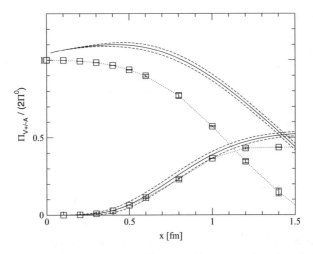

Fig. 9.8 Euclidean coordinate space correlation functions $\Pi_V(x) \pm \Pi_A(x)$ (upper and lower points and curves, respectively) normalized to free quark correlator. The solid lines show the correlation functions reconstructed from the ALEPH spectral functions and the dotted lines show the corresponding error band. The squares show the result of a random instanton liquid model

The situation is drastically different in the η' channel. Among the ~ 40 correlation functions calculated in the random ensemble, only the η' (and the isovector-scalar δ discussed in the next section) are completely unacceptable; The correlation function decreases very rapidly and becomes negative at $x \sim 0.4$ fm. This behavior is incompatible with the positivity of the spectral function. The interaction in the random ensemble is too repulsive, and the model "over-explains" the $U(1)_A$ anomaly.

The results in the unquenched ensembles (closed and open points) significantly improve the situation. This is related to dynamical correlations between instantons and anti-instantons (topological charge screening). The single instanton contribution is repulsive, but the contribution from pairs is attractive. Only if correlations among instantons and anti-instantons are sufficiently strong, the correlators are prevented from becoming negative. Quantitatively, the δ and η_{ns} masses in the streamline ensemble are still too heavy as compared to their experimental values. In the ratio ansatz, on the other hand, the correlation functions even shows an enhancement at distances on the order of 1 fm, and the fitted masses is too light. This shows that the η' channel is very sensitive to the strength of correlations among instantons.

In summary, pion properties are mostly sensitive to global properties of the instanton ensemble, in particular its diluteness. Good phenomenology demands $\bar{\rho}^4 n \simeq 0.03$, as originally suggested in Shuryak (1982b). The properties of the ρ meson are essentially independent of the diluteness, but show sensitivity to IA correlations. These correlations become crucial in the η' channel.

Let us add about dependence of the correlators on the number of colors N_c, studied in Schafer (2002). The results were obtained from simulations with $N = 128$ instantons in a Euclidean volume $V\Lambda^4 = V_3 \times 5.76$. V_3 was adjusted such that $(N/V) = (N_c/3)\Lambda^4$. In order to avoid finite volume artifacts the current quark mass was taken to be rather large, $m_q = 0.2\Lambda$. We observe that the rho meson correlation function exhibits almost perfect scaling with N_c and as a result the rho meson mass is practically independent of N_c. The scaling is not as good in the case of the pion. As a consequence there is some variation in the pion mass. However, this effect is consistent with $1/N_c$ corrections that amount to about 40% of the pion mass for $N_c = 3$. Finally, we study the behavior of the η' correlation function. There is a clear tendency toward $U(1)_A$ restoration, but the correlation function is still very repulsive for $N_c = 6$. It was also found that the η' correlation function only approaches the pion correlation for fairly large values of N_c. For example, the η' correlation function does not show intermediate range attraction unless $N_c > 15$.

After discussing the π, ρ, η' in some detail we only briefly comment on other correlation functions. The remaining scalar states are the isoscalar σ and the isovector δ (the f_0 and a_0 according to the notation of the particle data group). The sigma correlator has a disconnected contribution, which is proportional to $\langle \bar{q}q \rangle^2$ at large distance. In order to determine the lowest resonance in this channel, the constant contribution has to be subtracted, which makes it difficult to obtain reliable results. Nevertheless, we find that the instanton liquid favors a (presumably broad) resonance around 500–600 MeV. The isovector channel is in many ways similar to the η'. In the random ensemble, the interaction is too repulsive and the correlator becomes unphysical. This problem is solved in the interacting ensemble, but the δ is still very heavy, $m_\delta > 1$ GeV.

The remaining non-strange mesons are the a_1, ω and f_1. The a_1 mixes with the pion, which allows a determination of the pion decay constant f_π (as does a direct measurement of the $\pi - a_1$ mixing correlator). In the instanton liquid, disconnected contributions in the vector channels are small. This is consistent with the fact that the ρ and the ω, as well as the a_1 and the f_1 are almost degenerate.

Finally, we can also include strange quarks. $SU(3)$ flavor breaking in the 't Hooft interaction nicely accounts for the masses of the K and the η. More difficult is a correct description of $\eta - \eta'$ mixing, which can only be achieved in the full ensemble. The random ensemble also has a problem with the mass splittings among the vectors

ρ, K^* and ϕ (Shuryak and Verbaarschot 1993a). This is related to the fact that flavor symmetry breaking in the random ensemble is so strong that the strange and non-strange constituent quark masses are almost degenerate. This problem is improved (but not fully solved) in the interacting ensemble.

9.5.2 Baryonic Correlation Functions

As emphasized few times above, the existence of a strongly attractive interaction in the pseudo-scalar quark-anti-quark (pion) channel also implies an attractive interaction in the scalar quark-quark (diquark) channel. This interaction is phenomenologically very desirable, because it not only explains why the spin 1/2 nucleon is lighter than the spin 3/2 Delta, but also why Lambda is lighter than Sigma.

The vector components of the diagonal correlators receive perturbative quark-loop contributions, which are dominant at short distance. The scalar components of the diagonal correlators, as well as the off-diagonal correlation functions, are sensitive to chiral symmetry breaking, and the OPE starts at order $\langle \bar{q}q \rangle$ or higher. Instantons introduce additional, regular, contributions in the scalar channel and violate the factorization assumption for the 4-quark condensates. Similar to the pion case, both of these effects increase the amount of attraction already seen in the OPE.

The correlation function Π_2^N in the interacting ensemble is shown in Fig. 9.9. There is a significant enhancement over the perturbative contribution which corresponds to a tightly bound nucleon state with a large coupling constant. Numerically, we find[12] m_N = 1.019 GeV (see Table 9.4). In the random ensemble, we have measured the nucleon mass at smaller quark masses and found m_N = 0.960 ± 0.30 GeV. The nucleon mass is fairly insensitive to the instanton ensemble. However, the strength of the correlation function depends on the instanton ensemble. This is reflected by the value of the nucleon coupling constant, which is smaller in the interacting model.

Figure 9.9 also shows the nucleon correlation function measured in a quenched lattice simulation (Chu et al. 1994). The agreement with the instanton liquid results is quite impressive, especially given the fact that before the lattice calculations were performed, there was no phenomenological information on the value of the nucleon coupling constant and the behavior of the correlation function at intermediate and large distances.

The fitted position of the threshold is $E_0 \simeq 1.8$ GeV, larger than the mass of the first nucleon resonance, the Roper $N^*(1440)$, and above the $\pi \Delta$ threshold $E_0 = 1.37$ GeV. This might indicate that the coupling of the nucleon current to the Roper resonance is small. In the case of the $\pi \Delta$ continuum, this can be checked directly using the phenomenologically known coupling constants. The large value of the threshold energy also implies that there is little strength in the (unphysical) three-quark continuum. The fact that the nucleon is deeply bound can also be

---

[12]Note that this value corresponds to a relatively large current quark mass $m = 30$ MeV.

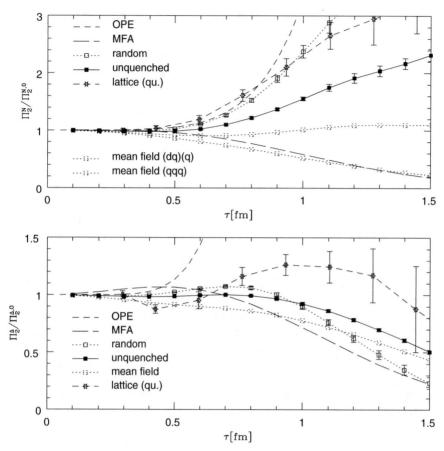

Fig. 9.9 Nucleon and Delta correlation functions Π_2^N and Π_2^Δ

demonstrated by comparing the full nucleon correlation function with that of three non-interacting quarks, see Fig. 9.9). The full correlator is significantly larger than the non-interacting (mean field) result, indicating the presence of a strong, attractive interaction.

Some of this attraction is due to the scalar diquark content of the nucleon current. This raises the question whether the nucleon (in our model) is a strongly bound diquark very loosely coupled to a third quark. In order to check this, we have decomposed the nucleon correlation function into quark and diquark components. Using the mean field approximation, that means treating the nucleon as a non-interacting quark-diquark system, we get the correlation function labeled (diq) in Fig. 9.9. We observe that the quark-diquark model explains some of the attraction seen in Π_2^N, but falls short of the numerical results. This means that while diquarks may play some role in making the nucleon bound, there are substantial interactions in the quark-diquark system. Another hint for the qualitative role of diquarks is

Table 9.4 Nucleon and delta parameters in the different instanton ensembles. All quantities are given in units of GeV. The current quark mass is $m_u = m_d = 0.1\Lambda$

	Unquenched	Quenched	RILM	Ratio ansatz (unqu.)
m_N	1.019	1.013	1.040	0.983
λ_N^1	0.026	0.029	0.037	0.021
λ_N^2	0.061	0.074	0.093	0.048
m_Δ	1.428	1.628	1.584	1.372
λ_Δ	0.027	0.040	0.036	0.026

provided by the values of the nucleon coupling constants $\lambda_N^{1,2}$. One can translate these results into the coupling constants $\lambda_N^{s,p}$ of nucleon currents built from scalar or pseudo-scalar diquarks. We find that the coupling to the scalar diquark current $\eta_s = \epsilon_{abc}(u^a C\gamma_5 d^b)u^c$ is an order of magnitude bigger than the coupling to the pseudo-scalar current $\eta_p = \epsilon_{abc}(u^a C d^b)\gamma_5 u^c$. This is in agreement with the idea that the scalar diquark channel is very attractive and that these configurations play an important role in the nucleon wave function.

The Delta correlation function in the instanton liquid is shown in Fig. 9.9. The result is qualitatively different from the nucleon channel, the correlator at intermediate distance $x \simeq 1$ fm is significantly smaller and close to perturbation theory. This is in agreement with the results of the lattice calculation (Chu et al. 1994). Note that, again, this is a quenched result which should be compared to the predictions of the random instanton model.

The mass of the delta resonance is too large in the random model, but closer to experiment in the unquenched ensemble. Note that similar to the nucleon, part of this discrepancy is due to the value of the current mass. Nevertheless, the Delta-nucleon mass splitting in the unquenched ensemble is $m_\Delta - m_N = 409$ MeV, still too large as compared to the experimental value 297 MeV. Similar to the ρ meson, there is no interaction in the Delta channel to first order in the instanton density. However, if we compare the correlation function with the mean field approximation based on the full propagator, see Fig. 9.9, we find evidence for substantial attraction between the quarks. Again, more detailed checks, for example concerning the coupling to the πN continuum, are necessary.

9.6   Comparison to Correlators on the Lattice

The study of hadronic (point-to-point) correlation functions on the lattice was pioneered by the MIT group (Chu et al. 1994) which measured correlation functions of the $\pi, \delta, \rho, a_1, N$ and Δ in quenched QCD. The correlation functions were calculated on a $16^3 \times 24$ lattice at $6/g^2 = 5.7$, corresponding to a lattice spacing of $a \simeq 0.17$ fm. We have already shown some of the results of the MIT group in Figs. 9.7, 9.8, and 9.9. The correlators were measured for distances up to ~ 1.5 fm. Using the parametrization introduced above, they extracted ground state masses and coupling constants and found good agreement with phenomenological results. What is even more important, they found the *full correlation functions* to agree with the predictions of the instanton liquid, even in channels (like the nucleon and delta) where no phenomenological information is available.

In order to check this result in more detail, they also studied the behavior of the correlation functions under cooling (Chu et al. 1994). The cooling procedure was monitored by studying a number of gluonic observables, like the total action, the topological charge and the Wilson loop. From these observables, the authors conclude that the configurations are dominated by interacting instantons after ∼25 cooling sweeps. Instanton-anti-instanton pairs are continually lost during cooling, and after ∼50 sweeps, the topological charge fluctuations are consistent with a dilute gas. The characteristics of the instanton liquid were already discussed above. After 50 sweeps the action is reduced by a factor ∼300 while the string tension (measured from 7×4 Wilson loops) has dropped by a factor 6.

The first comparison made between the instanton liquid results and those obtained on the lattice (Chu et al. 1994) are shown in Fig. 9.10a, for ρ vector (V) and π pseudoscalar (P) channels.

Even more direct comparison was between the correlator calculated on the "quantum" configurations, as compared to "cooled" or "semiclassical" lattice configurations (Chu et al. 1994). The behavior of the pion and nucleon correlation functions under cooling is shown in Fig. 9.11. The behavior of the ρ and Δ

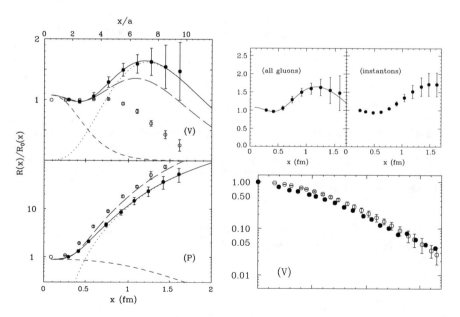

Fig. 9.10 The left panel shows the correlation functions for the vector (marked (V)) ρ channel and the pseudoscalar (marked (P)) π channel. The long-dashed lines are phenomenological ones, open and closed circles stand for RILM (Shuryak and Verbaarschot 1993b) and lattice calculation (Chu et al. 1994), respectively. The upper right panel compares vector correlators before and after "cooling". The lower part shows the same comparison for the ρ wave function; the closed and open points here correspond to "quantum" and "classical" vacua, respectively

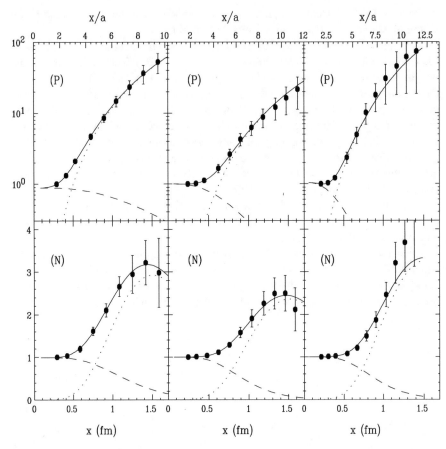

Fig. 9.11 Behavior of pion and proton correlation functions under cooling, from Chu et al. (1994). The left, center, and right panels show the results in the original ensemble, and after 25 and 50 cooling sweeps. The solid lines show fits to the data based on a pole plus continuum model for the spectral function. The dotted and dashed lines show the individual contributions from the pole and the continuum part

correlators (not shown) was quite similar. During the cooling process the scale was readjusted by keeping the nucleon mass fixed.[13]

---

[13] This introduces only a small uncertainty, the change in scale is $\sim 16\%$. We observe that the correlation functions are *stable under cooling*, they agree almost within error bars. This is also seen from the extracted masses and coupling constants. While m_N and m_π are stable by definition, m_ρ and g_ρ change by less than 2%, λ_π by 7% and λ_N by 1%. Only the delta mass is too small after cooling, it changes by 27%.

9.7   Gluonic Correlation Functions

One of the most interesting problems in hadronic spectroscopy is whether one can identify glueballs, bound states of pure glue, among the spectrum of observed hadrons. This question has two aspects. In pure glue theory, stable glueball states exist and they have been studied for a number of years in lattice simulations. In full QCD, glueballs mix with quark states, making it difficult to unambiguously identify glueball candidates.

Even in pure gauge theory, lattice simulations still require large numerical efforts. Nevertheless, a few results appear to be firmly established (1) The lightest glueball is the scalar 0^{++}, with a mass in the 1.5–1.8 GeV range. (2) The tensor glueball is significantly heavier $m_{2^{++}}/m_{0^{++}} \simeq 1.4$, and the pseudo-scalar is heavier still, $m_{0^{-+}}/m_{0^{++}} = 1.5$–1.8. (3) The scalar glueball is much smaller than other glueballs. The size of the scalar is $r_{0^{++}} \simeq 0.2$ fm, while $r_{2^{++}} \simeq 0.8$ fm (de Forcrand and Liu 1993). For comparison, a similar measurement for the π and ρ mesons gives 0.32 fm and 0.45 fm, indicating that spin-dependent forces between gluons are stronger than between quarks.

Gluonic currents with the quantum numbers of the lowest glueball states are the field strength squared ($S = 0^{++}$), the topological charge density ($P = 0^{-+}$), and the energy momentum tensors ($T = 2^{++}$);

$$j_S = (G^a_{\mu\nu})^2, \quad j_P = \frac{1}{2}\epsilon_{\mu\nu\rho\sigma}G^a_{\mu\nu}G^a_{\rho\sigma}, \quad j_T = \frac{1}{4}(G^a_{\mu\nu})^2 - G^a_{0\alpha}G^a_{0\alpha}. \quad (9.51)$$

The short distance behavior of the corresponding correlation functions is determined by the OPE

$$\Pi_{S,P}(x) = \Pi^0_{S,P}\left(1 \pm \frac{\pi^2}{192g^2}\langle f^{abc}G^a_{\mu\nu}G^b_{\nu\beta}G^c_{\beta\mu}\rangle x^6 + \ldots\right) \quad (9.52)$$

$$\Pi_T(x) = \Pi^0_T\left(1 + \frac{25\pi^2}{9216g^2}\langle 2\mathcal{O}_1 - \mathcal{O}_2\rangle \log(x^2)x^8 + \ldots\right) \quad (9.53)$$

where we have defined the operators $\mathcal{O}_1 = (f^{abc}G^b_{\mu\alpha}G^c_{\nu\alpha})^2$, $\mathcal{O}_2 = (f^{abc}G^b_{\mu\nu}G^c_{\alpha\beta})^2$ and the free correlation functions are given by

$$\Pi_{S,P}(x) = (\pm)\frac{384g^4}{\pi^4 x^8}, \qquad \Pi_T(x) = \frac{24g^4}{\pi^4 x^8}. \quad (9.54)$$

Power corrections in the glueball channels are remarkably small. The leading-order power correction $O(\langle G^2_{\mu\nu}\rangle/x^4)$ vanishes,[14] while radiative corrections of the form

---

[14]There is a $\langle G^2_{\mu\nu}\rangle\delta^4(x)$ contact term in the scalar glueball correlators which, depending on the choice of sum rule, may enter momentum space correlation functions.

$\alpha_s \log(x^2)\langle G_{\mu\nu}^2 \rangle/x^4$ (not included in (9.52)), or higher order power corrections like $\langle f^{abc} G_{\mu\nu}^a G_{\nu\rho}^b G_{\rho\mu}^c \rangle/x^2$ are very small.

On the other hand, there is an important low energy theorem that controls the large distance behavior of the scalar correlation function (Novikov et al. 1981)

$$\int d^4x \, \Pi_S(x) = \frac{128\pi^2}{b} \langle G^2 \rangle, \qquad (9.55)$$

where b denotes the first coefficient of the beta function. In order to make the integral well defined and we have to subtract the constant term $\sim \langle G^2 \rangle^2$ as well as singular (perturbative) contributions to the correlation function. Analogously, the integral over the pseudo-scalar correlation functions is given by the topological susceptibility $\int d^4x \, \Pi_P(x) = \chi_{top}$. In pure gauge theory $\chi_{top} \simeq (32\pi^2)\langle G^2 \rangle$, while in unquenched QCD $\chi_{top} = O(m)$. These low energy theorems indicate the presence of rather large non-perturbative corrections in the scalar glueball channels. This can be seen as follows; We can incorporate the low energy theorem into the sum rules by using a subtracted dispersion relation

$$\frac{\Pi(Q^2) - \Pi(0)}{Q^2} = \frac{1}{\pi} \int ds \, \frac{\mathrm{Im}\Pi(s)}{s(s + Q^2)}. \qquad (9.56)$$

In this case, the subtraction constant acts like a power correction. In practice, however, the subtraction constant totally dominates over ordinary power corrections. For example, using pole dominance, the scalar glueball coupling $\lambda_S = \langle 0|j_S|0^{++}\rangle$ is completely determined by the subtraction, $\lambda_S^2/m_S^2 \simeq (128\pi^2/b)\langle G^2 \rangle$.

For this reason, we expect instantons to give a large contribution to scalar glueball correlation functions. Expanding the gluon operators around the classical fields, we have

$$\Pi_S(x, y) = \langle 0|G^{2\,cl}(x)G^{2\,cl}(y)|0\rangle + \langle 0|G_{\mu\nu}^{a,cl}(x) \left[D_\mu^x D_\alpha^y D_{\nu\beta}(x, y) \right]^{ab}$$

$$\times G_{\alpha\beta}^{b,cl}(y)|0\rangle + \dots, \qquad (9.57)$$

where $D_{\mu\nu}^{ab}(x, y)$ is the gluon propagator in the classical background field. If we insert the classical field of an instanton, we find

$$\Pi_{S,P}^{SIA}(x) = \int \rho^4 dn(\rho) \frac{12288\pi^2 \rho^{-8}}{y^6(y^2+4)^5} \left[y^8 + 28y^6 - 94y^4 - 160y^2 - 120 \right.$$

$$\left. + \frac{240}{y\sqrt{y^2+4}} (y^6 + 2y^4 + 3y^2 + 2)\mathrm{asinh}\left(\frac{y}{2}\right) \right] \qquad (9.58)$$

with $y = x/\rho$.

There is no classical contribution in the tensor channel, since the stress tensor in the self-dual field of an instanton is zero. Note that the perturbative contribution in the scalar and pseudo-scalar channels have opposite sign, while the classical contribution has the same sign. To first order in the instanton density, we therefore find the three scenarios discussed in Sect. 9.5.1; *attraction* in the scalar channel, *repulsion* in the pseudo-scalar and *no* effect in the tensor channel. The single-instanton prediction is compared with the OPE in Fig. 9.12. We clearly see that classical fields are much more important than power corrections (Table 9.5).

Quantum corrections to this result can be calculated from the second term in (9.57) using the gluon propagator in the instanton field The singular contributions

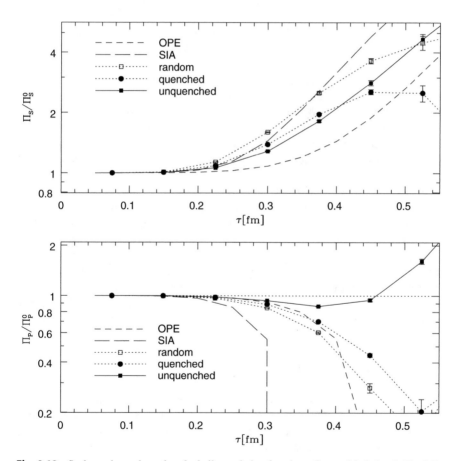

Fig. 9.12   Scalar and pseudo-scalar glueball correlation functions. Curves labeled as in Fig. 9.7

Table 9.5 Scalar glueball parameters in different instanton ensembles			Random	Quenched	Unquenched
	$m_{0^{++}}$	[GeV]	1.4	1.75	1.25
	$\lambda_{0^{++}}$	[GeV^3]	17.2	16.5	15.6

correspond to the OPE in the instanton field. There is an analog of the Dubovikov-Smilga result for glueball correlators; In a general self-dual background field, there are no power corrections to the tensor correlator. This is consistent with the result (9.53), since the combination $\langle 2\mathcal{O}_1 - \mathcal{O}_2 \rangle$ vanishes in a self-dual field. Also, the sum of the scalar and pseudo-scalar glueball correlators does not receive any power corrections (while the difference does, starting at $O(G^3)$).

Numerical calculations of glueball correlators in different instanton ensembles were performed in Schafer and Shuryak (1995). At short distances, the results are consistent with the single instanton approximation. At larger distances, the scalar correlator is modified due to the presence of the gluon condensate. This means that (like the σ meson), the correlator has to be subtracted and the determination of the mass is difficult. In the pure gauge theory we find $m_{0^{++}} \simeq 1.5\,\text{GeV}$ and $\lambda_{0^{++}} = 16 \pm 2 \,\text{GeV}^3$. While the mass is consistent with QCD sum rule predictions, the coupling is much larger than expected from naive calculations that do not enforce the low energy theorem.

In the pseudo-scalar channel the correlator is very repulsive and there is no clear indication of a glueball state. In the full theory (with quarks) the correlator is modified due to topological charge screening. The non-perturbative correction changes sign and a light (on the glueball mass scale) state, the η' appears. Non-perturbative corrections in the tensor channel are very small. Isolated instantons and anti-instantons have a vanishing energy momentum tensor, so the result is entirely due to interactions.

In Schafer and Shuryak (1995) we also evaluated the glueball wave functions. The most important result is that the scalar glueball is indeed small, $r_{0^{++}} = 0.2\,\text{fm}$, while the tensor is much bigger, $r_{2^{++}} = 0.6\,\text{fm}$. The size of the scalar is determined by the size of an instanton, whereas in the case of the tensor the scale is set by the average distance between instantons. This number is comparable to the confinement scale, so the tensor wave function is probably not very reliable. On the other hand, the scalar is much smaller than the confinement scale, so the wave function of the 0^{++} glueball may provide an important indication for the importance of instantons in pure gauge theory.

9.8   Wave Functions

One quantity which is close to *hadronic wave function* is the so called Bethe Salpeter amplitude (which is different from the so called light-cone one). Such Bethe-Salpeter amplitudes have been measured in a number of lattice gauge simulations. In the pion case this quantity is defined by

$$\psi_\pi(y) = \int d^4x \, \langle 0|\bar{d}(x) P e^{i\int_x^{x+y} A(x')dx'} \gamma_5 u(x+y)|\pi > . \qquad (9.59)$$

In practice, it is extracted from the three point correlator

$$\Pi_\pi(x, y) = \langle 0|T(\bar{d}(x)Pe^{i\int_x^{x+y} A(x')dx'}\gamma_5 u(x + y)\bar{d}(0)\gamma_5 u(0))|0\rangle$$
$$\sim \psi(y)e^{-m_\pi x} \tag{9.60}$$

where x has to be a large space-like separation in order to ensure that the correlation function is dominated by the ground state and y is the separation of the two quarks in the transverse direction $((x\dot{y}) = 0)$. In practice it is convenient to divide the 3-point by 2-point function, canceling the x-dependent part and the coupling constants.

Like the two point correlation functions, the Bethe-Salpeter amplitudes are calculated from the light quark propagator[15]

A qualitative understanding of the wave functions can be obtained using the single-instanton approximation. For small transverse separations y and $x \to \infty$ we get a very simple result

$$\psi_\pi(y) = 1 - y^2/(2\rho)^2 + \ldots \tag{9.61}$$

indicated that a pion radius (as determined by Bethe-Salpeter amplitude) is directly related to the instanton radius.

The wave functions in the random ensemble was calculated by Schafer and Shuryak (1995). Those for π, ρ, N and Δ are shown in Fig. 9.13. We observe that the pion and the proton as well as the rho meson and the delta resonance have very similar wave functions, but the sizes for pion and the proton are *smaller* than the rho meson and the delta resonance. We have already argued that the scalar diquark in the nucleon is linked with the instanton-induced attraction. The correlators for scalar (upper), tensor (middle) and pseudoscalar (lower) glueballs are shown in Fig. 9.14.

9.9 Brief Summary

- Definition of Euclidean correlation functions of two local operators at two different points separated by distance r. At small distances those can be calculated perturbatively, via free quark (or gluon) propagation from point x to point y. This leads to certain inverse powers of the distance $r = |x - y|$. At large distances Euclidean correlators are dominated by propagators of the lightest hadrons, posessing the same quantum numbers as the operators involved.
- In Minkowskian momentum representation the correlation functions have real and imaginary The imaginary part is also called "spectral density". From it one can find, by convolution with propagators, the real part using *dispersion relation*. This can be written both in Minkowskian and Euclidean formulations.

---

[15]In general, the inclusion of the Schwinger Pexp factor is expected to give an important contribution to the measured wave functions, since it corresponds to an additional string type potential, but not in the instanton model.

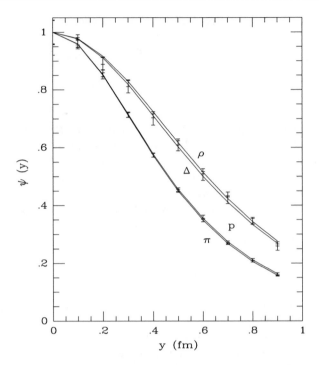

Fig. 9.13 Hadronic wave functions of the pion, rho meson, proton and delta resonance in the random instanton ensemble

- Using experimental spectral density one can calculate Euclidean correlator of vector currents with flavor quantum numbers ρ, ω, K^*, ϕ mesons. In all of them deviation from free quark propagation occurs only at surprisingly large distances $r > 1$ fm
- In contrast to that, in scalar and pseudoscalar channels deviation from free quark propagation occurs at much smaller distances $r \sim 1/4$ fm.
- Using combinations of correlators, one can identify instanton-induced 't Hooft $2N_f$-fermion operator. Its numerical value was obtained from lattice simulations, and found it to be rather large. It is shown that all observed strong deviations from pQCD at small r can be explained by this operator
- Correlators of all mesonic and baryonic local operators were calculated in the IILM. In spite of no confinement in this model, all of them agree with experimental and lattice data quite well, see e.g. Fig. 9.8 and Fig. 9.14 (upper).
- Even stronger splittings from pQCD and between different channels were found for gluonic correlation functions. The $J^P = 0^+$ scalar channel is attractive, while pseudoscalar 0^- is repulsive.

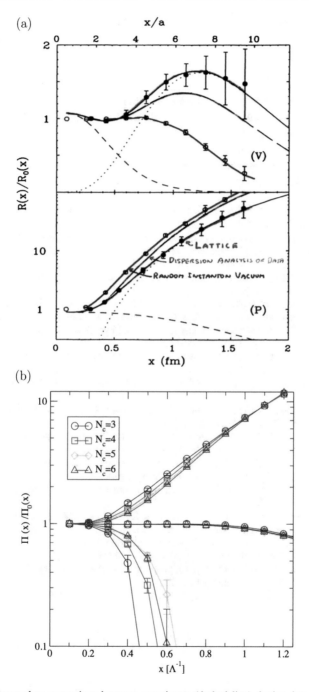

Fig. 9.14 Upper plot: comparison between experiment (dashed line), lattice data (circles) and random instanton liquid model (open squares). Lower plot: Gluonic correlation functions for different number of colors in the ILM, upper curves for scalar channel, middle for tensor, and lower set for pseudoscalar one

- One can generalize correlation functions to those of nonlocal operators. In particular, of a local one with that with two points split by spatial distance

$$\langle O_1(0) O_2(\tau, \vec{\delta}/2, -\vec{\delta}/2)\rangle$$

While correlator of two local objects provide hadronic wave functions at the (3d) origin, the nonlocal ones shed light on the spatial shape of hadronic wave functions, see Fig. 9.13.

References

Andrei, N., Gross, D.J.: The effect of instantons on the short distance structure of hadronic currents. Phys. Rev. **D18**, 468 (1978)

Chu, M.C., Grandy, J.M., Huang, S., Negele, J.W.: Evidence for the role of instantons in hadron structure from lattice QCD. Phys. Rev. **D49**, 6039–6050 (1994). hep-lat/9312071

de Forcrand, P., Liu, K.-F.: Glueball wave functions. Nucl. Phys. Proc. Suppl. **30**, 521–524 (1993)

Faccioli, P., DeGrand, T.A.: Evidence for instanton induced dynamics, from lattice QCD. Phys. Rev. Lett. **91**, 182001 (2003)

Novikov, V.A., Shifman, M.A., Vainshtein, A.I., Zakharov, V.I.: Are All Hadrons Alike? DESY-check = Moscow Inst. Theor. Exp. Phys. Gkae - Itef-81-048 (81,rec.jun.) 32 *P* and Nucl. Phys. B191 (1981) 301–369 and Moscow Inst. Theor. Exp. Phys. Gkae - Itef-81-042 (81,rec.apr.) 70 *P*. (104907). Nucl. Phys. **B191**, 301–369 (1981)

Nussinov, S., Lampert, M.A.: QCD inequalities. Phys. Rept. **362**, 193–301 (2002)

Schafer, T.: Instantons in QCD with many colors. Phys. Rev. **D66**, 076009 (2002). hep-ph/0206062

Schafer, T., Shuryak, E.V.: Glueballs and instantons. Phys. Rev. Lett. **75**, 1707–1710 (1995). hep-ph/9410372

Schafer, T., Shuryak, E.V.: The Interacting instanton liquid in QCD at zero and finite temperature. Phys. Rev. **D53**, 6522–6542 (1996). hep-ph/9509337

Schafer, T., Shuryak, E.V.: Implications of the ALEPH tau lepton decay data for perturbative and nonperturbative QCD. Phys. Rev. Lett. **86**, 3973–3976 (2001a)

Schafer, T., Shuryak, E.V., Verbaarschot, J.J.M.: Baryonic correlators in the random instanton vacuum. Nucl. Phys. **B412**, 143–168 (1994). hep-ph/9306220

Shuryak, E.V.: The role of instantons in quantum chromodynamics. 1. Physical vacuum. Nucl. Phys. **B203**, 93 (1982b)

Shuryak, E.V.: Pseudoscalar mesons and instantons. Nucl. Phys. **B214**, 237–252 (1983)

Shuryak, E.V.: Correlation functions in the QCD vacuum. Rev. Mod. Phys. **65**, 1–46 (1993)

Shuryak, E.V., Verbaarschot, J.J.M.: Mesonic correlation functions in the random instanton vacuum. Nucl. Phys. **B410**, 55–89 (1993a). hep-ph/9302239

Shuryak, E.V., Verbaarschot, J.J.M.: Quark propagation in the random instanton vacuum. Nucl. Phys. **B410**, 37–54 (1993b). hep-ph/9302238

Weingarten, D.: Mass inequalities for QCD. Phys. Rev. Lett. **51**, 1830 (1983)

Light-Front Wave Functions, Exclusive Processes and Instanton-Induced Quark Interactions

10

10.1 Quark Models of Hadrons

Historically, understanding of quark substructure of hadrons started in 1960's with two theoretical breakthroughs, both related with the corresponding symmetries.

One was related to properties of strange partners of better known particles, with K substituting pions, K^* substituting ρ mesons, Λ, Σ hyperons substituting nucleons, etc. It lead Gell-Mann and Neeman to the so called "eightfold way", their classification via the *octet* representation of the flavor $SU(3)$ group. Three flavors were, of course, u, d, s quarks, and spectra were well described via $SU(3)$-symmetric expression, perturbed by symmetry-breaking mass terms. This approach culminated by prediction of the baryon decuplet—another representation of $SU(3)$ group—with subsequent experimental discoveries of all its members, completed by the sss Ω^- baryon. Needless to say, with the discovery of c and then b quarks, nonrelativistic description of corresponding hadrons followed.

The description of masses were complemented by similar results on other properties, such as baryon magnetic moments. Note that Dirac particle has magnetic moment proportional to "magneton" unit $\sim e/m$, inversely proportional to mass. Baryon magnetic moments suggested that quarks have mass about 1/3 of the nucleon mass. Quark models of hadrons indeed use such "constituent quark masses".

Another insight into hadronic and quark properties was proposed even a bit earlier, in 1961, in Nambu and Jona-Lasinio (1961). This model, called NJL, pointed out the notion of chiral symmetry and its spontaneously breaking. The authors postulated existence of 4-fermion interaction, with some coupling G, strong enough to make a superconductor-like gap even in vacuum, at the surface of the Dirac sea. The second (and the last) parameter of the NJL model was the cutoff $\Lambda \sim 1\,\mathrm{GeV}$, below which their hypothetical attractive 4-fermion interaction operates.

Although there are many "quark models",[1] of different degree of sophistication, their common elements are:

(i) the constituent quark masses;
(ii) interquark potential, including confining linear $\sim r$ and Coulomb $\sim 1/r$ potentials
(iii) and various "residual interactions", mostly of 4-fermion type

The method of choice, like in atomic and nuclear physics, is selection of certain large basis of the wave functions, representing the Hamiltonian as a matrix in it, with its subsequent diagonalization.

While the constituent quark masses constitute most of the hadronic masses, the potentials (ii) contribute typically $O(1/10)$ of them, and the residual interactions are very important only in some "exceptional" hadrons, such as π, η' mesons.

The field of hadronic spectroscopy was, for a long time, reduced to mesons ($\bar{q}q$) and baryons (qqq) states, while larger systems—such as pentaquarks $q^4\bar{q}$ and dibaryons q^6, predicted by such models, were never seen. Only with inclusion of heavy quarks, pentaquarks were relatively recently discovered.

Completing this introductory section, let me also mention the "QCD cousins" with the number of colors $N_c \neq 3$. Obviously, there are two directions in which N_c can be changed, up and down, and both are rather interesting.

If $N_c \to \infty$, the baryons containing N_c quarks need to become very heavy. Skyrme's view on baryons, as classical solitons made of mesonic fields, become justified in this limit. This direction we will further discuss in chapter on holographic models 18.

One can also reduce N_c from 3 to 2, considering "two-color QCD". Due to specific properties of the $SU(2)$ group (two-index $\epsilon_{\alpha\beta}$) quarks and antiquarks are not really different, and it has additional Pauli-Gursay symmetry relating them. This symmetry has multiplets in which mesons and baryons enter together. In particular, with u, d quarks the chiral symmetry breaking leads to 5 massless Goldstone modes: 3 pions, diquark ud and its antiparticle.

One may wander how parameters of quark models depend on N_c, and whether extrapolation from $N_c = 2$ to the real world with $N_c = 3$ would be possible. It was pointed out Rapp et al. (1998) in connection to color superconductivity, that "residual forces", the Coulomb and instanton-induced ones, scale as

$$\frac{V_{qq}}{V_{\bar{q}q}} \sim \frac{1}{N_c - 1} \tag{10.1}$$

so they are: (1) negligible in the $N_c \to \infty$ limit; (2) equal to -1 in QED when $N_c = 0$, (3) equal when $N_c = 2$; and, finally, (4) for $N_c = 3$ they are different by

---

[1] In this chapter we will not discuss *glueballs*: we however will discuss them in later chapters, in connection to flux tubes and Pomerons.

a factor 1/2. So, the real world is equidistant, from small and large number of color limits!

10.2 Light-Front Observables

Unlike atomic or nuclear structure, the structure of hadrons is mostly experimentally probed in relativistic kinematics. Take as an example the pion form factor, at large momentum transfer. The most symmetric frame is in which initial pion has momentum $Q/2$ and outgoing one $-Q/2$. If $Q \sim few\,\mathrm{GeV} \gg m_\pi$ both momenta are large compared to the pion mass, so incoming and outgoing pions moves ultrarelativistically.

The iconic Deep Inelastic Scattering (DIS) of highly virtual photon on a nucleon gives us the so called Parton Distribution Functions (PDFs) $f_i(x, Q^2)$, where x is fraction of the total momentum carried by the quark absorbing the photon, and Q^2 is the photon virtuality, and $i =$ gluons, quarks and antiquarks of different flavors. Traditionally, DIS is compared to a microscope, the better is its spatial resolution $\sim 1/Q$ the more partons it can resolve.

Technical tool to describe the Q-dependence of PDFs was developed in pQCD and is known as Dokshitzer-Gribov-Lipatov-Altarelli-Parisi (DGLAP) evolution equation. It includes certain "splitting functions", describing probability of a parton splitting into two. Also important to note, as any other single-body Boltzmann equation, DGLAP is based on implicit assumption that higher correlations between bodies are small and can be neglected.

Let me only comment here that PDFs are, by their nature, single-parton observables, in a many-parton system. DGLAP has the form of a kinetic equation, and, surprisingly recently (Kharzeev and Levin 2017) noticed that, as for any kinetic equation, one can introduce the notion of *entropy*, its increase with time, with eventual approach of its maximum, the equilibrium state.

Entropy is property of the *single-parton density matrix*, but the hadrons themselves are of course pure quantum states, with wave functions. They too can be defined using light-front coordinates (LFWF) as $\psi(x_i, \vec{p}^i_\perp)$ with $i = 1..N$, N total number of partons. As we will discuss below, the wave function includes sectors with different N: e.g. baryons have the simplest sector with $N = 3$, and then $N = 5, 7 \ldots$ as extra quark pairs can be added.[2]

Completing this introductory section, let me mention several physics phenomena we will be discuss below in the chapter. Perhaps the most famous of them is the so called "spin crisis": related to the question how the total spin of the proton, of course equal to 1/2, is shared between different kinds of partons.

---

[2]Note that this feature is not new or exclusive to hadrons: e.g. atomic and nuclear states are also described as "closed shells", plus some number of "quasiparticles", plus any number of particle-hole pairs excited by the "residual" part of the Hamiltonian.

I would prefer to reformulate this puzzling issue into several: (1) What is the contribution of the valence d quark to the spin? (2) What is the contributions of the sea, of the different flavors of quarks and antiquarks? (3) Why is the sea of the polarized proton polarized at all? (4) Why is the sea polarized in the isospin?

10.3   Quark Models on the Light Front: Mesons in the $\bar{q}q$ Sector

Let me start with the general kinematics and normalization condition of the light front wave functions. They are defined in momentum representation, with x_i being momentum fractions along the beam direction $\vec{p}_{\parallel} = x\vec{P}$. The integration measure, which defines the orthogonality condition in the sector with $i = 1..N$ partons is defined as

$$\int \left(\prod_i dx_i \frac{d\vec{p}_{\perp}^i}{(2\pi)^2} \right) (\Pi_i x_i) \delta \left(\sum_i x_i - 1 \right) \delta \left(\sum_i \vec{p}_{\perp}^i \right)$$

$$\times \psi_n^*(x_i, \vec{p}_{\perp}^i) \psi_m(x_i, \vec{p}_{\perp}^i) = \delta_{mn} \tag{10.2}$$

Note that apart of two delta functions expressing momentum conservation, there is also a product of all momentum fractions, introduced for convenience. The so called "asymptotic" wave function, in this normalization, is independent of x_i. Note also that states in sectors with different N are assumed to be orthogonal to each other by definition.

The simplest case is the 2-body sector of the mesons. In this case the restriction on momentum fraction becomes just $x_1 + x_2 = 1$, and therefore the wave function is a function of a single longitudinal variable. It is convenient to select the *asymmetry* variable s defined via

$$s \equiv x_1 - x_2 \tag{10.3}$$

The inverse relations, expressing momentum fractions via it are

$$x_1 = \frac{1+s}{2}, \quad x_2 = \frac{1-s}{2}$$

The basic setting we will follow is due to Jia and Vary (2019). The Hamiltonian has four terms including (1) the effective quark masses coming from chiral symmetry breaking; (2) the longitudinal confinement; (3) the transverse motion and confinement; and, last but not least, (4) the NJL four-quark effective interaction

$$H = H_M + H_{conf}^{\parallel} + H_{conf}^{\perp} + H_{NJL} \tag{10.4}$$

$$H_M = \frac{M^2}{x_1} + \frac{\bar{M}^2}{x_2}$$

$$H^{\parallel}_{conf} = \frac{\kappa^4}{(M + \bar{M})^2} \frac{1}{J(x)} \partial_x J(x) \partial_x$$

$$H^{\perp}_{conf} = \vec{k}^2_{\perp} \left(\frac{1}{x_1} + \frac{1}{x_2} \right) + \kappa^4 x_1 x_2 \vec{r}^2_{\perp}$$

where $M, \bar{M}$ are masses of quark and antiquark, κ is the confining parameter, and the integration measure is $J(x) = x_1 x_2 = (1 - s)^2/4$, $\vec{k}_{\perp}, \vec{r}_{\perp}$ are transverse momenta and coordinates. If the masses are the same (no strange quarks) however, one can simplify it

$$\frac{1}{x_1} + \frac{1}{x_2} = \frac{1}{x_1 x_2} = \frac{4}{(1 - s^2)}$$

and therefore the matrix element of H_M simply lack the factor $(1 - s^2)$ normally present in the integration measure. The value of their parameters are

$$M_q = .337\,\text{GeV}, \quad \kappa = .227\,\text{GeV} \tag{10.5}$$

In Jia and Vary (2019) the basis is designed so that the part of Hamiltonian, other than H_{NJL}, is diagonal. In subsequent paper Shuryak (2019) this requirement is dropped and the basis is simply given by Jacobi polynomials $P_n^{1,1}(s)$. Other simplifications include dropping transverse momentum dependence and using only topology-induced 't Hooft 4-fermion operator. Therefore, it is absent in the ρ meson, and enter in effective Hamiltonian for π and η' with the opposite signs (we already seen it in the chapter on correlation functions).

The results are shown in Fig. 10.1. Note that the predicted PDF for mesons is just the wave function squared (there are no extra variables to integrate). The quark momentum distribution for ρ meson (in which the 4-fermion interaction is presumed to be absent) is peaked near the symmetric point $x = 1/2, s = 0$. The pion one, in contrast, has a completely different flat shape. The η' PDF is deviating to the opposite direction, getting even more concentrated in x.[3]

As one compares the lower plot to our result, one should keep in mind the fact that the PDF include also contribution from sectors with the quark number larger than 2, while ours (so far) do not.

10.4   Quark Models on the Light Front: Baryons as qqq States

The quark wave functions we will be discussing are defined as, for Delta baryon

$$|\Delta^{++}\rangle \sim \psi_{\Delta}(x_i)|u^{\uparrow}(x_1)u^{\uparrow}(x_2)u^{\uparrow}(x_3)\rangle \tag{10.6}$$

---

[3]Note that it is probability in momentum representation. The coordinate one does the opposite, making pion very compact and η', with a local repulsive potential at its core, of larger size.

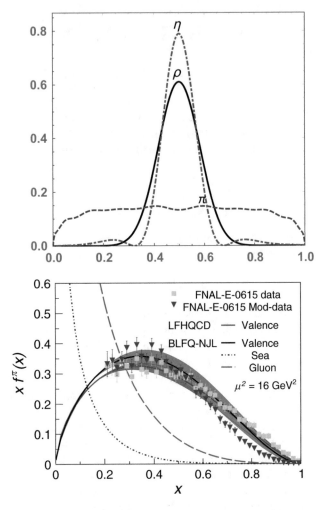

Fig. 10.1 Upper: momentum distribution for pion, rho and eta-prime mesons, calculated in the model. Lower (from Geesaman and Reimer (2019)) comparison between the measured pion PDF (points) and the JV model (lines)

and for the proton

$$|p \uparrow\rangle \sim \psi_p(x_i)\big(|u^\uparrow(x_1)u^\downarrow(x_2)d^\uparrow(x_3)\rangle \qquad (10.7)$$

$$-|u^\uparrow(x_1)d^\downarrow(x_2)u^\uparrow(x_3)\rangle\big)$$

and in what follows we will focus on the former component, in which the d quark has the last momentum fraction x_3. For a review see Braun (2006).

Here we follow Shuryak (2019) paper. Three x_i, $i = 1, 2, 3$ sums to one, so one needs two kinematic variables. The ones used in this case are defined as

$$s = \frac{x_1 - x_2}{x_1 + x_2}, \quad t = x_1 + x_2 - x_3 \tag{10.8}$$

in terms of which

$$x_1 = \frac{(1+s)}{2} \frac{(1+t)}{2}, \quad x_2 = \frac{(1-s)}{2} \frac{(1+t)}{2}, \tag{10.9}$$

and the integration measure

$$\int \left(\prod_i dx_i \right) \delta \left(\sum_i x_i - 1 \right) \left(\prod_i x_i \right) \ldots = \tag{10.10}$$

$$\frac{1}{2^5} \int_{-1}^{1} ds (1-s)(1+s) \int_{-1}^{1} dt (1-t)(1+t)^3 \ldots$$

is factorized. Therefore one can split it into two and select appropriate functional basis of two Jacobi functions

$$\psi_{n,m}(s, t) \sim P_n^{1,1}(s) P_m^{1,3}(t)$$

The Hamiltonian is simple generalization of that by Jia-Vary

$$H_{conf} = -\frac{\kappa^4}{J(s,t) M_q^2} \left[\frac{\partial}{\partial s} J(s,t) \frac{\partial}{\partial s} + \frac{\partial}{\partial t} J(s,t) \frac{\partial}{\partial t} \right] \tag{10.11}$$

with the measure function $J(s,t)$ appearing in the s, t, integration. Note that coefficient 4 in denominator is missing: this is cancelled by factor 4 coming from a difference between derivatives in x and s, t variables.

The third (and the last) effect we incorporate in this work is the topology-induced four-quark interactions. Note that topological 't Hooft Lagrangian is flavor antisymmetric. This means that it does not operate e.g. in baryons made of the same flavor quarks, like the $\Delta^{++} = uuu$. Another reason why the 't Hooft vertex should be absent is in any states in which all chiralities of quarks are the same, like LLL. For both these reasons, Δ^{++} is not affected by topology effects, therefore serving as a benchmark (like the ρ meson did in the previous section).

For discussion of the masses etc. see the original paper. The main results are the PDF and the wave functions. The distribution in momentum of the d quark is obtained by integration over s

$$d(x_3 = 1 - 2t) = \int_{-1}^{1} ds J(s,t) \Psi_N^2(s,t) \tag{10.12}$$

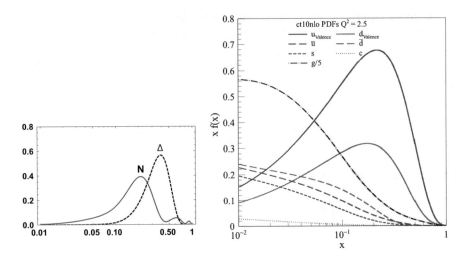

Fig. 10.2 Left: our calculation of the d quark distribution in the Nucleon times x, $xd(x)$ (red, solid) and Delta (black, dashed) states. For comparison, the right plot shows empirical structure functions (copied from Geesaman and Reimer (2019)), where the valence $xd(x)$ is also shown in red

and it is shown in Fig. 10.2 (upper), for Delta and Nucleon wave functions we obtained. Two comments: (1) the peak in the Nucleon distribution moves to lower x_d, as compared to $x = 1/3$ expected in Δ and non-interacting three quarks; (2) there appears larger tail toward small x_d in the nucleon, but also some peaks at large x_d. Both are unmistakably the result of local ud pairing (strong rescattering) in a diquark cluster. The Fig. 10.2 (lower), shown for comparison, includes the empirical valence $xd(x)$ distribution, also shown by red solid line. The location of the peak (1) roughly corresponds to data, and (2) the presence of small-x_d tail is also well seen (recall that what is plotted is distribution times x). The experimental distribution is of course much smoother than ours. It is expected feature: our wave function is expected to be "below the pQCD effects", at resolution say $Q^2 \sim 1\,\mathrm{GeV}^2$, while the lower plot is at $Q^2 = 2.5\,\mathrm{GeV}^2$, ant it includes certain amount of pQCD gluon radiation. contains higher order correlations between quarks.

In order to reveal the structure, one can of course compare the wave functions without any integrations, as they depend on two variables s, t only. Such plots, of $J(s, t)\Psi^2(s, t)$, we show In Fig. 10.3 for the Delta, the nucleon, and some model discussed in Chernyak et al. (1989). While the Delta shows a peak near the symmetry point $x_1 = x_2 = x_3$ as expected, without any other structures, our Nucleon WF indicate more complicated dynamics. Indeed, there appear several bumps, most prominent near $s \approx 1$, $t \approx 1$ which is $x_1 \approx 1$. Such strong peaking corresponds to large momentum transfer inside the ud diquark clusters. Yet there is also the peak in the middle, roughly corresponding to that in Delta. So, the nucleon wave function is a certain coherent mixture of a three-quark and quark-diquark components.

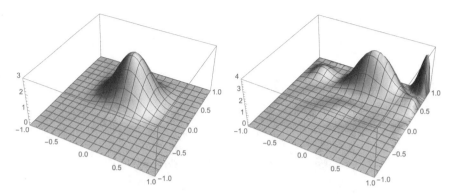

Fig. 10.3 The probability distribution $J(s,t)\Psi^2(s,t)$ in s,t variables, for the Delta and Nucleon lowest states, as calculated from the model

Additional information about the nucleon wave function may be obtained from the amplitudes corresponding to particular basis states. The wave function coefficients C_n^α, defining expansion in the basis functions

$$|\alpha> = \sum_n C_n^\alpha |n>$$

are shown in Fig. 10.4. The upper plot compares those for the ground state Delta and Nucleon channels. Note first that the largest coefficients are the first (corresponding to trivial $\psi_{0,0}(s,t) \sim const(s,t)$). Furthermore, for the Delta it is close to one, while it is only $\sim 1/2$ for the nucleon. The fraction of "significant coefficients" is much larger for N. The nucleon wave function has a nontrivial tail toward higher n, l harmonics which does not show any decreasing trend. One may in fact conclude that convergence of the harmonic expansion is not there. This can be traced to apparent peak near $x = 1$, perhaps the pointlike residual interaction leads to true singularity there.

Two lower plots of Fig. 10.4 address the distribution of the wave function coefficients C_n^α in these two channels, without and with the residual four-quark interaction. It includes not the ground state but the lowest 25 states in each channel. As it is clear from these plot, in the former (Delta) case the distribution is very non-Gaussian, with majority of coefficients being small. The latter (Nucleon) case, on the other hand, is in agreement with Gaussian. In other quantum systems, e.g. atoms and nuclei, Gaussian distribution of the wave function coefficients C_n^α is usually taken as a manifestation of "quantum chaos". In this language, we conclude that our model calculation shows that the residual four-quark interaction leads to chaotic motion of quarks, at least inside the Nucleon resonances. (If this conclusion surprises the reader, we remind that the same interaction was shown to produce chaotic quark condensate in vacuum. In particular, numerical studies of Interacting Instanton Liquid in vacuum has lead to Chiral Random Matrix theory of the vacuum Dirac eigenstates near zero, accurately confirmed by lattice studies.)

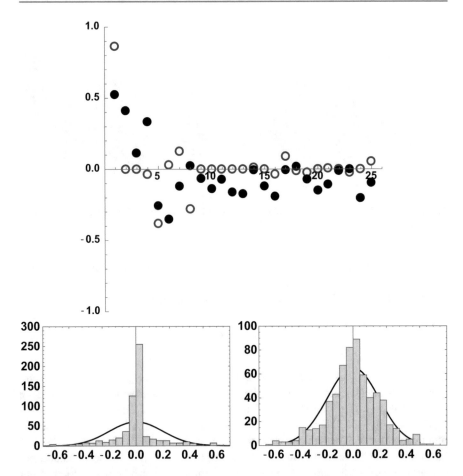

Fig. 10.4 Upper: The coefficients of the wave function C_n^α (the order of basis functions is specified in Appendix), for the ground state Delta (open points) and the Nucleon (closed points) states. Two lower plots contain histograms of these coefficients for 25 lowest Delta and Nucleon states, respectively. The black lines in the background are the Gaussians with the appropriate width

10.5   Quark Models on the Light Front: Pentaquarks and the Five-Quark Sector of Baryons

Here we continue to follow Shuryak (2019) paper. The kinematics in the **5-particle sector** has four variables for longitudinal momenta s, t, u, w, collectively called

$z_i, i = 1, 2, 3, 4$ are defined by

$$s = \frac{x_1 - x_2}{x_1 + x_2}, \quad t = \frac{x_1 + x_2 - x_3}{x_1 + x_2 + x_3}, \tag{10.13}$$

$$u = \frac{x_1 + x_2 + x_3 - x_4}{x_1 + x_2 + x_3 + x_4}, \quad w = x1 + x2 + x3 + x4 - x5$$

The principle idea can also be seen from the inverse relations

$$x_1 = \frac{1}{2^4}(1 + s)(1 + t)(1 + u)(1 + w)$$

$$x_2 = \frac{1}{2^4}(1 - s)(1 + t)(1 + u)(1 + w)$$

$$x_3 = \frac{1}{2^3}(1 - t)(1 + u)(1 + w),$$

$$x_4 = \frac{1}{2^2}(1 - u)(1 + w) \tag{10.14}$$

$$x_5 = 1 - x_1 - x_2 - x_3 - x_4 = \frac{1}{2}(1 - w)$$

The integration measure follows the previous trend, has the form of product of factors depending on single variables of the set, namely

$$\int_{-1}^{1} \frac{dsdtdudw}{16777216}(1 - s)(1 + s)(1 - t)(1 + t)^3(1 - u)(1 + u)^5(1 - w)(1 + w)^7 \ldots \tag{10.15}$$

The orthonormal polynomial basis to be used is then provided by the product of appropriate Jacobi polynomials

$$\psi_{lmnk}(s, t, u, w) \sim P_l^{1,1}(s) P_m^{1,3}(t) P_n^{1,5}(u) P_k^{1,7}(w) \tag{10.16}$$

with normalization constants defined to satisfy (10.2).

Without inclusion of four-quark residual interaction we get multiple pentaquark states, with the lowest mass

$$M_{min.penta} = 2.13 \, \text{GeV} \tag{10.17}$$

Fig. 10.5 The only diagram
in which four-quark
interaction connects the three-
and five-quark sectors,
generating the $\bar{u}$ sea

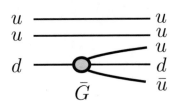

To get this number in perspective, let us briefly remind the history of the light pentaquark search. In 2003 LEPS group reported observation of the first pentaquark called $\Theta^+ = u^2 d^2 \bar{s}$ with surprisingly light mass, of only 1.54 GeV, 0.6 GeV lower than our calculation (and many others) yield. Several other experiments were also quick to report observation of this state, till other experiments (with better detectors and much high statistics) show this pentaquark candidate does not really exist. Similar sad experimental status persists for all 6-light-quark dibaryons , including the flavor symmetric $u^2 d^2 s^2$ spin-0 state much discussed in some theory papers. Yet neither instanton liquid nor lattice studies never found any indication for light pentaquarks or dibaryons: in fact ud, us, sd diquarks strongly repel each other.

The four-quark interaction relates the three-quark of the baryons with the five-quark pentaquark states.

The Hamitonian matrix element corresponding to the diagram shown in Fig. 10.5 we calculated between the nucleon and each of the pentaquark wave functions, defined above, by the following $2 + 4$ dimensional integral over variables in three- and five-quark sectors, related by certain delta functions

$$\langle N|H|5q, i \rangle = \bar{G} \int ds\,dt\,J(s, t)\,ds'\,dt'\,du'\,dw'\,J(s', t', u', w') \qquad (10.18)$$

$$\psi_N(s, t)\delta(x_1 - x_1')\delta(x_2 - x_2')\psi_i(s', t', u', w')$$

The meaning of the delta functions is clear from the diagram, they are of course expressed via proper integration variables and numerically approximated by narrow Gaussians. After these matrix elements are calculated, the five-quark "tail" wave function is calculated via standard perturbation theory expression

$$\psi_{tail}(s', t', u', w') = -\sum_i \frac{\langle N|H|5q, i \rangle}{M_i^2 - M_N^2} \psi_i(s', t', u', w') \qquad (10.19)$$

The typical value of the overlap integral itself for different pentaquark state is $\sim 10^{-3}$, and using for effective coupling $\bar{G}$ the same value as we defined for G from the nucleon, namely ~ 17 GeV^2, one finds that admixture of several pentaquarks to the nucleon is at the level of a percent. The normalized distribution of the 5-th body, namely $\bar{u}(x)$, over its momentum fraction is shown in Fig. 10.6. One can see a peak at $x_{\bar{u}} \sim 0.05$, which looks a generic phenomenon. The oscillations at large $x_{\bar{u}}$ reflect

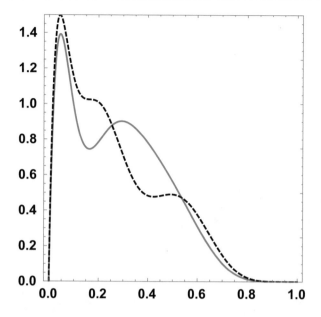

Fig. 10.6 The momentum fraction distribution of antiquarks $\bar{u}(x)$, from the 5-quark sector, for the Proton and Delta baryons (solid and dashed lines, respectively)

strong correlations in the wave function between quarks, as well as perhaps indicate the insufficiently large functional basis used. This part of the distribution is perhaps numerically unreliable.

As originally emphasized by Dorokhov and Kochelev (1993), The 't Hooft topology-induced four-quark interaction leads to processes

$$u \rightarrow u(\bar{d}d), \quad d \rightarrow d(\bar{u}u)$$

but not

$$u \rightarrow u(\bar{u}u), \quad d \rightarrow d(\bar{d}d)$$

which are forbidden by Pauli principle applied to zero modes. Since there are two u quarks and only one d in the proton, one expects this mechanism to produce twice more $\bar{d}$ than $\bar{u}$.

The available experimental data, for the *difference* of the sea antiquarks distributions $\bar{d} - \bar{u}$ (from Geesaman and Reimer (2019)) is shown in Fig. 10.7. In this difference the symmetric gluon production should be cancelled out, and therefore it is sensitive only to a non-perturbative contributions.

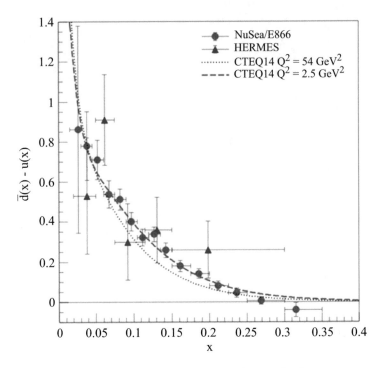

Fig. 10.7 The difference of sea antiquarks momentum distributions, $\bar{d} - \bar{u}$

Few comments:

(i) First of all, the sign of the difference is indeed as predicted by the topological interaction, there are more anti-d than anti-u quarks;

(ii) Second, since $2 - 1 = 1$, this representation of the data directly give us the nonperturbative antiquark production per valence quark, e.g. that of $\bar{u}$. This means it can be directly compared to the distribution we calculated from the five-quark tail of the nucleon and Delta baryons, Fig. 10.6.

(iii) The overall shape is qualitative similar, although our calculation has a peak at $x_{\bar{u}} \sim 0.05$ while the experimental PDFs do not indicate it. Of course, there exist higher-quark-number sectors with 7 and more quarks in baryons, which our calculation does not yet include: those should populate the small x end of the PDFs.

(iv) The data indicate much stronger decrease toward large $x_{\bar{u}}$ than the calculation.

(v) There are other theoretical models which also reproduce the flavor asymmetry of the sea, e.g. those with the pion cloud. In principle, one should be able to separate those and topology-induced mechanism (we focused here) by further combining flavor and spin asymmetry of the sea. In particular, as also noticed in Dorokhov and Kochelev (1993), if d quark producing $u\bar{u}$ pair has positive helicity, the sea quark and antiquark from 't Hooft four-quark operators must have the *opposite* (that is negative) helicity. The spin-zero pion mechanism, on the other hand, cannot transfer spin and would produce flavor but not spin sea asymmetry.

Summarizing this chapter, let me first comment that application of some model Hamiltonians to light front wave functions of hadrons should have been done long time ago. Yet it is just starting, with the model presented in this paper still being (deliberately) rather schematic, but at least including constituent quark masses, confinement and some residual interactions due to gauge topology. We have shown that local 4-fermion residual interaction does indeed modify meson and baryon wave in a very substantial way. In Jia and Vary (2019) it was shown that the π and ρ meson have very different wave functions. We now show that it is also true for baryons: the proton and the Δ have qualitatively different wave functions. The imprint left by strong diquark (ud) correlations on the light-front wave functions is now established. We furthermore found evidences that Nucleon resonances show features of "quantum chaos" in quark motion. We also seen the first steps toward the "un-quenching" the light-front wave functions, estimating multiple matrix elements of the mixings between three-quark and five-quark (baryon-pentaquark) components of the nucleon wave function. This puts mysterious spin and flavor asymmetries of the nucleon sea inside the domain of consistent Hamiltonian calculations.

10.6 Hard and Semihard Exclusive Processes

So far we discussed one use of the light-front wave functions, for parton distribution functions (PDFs) in hard processes like deep inelastic lepton-hadron scattering (DIS). In those only one of the quarks of the target hadron is "probed" by a near-pointlike probe because the spatial resolution is much smaller, $\delta x \sim 1/Q \ll 1\,\mathrm{fm}$, than a femto-meter, taken as typical hadronic size. What happens with all other hadron constituents is not considered.

Exclusive processes are the opposite case: they must lead to a well-defined final state. Out of many examples we will focus on the simplest cases of *elastic* collision of mesons, e.g. π and ρ. By "hard" processes we will mean those in which momentum transfer is very large, specifically $Q^2 > 10\,\mathrm{GeV}^2$, and "semihard" are those in the important window

$$2\,\mathrm{GeV}^2 < Q^2 < 10\,\mathrm{GeV}^2$$

In the former case the "hard block" can be described by perturbative gluon exchanges, while in the latter case perturbative and nonperturbative contributions to scattering are in general comparable. It is important that "internal" hadronic momenta and quark effective masses have a scale

$$\langle \vec{p}_\perp^2 \rangle \sim M^2 \sim 0.1\,\mathrm{GeV}^2 \ll Q^2 \qquad (10.20)$$

smaller than momentum transfer. This allows us to assume *factorization of the amplitude* into initial and final wave functions, with the "hard block" in between.

Consider elastic scattering of π meson, struck by virtual photon with large momentum. In the so-called Breit frame, the space-like photon carries $q = (0, 0, Q, 0)$ (we here put energy in the fourth place), the incoming pion carries $p = (0, 0, Q/2, \sqrt{m^2 + (Q/2)^2})$, and the outgoing pion carries $p' = (0, 0, -Q/2, \sqrt{m^2 + (Q/2)^2})$. We will however ignore the pion mass in energy, approximating it by just $Q/2$, in the ultrarelativistic "hard momentum limit". The quark momenta inside the pion are not strictly along the three axes, so there are some nonzero transverse momenta $\vec{k}_\perp \neq 0$: we however will also ignore those and write $k_q^\mu \approx x p^\mu$, with x fraction of the momentum. In the sector of only quark and antiquark in the pion, $x_q + x_{\bar{q}} = 1$.

The diagrams shown in Fig. 10.8 from Shuryak and Zahed (2020) we follow in this section indicate one-gluon exchange (a), Born-style nonperturbative contributions (b) in which a gluon is substituted by (Fourier transform of) nonperturbative fields, and diagrams with "dressed" propagators in the instanton field (c, d).

In (a) one finds the conventional Feynman diagram in empty space, so momenta are conserved along the lines and those are as indicated in the figure. The photon

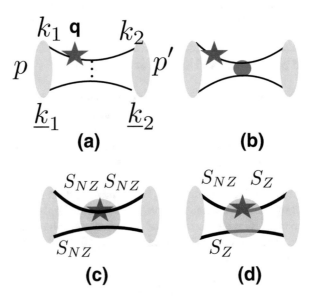

Fig. 10.8 The perturbative one-gluon exchange diagram (**a**) explains notations for momenta of quarks and mesons. The thin solid lines are free quark propagators, the red star indicates the virtual photon (or scalar) vertex, bringing in large momentum q^μ, and shaded ovals represent the (light-front) mesonic wave functions. The diagram (**b**) is a "Born-style contribution", in which the gluon propagator is substituted by the Fourier transform of the instanton field. The diagrams (**c, d**) contain three propagators in the instanton background (thick lines). In (**c**) all of them are S_{NZ} made of nonzero Dirac modes, while in (**d**) two of them are S_Z made of quark zero modes. This last contribution we will refer to as "'t Hooft-style term"

strikes a quark (star) that then turns around and strikes the antiquark with its gluon field (dashed line), turning the antiquark around as well. In contrast to DIS, *both* quarks need to be nearly at the same point in the transverse plane, because the virtuality of intermediate quark and of the gluon are both large $O(Q^2)$, and therefore they cannot travel far. Diagram (a) was calculated in the early days of QCD (Brodsky and Farrar 1973; Chernyak and Zhitnitsky 1977) with the conclusions that elastic meson form factors are such that at large Q

$$Q^2 F(Q^2) \to const(Q) \tag{10.21}$$

The gray ellipses in Fig. 10.8 stand for *light-front distribution* amplitudes depending on quark momenta

$$\Phi^\pi(x, y) = \langle 0 | T^*(q(x)[x, y]\overline{q}(y)) | \pi(p) \rangle \tag{10.22}$$

with $[x, y]$ a Wilson line. Its generic form is

$$\varphi_{\pi^+}(x) = \frac{1}{if_\pi} \int_{-\infty}^{+\infty} \frac{dz^-}{2\pi} e^{ixp\cdot z} \langle 0 | \overline{d}(0)\gamma^+\gamma_5[0, z]u(z) | \pi^+(p) \rangle$$

$$\varphi_{\pi^+}^P(x) = \frac{p^+}{f_\pi \chi_\pi} \int_{-\infty}^{+\infty} \frac{dz^-}{2\pi} e^{ixp\cdot z} \langle 0 | \overline{d}(0)i\gamma_5[0, z]u(z) | \pi^+(p) \rangle \tag{10.23}$$

$$\frac{\varphi_{\pi^+}^{T\prime}(x)}{6} = \frac{1}{f_\pi \chi_\pi} \frac{p^\mu p^{\prime\nu} p^+}{p \cdot p^\prime} \int_{-\infty}^{+\infty} \frac{dz^-}{2\pi} e^{ixp\cdot z} \langle 0 | \overline{d}(0)\sigma_{\mu\nu}\gamma_5[0, z]u(z) | \pi^+(p) \rangle$$

with all DAs normalized to 1. Note that all terms are explicitly odd under P parity, as we define the axial, pseudoscalar, and tensor terms. The first *chirally diagonal* term has a coefficient $f_\pi \approx 133$ MeV, the pion decay constant, fixed by the semi-leptonic decay amplitude. The second and third terms are *chirally non-diagonal*, and in principle they have independent DAs. Those are in fact the primary ones, surviving when chiral symmetry is restored at $T > T_c$. They also have larger coupling constant (wave function at the origin). The value of the constant χ_π can in fact be fixed by the divergence of the axial current and the PCAC relation

$$-(m_u + m_d) \langle 0 | \overline{d}(0)i\gamma^5 u(0) | \pi^+(p) \rangle \tag{10.24}$$

$$= (m_u + m_d) \operatorname{Tr}\left(i\gamma^5 \left(-\frac{if_\pi}{4}\gamma^5 \chi_\pi \right) \right) \int_0^1 dx\, \varphi^P(x)$$

$$= (m_u + m_d) f_\pi \chi_\pi \equiv f_\pi m_\pi^2$$

Using the Gell-Mann-Oakes-Renner relation,

$$f_\pi^2 m_\pi^2 = -2(m_u + m_d)\langle\bar\psi\psi\rangle \tag{10.25}$$

with $|\langle\bar\psi\psi\rangle| \approx (240\,\text{MeV})^3$, which yields

$$\chi_\pi = \frac{m_\pi^2}{(m_u + m_d)} \approx 1.2\text{--}1.5\,\text{GeV} \tag{10.26}$$

depending on what normalization point is used for quark masses.

Note that chirally non-diagonal DAs contribute formally subleading "higher twist" corrections containing χ_π^2/Q^2. However, as we will see soon, their contribution is far from being small in the kinematic region of interest. The same reasoning applies to the vector mesons, e.g. ρ^+ that we will discuss. More specifically, the (transversely polarized) rho meson density matrix reads

$$\varphi_{\rho_\perp}(x) = g_\perp^{(v)}(x) = \frac{p^+}{f_\rho m_\rho} \int_{-\infty}^{+\infty} \frac{dz^-}{2\pi} e^{ixp\cdot z} \langle 0|\bar d(0)\slashed\epsilon_\perp u(z)|\rho^+(p)\rangle$$

$$\varphi_{\rho_\perp}^T(x) = \phi_\perp(x)$$
$$= \frac{p^+}{2f_\rho^T p\cdot p'} \int_{-\infty}^{+\infty} \frac{dz^-}{2\pi} e^{ixp\cdot z} \langle 0|\bar d(0)(\slashed\epsilon_\perp\slashed p' - \epsilon_\perp\cdot p)u(z)|\rho^+(p)\rangle$$

$$\varphi_{\rho_\perp}'^A(x) = g_\perp'^{(a)}(x)$$
$$= \frac{2ip^+}{3f_\rho m_\rho p\cdot p'}\epsilon^{\mu\nu\rho\sigma}\epsilon_{T\mu}p'_{\rho}p_\sigma \int_{-\infty}^{+\infty} \frac{dz^-}{2\pi} e^{ixp\cdot z} \langle 0|\bar d(0)\gamma_\mu\gamma_5 u(z)|\rho^+(p)\rangle \tag{10.27}$$

with non-vanishing $\epsilon_\perp^\mu \neq 0$ only for $\mu = 1, 2$.

We note that both contributions $\slashed\epsilon_T m_\rho$ and $\slashed\epsilon_T\slashed p = \epsilon_{T\mu}p_\nu\sigma^{\mu\nu}$ (because $\epsilon_T^\mu p_\mu = 0$) yield a parity even rho meson vertex. Two components are, again, chirality conserving and chirality flipping, respectively. Again, $f_\rho \approx 210\,\text{MeV}$ is the rho electromagnetic decay constant. The second constant we, generally speaking, not fixed, but in the "results" section plots we assumed $\tilde f_\rho = f_\rho$. Both wave functions are normalized to 1, with

$$\langle 0|\bar d(0)\gamma^\mu u(0)|\rho_T^+(p)\rangle$$
$$= -\int_0^1 dx\,\text{Tr}\left(\gamma^\mu\left(-\frac{i}{4}\slashed\epsilon_T(f_\rho m_\rho\varphi_{\rho_T^+}(x) + \tilde f_\rho\slashed p\,\tilde\varphi_{\rho_T^+}(x))\right)\right) \equiv if_\rho m_\rho\epsilon_T^\mu \tag{10.28}$$

Using these distributions, one can calculate the diagram (a): for the pion, the result is

$$
\begin{aligned}
V_a^\pi(Q^2) = {}& \epsilon_\mu(q)(p^\mu + p'^\mu)(e_u + e_{\bar{d}}) \left[\left(\frac{2C_F \pi \alpha_s f_\pi^2}{N_c Q^2} \right) \right. \\
& \times \int dx_1 dx_2 \left(\frac{1}{\bar{x}_1 \bar{x}_2 + m_{\text{gluon}}^2/Q^2} \right) \left(\varphi_\pi(x_1)\varphi_\pi(x_2) \right. \\
& + 2\frac{\chi_\pi^2}{Q^2} \left(\varphi_\pi^P(x_1)\varphi_\pi^P(x_2) \left(\frac{1}{\bar{x}_2 + E_\perp^2/Q^2} - 1 \right) \right. \\
& \left. \left. \left. + \frac{1}{6}\varphi_\pi^P(x_1)\varphi_\pi'^T(x_2) \left(\frac{1}{\bar{x}_2 + E_\perp^2/Q^2} + 1 \right) \right) \right) \right]
\end{aligned}
\tag{10.29}
$$

Here we show explicitly electromagnetic charges $e_u = 2/3$, $e_{\bar{d}} = 1/3$, although of course the total charge of positive pion is $e_u + e_{\bar{d}} = 1$. The color matrices give the factor $C_F = (N_c^2 - 1)/2N_c = 4/3$ with $N_c = 3$ number of colors. Large spacelike photon momentum is q^μ, and $q_\mu q^\mu = -Q^2 < 0$. Photon polarization vector is $\epsilon_\mu(q)$, with $\epsilon_\mu q^\mu = 0$.

Historically, only the first "leading twist" DA ϕ_π was used in the pioneering papers. Furthermore, higher order diagrams including gluon radiation were calculated, and it was found that, in much stricter limit $\log(Q/\Lambda_{QCD}) \gg 1$, the "asymptotic wave function" is the only one surviving

$$
\phi_\pi \to \phi_\pi^{asymptotic}(\xi) = (3/4)(1 - \xi^2)
\tag{10.30}
$$

(Needless to say, this limit is very far from realistic kinematic range we discuss.) Since f_π^2 (describing quark density in the pion at the origin, in the transverse plane) is known, the formula above gave asymptotic value of the limit (10.21). While experiments and some lattice result show $Q^2 F(Q^2)$ to be approximately constant in semihard domain, yet the value remains significantly above this asymptotic prediction. In 1980s Chernyak and collaborators (Chernyak and Zhitnitsky 1984) proposed pion wave function

$$
\phi_{CZ}(x) = 30x(1 - x)(2x - 1)^2
\tag{10.31}
$$

(known as "the double-hump" one), which increases the integral as compared to the asymptotic wave function.

The integral over momentum fraction can have divergences, and therefore we kept "regulators" in denominator, the gluon mass and the quark transverse energy $E_\perp^2 = k_\perp^2 + M^2$. Note that naively the second part (with $\tilde{\phi}$) is suppressed by chi^2/Q^2 in front of the second integral, but if the wave function is finite at the edges, regulators give extra Q^2 and it is not small. As we found in Shuryak and Zahed (2020), the second term increases the perturbative answer (a) by about factor 3: and yet it is not enough! The missing contribution is the *nonperturbative contributions to the hard block* we suggest.

Naively, one can either use fields of nonperbative objects (instantons) instead of a one-gluon exchange, or put in NJL 4-fermion operators, as indicated by the diagram (b). Specific calculations performed in Shuryak and Zahed (2020) use propagators in the instanton field: those are too technical to be presented here.

Below we present some of the results, with corresponding plots for two types of elastic mesonic form factors: (a) the *vector* ones, associated with hard scattering of a photon; and (b) the *scalar* ones, associated with scattering via Higgs boson exchange. (Of course, only the photon scattering is available for laboratory experiments, while the scalar form factors are calculated in numerical lattice gauge theory simulations.) The hard block operators are different for each probe type, so for each meson we have two hard block operators, vector and scalar, with three contributions of diagrams (a, c, d) in each. In the paper Shuryak and Zahed (2020) one can find also form factors related with graviton and dilaton scattering, and also discussion of η' and its chiral partner $\delta = a_0(1540)$ scalar meson.

10.6.1 Vector Form Factors of the Pseudoscalar Mesons

We will keep the notations of the contributions as explained in Fig. 10.8. For example, the (photon-induced) *vector scattering amplitude* on the pion, with perturbative one-gluon exchange will be referred to as V_a^π, already given above in (10.29).

Here we show explicitly electromagnetic charges $e_u = 2/3, e_{\bar{d}} = 1/3$, although of course the total charge of positive pion is $e_u + e_{\bar{d}} = 1$. The color matrices give the factor $C_F = (N_c^2 - 1)/2N_c = 4/3$ with $N_c = 3$ number of colors. Large spacelike photon momentum is q^μ, and $q_\mu q^\mu = -Q^2 < 0$. Photon polarization vector is $\epsilon_\mu(q)$, with $\epsilon_\mu q^\mu = 0$. Momenta of the initial and final mesons are p and p', respectively. The pion decay constant is $f_\pi \approx 133$ MeV, which represents the wave function at the origin in transverse plane. For pion distribution we use expression (10.23) which includes not only the chirally diagonal part of the distribution $\phi_\pi(x)$ but also the chirally non-diagonal one $\tilde{\phi}_\pi(x)$. Both depend only on the longitudinal momentum fraction x of one of the quarks only, as they assumed to correspond to a two-body sector of full wave function. Here the bar indicates that the momentum fractions are those of antiquarks, $\bar{x}_i \equiv 1 - x_i$. Using asymmetry parameters, these variables are written as $x_i = (1 + \xi_i)/2, \bar{x}_i = (1 - \xi_i)/2$. The regulators are the gluon mass and quark "transverse energy".

The contribution we call the *Born-like instanton contribution* V_b^π has the same Dirac traces and can be obtained by a substitution to V_a^π by the Fourier transforms of the instanton gauge field instead of gluon propagator

$$\pi\alpha_s(Q/2) \rightarrow \kappa \left\langle \mathbb{G}^2(Q\rho\sqrt{\bar{x}_1\bar{x}_2}) \right\rangle \qquad (10.32)$$

and is therefore

$$
V_b^\pi(Q^2) = \epsilon_\mu(q)(p^\mu + p'^\mu)(e_u + e_{\bar{d}}) \left[\left(\frac{2C_F \kappa f_\pi^2}{N_c Q^2} \right) \right.
$$

$$
\times \int dx_1 dx_2 \left\langle \mathbb{G}^2(Q\rho\sqrt{x_1 x_2}) \right\rangle \left(\frac{1}{\bar{x}_1 \bar{x}_2 + m_{\text{gluon}}^2/Q^2} \right) \left(\varphi_\pi(x_1)\varphi_\pi(x_2) \right.
$$

$$
+ 2\frac{\chi_\pi^2}{Q^2} \left(\varphi_\pi^P(x_1)\varphi_\pi^P(x_2) \left(\frac{1}{\bar{x}_2 + E_\perp^2/Q^2} - 1 \right) \right.
$$

$$
\left. \left. \left. + \frac{1}{6}\varphi_\pi^P(x_1)\varphi_T'(x_2) \left(\frac{1}{\bar{x}_2 + E_\perp^2/Q^2} + 1 \right) \right) \right) \right] \tag{10.33}
$$

The contribution V_c^π, with three *non*zero mode propagators, is

$$
V_c^\pi = \epsilon_\mu(q)(p^\mu + p'^\mu)(e_u + e_{\bar{d}}) \left[\frac{\kappa\pi^2 f_\pi^2 \chi_\pi^2}{N_c M_Q^2} \langle \rho^2 \mathbb{G}_V(Q\rho) \rangle \right.
$$

$$
\left. \times \int dx_1 dx_2 \bar{x}_1 \left(\varphi_\pi^P(x_1)\varphi_\pi^P(x_2) - \frac{1}{36}\varphi_\pi'^T(x_1)\varphi_\pi'^T(x_2) \right) \right] \tag{10.34}
$$

The function G_V is averaged over the instanton size distribution, as explained in Shuryak and Zahed (2020). There is only a single integral over distribution $\tilde{\phi}$, since the other one coincides with the normalization and is just 1.

The contribution of the zero mode ('t Hooft vertex) part to the vector pion form factor is zero

$$
V_d^\pi = -\epsilon_\mu(q)(p^\mu + p'^\mu)(e_u + e_{\bar{d}}) \tag{10.35}
$$

$$
\times \left[\left(\frac{1}{N_c^2(N_c+1)} \right) \frac{4\kappa\pi^2 f_\pi^2 \chi_\pi^2}{3M_Q^2} \left\langle \rho^2 \frac{K_1(Q\rho)}{Q\rho} \right\rangle \int dx_1 dx_2 \varphi_\pi^P(x_1)\varphi_\pi'^T(x_2) \right]
$$

This contribution vanishes after the x integration is carried.

The summary plot of the vector pion form factor is shown in Fig. 10.9, for flat distributions $\phi_\pi(x) = \tilde{\phi}_\pi(x) = 1$. The perturbative contributions V_a^π (closed circles) is the *sum of both* chiral structures of the pion density matrices.

The instanton Born-style contributions to V_b^π is relatively close to V_a^π if the instanton diluteness parameter $\kappa = 1$. To avoid misunderstanding, we note that V_b amplitude does *not* really constitute a consistent account for instanton effects, as are V_c, V_d, and it is *not* shown in the summary plot.

The instanton-induced V_c^π (squares) at $\kappa = 1$ is comparable to perturbative V_a^π in magnitude but has a different dependence on Q^2. Taken together (dots) they predict the pion form factor for corresponding values of Q^2, reasonably well joining the experimental data at the lower end. We remind the reader that it is not a fit: no parameters were specially tuned for this to happen.

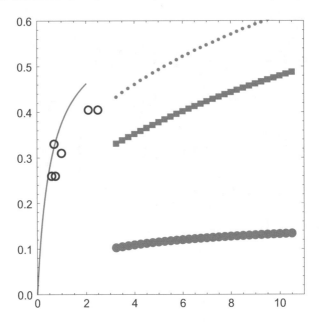

Fig. 10.9 The vector form factors of the pion times squared momentum transfer, $Q^2 F_\pi(Q^2)$ (GeV^2) versus Q^2(GeV^2). Closed discs show the total perturbative contribution. Squares correspond to the instanton contribution from nonzero mode propagators. The dotted line above is their sum. A curve in the l.h.s. is the usual dipole formula, and open points are from experimental measurements. We do not show data points at smaller Q^2, where they do agree with the dipole formula curve

10.6.2 Scalar Form Factors of the Pseudoscalar Mesons

One can think of point-like scalar quantum, hitting one of the quarks with momentum transfer q^μ to be the Higgs boson. If so, the corresponding couplings are Yukawa couplings λ_q of the standard model: but, of course, it is unimportant for form factors. (For example, lattice groups use for convenience $\lambda_u = 1, \lambda_d = 0$.).

The amplitude of elastic scattering of a virtual scalar on a pion, with a perturbative one-gluon exchange between quarks, leads to the following scattering amplitude:

$$S_a^\pi = -(\lambda_u + \lambda_d)M_Q\left[\left(\frac{2\pi C_F \alpha_s f_\pi^2 \chi_\pi}{N_c Q^2 M_Q}\right)\int_0^1 \frac{dx_1 dx_2}{\bar{x}_1 \bar{x}_2 + m_{\text{gluon}}^2/Q^2}\right.\tag{10.36}$$

$$\left.\times\left(\varphi_\pi(x_1)\varphi_\pi^P(x_2)\left(1+\frac{2}{\bar{x}_2 + E_\perp^2/Q^2}\right)+\frac{1}{6}\varphi_\pi(x_1)\varphi_\pi^{T'}(x_2)\right)\right]$$

Note that scalar amplitudes have negative overall sign, which does not matter as couplings λ_q are arbitrary. This sign of course does not affect contribution to the form factor, the square bracket.

In the previous section, on the vector form factor, it was obvious that the factors
with charges and momenta in the amplitude do not belong to form factors, as they
are also present in the forward scattering at $Q = 0$. The situation with the scalar
form factors is a bit different. The forward scattering amplitude on a hadron h is
proportional to

$$\sum_q \lambda_q \langle h | \bar{q} q | h \rangle = - \sum_q \lambda_q \frac{\partial M_h^2}{\partial m_q} \qquad (10.37)$$

thanks to the Feynman-Hellman theorem. The derivative appearing here is known
for the pion from the Gell-Mann-Oaks-Renner relation, and for most hadrons from
lattice chiral extrapolations.

In the scalar plots to follow, we will the show square brackets in the amplitudes
times Q^2 without the factors in front, as we did for the vector cases. Indeed, our main
focus is on the relative magnitude of different contributions. However the reader
should be cautioned that the true scalar form factor $F_S(Q^2)$ requires multiplication
by an additional factor

$$K_S = \frac{\sum_q \lambda_q \partial M_h^2 / \partial m_q}{M_Q \sum_q \lambda_q} \qquad (10.38)$$

to enforce the standard form factor normalization $F_S(Q = 0) = 1$.

The contribution of the instanton-induced diagram (c) (three nonzero mode
propagators) is

$$S_c^\pi (Q^2) = -(\lambda_u + \lambda_{\bar{d}}) M_Q \left[\left(\frac{\kappa \pi^2 \chi_\pi f_\pi^2}{N_c M_Q^3} \right) \left\langle (Q\rho)^2 \mathbb{G}_S(Q\rho) \right\rangle \right.$$

$$\left. \times \int_0^1 dx_1 dx_2 \, \bar{x}_2 \varphi_\pi (x_1) \left(\varphi_\pi^P (x_2) - \frac{1}{6} \varphi_\pi^{T'}(x_2) \right) \right] \qquad (10.39)$$

Note that unlike V_c^π, here there is another form factor $\mathbb{G}_S$ that is a part of $\mathbb{G}_V$.
The contribution from the zero modes and the 't Hooft vertex

$$S_d^\pi (Q^2) = -(\lambda_u + \lambda_{\bar{d}}) M_Q \left[\left(\frac{1}{N_c^2(N_c + 1)} \right) \left(\frac{\kappa \pi^2 f_\pi^2 \chi_\pi}{M_Q^3} \right) \langle Q\rho K_1(Q\rho) \rangle \right.$$

$$\times \int_0^1 dx_1 dx_2 \left(x_1 \varphi_\pi (x_2) \left(\varphi_\pi^P (x_1) + \frac{1}{6} \varphi_\pi^{T'}(x_1) \right) \right.$$

$$\left. \left. + x_2 \varphi_\pi (x_1) \left(\varphi_\pi^P (x_2) + \frac{1}{6} \varphi_\pi^{T'}(x_2) \right) \right) \right] \qquad (10.40)$$

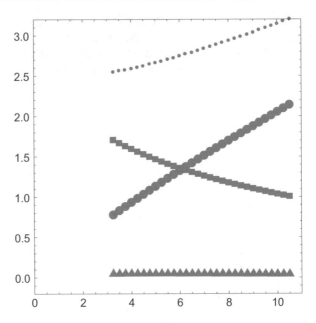

Fig. 10.10 The scalar form factors of the pion times momentum transfer squared, $Q^2 F_\pi^S(Q^2)$ (GeV^2) versus Q^2(GeV^2). Black closed points is the perturbative contribution, and black squares correspond to the instanton contribution of three non-zero mode propagators, black triangles are due to diagram d with two zero mode propagators

Perturbative and instanton contributions to scalar form factor of the pion are plotted versus Q^2 in Fig. 10.10. Again, one finds them to be comparable in magnitude but quite different in Q dependence.

10.6.3 Form Factors of Transversely Polarized Vector Mesons

The transversely polarized rho meson form factors are both electric and magnetic (see below). For simplicity, we quote here the contribution to the electric or charge form factor by choosing the transverse polarization $\epsilon_T(p, p')$ of the ρ_T with momentum p, p' to be also transverse to q, or $\epsilon_T(p, p') \cdot q = 0$.

The perturbative contribution (a) for the transversely polarized rho vector form factor is formally subleading (containing extra m_ρ^2/Q^2), like the χ_π^2/Q^2 in the second term of V_a^π:

$$V_a^\rho(Q^2) = \epsilon_\mu(q)(p^\mu + p'^\mu)(e_u + e_{\bar{d}})(-\epsilon_T'^* \cdot \epsilon_T)\left(\frac{2\pi C_F \alpha_s f_\rho^2 m_\rho^2}{N_c Q^4}\right)$$

$$\times \int_0^1 \frac{dx_1 dx_2}{\bar{x}_1 \bar{x}_2 + m_{\text{gluon}}^2/Q^2}$$

$$\times\left[\left(\varphi_\rho(x_1)\varphi_\rho(x_2) - \frac{1}{16}\varphi_\rho^{A'}(x_1)\varphi_\rho^{A'}(x_2)\right)\right.$$

$$\times\left(\frac{1}{\bar{x}_1 + E_\perp^2/Q^2} + \frac{1}{\bar{x}_2 + E_\perp^2/Q^2} - 2\right)$$

$$\left.+\frac{1}{2}\varphi_\rho^{A'}(x_1)\varphi_\rho(x_2)\left(\frac{1}{\bar{x}_1 + E_\perp^2/Q^2} - \frac{1}{\bar{x}_2 + E_\perp^2/Q^2}\right)\right] \quad (10.41)$$

Note that the overall minus sign here does not mean that this amplitude is negative, since the product of two polarization vectors is negative $(\epsilon_T^* \cdot \epsilon_T') < 0$ in Minkowski metrics used (Fig. 10.11).

The one-gluon exchange to the scalar form factor of the transverse rho meson is

$$S_a^\rho(Q^2) = -(\lambda_u + \lambda_{\bar{d}}) M_Q (-\epsilon_T'^* \cdot \epsilon_T)\left[\left(\frac{\pi C_F \alpha_s}{N_c}\frac{m_\rho f_\rho f_\rho^T}{M_Q Q^2}\right)\right.$$

$$\times \int \frac{dx_1 dx_2}{\bar{x}_1\bar{x}_2 + m_{\text{gluon}}^2/Q^2}\left(\frac{\varphi_\rho^T(x_1)(\varphi_\rho(x_2) - \varphi_\rho^{A'}(x_2)/4)}{\bar{x}_1 + E_\perp^2/Q^2}\right.$$

$$\left.\left.+\frac{(\varphi_\rho(x_1) - \varphi_\rho^{A'}(x_1)/4)\varphi_\rho^T(x_2)}{\bar{x}_2 + E_\perp^2/Q^2}\right)\right] \quad (10.42)$$

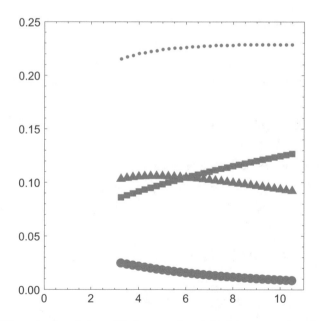

Fig. 10.11 The vector form factors of the transversely polarized rho meson, times the momentum transfer squared, $Q^2 F_\pi^S(Q^2)$ (GeV^2) versus Q^2(GeV^2). Black closed points show the perturbative contribution, black triangles correspond to instanton zero mode ('t Hooft vertex) contribution, and squares are for contribution of nonzero mode propagators

Fig. 10.12 The scalar form factors of the transversely polarized rho meson times momentum transfer squared, $Q^2 F_\pi^S(Q^2)$ (GeV^2) versus Q^2(GeV^2). Black closed points stand for the perturbative contribution, Black squares are for instanton contribution of three nonzero mode propagators, black triangles are due to diagram d with two zero mode propagators

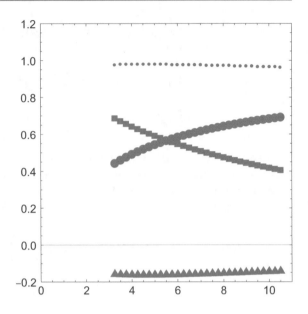

The *Born-like instanton contribution* to the rho vector form factor V_b^ρ is also given by the substitution of (10.32) to V_a^ρ and is thus not given. Perturbative and nonperturbative contributions to scalar propagator of transversely polarized ρ are shown in Fig. 10.12.

The contribution of the *nonzero mode* propagators to rho meson vector form factor is

$$V_c^\rho(Q^2) = \epsilon_\mu(q)(p^\mu + p'^\mu)(e_u + e_{\bar{d}})(-\epsilon_T'^* \cdot \epsilon_T)\left[\frac{\kappa\pi^2 f_\rho^2 m_\rho^2}{N_c M_Q^2}\left\langle \rho^2 \mathbb{G}_V(Q\rho)\right\rangle\right.$$

$$\left.\times \int_0^1 dx_1 dx_2\, \bar{x}_1\left(\varphi_\rho(x_1) - \frac{\varphi_\rho^{A'}(x_1)}{4}\right)\left(\varphi_\rho(x_2) + \frac{\varphi_\rho^{A'}(x_2)}{4}\right)\right] \quad (10.43)$$

and the contribution to the rho meson scalar form factor is

$$S_c^\rho(Q^2) = -(\lambda_u + \lambda_{\bar{d}}) M_Q (-\epsilon_T'^* \cdot \epsilon_T)\left[\left(\frac{\kappa\pi^2 f_\rho f_\rho^{T2} m_\rho}{4N_c M_Q^4}\right)\left\langle (Q\rho)^2 \mathbb{G}_S(Q\rho)\right\rangle\right.$$

$$\times \int dx_1 dx_2\left(\bar{x}_1\left(\varphi_\rho(x_1) - \frac{\varphi_\rho^{A'}(x_1)}{4}\right)\varphi_\rho^T(x_2)\right.$$

$$\left.\left.+ \bar{x}_2\left(\varphi_\rho(x_2) - \frac{\varphi_\rho^{A'}(x_2)}{4}\right)\varphi_\rho^T(x_1)\right)\right] \quad (10.44)$$

The contribution of the *'t Hooft vertex* to the vector form factor of the transversely polarized rho is

$$V_d^\rho(Q^2) = -(e_u + e_{\bar{d}})\left(\epsilon_\mu(q)(p^\mu + p'^\mu)(\epsilon_T'^* \cdot \epsilon_T)\right)$$

$$\left[\langle Q\rho\, K_1(Q\rho)\rangle\left(\frac{2\kappa\pi^2}{N_c^2(N_c+1)}\frac{f_\rho^{T2}}{M_Q^2}\right)\right] \tag{10.45}$$

The contribution of the *'t Hooft vertex* to the scalar form factor is

$$S_d^\rho(Q^2) = +(\lambda_u + \lambda_{\bar{d}})\, M_Q\,(-\epsilon_T'^* \cdot \epsilon_T)$$

$$\times\left[\left(\frac{1}{N_c^2(N_c+1)}\right)\left(\frac{\kappa\pi^2 f_\rho f_\rho^T m_\rho}{M_Q^3}\right)\langle Q\rho K_1(Q\rho)\rangle\right.$$

$$\times \int_0^1 dx_1 dx_2\left(x_1\varphi^T(x_2)\left(\varphi_\rho(x_1) + \frac{1}{4}\varphi_\rho^{A'}(x_1)\right)\right.$$

$$\left.\left.+x_2\varphi_\rho^T(x_1)\left(\varphi_\rho(x_2) + \frac{1}{4}\varphi_\rho^{A'}(x_2)\right)\right)\right] \tag{10.46}$$

10.7 Brief Summary

- The chapter starts with brief discussion of hadron spectroscopy and related quark models, formulated in the rest frame.
- Then we discuss the so called light-front observables, such as parton distribution functions (PDFs). Since for interacting systems boost is a complicated operator, instead of using boosted rest frame wave functions one may diagonalize instead a Hamiltonian defined directly on the light-front. Those should not be confused with 3d wave functions discussed in the previous chapter: the light-front ones have separate dependence on 2d transverse and longitudinal coordinates.
- We discussed such calculations for light mesons (π, ρ, η'), baryons Δ, N and pentaquarks.
- we further have shown how one can calculate admixture of five-quark states to the leading three-quark configuration of the nucleons. For example, a proton can be viewed with admixture $uud + uud(\bar{u}u)$, from which PDF of $\bar{u}$ can be calculated and compared with the experiment.
- Light-front distributions appear in exclusive processes, provided the so called *factorization* of the process can be defined, as a convolution between *hard block* at scale Q and the wave functions.
- As shown in last 4 plots, the instanton-induced contributions to hard block are indeed *comparable* to perturbative ones in magnitude.
- They play quite different roles in the wave functions and form factors. Indeed, the wave function of the pions are dominated by 't Hooft-like interactions, and

not so for ρ mesons. But in form factors this contribution to hard block vanishes for the pion, and is present for ρ.

- Even perturbative one-gluon exchange contribution must include *chiral-nondiagonal* part of the wave functions. Formally higher twist, they are not at all small in the semihard region discussed. Their presence is e.g. crucial for the very existence of *scalar* form factors.
- The instanton-induced and perturbative contributions have quite different Q dependence. So, once the (lattice) data for these form factors be known, one would be able to tell them apart. At the moment, in semihard domain $Q^2 = 2$–$10\,\mathrm{GeV}^2$ there are only results from Davies et al. (2018) collaboration, for charm and strange quarks.
- One may wander if one can somehow generalize the instanton-induced effects to arbitrary NJL-type quark-quark local forces. It cannot strictly speaking be done, unless NJL provide well defined propagators, like the one used for instantons. Yet some approximations may perhaps be developed.

References

Braun, V.M.: Nucleons on the light-cone: theory and phenomenology of baryon distribution amplitudes. In: Continuous Advances in QCD. Proceedings, 7th Workshop, QCD 2006, Minneapolis, May 11–14, 2006, pp. 42–57 (2006)

Brodsky, S.J., Farrar, G.R.: Scaling laws at large transverse momentum. Phys. Rev. Lett. **31**, 1153–1156 (1973)

Chernyak, V., Zhitnitsky, A.: Asymptotic behavior of hadron form-factors in quark model (In Russian). JETP Lett. **25**, 510 (1977)

Chernyak, V.L., Zhitnitsky, A.R.: Asymptotic behavior of exclusive processes in QCD. Phys. Rept. **112**, 173 (1984)

Chernyak, V.L., Ogloblin, A.A., Zhitnitsky, I.R.: Calculation of exclusive processes with baryons. Z. Phys. **C42**, 583 (1989). [Sov. J. Nucl. Phys.48,889(1988)]

Davies, C., Koponen, J., Lepage, P.G., Lytle, A.T., Zimermmane-Santos, A.C.: Meson electromagnetic form factors from lattice QCD. PoS Lattice **2018**, 298 (2018)

Dorokhov, A.E., Kochelev, N.I.: Instanton induced asymmetric quark configurations in the nucleon and parton sum rules. Phys. Lett. **B304**, 167–175 (1993)

Geesaman, D.F., Reimer, P.E.: The sea of quarks and antiquarks in the nucleon. Rept. Prog. Phys. **82**(4), 046301 (2019)

Jia, S., Vary, J.P.: Basis light front quantization for the charged light mesons with color singlet Nambu–Jona-Lasinio interactions. Phys. Rev. **C99**(3), 035206 (2019)

Kharzeev, D.E., Levin, E.M.: Deep inelastic scattering as a probe of entanglement. Phys. Rev. **D95**(11), 114008 (2017)

Nambu, Y., Jona-Lasinio, G.: Dynamical model of elementary particles based on an analogy with superconductivity. 1. Phys. Rev. **122**, 345–358 (1961) [127(1961)]

Radyushkin, A.: Deep Elastic Processes of Composite Particles in Field Theory and Asymptotic Freedom (1977)

Rapp, R., Schafer, T., Shuryak, E.V., Velkovsky, M.: Diquark bose condensates in high density matter and instantons. Phys. Rev. Lett. **81**, 53–56 (1998). hep-ph/9711396

Shuryak, E.: Light-front wave functions of mesons, baryons and pentaquarks, with topology-induced local 4-quark interaction. Phys. Rev. D **100**, 114018 (2019)

Shuryak, E., Zahed, I.: The nonperturbative quark-antiquark interactions in mesonic formfactors (2020)

The Topological Landscape and the Sphaleron Path

11

11.1 The Sphalerons

Let me begin with a qualitative picture. Like monopoles, sphalerons are 3-d magnetic objects made of $SU(2)$ non-Abelian gauge fields. They are "solitons", solutions of classical Yang-Mills eqn. Unlike monopoles, their magnetic field lines do not go radially (in a "hedgehog way") but form circles around the center, see Fig. 11.1. Three colors (red, blue and green) on the picture depict not the number of colors (which is two) but three adjoint fields (due to three generators of $SU(2)$). As the picture suggests, each field type rotates around a different axis.

Historically existence of such solution have been suggested by Dashen et al. (1974) as an extended soliton for the Young-Mills-Higgs equations of motion in the electroweak sector of the Standard Model (Weinberg-Salam theory), as a 3d soliton alternative to the t'Hooft-Polyakov monopole in Georgi-Glashow model. A decade later (Klinkhamer and Manton 1984) (KM) have looked for an explicit solution, variationally and numerically (in note-added-in-proofs). As all other non-perturbative solitons, the field is $O(1/g)$ and the mass $O(1/g^2)$ times the relevant scale. The electroweak coupling is small and the mass of the KM sphaleron solution turned out to be as large as[1]

$$M_{KM} \approx 8\,TeV \tag{11.1}$$

The fact that only numerical solution was found is related to the fact that it is not represented by any simple spherically symmetric shape, as it gets deformed by

---

[1] Since it is smaller than current $\sqrt{s}$ of LHC one may ask whether one can expect its production there. Unfortunately, even if one would build collider with much higher energy available, there are no chances to produce this object experimentally, due to vanishingly small overlap with two colliding proton state.

© The Author(s), under exclusive license to Springer Nature Switzerland AG 2021
E. Shuryak, *Nonperturbative Topological Phenomena in QCD and Related Theories*, Lecture Notes in Physics 977,
https://doi.org/10.1007/978-3-030-62990-8_11

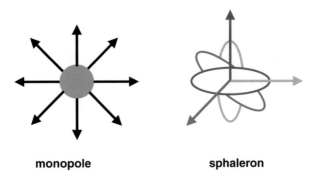

monopole   **sphaleron**

Fig. 11.1 Qualitative distinction between the magnetic field in monopole and spalerons

a specific direction taken by the Higgs field in broken electroweak vacuum with nonzero Higgs VEV. Indeed one of the components, Z boson, does not have the same mass as the other two, $W^{\pm}$, which makes it "elliptic". For early review consult e.g. McLerran (1989).

It has been subsequently found that, while this configuration solves the equation of motion, it is unstable. Unlike other solitons discussed so far—fluctons, monopoles and instantons—which are *minima* of the action, the sphalerons are *saddle points* in a space of all configurations. While for fluctons the quadratic operator of the fluctuations had only positive modes, and for instantons positive and zero modes, for sphalerons one (or more) *negative* mode appears in the spectrum. Thus their name, *sphaleron*, which (according to its discoverers) means in Greek "ready to fall".

In this chapter we will not follow these original papers. First of all, our main focus will be on QCD applications, where complications related with Higgs and its VEV are absent. Even in section devoted to electroweak sphalerons and cosmological baryogenesis, it would be sufficient for us to consider a much simpler analytic solution found later for pure gauge sector (Ostrovsky et al. 2002) and known as the "COS sphaleron". The way toward it was historically related to the instanton-antiinstanton configurations, which we need to consider first.

11.2   Instanton-Antiinstanton Interaction and the "Streamline" Set of Configurations

In the Chap. 6 devoted to interacting instanton ensembles we skipped the discussion of the instanton-antiinstanton interaction, which we will discuss now.[2]

---

[2]Interaction between instantons does not exist at the classical level, as is clear from the relation between the action and the topological charge.

Returning to the double-well potential, let us start with a simple "sum ansatz"

$$q_{sum}(\tau) = \frac{1}{g}\left(\frac{1}{\tanh(\tau - \tau_I)} - \frac{1}{\tanh(\tau - \tau_A)} - 1\right). \tag{11.2}$$

This path has the action $S_{IA}(T) = 1/g^2(1/3 - 2e^{-T} + O(e^{-2T}))$, where $T = |\tau_I - \tau_A|$. It is qualitatively clear that if the two instantons are separated by a large time interval $T \gg 1$, the action $S_{IA}(T)$ is close to $2S_0$. In the opposite limit $T \to 0$, the instanton and the antiinstanton annihilate and the action $S_{IA}(T)$ should tend to zero. In that limit, however, the IA pair is at best an approximate solution of the classical equations of motion and it is not clear how the path should be chosen.

The best way to deal with this problem is the "streamline" method (Balitsky and Yung 1986). To define them, one starts with an extremum of the action (in this case, infinitely separated IA pair) and lets the system evolve with the "gradient flow" to smaller action. The "force" is defined as $f(\tau) = \partial S[x]/\partial x(\tau)$ and the "streamline equation" is sliding along the direction of the force

$$\frac{dx(\tau)}{dt} = f(\tau) \tag{11.3}$$

Here t is extra "sliding time", not to be confused with the Euclidean time τ on which all paths depend. At zero t there are well-separated instanton and antiinstanton, at $t \to \infty$ they annihilate each other to a trivial path $x(\tau) = 0$. A set of such paths for the double-well instantons were obtained numerically in my paper (Shuryak 1988).

For the gauge theory instantons one can start with the simplest *sum ansatz*

$$\frac{g}{2}A_\mu^a = \frac{\bar{\eta}_{a\mu\nu}y_I^\nu\rho^2}{y_I^2(y_I^2+\rho^2)} + \frac{\eta_{a\mu\nu}y_{\bar{I}}^\nu\rho^2}{y_{\bar{I}}^2(y_{\bar{I}}^2+\rho^2)} \tag{11.4}$$

where, for simplicity, we selected the same radii and orientation of both solitons. Note however that while for a single instanton the pure gauge singularity at the origin cancels in the expression for the fields, it does not do so for the sum ansatz. This can be cured in the following *ratio ansatz*

$$\frac{g}{2}A_\mu^a = \frac{\bar{\eta}_{a\mu\nu}y_I^\nu\rho^2/y_I^2 + \eta_{a\mu\nu}y_{\bar{I}}^\nu\rho^2/y_{\bar{I}}^2}{1 + \rho^2/y_{\bar{I}}^2 + \rho^2/y_I^2} \tag{11.5}$$

In search for better approximation (Verbaarschot 1991), using conformal invariance of classical Yang-Mills equation, mapped this problem into a $co-central$ instanton and antiinstanton with different radii ρ_1, ρ_2. This introduced a notion that the action dependence on the 4-d distance between the centers and these two sizes may only come in a conformal invariant dimensionless combination[3]

$$X^2_{conf} = \frac{(z^I_\mu - z^{\bar{I}}_\mu)^2 + (\rho_1 - \rho_2)^2}{\rho_1 \rho_2} \tag{11.6}$$

The gradient flow (streamline) equation has the same meaning as in QM: the force is substitute by the current $j^a_\mu = \partial S / \partial A^a_\mu$ and then

$$\frac{d A^a_\mu(x)}{dt} = j^a_\mu \tag{11.7}$$

The "streamline" set of configurations, going along the gradient, is known in mathematics as *Lefschetz thimbles*, special lines connecting extrema in the complex plain.

In co-centrical setting it has been reduced to a single-variable and solved numerically. Verbaarschot also discovered that the so called *Yung ansatz*

$$
\begin{aligned}
ig A^{Yung}_\mu(x) &= ig A^{Yung}_{a\mu}(x) \frac{\tau^a}{2} \\
&= \frac{\bar{\tilde{y}}_2}{\sqrt{\tilde{y}_2}} \frac{R}{\sqrt{R^2}} \frac{(\bar{\sigma}_\mu y_1 - y^\mu_1) \rho^2_1}{y^2_1 (y^2_1 + \rho^2_1)} \frac{\bar{R}}{\sqrt{R^2}} \frac{\tilde{y}_2}{\sqrt{\tilde{y}_2}} \\
&\quad + \frac{(\bar{\sigma}_\mu y_2 - y^\mu_2) \rho^2_2}{y^2_2 + \rho^2_2} + \frac{\rho_1 \rho_2}{z y^2_1 (y^2_2 + \rho^2_2)} \\
&\quad \times \left[(\bar{\sigma}_\mu y_1 - y^\mu_1) - \frac{\bar{\tilde{y}}_2}{\sqrt{\tilde{y}_2}} \right. \\
&\quad \left. \times \frac{R}{\sqrt{R^2}} (\bar{\sigma}_\mu y_1 - y^\mu_1) \frac{\bar{R}}{\sqrt{R^2}} \frac{\tilde{y}_2}{\sqrt{\tilde{y}_2}} \right],
\end{aligned}
\tag{11.8}
$$

is numerically close to his streamline solution.

A word of explanation of notations here: all vectors without an indicative index are SU(2) matrices obtained by their contraction with the vector $\sigma_\mu = (1, -i\vec{\tau})$, for example $R = x_1 - x_2 = R_\mu \sigma_\mu$. An overbar similarly denotes contraction with $\bar{\sigma} = (1, i\vec{\tau})$. Note that barred and unbarred matrices always alternate, in all terms; this is because one index of each matrix is dotted and the other not, in spinor notation. The additional coordinate with tilde is

$$\tilde{y}_2 = x_2 - \frac{R \rho_2}{z \rho_1 - \rho_2}. \tag{11.9}$$

---

[3] Which has the meaning of the geodesic distance between points in the AdS_5 space, with the sizes ρ_i identified with the 5-th coordinates. We will return to this point in discussion of AdS/CFT correspondence.

The u here stands for relative orientation SU(2) color matrix parameterized by $u_\mu \sigma_\mu$, and $u \cdot \hat{R}$ is its projection to unit relative distance vector. For same orientation of the instanton and antiinstanton $u = (0, 0, 0, 1)$.

This expression represents a very good approximation to the streamline equation not only at large distances $(z_\mu^I - z_\mu^{\bar{I}})^2 \gg \rho^2$, as claimed by Yung in the original paper, but also at *all* distances as well. In fact at distance zero the formula produces a very complicated field A_μ, which however after inspection was found to be a pure gauge, with zero field strength!

The interaction for this ansatz (Verbaarschot 1991) is

$$
S_{IA} = \frac{8\pi^2}{g^2} \frac{1}{(\lambda^2 - 1)^3} \left\{ -4 \left(1 - \lambda^4 + 4\lambda^2 \log(\lambda)\right) \left[|u|^2 - 4|u \cdot \hat{R}|^2\right] \right.
\tag{11.10}
$$

$$
\left. +2 \left(1 - \lambda^2 + (1 + \lambda^2) \log(\lambda)\right) \left[(|u|^2 - 4|u \cdot \hat{R}|^2)^2 + |u|^4 + 2(u)^2(u^*)^2\right] \right\},
$$

where

$$
\lambda = \frac{R^2 + \rho_1^2 + \rho_2^2}{2\rho_1 \rho_2} + \left(\frac{(R^2 + \rho_1^2 + \rho_2^2)^2}{4\rho_1^2 \rho_2^2} - 1\right)^{1/2}.
\tag{11.11}
$$

is related to the conformal distance parameter defined above. Large distance $R \gg \rho$ is large λ, zero distance at $\rho_1 = \rho_2$ is $\lambda = 1$.

11.3 From the Instanton-Antiinstanton Configurations to the Sphaleron Path

In the previous section we defined some set of configurations describing an instanton and an antiinstanton, placed at a certain (Euclidean time) positions $\tau = -T/2$ and $\tau = T/2$ (see Fig. 11.2).

Fig. 11.2 The instanton and antiinstanton are placed at $\tau = -T/2$ and $\tau = T/2$ along the Euclidean time axes. The sphaleron path configurations are located at the middle plane $\tau = 0$

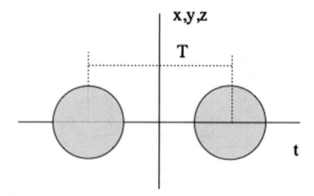

Using those, we will deduce some special set of 3d configurations possessing only *magnetic* fields. Indeed, the instanton fields are self-dual, $\vec{E} = \vec{B}$, while the antiinstanton ones are anti-selfdual, $\vec{E} = -\vec{B}$. By symmetry, at the 3-d plane $\tau = 0$ in between, the electric field must vanish. Therefore, at this 3-d plane the field is purely magnetic.

Furthermore, these configurations make a one-dimensional set, parameterized by the time distance T between the instanton and the anti-instanton. For each of those, one can define their energy normalized by size

$$ E R = \frac{1}{2} \left[\int d^3 r \, r^2 \mathcal{B}^2 \times \int d^3 r \mathcal{B}^2 \right]^{1/2}. \tag{11.12} $$

and the Chern–Simons number $N_{CS}(T)$. It has been calculated by Ostrovsky et al. (2002) for the "streamline" Yung ansatz, and shown in Fig. 11.3. Note that at $N_{CS} = 0$ (corresponding to large distance T) the energy is zero, and it is also tend to zero at $N_{CS} = 1$. The maximum is in the middle, at $N_{CS} = 1/2$, when the instanton and anti-instanton "half-overlap". These configurations constitute the so called *sphaleron path*, leading from one classical vacuum to the next.

Let us now look at this problem from a different point of view. In the $A_0 = 0$ gauge the electric field is given by the time derivative of A_m. Therefore the electric field has a meaning of momenta, conjugated to coordinates A_m, and E^2 term in

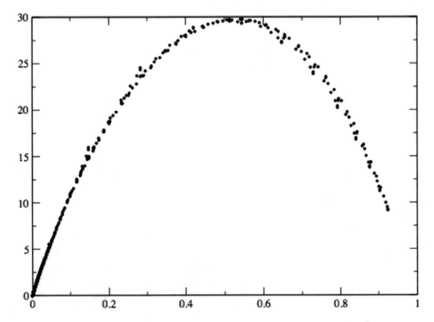

Fig. 11.3 The energy times the r.m.s, radius $E \cdot R$ versus Chern–Simons number $\tilde{N}_{CS}$, for the 3-d configurations obtained from the instanton-antiinstanton Yung ansatz as described in the text. The curve should be compared to the sphaleron path we soon obtain from constrained minimization

energy interpreted as field kinetic energy. Pure magnetic field configurations with zero electric field, have thus only the potential energy.

One can view those as some *turning points*, at which momentum is zero. Indeed, such points connect the tunneling part of the path at $E < V$ with the real time evolution at $E > V$. In fact the action on the plane can be viewed as given the *probability* (rather than the amplitude) of such forced tunneling, with a vertical line in the Fig. 11.2 literally interpreted as the *unitarity cut* describing the field configuration ready to propagate into the Minkowski time.

11.4 The Sphaleron Path from a Constrained Minimization

By "sphaleron path" we mean a one-parameter set of configurations, interpolating between the nearest values of Chern–Simons number, e.g. from $N_{CS} = 0$ to $N_{CS} = 1$. Because of symmetry of the barrier we are trying to calculate, it is expected that the maximum—the sphaleron – corresponds to $N_{CS} = 1/2$.

Naively, what needs to be done is to fix N_{CS} to some value, and then find among such configurations the one with *minimal possible energy*. This is indeed what was done in electroweak theory. The expected mass scale is $\sim v/\alpha_{EW}$ is defined by the Higgs VEV $v \approx 1/4\,TeV$ and the electroweak coupling constant.

We however mentioned above, that in all application we will discuss the Higgs VEV, or even existence of the Higgs fields itself, will not be important. The problems we want to address are dominated by the gauge fields: thus it is natural to ask whether one can identify sphaleron path in *pure gauge theory*.

Unlike electroweak theory, classical pure gauge theory is scale invariant, it has no dimensional parameters. Naively this would imply that all minima are at zero value, because one can always reduce the energy by rescaling the size of the configuration upward. To break this unwanted scale symmetry, one needs to set an additional requirement, basically fixing the size of the soliton in question. It can e.g. be defined as r.m.s. radius

$$< r^2 > = \frac{\int d^3 x\, r^2 \mathcal{B}^2}{\int d^3 x \mathcal{B}^2} \tag{11.13}$$

If it is fixed, there will be a particular static solution with the minimal energy we are looking for. The product of the energy times r.m.s. size is dimensionless, and that will be what we will evaluate below. We will follow in this section the work by Ostrovsky et al. (2002), and will find the shape of the barrier.

Since we are interested in static 3d configurations, those would be purely made up of magnetic fields, since the electric fields have negative T-parity and thus not allowed. As a starting simplifying assumption, we will consider a spherically symmetric 3-d configuration of the gauge field. Indeed, often the field configurations with the minimal energy have the maximal possible symmetry. Of course, we expect the energy density $(\vec{B}^a)^2$—and not the gauge fields themselves—be spherically symmetric.

For the SU(2) color subgroup in which we are interested, configurations of the gauge field $\mathcal{A}_\mu^a$ can be expressed through the following four space-time $(0, j = 1..3)$ and color $(a = 1..3)$ structures

$$\mathcal{A}_j^a = A(r, t)\Theta_j^a + B(r, t)\Pi_j^a + C(r, t)\Sigma_j^a$$

$$\mathcal{A}_0^a = D(r, t)\frac{x^a}{r} \tag{11.14}$$

with three mutually orthogonal projectors

$$\Theta_j^a = \frac{\epsilon_{jam}x^m}{r}, \quad \Pi_j^a = \delta_{aj} - \frac{x_a x_j}{r^2}, \quad \Sigma_j^a = \frac{x_a x_j}{r^2}. \tag{11.15}$$

While for the sphaleron path problem the four functions should be static (independent on time t), the attentive reader would notice that we included the time. The reason for it is that later on we will also discuss a dynamical problem of the *sphaleron explosion*.

One may rewrite a problem with r and t- dependent functions as some 1+1 dimensional Lagrangian. In fact this is true for the problem at hand, and our four functions of Eq. (11.14) can be rewritten as four fields of the Abelian gauge-Higgs model ($A_{\mu=0,1}$, ϕ, α) on a hyperboloid :

$$A = \frac{1 + \phi\sin\alpha}{r}, \quad B = \frac{\phi\cos\alpha}{r}, \quad C = A_1, \quad D = A_0. \tag{11.16}$$

One can express the field strengths in these terms as

$$\begin{aligned}\mathcal{E}_j^a = \mathcal{G}_{0j}^a &= \frac{1}{r}[\partial_0\phi\sin\alpha + \phi\cos\alpha(\partial_0\alpha - A_0)]\Theta_j^a \\ &+ \frac{1}{r}[\partial_0\phi\cos\alpha - \phi\sin\alpha(\partial_0\alpha - A_0)]\Pi_j^a \\ &+ (\partial_0 A_1 - \partial_1 A_0)\Sigma_j^a \end{aligned} \tag{11.17}$$

and

$$\begin{aligned}\mathcal{B}_j^a = \frac{1}{2}\epsilon_{jkl}\mathcal{G}_{kl}^a &= \frac{1}{r}[-\partial_1\phi\cos\alpha + \phi\sin\alpha(\partial_1\alpha - A_1)]\Theta_j^a \\ &+ \frac{1}{r}[\partial_1\phi\sin\alpha + \phi\cos\alpha(\partial_1\alpha - A_1)]\Pi_j^a \\ &+ \frac{1 - \phi^2}{r^2}\Sigma_j^a, \end{aligned} \tag{11.18}$$

where $\partial_0 \equiv \partial_t$ and $\partial_1 \equiv \partial_r$. Putting those expression into the usual 3+1 dimensional Minkowski action and integrating over angles one find reduced 1+1d action

$$S = \frac{1}{4g^2} \int d^3x dt \left[\left(B_j^a \right)^2 - \left(\mathcal{E}_j^a \right)^2 \right]$$

$$= 4\pi \int dr dt \left[(\partial_\mu \phi)^2 + \phi^2 (\partial_\mu \alpha - A_\mu)^2 \right.$$

$$\left. + \frac{(1 - \phi^2)^2}{2r^2} - \frac{r^2}{2} (\partial_0 A_1 - \partial_1 A_0)^2 \right], \qquad (11.19)$$

with the summation now over the 1+1 dimensional indices. The t, r space is with the $(-, +)$ metric. Note that ϕ and α now have a meaning of the modulus and phase of some charged scalar. The charge is Abelian, as seen from the last term containing the "field strength" squared.

What exactly does this elegant re-writing of the Lagrangian in new fields give us? Well, it helps to understand better the remaining symmetries of the model. The spherical ansatz is preserved by a set of gauge transformations generated by unitary matrices of the type

$$U(r, t) = \exp \left(i \frac{\beta(r, t)}{2r} \tau^a x^a \right). \qquad (11.20)$$

These transformations naturally coincide with the gauge symmetry of the corresponding abelian Higgs model:

$$\phi' = \phi, \quad \alpha' = \alpha + \beta, \quad A'_\mu = A_\mu + \partial_\mu \beta. \qquad (11.21)$$

This freedom can be used to gauge out, for example, one component of A_μ: we will use the gauge $\mathcal{A}_0 = 0$ from now on.

Before proceeding any further, let us express the topological current

$$K_\mu = -\frac{1}{32\pi^2} \epsilon^{\mu\nu\rho\sigma} \left(\mathcal{G}_{\nu\rho}^a A_\sigma^a - \frac{g}{3} \epsilon^{abc} A_\nu^a A_\rho^b A_\sigma^c \right). \qquad (11.22)$$

in the reduces form, in the $\mathcal{A}_0 = 0$ gauge:

$$K^0 = \frac{1}{8\pi^2 r^2} \left[(1 - \phi^2)(\partial_1 \alpha - A_1) - \partial_1 (\alpha - \phi \cos \alpha) \right]$$

$$K^i = \frac{x^i}{8\pi^2 r^3} \left[(1 - \phi^2) \partial_0 \alpha - \partial_0 (\alpha - \phi \cos \alpha) \right], \qquad (11.23)$$

while the topological charge becomes

$$\partial_\mu K^\mu = \frac{1}{8\pi^2 r^2} \left\{ -\partial_0 \left[(1 - \phi^2)(\partial_1 \alpha - A_1) \right] \right.$$
$$\left. + \partial_1 \left[(1 - \phi^2)(\partial_0 \alpha - A_0) \right] \right\} . \tag{11.24}$$

Note that only gauge-invariant combinations of field derivatives appear here.

As a "topological coordinate" marking the tunneling paths and the turning states one can use the Chern–Simons number

$$N_{CS} = \int d^3 x K_0 = -\frac{1}{2\pi} \int dr (1 - \phi^2)(\partial_1 \alpha - A_1)$$
$$+ \frac{1}{2\pi} (\alpha - \cos\alpha)|_{r=0}^{r=\infty} \tag{11.25}$$

The first, gauge-invariant term is sometimes called the *corrected* or *true* Chern–Simons number $\tilde{N}_{CS}$, while the second (gauge-dependent) term is referred to as the *winding number*. It is the change in $\tilde{N}_{CS}$ which is equivalent to the integral over the local topological charge.

Now we are done with the digression of re-writing spherically symmetric 3+1 problem into a 1+1 form, and return to the static sphaleron path. To keep both *the Chern-Symons number* and *the mean radius* constant, we introduce two Lagrange multipliers $1/\rho^2$, η and search for the minimum of the following functional

$$\tilde{E} = \frac{4\pi}{g^2} \int dr \left(1 + \frac{r^2}{\rho^2} \right) \left[(\partial_r \phi)^2 + \phi^2 (\partial_r \alpha)^2 + \frac{(1 - \phi^2)^2}{2r^2} \right]$$
$$+ \frac{\eta}{2\pi} \int dr (1 - \phi^2) \partial_r \alpha \tag{11.26}$$

It is convenient to introduce new variable $\xi = 2 arctan(r/\rho) - \pi/2$. Then

$$\tilde{E} = \frac{8\pi}{g^2} \left\{ \int_{-\pi/2}^{\pi/2} d\xi \left[(\partial_\xi \phi)^2 + \phi^2 (\partial_\xi \alpha)^2 + \frac{(1 - \phi^2)^2}{2 \cos^2 \xi} + \kappa (1 - \phi^2) \partial_\xi \alpha \right] \right\} \tag{11.27}$$

where $\kappa = \eta \rho g^2/(32\pi^2)$ The Euler–Lagrange equations are

$$\partial_\xi^2 \phi - \phi(\partial_\xi \alpha)^2 + \frac{(1 - \phi)^2 \phi}{cos^2 \xi} + 2\kappa \phi \partial_\xi \alpha = 0 \tag{11.28}$$

$$\partial_\xi (\phi^2 \partial_\xi \alpha) + \kappa \partial_\xi (1 - \phi^2) = 0$$

Finiteness of the energy demands the following boundary conditions $\phi^2(\xi = -\pi/2) = \phi^2(\pi/2) = 1$

The second Eq. (11.28) gives

$$\partial_\xi \alpha = -\kappa \frac{1 - \phi^2}{\phi^2} \tag{11.29}$$

with integration constant equals 0 as it follows from the form of energy. After substitution $\partial_\xi \alpha$ to the Eq. (11.28) one has

$$\partial_\xi^2 \phi + \frac{(1 - \phi^2)\phi}{cos^2\xi} = \kappa^2 \frac{1 - \phi^4}{\phi^3} \tag{11.30}$$

The solution to this equation exists for $-1 < \kappa < 1$, it is $\phi^2 = 1 - (1 - \kappa^2)cos^2\xi$. Assuming ϕ to be positive one finds finally

$$\phi(r) = \left(1 - (1 - \kappa^2) \frac{4\rho^2 r^2}{(r^2 + \rho^2)^2} \right)^{1/2} \tag{11.31}$$

$$\partial_r \alpha(r) = -2\kappa \frac{1 - \phi^2}{\phi^2} \frac{\rho}{r^2 + \rho^2}. \tag{11.32}$$

For any κ mean radius of the solution is the same $< r^2 >= \rho^2$, and the energy density, the total energy, and the (corrected) Chern-Symons number are respectively

$$B^2/2 = 24 \left(1 - \kappa^2 \right)^2 \rho^4 / \left(r^2 + \rho^2 \right)^4 \tag{11.33}$$

$$E_{stat} = 3\pi^2 \left(1 - \kappa^2 \right)^2 / \left(g^2 \rho \right)$$

$$\tilde{N}_{CS} = \text{sign}(\kappa)(1 - |\kappa|)^2 (2 + |\kappa|)/4$$

Two last equations define the parametric form of the potential, see Fig. 11.4. The same profile obviously continues from 1/2 to 1, and so on, as a periodic potential with zeros at all integer values of N_{CS}, as a chain of mountains separated by valleys. In fact there are mountains of any hight, but tall ones are narrow. If energy is expressed in units of $1/g\rho$, it becomes unique. Note that the maximum is about parabolic but near zero energy the behavior is linear; so valleys are actually more like a deep canyons. The maximum is the *sphaleron solution* corresponding to $\kappa = 0$

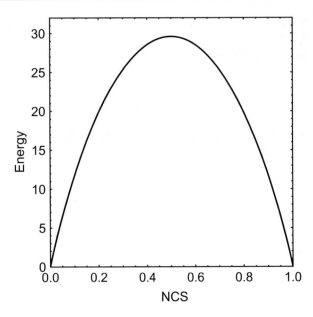

Fig. 11.4 The potential energy E (in units of $1/g^2\rho$) versus the Chern–Simons number $\tilde{N}_{CS}$, for the "sphaleron path" solution to be derived in Sphaleron chapter

and $N_{CS} = 1/2$

$$\phi = \frac{|r^2 - \rho^2|}{r^2 + \rho^2}, \qquad \alpha = \pi\theta(r - \rho). \tag{11.34}$$

11.5   Sphaleron Explosions

As we already mentioned above, the sphalerons are saddle point solution which are unstable. Add a small perturbation to a ball placed on the top of the mountain pass, and will start rolling down into one or the other valley.

The same paper Ostrovsky et al. (2002) came up with an analytic solution for this problem as well. The static sphaleron field configuration, found in previous section, is used as the initial condition for real-time, Minkowski evolution of the gauge field. Let us first consider the equations of motion in the 1+1 dimensional dynamical system. Variation of the action, Eq. (11.19), gives

$$\partial_\mu\partial^\mu\phi + \phi(\partial_\mu\alpha - A_\mu)^2 + \frac{(1 - \phi^2)\phi}{r^2} = 0 \tag{11.35}$$

$$\partial^\mu\left[\phi^2\left(\partial_\mu\alpha - A_\mu\right)\right] = 0 \tag{11.36}$$

$$\phi^2(\partial_1\alpha - A_1) - \partial_0\left[\frac{r^2}{2}(\partial_0 A_1 - \partial_1 A_0)\right] = 0$$

$$\phi^2(\partial_0\alpha - A_0) - \partial_1\left[\frac{r^2}{2}(\partial_0 A_1 - \partial_1 A_0)\right] = 0. \tag{11.37}$$

The solution of Eq. (11.36) has the form

$$\phi^2(\partial_0\alpha - A_0) = -\partial_1\psi$$

$$\phi^2(\partial_1\alpha - A_1) = -\partial_0\psi, \tag{11.38}$$

where $\psi(r, t)$ is an arbitrary smooth function. Equations (11.37) are consistent with this solution if

$$\partial_0 A_1 - \partial_1 A_0 = -\frac{2\psi}{r^2} \tag{11.39}$$

Now, combining Eqs. (11.36) and (11.37) one has

$$\partial^\mu\left(\frac{\partial_\mu\psi}{\phi^2}\right) = \partial_0 A_1 - \partial_1 A_0 = \frac{2\psi}{r^2}, \tag{11.40}$$

which can be viewed as a necessary and sufficient condition for ψ to be a solution for Eqs. (11.36) and (11.37) simultaneously. Equation (11.35) is now

$$\partial_\mu\partial^\mu\phi - \frac{(\partial_\mu\psi)^2}{\phi^3} + \frac{(1-\phi^2)\phi}{r^2} = 0. \tag{11.41}$$

The initial conditions for Eqs. (11.40) and (11.41) are

$$\phi(r, 0) = \phi(r),$$

$$\partial_0\phi(r, t)|_{t=0} = 0,$$

$$\partial_1\psi(r, 0) = -\phi(r)^2\partial_0\alpha(r) = 0 \Rightarrow \psi(r, 0) = 0,$$

$$\partial_0\psi(r, t)|_{t=0} = -\phi(r)^2\partial_1\alpha(r),$$

where the t-independent fields on the right sides of the equations are the static solutions of ϕ and α from the previous section.

As with static solutions, it is more convenient to discuss the time-evolution equations in hyperbolic coordinates. Let us choose ω and τ such that

$$r = \frac{\rho\cos\omega}{\cos\tau - \sin\omega}, \quad t = \frac{\rho\sin\tau}{\cos\tau - \sin\omega}. \tag{11.42}$$

The physical domain of $0 < r < \infty$ and $-\infty < t < \infty$ is covered by $-\pi/2 < \omega < \pi/2$ and $-\pi/2 + \omega < \tau < \pi/2 - \omega$. For $t > 0$, the corresponding domain is $-\pi/2 < \omega < \pi/2$ and $0 < \tau < \pi/2 - \omega$. This change of variables (11.42) is a conformal one.

In the new variables Eqs. (11.40) and (11.41) become

$$ -\partial_\tau^2 \phi + \partial_\omega^2 \phi - \frac{(\partial_\tau \psi)^2 - (\partial_\omega \psi)^2}{\phi^3} + \frac{(1 - \phi^2)\,\phi}{\cos^2 \omega} = 0 $$

$$ -\partial_\tau \frac{\partial_\tau \psi}{\phi^2} + \partial_\omega \frac{\partial_\omega \psi}{\phi^2} - \frac{2\psi}{\cos^2 \omega} = 0. \qquad (11.43) $$

Before solving these equations let us note that it is possible to predict the large-t behavior of gauge field from the form of the conformal transformation (11.42). Indeed, the $t \to \infty$ limit corresponds to the line $\tau = \pi/2 - \omega$ on the (ω, τ) plane. If one now takes the limit $|r - t| \to \infty$ (regardless of the limit for $|r - t|/t$), the position on (ω, τ) plane is either $\omega \to -\pi/2$, $\tau \to 0$ or $\omega \to \pi/2$, $\tau \to \pi$. This means that the entire line $\tau = \pi/2 - \omega$ corresponds to space-time points with finite differences between r and t and, therefore, if ϕ and ψ are smooth functions of ω and τ, then for asymptotic times the field is concentrated near the $r = t$ line. This corresponds to the fields expanding as a thin shell in space.

We must now supply Eqs. (11.43) with initial conditions, which are

$$ \phi(\omega, \tau = 0)^2 = 1 - (1 - \kappa^2)\cos^2 \omega $$

$$ \partial_\tau \phi(\omega, \tau)|_{\tau=0} = 0 $$

$$ \psi(\omega, \tau = 0) = 0 $$

$$ \partial_\tau \psi(\omega, \tau)|_{\tau=0} = \frac{\rho}{1 - \sin \omega} \partial_t \psi(\omega, \tau)|_{t=0} $$

$$ = \kappa(1 - \kappa^2)\cos^2 \omega. \qquad (11.44) $$

One of the solutions of Eqs. (11.43), first found in 1977 by Lüscher and Schechter, is

$$ \phi(\omega, \tau)^2 = 1 - \left(1 - q^2(\tau)\right)\cos^2 \omega $$

$$ \psi(\omega, \tau) = \frac{\dot{q}(\tau)}{2}\cos^2 \omega, \qquad (11.45) $$

with a function $q(\tau)$ that satisfies

$$ \ddot{q} - 2q(1 - q^2) = 0. \qquad (11.46) $$

This is the equation for a one-dimensional particle moving in double-well potential of the form $U(q) = (1 - q^2)^2/2$.

We now have to check that the solution satisfies the initial conditions, (11.44). This is indeed the case if one identifies $q(0) = \kappa$ and takes $\dot{q}(0) = 0$. For the initial condition of this type (i.e. for energy $\varepsilon = \dot{q}^2/2 + U(q) < 1/2$), the solution of Eq. (11.46) is

$$q(\tau) = \tilde{q}\, dn\, (\tilde{q}(\tau - \tau_0), k)\,, \qquad (11.47)$$

where dn is Jacobi's function and $\tilde{q} = \sqrt{2 - \kappa^2}$ is the second stopping point for a particle in the potential $U(q)$. We have also defined

$$k^2 = 2\frac{1 - \kappa^2}{2 - \kappa^2} \quad \text{and} \quad \tau_0\tilde{q} = \frac{T}{2}\,,$$

where T, the period of oscillations in the potential $U(q)$, is $T = 2K(k)$, with $K(k)$ being the complete elliptic integral of the first kind. The idea is, of course, that "oscillations" in τ begin from the rest point, close to $\tau = 0$.

Let us now look at several properties of the solution for large times. The solution (11.47) is apparently regular in the (ω, τ) plane, and therefore for large times the field is concentrated near $r = t$. At asymptotic times the energy density, $e(r, t)$, is given by

$$4\pi e(r, t) = \frac{8\pi}{g^2\rho^2}\left(1 - \kappa^2\right)^2 \left(\frac{\rho^2}{\rho^2 + (r - t)^2}\right)^3. \qquad (11.48)$$

The change in topological charge is

$$\begin{aligned}
\Delta Q &= \int_0^{\infty} d^3x\, dt\, \partial_\mu K^\mu \\
&= \frac{1}{2\pi}\int dr\, dt\left[-\partial_t^2\psi + \partial_r^2\psi - \frac{2\psi}{r^2}\right] \\
&= \frac{\pi}{2}\kappa(3 - \kappa^2) - \text{sign}(\kappa)\arccos\left(\frac{cn(\tilde{q}\pi, k)}{dn(\tilde{q}\pi, k)}\right). \qquad (11.49)
\end{aligned}$$

The evolution of $\tilde{N}_{CS}$ begins from time $t = 0$, where

$$\tilde{N}_{CS}(0) = \frac{1}{4}\text{sign}(\kappa)(1 - |\kappa|)^2(2 + |\kappa|)\,, \qquad (11.50)$$

and as $t \to \infty$ its limit is $\tilde{N}_{CS}(\infty) = \tilde{N}_{CS}(0) + \Delta Q$.

We now estimate number of gluons produced by the described evolution. In ϕ, ψ language the chromoelectric and chromomagnetic fields are

$$E_j^a = \frac{1}{r}\left(\partial_t\phi\sin\alpha - \frac{\partial_r\psi\cos\alpha}{\phi}\right)\Theta_j^a$$
$$+ \frac{1}{r}\left(\partial_t\phi\cos\alpha + \frac{\partial_r\psi\sin\alpha}{\phi}\right)\Pi_j^a + \frac{2\psi}{r^2}\Sigma_j^a, \qquad (11.51)$$

$$B_j^a = -\frac{1}{r}\left(\partial_r\phi\cos\alpha + \frac{\partial_t\psi\sin\alpha}{\phi}\right)\Theta_j^a$$
$$+ \frac{1}{r}\left(\partial_r\phi\sin\alpha - \frac{\partial_t\psi\cos\alpha}{\phi}\right)\Pi_j^a + \frac{1-\phi^2}{r^2}\Sigma_j^a. \qquad (11.52)$$

Terms proportional to Σ_j^a are longitudinal and die out as $t \to \infty$. The remainder is a purely transverse field. The main result becomes apparent when we choose a gauge where

$$\phi\partial_r\phi\cos\alpha + \partial_r\psi\sin\alpha = 0,$$

in which

$$E_j^a \to \frac{1}{r}\sqrt{\frac{(\partial_r\psi)^2}{\phi^2} + (\partial_r\phi)^2}\,\Theta_j^a$$
$$\to \frac{1-\kappa^2}{r\rho}\left(\frac{\rho^2}{\rho^2 + (r-t)^2}\right)^{3/2}\Theta_j^a, \qquad (11.53)$$

$$B_j^a \to \frac{1-\kappa^2}{r\rho}\left(\frac{\rho^2}{\rho^2 + (r-t)^2}\right)^{3/2}\Pi_j^a. \qquad (11.54)$$

We now perform a Fourier transform, finding

$$E_j^a(\vec{k}) = 4\pi\rho(1-\kappa^2)K_1(\omega\rho)\Theta_j^a$$
$$B_j^a(\vec{k}) = 4\pi\rho(1-\kappa^2)K_1(\omega\rho)\Pi_j^a, \qquad (11.55)$$

where Θ_j^a and Π_j^a are the color/space projectors in momentum space analogous to those in coordinate space (11.15), the frequency $\omega = |\vec{k}|$, and K_1 is a Bessel function. One can easily verify that $B_j^a = \epsilon_{jlm}k_l E_m^a/k$, as is required for a radiation field. Time evolution of electric and magnetic fields is shown in Fig. 11.5.

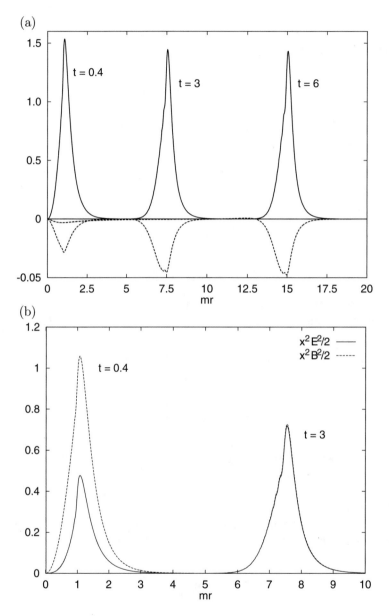

Fig. 11.5 (**a**) The energy (solid lines) and Chern–Simons number densities (dashed lines) for three times during the explosion, $t = 0.4$, 3, and 6 fm. (**b**) the electric and magnetic fields squared

Completing this section, let us mention that alternative derivation of the solution for the sphaleron explosion has been found by Shuryak and Zahed (2003a). It starts with Euclidean 4-d symmetric ansatz

$$gA_\mu^a = \eta_{a\mu\nu}\partial_\nu F(y), \quad F(y) = 2\int_0^{\xi(y)} d\xi' f(\xi') \tag{11.56}$$

with $\xi = ln(x^2/\rho^2)$ and η the 't Hooft symbol. Upon substitution of the gauge fields in the gauge Lagrangian $G_{\mu\nu}^2$ one finds that the effective Lagrangian has the form

$$L = \int d\xi \left[\frac{\dot{f}^2}{2} + 2f^2(1-f)^2\right] \tag{11.57}$$

corresponding to the motion of a particle in a double-well potential.

The off-center conformal transformation in question has the form

$$(x+a)_\mu = \frac{2\rho^2}{(y+a)^2}(y+a)_\mu \tag{11.58}$$

with $a_\mu = (0,0,0,\rho)$. While the original solution depends only on the radial coordinate in 4 dimensions y^2, in terms of x_μ this symmetry is broken and there is separate dependence on x_4 and the 3-dimensional radius $r = \sqrt{x_1^2 + x_2^2 + x_3^2}$.

The last step in getting the final solution is the analytic continuation to Minkowski time t, via $x_4 \to it$. It has the explicit form

$$gA_4^a = -f(\xi)\frac{8t\rho x_a}{\left[(t-i\rho)^2 - r^2\right]\left[(t+i\rho)^2 - r^2\right]}$$

$$gA_i^a = 4\rho f(\xi)\frac{\delta_{ai}(t^2 - r^2 + \rho^2) + 2\rho\epsilon_{aij}x_j + 2x_i x_a}{\left[(t-i\rho)^2 - r^2\right]\left[(t+i\rho)^2 - r^2\right]} \tag{11.59}$$

which includes i but still is manifestly real. Expressions for the gauge fields can be easily generated from it.

The only comment worth making is about the original parameter ξ in terms of these Minkowskian coordinates, which we still call x_μ, has the form

$$\xi = \frac{1}{2}\log\frac{y^2}{\rho^2} = \frac{1}{2}\log\left(\frac{(t+i\rho)^2 - r^2}{(t-i\rho)^2 - r^2}\right) \tag{11.60}$$

which is pure imaginary. To avoid carrying the extra i, we use the real

$$\xi_E \to -i\xi_M = arctan\left(\frac{2\rho t}{t^2 - r^2 - \rho^2}\right) \tag{11.61}$$

and in what follows we will drop the suffix E. Switching from imaginary to real ξ corresponds to switching from the Euclidean to Minkowski spacetime solution. It changes the sign of the acceleration in equation of motion , or, equivalently, the sign of the effective potential $V_M = -V_E$, to that of the normal double-well problem. The static sphaleron solution corresponds to the particle standing on the potential maximum at $f = 1/2$, and the needed sphaleron decay to "tumbling" paths. Since the start from exactly the maximum takes a divergent time, we will start nearby the turning point.

11.6 Chiral Anomaly and Sphaleron Explosions

We had already discussed the relation between the 3d and 4d topological charges: the divergence of the current containing Chern–Simons number is equal to

$$\partial_\mu K_\mu = \frac{1}{32\pi^2} G^a_{\mu\nu} \tilde{G}^a_{\mu\nu} \tag{11.62}$$

the 4d topological charge

$$Q_T = \frac{1}{32\pi^2} \int d^4x\, G^a_{\mu\nu} \tilde{G}^a_{\mu\nu} \tag{11.63}$$

We used Gauss theorem connecting the volume integral of the r.h.s., the 4d topological charge, to the change of the 3d topological charge, the Chern–Simons number:

$$N_{CS}(\tau \to \infty) - N_{CS}(\tau \to -\infty) = Q \tag{11.64}$$

The r.h.s. of this relation also appears in another important relation, known in QCD-like theories as Adler–Bell–Jackiw (ABJ) or *axial anomaly*:

$$\partial_\mu j^5_\mu = \frac{1}{32\pi^2} \epsilon^{\alpha\beta\gamma\delta} G_{\alpha\beta} G_{\gamma\delta} \tag{11.65}$$

including divergence of the axial quark current

$$j^5_\mu = \bar{q}\gamma_\mu\gamma_5 q \tag{11.66}$$

It was historically obtained from triangular diagrams with one axial and two vector currents: its derivation we will not discuss, yet the physics related to it we will discuss in next chapters.

Since the r.h.s. of both equations is the same topological charge, their change (in appropriate units) is the same. Say, one instanton $Q = 1$ produces change in axial density $\Delta j^5_0 = 2N_f$. The classical minima with different Chern–Simons number are associated with different axial charge. The difference between them is gauge invariant and given by Q. Another combination of two currents *is* in fact conserved, but not gauge invariant.

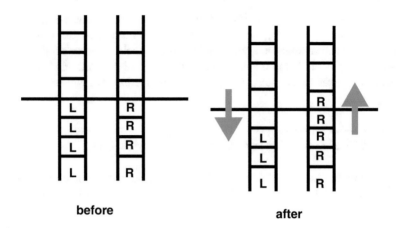

before   **after**

Fig. 11.6 Occupancy of the "Dirac sea hotel", before and after the earthquake

Integrating the zeroth component of the currents over space one finds that, in QCD with N_f light quarks

$$\Delta Q_5 = (2N_f)\Delta N_{CS} \quad (QCD) \tag{11.67}$$

Therefore the Chern–Simons number of the gauge field configuration is *rigidly locked to the axial charge*, the number of left minus right-polarized fermions! For example, if $N_f = 3$, a transition over the sphaleron barrier from $N_{CS} = 0$ to $N_{CS} = 1$ must be accompanied by 6 units of the axial charge.

Let me translate this into a small story, so you better remember it. There was a sea-side hotel called the Dirac Sea Hotel. It was so big that we think of it extending indefinitely, both above the ground and below. It has a bit strange policy to put occupants at as low level as possible, and to mark which of the occupants are right and left-handed, which were put in two separate towers, see the left figure. A strange earthquake happened one day, leading to the shift as indicated in the right figure: all lefties went down by one floor, and all righties one up, see Fig. 11.6.

The same anomaly relation in electroweak theory leads to even more drastic consequences. Since only the left-polarized quarks and leptons interact with the electroweak gauge field, there are no right-hand fields in the equation. Furthermore, there is no distinction between vector and axial current, as only left-handed part matters. Therefore, the *number* (rather than just chiralities) of quarks and leptons are changed in electroweak theory. A travel from $N_{CS} = 0$ to $N_{CS} = 1$ must lead to simultaneous production of 9 quarks and 3 leptons! If the polarization of the sphaleron field is "up", these 12 fermions are $t_r t_b t_g c_r c_b c_g u_r u_b u_g \tau \mu e$ where $r = red, b = blue, g = green$ are three colors of the quarks. The baryon and lepton numbers increase by 3 units, and their difference is zero

$$\Delta B = \Delta L = 3\Delta N_{CS} \quad (electroweak\ part\ of\ the\ SM) \tag{11.68}$$

Can those drastic statement be put to experimental tests?

As far as the electroweak theory is concern, the chances to do so are not there. In the Higgs-broken vacuum we live today one has to go over the barrier with the hight $M_{KM} \approx 8\,\mathrm{TeV}$. And the energy is not the main problem: one needs to produce $O(100)$ W bosons, fit all of them in a small volume of an electroweak scale, keeping all of them in a form of coherent configurations of the sphaleron path. Electroweak sphaleron transitions may however happen above and near the electroweak phase transition, because in this case Higgs VEV is absent or small.

In QCD sphalerons have energy of only several GeV, so one might think those are well studied experimentally. Unfortunately, this is not the case: it still remains to be done. One proposal (Shuryak and Zahed 2003b) to do that in hadron-hadron collisions is via double-diffractive production of hadronic clusters with strong left-right quark asymmetry, say six units of the axial charge. These clusters should resemble hadronic channels which we discussed in connection to η_c decays.

In heavy ion collisions one expect multiple sphaleron processes at the initial stage of the collision, leading to fireballs with a disbalanced chirality. Observation of this can be done with the help of Chiral Magnetic Effect (CME), to be discussed at the end of this chapter.

While we have shown that general Adler–Bell–Jackiw (ABJ) anomaly relation require locking of the Chern–Simons number to the (axial) charge of the fermions, it has not yet been explained how exactly it happens. In fact one can follow this phenomenon using the "exploding sphaleron" solution derived previously.

Let us start at time $t = 0$, from the static (but unstable) sphaleron. Dirac equation in the field of the sphaleron has a fermionic zero mode. Like we previously discussed it for the monopole, it should be i interpreted as a zero energy bound state of a massless fermion, which can be occupied or empty.

Sphaleron explosion solution includes some radial electric fields (at intermediate time only, at the late time it is transverse). This field can accelerate a fermion, from the initial zero energy state to some excited state with positive energy. Thus a qualitative picture of fermion production does not mean their produciton "from nowhere", but includes level motion, out of occupied levels in the "Dirac sea" to physical states with positive energy.

What about antifermions? In order to have *the same* zero mode (or other) solution as fermions, one needs the same form of the Dirac equation, with the same colormagnetic moment. Since the charge of antifermion is opposite to that of the fermion, one has to flip the spin to compensate.

In QCD there is no problem with that, just chirality of the produced positive energy antiquark would be opposite to that of the quark. As a result, one has production of a particle-hole pair. The baryon charge is not changed, but the axial one—the difference between left and right-handed fermions—is changed by two units.

In electroweak theory there is a problem: only left-handed fermions interact, while right-handed do not. As a result, there is no antifermion solution mirroring the fermion one, and the sole fermion is produced. Both axial and baryon (or lepton) charge is changed.

This reasoning is simple and consistent with what one expects from the anomaly, so there is no doubts in its validity. However attempts to follow dynamically the outlined scenario had encounter the following problem. Physical fermions are either produced or not, so change in the axial (or total) charge can only be an integer. The change of the Chern–Simons number during the sphaleron explosion is expected to be 1/2, and complemented by a similar process of sphaleron formation it is expected to give 1, also an integer.

However, the first numerical solutions of the KM electroweak sphaleron explosion revealed a trouble: the Chern–Simons number apparently refused to settle at the expected change of 1/2. It looked like a ball rolling down from the saddle point of the potential does not roll all the way to the valley's bottom! It was first taken as numerical problem, but the same feature was confirmed in subsequent numerical solutions. Moreover, it is there in the COS analytic solution presented above: the Chern–Simons number stabilizes at late times is $N_{CS} \simeq 0.12$. The not-quite completed transition in terms of N_{CS} is consequence of the classical approximation, in which the gauge boson mass is neglected.

Fortunately, the analytic solution to the Dirac equation in the background of the exploding sphaleron was found in Shuryak and Zahed (2003a)

The solution to the massless Dirac equation in the Minkowski background field of exploding sphalerons can also be obtained by the same conformal mapping, from the O(4) Euclidean zero modes. We explicitly construct these states and show that at the initial Minkowski time t=0 those are zero energy states, while at asymptotically large time they reduce to a free quark or free antiquark of specific chirality. We also calculate the spectrum of the produced fermions.

The solution itself is obtained by the following inversion formula

$$Q_+(x) = \gamma_4 \frac{\gamma_\mu \, (y+a)_\mu}{1/(y+a)^2} \, \Psi_+(y) \tag{11.69}$$

and it solves the ($\gamma_4\times$) Dirac equation in the (Euclidean) 4-d spherically symmetric gauge configuration.

Omitting the technical details, let us proceed to the results. The quark spectrum shown in Fig. 11.7 is close to black-body Planck distribution, with an effective temperature $T = 2/\rho$. The distribution integrates exactly to *one* produced fermion of each kind in electroweak theory (with also antiquarks in QCD) as the anomaly relation requires.

11.7   Brief Summary

* *Sphalerons* are 3d (static magnetic) solutions of classical equations of motion, Yang-Mills for gauge fields plus those for scalar field. They possess semi-integer values of the Chern–Simons number $N_{CS} = 1/2 + integer$. They are therefore extrema of the action, but instead of being minima they actually are *saddle points*, unstable against decays along the so called *sphaleron path*.

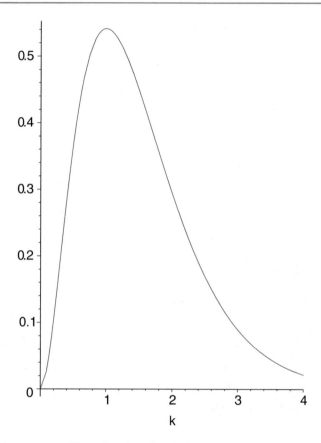

Fig. 11.7 The spectrum of **R** quarks released at the Sphaleron point versus its momentum k in units of $1/\rho$

- There are three different ways to derive them (in pure gauge settings): (1) as 3d configuration on a plane equidistant between instanton and antiinstanton; (2) energy minimization, constrained by fixed N_{CS} and r.m.s. size ρ; (3) conformal off-center transformation from certain 4d spherical solution in Euclidean space.
- The energy of the sphaleron path states is the same for all approaches, and is shown in Fig. 11.4 (recall that without derivation we discussed it already in Chap. 6).
- *Sphaleron explosions* are time-dependent solutions of equation of motion, which start from the sphaleron path configurations. At time zero there is no electric field, but it appears in the process. At late time solutions take form of expanding spherical wave with electric and magntic fields normal to each other and to the radial direction.
- Since chiral anomaly locks together the change in N_{CS} and in fermionic axial charge (the difference between number of left and right polarized quanta),

sphaleron explosions in QCD must be accompanied by generation of $2N_f = 6$ units of chirality.

- In electroweak theory sphaleron explosions produce 12 fermions, 9 quarks and 3 leptons. Their specific flavors are defined by orientation of the original state in weak $SU(2)$. These fermions are accelerated to positive energies by radial electric field, and their spectra depend on the size of the exploding initial state.

References

Balitsky, I.I., Yung, A.V.: Collective - coordinate method for quasizero modes. Phys. Lett. **B168**, 113–119 (1986). [273(1986)]

Dashen, R.F., Hasslacher, B., Neveu, A.: Nonperturbative methods and extended hadron models in field theory. 3. Four-dimensional nonabelian models. Phys. Rev. **D10**, 4138 (1974)

Klinkhamer, F.R., Manton, N.S.: A saddle point solution in the Weinberg-Salam theory. Phys. Rev. **D30**, 2212 (1984)

McLerran, L.D.: Anomalies, Sphalerons and Baryon number violation in electroweak theory. Acta Phys. Polon. **B20**, 249–286 (1989)

Ostrovsky, D.M., Carter, G.W., Shuryak, E.V.: Forced tunneling and turning state explosion in pure Yang-Mills theory. Phys. Rev. **D66**, 036004 (2002). hep-ph/0204224

Shuryak, E.V.: Toward the quantitative theory of the 'Instanton Liquid' 4. Tunneling in the double well potential. Nucl. Phys. **B302**, 621–644 (1988)

Shuryak, E., Zahed, I.: Prompt quark production by exploding sphalerons. Phys. Rev. **D67**, 014006 (2003a). hep-ph/0206022

Shuryak, E., Zahed, I.: Semiclassical double pomeron production of glueballs and eta-prime. Phys. Rev. **D68**, 034001 (2003b). hep-ph/0302231

Verbaarschot, J.J.M.: Streamlines and conformal invariance in Yang-Mills theories. Nucl. Phys. **B362**, 33–53 (1991). [Erratum: Nucl. Phys.B386,236(1992)]

Sphaleron Transitions in Big and Little Bangs 12

12.1 Electroweak Sphalerons and Primordial Baryogenesis

12.1.1 Introduction to Cosmological Baryogenesis

The observed Baryonic Asymmetry of the Universe (BAU) is usually expressed as the ratio of the baryon density to that of the photons. This key parameter enters calculations of primordial nucleosynthesis of such nuclei as d, t, He^4, Li^7: all of them agree that its value is

$$n_B/n_\gamma \sim 6 \times 10^{-10} \tag{12.1}$$

The question how it was produced is among the most difficult open questions of physics and cosmology. Sakharov had formulated three famous *necessary conditions* for its generation:

(1) The baryon number violation
(2) The CP violation
(3) Deviations from thermal equilibrium

Although all of them are formally satisfied by the standard model (SM) and standard Big Bang cosmology, we do not yet know specific mechanisms creating the baryon asymmetry. Smooth crossover electroweak transition does not generate large deviations from thermal equilibrium, the baryon number violation is suppressed by huge sphaleron barrier, and CP violation via CKM quark mixing matrix is quite small. It is therefore widely accepted that BAU is due to some so far hypothetical mechanisms, e.g., heavy neutrino decays or large CP violation in Higgs sector (for review of the field, see, for example, Dine and Kusenko 2003). This common view

© The Author(s), under exclusive license to Springer Nature Switzerland AG 2021
E. Shuryak, *Nonperturbative Topological Phenomena in QCD and Related Theories*, Lecture Notes in Physics 977,
https://doi.org/10.1007/978-3-030-62990-8_12

is from time to time challenged by some suggested scenarios: one of them we will discuss later in the chapter.

In the electroweak theory, semiclassical description of the tunneling through the barrier, separating topologically distinct gauge fields, is given by the (by now much discussed) *instanton* solution. However, unlike in QCD, the coupling is weak, the action is large, and the tunneling probability in the broken phase we live in is extremely low:

$$\Gamma_{tunneling}/T^4 \sim \exp(-4\pi/\alpha_w) \sim 10^{-170} \qquad (12.2)$$

What about thermal excitation of electroweak sphalerons? In electroweak theory the coupling is small; thus the sphaleron energy large is $E \sim v/\alpha_w \sim O(10\,\text{TeV})$. Naive comparison with the highest temperature at which the broken phase exists, namely, the electroweak critical temperature $T_{EW} \approx 0.16\,\text{TeV}$ give Boltzmann factor, is not as small as the probability of tunneling just mentioned, but it is still very small, $O(exp(-100))$. However, close to T_{EW}, the Higgs VEV is small, and the sphaleron rate should not be suppressed that much. We will return to this issue shortly.

However, this estimate is too naive, as the parameters of the electroweak theory are strongly renormalized near T_c. The main reason of the increase of the rate, from the KM sphaleron to revised one, is of course the reduction of the Higgs VEV from its vacuum value $v = 246\,\text{GeV}$ to near zero.

More accurate calculation of the equilibrium sphaleron rates at the electroweak crossover region give much larger rates. For an update see, for example, Burnier et al. (2006) who estimated those in the range

$$\Gamma/T^4 \approx 10^{-20} \qquad (12.3)$$

including rather large pre-exponent calculated semiclassically in Arnold and McLerran (1987),Carson et al. (1990), and Moore (1996). Such rates are, however, still too small for the solution of the baryogenesis puzzle.

On the other hand, in the symmetric phase at $T > T_c$, in which there is no-Higgs VEV, the sphaleron size ρ can be much larger than the electroweak scale $T_c \sim 100\,\text{GeV}$, with respectively much higher sphaleron rates. There is a limit to ρ set by the inverse magnetic screening mass:

$$\rho < \frac{1}{m_{mag}} \sim \frac{1}{g^2 T} \qquad (12.4)$$

and the dimensional arguments thus put the sphaleron transition rate to $\Gamma \sim \alpha_w^4 T^4$. More complicated analysis of the problem (Arnold et al. 1997) shows that it is suppressed by one more power, so

$$\frac{\partial(\Delta N_{CS})^2}{\partial t \partial V} = \Gamma \sim \alpha_w^5 T^4 \qquad (12.5)$$

Although there appears a rather high power of weak coupling, the exponential suppression is gone, and so in the symmetric phase, the sphaleron rates obtained are *larger* than the expansion rate of the Universe at the corresponding era. This means appearance of another problem: all asymmetries which may be primordially generated would then be wiped out above the electroweak critical temperature! These considerations force us to search for resolution of the baryon asymmetry puzzle at narrow temperature interval at or right below the electroweak phase transition.

But before we return to recent works on electroweak sphalerons, let us briefly describe the history of cosmological electroweak transition.

12.1.2 Electroweak Phase Transition

Originally, this transition was assumed to be first order, in which the broken phase (in which Higgs field obtains a nonzero VEV) first appears in the form of bubbles. Their coalescence and existence of bubble walls, moving in the process, have created hopes in the 1980s–1990s for large local deviations from thermal equilibrium.

However, further studies of the transition have shown that the first-order transition can only happen if the Higgs mass is relatively small, less than 70 GeV or so. After CERN discovery of Higgs particle with a mass of 125 GeV, it became obvious that SM only predicts relatively smooth crossover transition. After this fact has been acknowledged, people looked at various phenomena *beyond* the SM, such as its supersymmetric extensions, which may still allow the first-order transitions. We will also briefly discuss the so-called "cold" or "hybrid" scenario, in which electroweak transition starts right after inflation, with strong deviation from equilibrium.

Let us, however, return to standard cosmology and SM framework. The best value of electroweak transition today is calculated on the lattice; e.g., D'Onofrio et al. (2014) produce the following value:

$$T_{EW} = (159 \pm 1)\,\text{GeV} \tag{12.6}$$

The temperature of the Universe today is $T_{\text{now}} = 2.73\,\text{K}$. The ensuing redshift z-factor is

$$z_{EW} = \frac{T_{EW}}{T_{\text{now}}} \approx 6.8 \cdot 10^{14} \tag{12.7}$$

During the radiation-dominated era, the relation of time to temperature is given by Friedmann relation:

$$t = \left(\frac{90}{32\pi^3 N_{\text{DOF}}(t)}\right)^{\frac{1}{2}} \frac{M_P}{T^2} \tag{12.8}$$

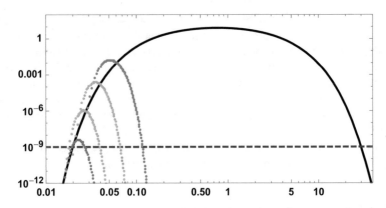

Fig. 12.1 The sphaleron suppression rates as a function of the sphaleron size ρ in GeV^{-1}. The solid curve corresponds to the unbroken phase $v = 0$ at $T = T_{EW}$. Four sets of points, top to bottom, are for well-broken phase, at $T = 155, 150, 140, 130$ GeV. They are calculated via Ansatz B described in Appendix C and normalized to lattice-based rates. The horizontal dashed line indicates the Hubble expansion rate relative to these rates

Inserting the Planck mass $M_P = 1.2 \cdot 10^{19}$ GeV, the transition temperature, and the effective number of degrees of freedom $N_{\mathrm{DOF}} = 106.75$, we find the time after the Big Bang to be

$$t_{EW} \sim 0.9 \cdot 10^{-11}s, \quad ct_{EW} \approx 2.7 \, \mathrm{mm} \tag{12.9}$$

Quite amusingly, these numbers are nearly exactly 1000 times larger than critical and freeze-out temperatures in heavy ion collisions, $T_c \approx 155$ MeV and hadronic freeze-out $T_{FO} = 117 - 140$ MeV, depending on the collision energy.

12.1.3 Sphaleron Size Distribution

The details can be found in our original paper (Kharzeev et al. 2019); let me just show the result in Fig. 12.1. In short, at $T > T_{EW}$ (as shown by solid line), the sphaleron tends to be rather large size, on a scale of electroweak magnetic screening length. But as the temperature shifts below T_{EW} and nonzero Higgs VEV appears, their sizes rapidly shrink (as shown by the colored points). The mass of the sphaleron grows; as with it, the Boltzmann factor $exp(-M/T)$ rapidly falls till at $T < T_{fo} \approx 130$ GeV, and the rate goes smaller than the Universe expansion rate (the horizontal dashed line), and sphalerons become unimportant.

12.1.4 The Hybrid (Cold) Cosmological Model and Sphalerons

A scenario which we will discuss would then be the so-called hybrid (cold) cosmological scenario in which the end of inflation coincides with the electroweak transition, so that equilibration happens at $T < T_c$. This ensures *large deviations from equilibrium*. While based on some fine-tuning of the unknown physics of the inflation, it avoids many pitfalls of the standard cosmology, such as "erasure" of asymmetries generated before the electroweak scale.

The *baryon number violation* in it is due to the sphaleron transitions that occur inside the bubbles of certain size ρ: and those can be studied numerically and semiclassically in detail. The *CP violation* is not yet calculated accurately, but its magnitude near the sphalerons will be estimated.

In such a scenario, there are coherent oscillations of the gauge/scalar fields studied in detail in real-time lattice simulations (Garcia-Bellido et al. 2004; Tranberg and Smit 2003). The simulated models include two scalars—the inflaton and the Higgs boson—and the electroweak gauge fields of the SM, in the approximation that the Weinberg angle is zero (Z is degenerate with W). All fermions of the SM are ignored: the effect of the top quark in particular is the subject of the present paper. After inflation ends, all bosonic fields are engaged in damped oscillations for relatively short time, at the end of which the Higgs VEV and gauge fields stabilize to their equilibrium values, with the bulk temperature $T_{bulk} \sim 50\,\text{GeV}$, well below the critical (crossover) temperature:

(1) One important finding of the simulations is that the initial coherent oscillations of scalars soon give way to the usual broken phase. The most important feature is persistence of "no-Higgs spots" in which Higgs VEV is very far from the equilibrium value v and is instead close to zero. The gauge fields in them have, however, rather high magnitude. Figure 12.2 (from Garcia-Bellido et al. 2004) shows an example of a snapshot of the Higgs field modulus. Typically the volume fraction occupied by such "no-Higgs spots" is of the order of several percents and is decreasing with time.

(2) The second important findings is that of topologically nontrivial fluctuations of the gauge fields. As shown in the lower Fig. 12.2, those are well localized only *inside* the "no-Higgs spots" mentioned above. Indeed, this becomes apparent from the distribution of the topological charge shown in the lower part of Fig. 12.2, for the same time configuration of the Higgs as shown in the upper part of the Fig. 12.2. Of course, it is only one snapshot, but the authors found from the simulations that it is true for the whole sample.

The fraction of "hot spots" (no-Higgs-VEV) which induced topological transitions is also in the range of few percent. More precise measure is the so-called sphaleron rate which is defined by the mean square deviation from zero of the Chern-Simons number

$$\Gamma(t) = \frac{1}{m^4 V} \frac{d\Delta N_{CS}^2}{dt} \tag{12.10}$$

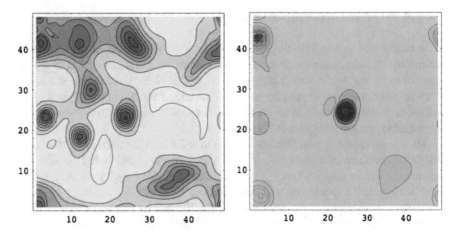

Fig. 12.2 The contour plots from Garcia-Bellido et al. (2004) of the modulus of the Higgs field $|\phi(x)|^2$ (upper plot) and the topological charge density $Q(\vec{x}, t)$ at time $mt = 19$, for the model A1, $N_s = 48$. Red (dark) areas in the upper plot correspond to small VEV, while yellow (light) bulk corresponds to the broken phase. On the lower plot lumps of the topological charge density appear as red regions (dark in black and white display). While most of the no-Higgs spots do not have the topological transitions; all transitions seem to be inside the spots

(here and below all quantities are defined via one characteristic mass parameter m: for the simulations its value is about 264 GeV, close to v).

Since the process only exists for finite period of time—too early there are no gauge fields, and too late there are no no-Higgs spots—its time integral is well converging. Its value

$$I(mt) = \int_{t_i}^{t} d(mt)\Gamma(t) \tag{12.11}$$

is in the range

$$I \sim 10^{-4} \tag{12.12}$$

(for more details about different parameter sets, see Table II of Garcia-Bellido et al. 2004). This quantity, as well as of course snapshots like those shown in Fig. 12.2, directly gives the spatial distance between the topological fluctuations $R_{sph}m = 20 \ldots 30$.

(3) Space-time evolution of the topological charge Q is shown in Fig. 12.3 as two snapshots. One can see that the fluctuation at some time moment is very much concentrated in a small spherical cluster (the upper one in Fig. 12.3) and is followed by an expanding spherical shell (the lower one in Fig. 12.3) which gets near-empty inside.

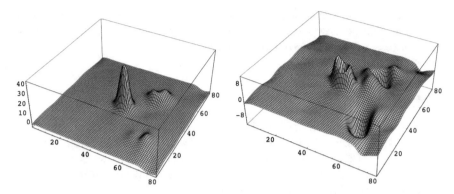

Fig. 12.3 From Garcia-Bellido et al. (2004). Two snapshots of the topological charge, at times $mt = 18$ and 19

The decomposition into electric and magnetic components of the field shows that the fluctuations start as nearly (90%) magnetic object at $mt \sim 17$, with the electric field and thus the topological charge $\sim \vec{E}\vec{B}$ peaking some time later. Then there appears an expanding shell, followed by the magnetic field rebounding to its secondary maximum of smaller amplitude. We will return to discussion of all those features in the next chapter.

Before discussing these numerical results, let us point out an important issue which is discussed in Flambaum and Shuryak (2010) but is outside of our current interest: the "no-Higgs spots" seen in numerical simulations can be identified with the *non-topological* solitons of the electroweak theory known as the "WZ-top bags" (Crichigno et al. 2010).

What we will consider now is the role of the quarks, especially the top quarks, in baryogenesis: due to the high cost of inclusion of the fermions, those up to now were not included in lattice numerical simulations.

Now we turn to the next question: how much the *sphaleron rate* can be affected by their presence, as compared to purely bosonic ones in the simulations (Garcia-Bellido et al. 2004)?

The Adler-Bell-Jackiw anomaly requires 12 fermions to be produced. Particular fermions depend on the orientation of the gauge fields in the electroweak $SU(2)$: since we are interested in utilization of top quarks, we will assume it to be "up." In such case the produced set contains $t_r t_b t_g c_r c_b c_g u_r u_b u_g \tau \mu$, e, where $r, b, and g$ are quark colors. We refer to it below as the $0 \rightarrow 12$ reaction. Of course, in matter with a nonzero fermion Density, many more reactions of the type $n \rightarrow (12 - n)$ are allowed, with n (anti)fermions captured from the initial state.

Evaluating the back reaction of the fermions on gauge field, the (analytic) solution to the Dirac equation of the "expansion stage" (Shuryak and Zahed 2003a) we discussed above is very useful. A new element we are adding now is that its time reflection can also describe the compression stage, from free fermions captured by a convergent spherical wave of gauge field at $t \rightarrow -\infty$ and ending at the sphaleron zero mode at $t = 0$. The details will be omitted, but the main result is that by "eating the top quarks" already present in the bags, one effectively lowers the barrier and thus reaches further increase of the sphaleron rate.

12.1.5 Effective Lagrangian for CP Violation

The only known[1] source of CP violation in the standard model is that induced by the phase of the of the Cabibbo-Kobayashi-Maskawa (CKM) quark mixing matrix. CP violation, originally discovered in decays of neutral K mesons, is now studied mostly by LHCb CERN experiment at LHC in B meson decays. It is a very large and involved field, into which we have no time to go.

Still, we need some brief introduction here. Let me start with QED and remind that fourth-order "electron box" diagram generates effective Lagrangian $L = O(F_{\mu\nu}^4)$, originally calculated by Heisenberg in Euler in the 1930s. Similarly, one can imagine some "background electroweak fields" W_μ, Z_μ and effective Lagrangians generated by quark loop. On general ground, the CP-odd effects require at least four CKM matrices, so the effect may in principle appear starting from the fourth-order $O(W^4)$ "quark box" diagram. So, one may think that the calculation of CP-violating effective Lagrangian is just an exercise, most likely done half century ago when electroweak theory was found.

I am sorry to tell that it is rather involved calculation. For some reason, plenty of field theorists who do diagrams of all possible kind in all possible theories failed to solve this important issue.

According to one explicit calculation of the effective CP-odd Lagrangian Hernandez et al. (2009) in the leading W^4 order, there is no CP violation (and we will soon check that it is true in our approach). These authors then observe that the effect appeared in next-to-leading order diagram leads to the following dimension 6 operator:

$$L_{CP} = C_{CP}\epsilon^{\mu\nu\lambda\sigma}\left[Z_\mu W_{\nu\lambda}^+ W_\alpha^- \left(W_\sigma^+ W_\alpha^- + W_\alpha^+ W_\sigma^-\right) + \text{c.c.}\right] \qquad (12.13)$$

containing four charged gauge boson W fields and one neutral Z. Yet subsequent investigations in Garcia-Recio and Salcedo (2009) have not confirmed a nonzero coefficient for this operator, but came up instead with a set operators of dimension 6 possessing with a completely different structure. Another group (Brauner et al. 2012) confirmed their finding. Remarkably, all the 13 operators O_i found are C-odd and P-even, while the above-given (12.13) is P-odd and C-even.

---

[1]Perhaps similar phase of the neutrino mass mixing matrix will be discovered soon.

Shuryak and Zahed (2016) proposed to simplify the problem, by splitting it into two separate parts. Quark in the loop moves in the background field, so the natural basis in which one should approach the problem is that of Dirac eigenstates:

$$i\not{D}\psi_\lambda(x) = \lambda\psi_\lambda(x) \tag{12.14}$$

for the Dirac operator in this field. The propagators are in this representation diagonal, as they are inverse to the Dirac operator, and contain quark masses in a standard manner. The vertices are matrix elements of fields projected on currents made of $\psi_\lambda(x)$. One may start with diagonal vertices.

The heart of the problem is in performing flavor summation, with four CKM matrices and needed number of propagators. The lowest-order box is given by

$$\int \prod_i^4 d^4x_i \, \mathrm{Tr}\left(\not{W}(x_1)\hat{V}\hat{S}_u(x_1,x_2)\not{W}(x_2)\hat{V}^\dagger\hat{S}_d(x_2,x_3) \right.$$

$$\left. \not{W}(x_3)\hat{V}\hat{S}_u(x_3,x_4)\not{W}(x_4)\hat{V}^\dagger\hat{S}_d(x_4,x_1) \right)$$

Here hats indicate that the CKM matrix V and the propagators are 3×3 matrices in flavor space. The propagator also has labels u, d, indicating up- or down-type quarks.

The trace is over Dirac and color indices as well: slash with W field as usual means convolution of vector potential with gamma matrices. The point is one should look first at the flavor trace, multiplying eight matrices explicitly (in Mathematica). And indeed, the leading box diagram produces no CP violation! We also found that if one adds one Z field, we still get zero. Adding two Z, one on u and one on d quark line, does produce the expression

$$F(\lambda) = \lambda^4 \mathrm{Tr}\left(\hat{V} S_d \hat{V}^\dagger S_u \hat{V} S_d \hat{V}^\dagger S_u \right) \tag{12.15}$$

Let us use representation in which the main operator is diagonalized, so that

$$i\not{D}\psi_\lambda(x) = \lambda\psi_\lambda(x) \tag{12.16}$$

where the notations with the slash here and below mean the convolution with the Dirac matrices $\not{D} = D_\mu\gamma_\mu$. The corresponding (Euclidean time) propagator—describing a quark of flavor f propagating in the background—can thus be written as the sum over modes:

$$S(x,y) = \sum_\lambda \frac{\psi_\lambda^*(y)\psi_\lambda(x)}{\lambda + M\not{\partial}^{-1}M^+} \tag{12.17}$$

The generic fourth-order diagram in the weak interaction, containing necessary four CKM matrices, takes in the coordinate representation the form

$$\int d^4x_1 d^4x_2 d^4x_3 d^4x_4 \mathrm{Tr}[\mathcal{W}(x_1)\hat{V}\hat{S}_u(x_1, x_2)\mathcal{W}(x_2)$$

$$\hat{V}^+\hat{S}_d(x_2, x_3)\mathcal{W}(x_3)\hat{V}\hat{S}_u(x_3, x_4)\mathcal{W}(x_4)\hat{V}^+\hat{S}_d(x_4, x_1)]$$

where V is the CKM matrix; the hats on them and the propagators with u, d subscripts indicate that they are the 3×3 matrices in flavor subspace; and the trace is implied to be over flavor indices. If one considers next order diagrams, with Z, ϕ field vertices, the expressions are generalized straightforwardly.

Spin-Lorentz structure of the resulting effective action is very complicated. To understand the scale dependence, we will now we make strong simplifying assumptions. First, we will focus on the diagonal matrix elements of the operator $\mathcal{W}$ and assume it to be approximately proportional to λ (with some coefficient ξ):

$$< \lambda|\mathcal{W}|\lambda' > \approx \xi\lambda\delta_{\lambda\lambda'} \qquad (12.18)$$

Second, we assume that right-handed operator $i\partial\!\!\!/$ can similarly be represented by diagonal matrix element we will call $\not{p}$. If so, one can use the orthogonality condition of different modes and perform the integration over coordinates, producing much simplified expression, with a single sum over eigenvalues $\sum_\lambda F(\lambda)$ where

$$F(\lambda) = \lambda^4 \mathrm{Tr}\left(\hat{V}S_d\hat{V}^+ S_u \hat{V}S_d\hat{V}^+ S_d\right) \qquad (12.19)$$

This is the diagram in the λ-representation, which generalizes the momentum representation valid only for constant fields. Unlike momenta, the spectrum of Dirac eigenvalues λ may have various spectral densities. In particular, there is a zero mode, corresponding to zero mode in the original four-dimensional symmetric case. This describes the fermion production on various backgrounds, such as the exploding sphaleron.

Independent of what physical meaning and spectrum of λ are, the point is that one can perform multiplication of flavor matrices and extract universal function of λ, describing dependence of CP violation on the scale.

Using the standard form of the CKM matrix $\hat{V}$, in terms of known three angles and the CP-violating phase δ, and also six known quark masses, one can perform the multiplication of these eight-flavor matrices and identify the lowest-order CP-violating term of the result. Performing the multiplication in the combination above, one finds a complicated expression which does *not* have $O(\delta)$ term, so there is no lowest-order CP violation. This agrees with a statement from earlier works that the leading fourth-order diagram generates no operators.

Higher-order diagrams, however, do have such contributions. For example, if on top of four W vertices with CKM matrices there are also two Z, flavor trace looks as follows:

$$F_{ZZ}(\lambda) = \lambda^6 \text{Tr} \left(\hat{V} S_d \hat{V}^+ S_u \hat{V} S_d Z S_d \hat{V}^+ S_u Z S_u \right) \tag{12.20}$$

Now the flavor trace has the lowest-order CP violation described by the following symmetric expression:

$$
\begin{aligned}
&Im\, F_{ZZ}(\lambda) \\
&= 2\lambda^6 \frac{J\left(m_b^2 - m_d^2\right)\left(m_b^2 - m_s^2\right)\left(m_d^2 - m_s^2\right)\left(m_c^2 - m_t^2\right)\left(m_c^2 - m_u^2\right)\left(m_t^2 - m_u^2\right)}{\Pi_{f=1\ldots6}\left(\lambda^2 + m_f^2\right)^2}
\end{aligned}
$$

The numerator is the Jarlskog combination of the CKM angles and differences of masses squared. As expected, the effect vanishes when the mass spectrum of either u-type or d-type quarks gets degenerate.

A plot of this function[2] is shown in Fig. 12.4. Because of the cancelation between different quark flavors, it is very small at large λ, about 10^{-19} at the electroweak scale and $\lambda \sim 100\,\text{GeV}$ at the r.h.s. of the plot, which can be called "the Jarlskog regime." Yet at the scale near and below $\lambda = 1\,\text{GeV}$—so to say "the Smith regime"—it is 12 orders of magnitude larger!

These calculations show that for sphalerons produced by cold electroweak scenario, with size at the scale $\rho \sim 1/80\,\text{GeV}$, the CP-odd effect is way too small to explain the baryon asymmetry. If some mechanism will be found, which can generate the sphaleron transitions with much large sizes, say 1 GeV, the CP violation in the CKM matrix would be sufficient!

12.1.6 The CP Violation in the Background of Exploding Sphalerons

The paper Kharzeev et al. (2019) performed evaluation of the CP violation in the background of exploding sphaleron, which we will here follow. But first, a general discussion of why exploding sphalerons are different from arbitrary background field is discussed in the previous subsection.

One obvious (but important) point is that as any classical object, sphaleron has strong fields $A \sim 1/g_{EW}$, and therefore in the expansion over the field operators,

---

[2]Quark masses are all taken as they are in our world, that is, at Higgs field equal v. However, in a "hot spot" in which the sphaleron transition happens, the Higgs expectation value is smaller than in the broken phase, $\phi < v$, by certain factor. One can take care of this by rescaling all $\tilde{\lambda}$ by this ϕ/v factor. Since the function is dimensionless, its values are preserved, and the plot just moves horizontally as a whole by this factor.

Fig. 12.4 The CP-violating
part of the $W^4 Z^2$ diagram
$Im F_{ZZ}(\lambda)$ versus λ (GeV)

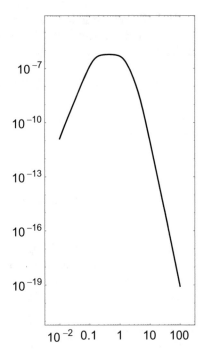

there is no coupling constant. So, for diagrams with 4 W and 2 Z, we speak about
the absence of α_{EW}^6 factor, or increase by say 10 orders of magnitude.

It is, however, still true that one can have perturbative expansion, but with a
parameter being deviation of CKM matrix from the unit one. It is still true that CP
violation requires at least four CKM matrices.

The nontrivial point made by Kharzeev et al. (2019) is the so-called *topological
stability*: the Dirac operator in the background of a sphaleron explosion still
possesses a topological zero mode,[3] $\lambda = 0$, surviving any gluon rescattering.
For this mode the formulae from the previous section cannot be applied, and CP
violation needs to be calculated in a different way.

Not going into detail, we emphasize the physics. In the broken phase $T < T_{EW}$,
six different quarks obtain six different masses, proportional to now nonzero Higgs
VEV $v(T)$. Suppose sphalerons produce first a particular quark, say u: in a CKM
vertex, it can morph into a different ones and then again. Combining amplitude
and conjugated amplitudes one can single out interference diagrams with (say) four
CKM vertices. When quark propagates between those, it gets a phase proportional

---

[3]Note that this scenario is different from earlier ones, based on small *momenta* $\vec{p} \approx 0$, which are
not topologically protected from rescatterings.

to time and m_Q^2/M_{KW}, where m_Q is its mas and M_{KW} "thermal Klimov-Weldon" quark mass

$$M_{KW} = \frac{g_s T}{\sqrt{6}} \sim 50\,\text{GeV} \tag{12.21}$$

induced by the real part of the forward scattering amplitude of a gluon on a quark. Different possible options with different intermediate flavors interfere with each other and produce CP-violating imaginary part. This means that probability to produce quarks and antiquarks are not the same, which leads to BAU.

Omitting details, let me go to (preliminary) answer from Kharzeev et al. (2019): for a single light quark (u, d) production, the CP asymmetry is

$$A_{CP} \sim J \frac{\left(m_b^2 - m_s^2\right)\left(m_c^2 - m_u^2\right)\rho^2}{M_{KW}^2} \sim 10^{-9} \tag{12.22}$$

where the sphaleron size ρ stands for typical distance between vertices.

Is it enough for BAU observed? In a very crude way, very small sphaleron rate near freeze-out $\sim 10^{-16}$ is basically canceled by very long time $Tt \sim 10^{16}$ in units of microscale. Therefore, up to a couple of orders of magnitude, the CP asymmetry is the BAU ratio. So to my opinion, the scenario *may* work, although the real calculation for a total CP asymmetry is still needed. Needless to say, there is and will be long process of its scrutiny.

12.1.7   Electroweak Sphaleron Explosion: Other Potential Observables

Sounds and gravity waves are naturally emitted at explosions, and sphaleron explosions are no exception. In a symmetric phase, $T > T_{EW}$, we have explicit time-dependent solution of the classical Yang-Mills equations (Ostrovsky et al. 2002; Shuryak and Zahed 2003a), from which one can calculate the stress tensor components and see that the word "explosion" is not a metaphor here. There is, however, a problem: in the symmetric phase, the sphaleron explosion is spherically symmetric and therefore cannot radiate direct gravitational waves. However, the indirect gravitational waves can still be generated at this stage, through the process

$$\text{sound} + \text{sound} \rightarrow \text{gravity   wave}$$

pointed out in Kalaydzhyan and Shuryak (2015). After the EWPT, at $T < T_{EW}$, the nonzero VEV breaks the symmetry, and the sphalerons (and their explosions) are no longer spherically symmetric but elliptic. With a nonzero and time-dependent quadrupole moment, they generate direct gravitational radiation. The corresponding matrix elements of the stress tensor were evaluated in Kharzeev et al. (2019).

Magnetic clouds and their linkage: In the broken phase, at $T < T_{EW}$, the original third component of the $SU(2)$ non-Abelian gauge field A_μ^3 gets mixed with electromagnetism, producing physical massive Z_μ and a massless a_μ. This mixing is controlled by the so-called Weinberg angle θ_W. Therefore sphaleron explosions in this phase create electromagnetic magnetic clouds. Since there are no electromagnetic magnetic monopoles, they are not screened but survive till today, as the so-called *intergalactic magnetic clouds*.

The interesting feature of those is that chirality of fermions (electrons) produced by sphalerons can be directly transported to chirality of the magnetic field, known as "magnetic linkage" or Abelian Chern-Simons number:

$$\int d^3x\, AB \sim B^2 \xi^4 \sim const \tag{12.23}$$

The configurations with nonzero (12.23) are called *helical*.

Multiple sphaleron and anti-sphaleron transitions create large number of left- and right-linked objects, which would annihilate each other (as quarks and antiquarks do). But if there is a (CP-violating) *asymmetry* between sphaleron and anti-sphaleron transitions, there would be remaining linkage which will not be able to annihilate (as remaining quarks do). Therefore, potential observation of helical misbalance in magnetic intergalactic fields today can be related to BAU.

Of course, many magnetic phenomena may happen, between electroweak transition and now. One of them is the so-called *inverse cascade* effect, following from Maxwell equation in matter with chiral unbalance and current proportional to magnetic field (CME). Discussion of them goes well beyond these lectures.

12.2   QCD Sphalerons

12.2.1   Sphaleron Transitions at the Initial Stage of Heavy Ion Collisions

In these lectures we do not systematically discuss heavy ion collisions, often referred to as "the little Bangs." For our current purposes, it is enough to point out few similarities and important differences from cosmology, the Big Bang. Similarly, both are explosions of systems much larger than the microscale of the matter involved, expanding slowly enough to warrant adiabatic approximation and entropy conservation. In both cases, we are interesting in crossing the phase transitions, electroweak or QCD ones, and phenomena associated with those. It is assumed that most of excitations from the initial time have been relaxed to thermal equilibrium.

Yet, in both cases, there are certain long-wavelength acoustic modes, with their relaxation time being so long that they do *not* die out from the initiation (Bang) time. Both in cosmology and heavy ion collisions, those are visible in final matter distribution: as acoustic microwave background temperature fluctuations in cosmology and as azimuthal correlations in the Little Bang. Their physics is rather similar: but in Universe we speak about angular harmonics up to few thousands and heavy ion collisions up to ten or so. Both in principle tell us something about the "initial state." For Big Bang it is amplitude and statistical properties of initial quantum oscillations of whatever fields there were at Big Bang;[4] for Little Bang it is (much less mysterious) initial positions of the nucleons at the moment of the collision.

Of course, in these lectures we are not interested in just initial distribution of matter densities in space, resulting in such acoustic mode excitation. Our interest is in the initial distribution in relevant *topological coordinates*, the Chern-Simons number, and its relaxation to equilibrium via certain instanton/sphaleron transitions.

High-energy collisions of nuclei start from highly excited out-of-equilibrium initial stage of mostly gluonic field. Unfortunately, it is not easy to make reliable ab initio calculations of it. All we experimentally know are the one-body density matrix or gluon density functions (PDFs) as a function of their momentum fraction x, $g(x)$. Very little is known about two-gluon correlations or coherency of those gluons. One popular view of it is *Color Glass Condensate* model, suggested by McLerran and Venugopalan (1994) which views it as high occupancy classical gluonic field. If so, it is calculated via classical Yang-Mills equations with certain sources. After collision, the field gets modified to (still out-of-equilibrium and classical) state of glue called GLASMA. With system expansion and decreasing density, the occupation numbers eventually reach $O(1)$, and thermal state of the gluon plasma is established.

Studies of GLASMA are done by numerical real-time solution of classical Yang-Mills equation, technically similar to simulations of out-of-equilibrium electroweak stage of the Big Bag we discussed above. In this section we will follow work by Mace et al. (2016) (in which you can also find references on the previous works on the subject).

During the evolution of GLASMA, there is diffusive spread over the topological landscape, so that measuring the Chern-Simons number as a function of time one observes a typical diffusive behavior (see Fig. 12.5 from Mace et al. 2016). "Cooled" configurations, by various methods, descend to the bottom of the topological landscape, that is, toward zero energy and integer values of ΔN_{CS}, shown by (violet) rectangular curve in the upper figure.

The lower figure shows the histogram of the distribution over ΔN_{CS}. Since in QCD there is no CP violation, the curve is left-right symmetric, that is, mean $< \Delta N_{CS} >= 0$. The width of the distribution indicates the magnitude of the additional chiral imbalance in the typical events created between the time $Q_s t = 10$ during the time interval $Q_s \delta t = 10$. Taking say the characteristic parton momenta in GLASMA $Q_s \sim 2\,\text{GeV}$, one finds that both the absolute time and time interval are about 1 fm/c. Taking this histogram as an example of that, we see than r.m.s. deviation corresponds to the diffusive motion adding $\Delta N_{CS}^{r.m.s.} \approx 3$ new transitions. For three

---

[4]In his papers and 2020 Nobel talk, Penrose speaks about observed multiple circles with common center in microwave backgrounds, hinting to some events happened *prior* to Big Bang!

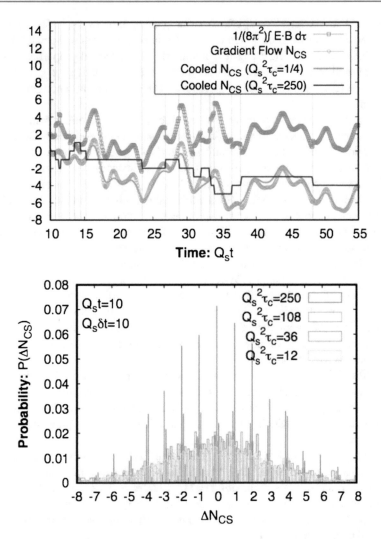

Fig. 12.5 Upper: evolution of the Chern-Simons number for a 0.35 single nonequilibrium configuration. Different curves correspond to different extraction procedures. Lower: histograms of the distribution over N_{CS} at time $Q_s t = 10$

light quark flavors, this corresponds to the added chiral charge of the configurations to be about

$$Q_5^{r.m.s.} = 2N_f * \Delta N_{CS}^{r.m.s.} \approx 20 \tag{12.24}$$

The rate of sphaleron transition decreases with time, so the first fm/c is dominant in such calculations.

Note that we call diffusive component ΔN_{CS}, not just N_{CS}, because this is the amount of sphaleron transition generated in GLASMA, on top of what it is at the time of the collision $N_{CS}(t = 0)$. This initial value is the subject of the next subsection.

12.2.2 Sphalerons from Instant Perturbations

As we argue in most chapters of this book, the QCD vacuum contains virtual fields which are topologically nontrivial. Given certain amount of energy, these virtual fields may become real excitations observable experimentally. Before we discuss specific proposal for experiments, let us first discuss some solvable quantum-mechanical example explaining the principle on which they are based.

We will show that basically if one wants to make some virtual fields of the vacuum real, all one has to do is to clap the palms of one's hands strong enough! Indeed, strong enough instantaneous perturbations applied to a system at some coordinate *under the barrier* will localize it in a state *near the top of the barrier* at the same value of the coordinate. Schematic picture of such excitation is indicated in Fig. 12.6 by vertical red arrow.

In QCD setting we will study production of sphalerons in high-energy collisions by "instanton-induced processes" in the next chapter. Here we only outline the argument by a quantum mechanical example from my paper (Shuryak 2003). Its main idea is that near-instantaneous perturbation does not leave time to move—change any coordinates, including the topological ones—while the energy can be changed by the amount determined by energy-time uncertainty relation $\Delta E \sim 1/\Delta t$. As we will see, under rapid perturbation, the system jumps mostly *on the barrier*, into the analogs of the "sphaleron path" states we discuss in this chapter.

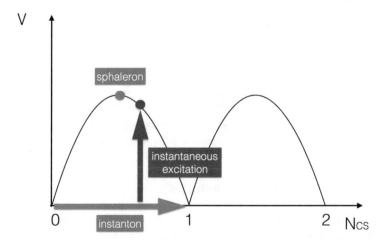

Fig. 12.6 Schematic representation of the potential as a function of Chern-Simons number. Horizontal blue arrow marked "instanton" indicates the tunneling process in vacuum, while vertical red arrow shows the direction of instantaneous excitation

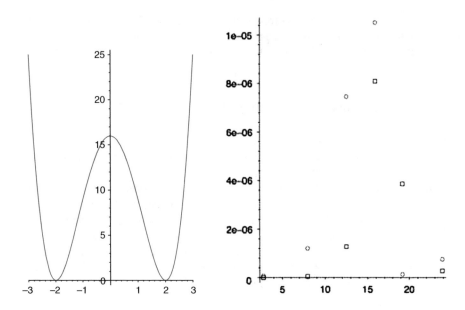

Fig. 12.7 (Left) The double-well potential used. (Right) The excitation probability P_n of the double-well system versus the excitation energy. Two sets of points are for two excitation functions mentioned in the text

Let us consider a double-well potential (shown in Fig. 12.7) and try to detect a particle's presence under the barrier, near its middle $x = 0$ point. To do so one may introduce an external perturbation $\delta V \approx \delta(x) f(t)$ localized near $x = 0$ and some time dependence $f(t)$. Standard solution is expanding $f(t)$ into Fourier integral and observing that frequencies tuned to the transition from the ground to the n-th level $\omega = E_n - E_0$ would create real transitions. One can calculate the probability of the transition $P_n = < 0|\delta V|n >|^2$. The results are plotted versus n as points in Fig. 12.7 Note that the peak excitation energy corresponds to the "sphaleron excitation," the maximum of the potential, $V \approx 16$ in this example.

The reason for this is the following: While there can be enough energy to excite the system to even higher states, the overlap matrix element to those higher states with the ground wave function rapidly decreases with excitation energy. Only near the maximum of the potential the final spatial wave function is smooth enough, and thus the overlap with the ground state wave function is large.

12.2.3 QCD Sphalerons in Experiments

As we already discussed above, the sphalerons were discovered in the context of electroweak theory, but—because there is no hope to produce an electroweak sphaleron experimentally—it has been mostly discussed in relation to cosmological applications. Now, returning to QCD, with its energy scale a factor 1000 lower, one might think that sphaleron production happens routinely in any hadronic or heavy ion collisions and is therefore studied throughout in experiments. Yet, as we will now discuss, only the former part of the previous sentence is true.

Three suggestions of how one can observe the QCD sphalerons experimentally were discussed in literature. We will go over these ideas briefly here, as discussion of real experiments would take us too far from the main goals of this book.

Discussion of sphaleron excitation in high-energy collisions has historically started in the setting of electroweak theory. If observed, it would be a spectacular demonstration of baryon and lepton number violation in standard model: but *as* we already mentioned several times, there are no prospects to do so experimentally. Still there were significant theoretical efforts made, and important insights were gained: for a good review, see Mattis (1992).

Schrempp, Ringwald, and collaborators applied this theory to lepton-hadron deep inelastic scattering (see Moch et al. 1997 and subsequent works). The idea was to consider deep inelastic *ep* collisions with large momentum transfer Q, in which small-size instantons will excite large mass $M \sim Q$ sphalerons. Unlike perturbative processes, resulting in a single-quark jet (or few jets, with radiated hard gluons), the sphaleron is expected to decay into a high multiplicity near-isotropic cluster of particles. The idea was to keep the scale in the weak coupling domain, so that the semiclassical calculation be well controlled, say $M \sim 10\,\text{GeV}$. Unfortunately, in this regime the cross-section is too small, compared to various perturbative processes, and the project eventually collapsed since it was not possible to tell the "signal" from the "background."

Another direction for experiments, which can be called "soft," has been proposed by Shuryak and Zahed (2003b). Instead of hard gluons, we propose to use soft Pomerons, focusing on the double-diffractive pp or $\gamma\gamma$ collisions. The sphaleron is expected to be found among various gluonic clusters which "jump out of the vacuum" at mid-rapidity. Observations of the two scattered protons provide very constrained kinematics and define the mass of the clusters. There are some old experiments which observed few GeV clusters of hadrons, with an isotropic decay: but theory quantum numbers were never controlled. Our suggestion was to look for certain exclusive channels related to the anomaly, which tells us that quark state produced must be of the particular flavor-spin structure:

$$\left(\bar{u}_R u_L\right)\left(\bar{d}_R d_L\right)\left(\bar{s}_R s_L\right) + (L \leftrightarrow R) \tag{12.25}$$

which would confirm or reject their topological origin. Similar argument was quite successful in the decays of η_c, which we will discuss in the next chapter. The scale of the sphaleron mass there should correspond to the average instanton size $\rho \sim 0.3\,fm$, corresponding to the sphaleron mass of about 3 GeV (incidentally, close to the mass of η_c.) The proposal to do these experiments at RHIC STAR detector in the pp mode has been made, but not yet done.

The third approach was suggested by Kharzeev et al. (2008). Instead of looking for individual sphaleron transitions, they proposed to observe total global chiral

$L - R$ disbalance of the fireball produced in heavy ion collision by fluctuating multiple sphaleron transitions. Note that the chiral charge is CP-odd quantity, and so its average value is of course zero in strong interactions.

Its observation was proposed to do based on the so-called chiral magnetic effect (CME) (Fukushima et al. 2008), according to which the CP-odd chiral disbalance leads to an *electric* current along the applied *magnetic* field.[5]

It is important that ambient matter should be quark-gluon plasma, at $T > T_c$, in which chiral symmetry is unbroken and thus chiral disbalance remains conserved. This suggestion has led to significant experimental activity. We of course cannot describe it here in detail: the effect is clearly seen, but possible backgrounds are not yet completely understood.

The experimental program continues, at both RHIC and LHC. Eventually we will learn the sphaleron rates, both the "primordial one," from nonzero topology in the vacuum wave function as well as that in GLASMA. Experimental check of the sphaleron theory in QCD will, no doubt, strengthen also our understanding of electroweak sphalerons in the Big Bang.

12.2.4 Diffractive Production of Sphalerons

After the correspondence between the instanton-anti-instanton configurations and sphalerons has been elucidated, let us return to high-energy collisions, following Nowak et al. (2001).

Calculations of the instanton-induced scattering proceeded gradually. First, one can consider a single gluon exchange between the vacuum field of the instanton and the quarks: in this case a gluon is of course just a perturbative tail of an instanton, a weak dipole-shaped potential. Then came realization that any number of exchanged gluons are summed up in the Wilson line, which can be easily analytically evaluated in the background field of the instanton. The probability is the amplitude squared, so it can be represented by the upper picture in Fig. 12.8, in which it is assumed that the vertical line—the unitarity cut—is very much removed from the instanton, so that its field at the cut can be ignored. The two parts of the plot are shown together, but they are in fact independent matrix elements with a single soliton.

Next came discussion of inelastic processes, in which first a gluon or few were passing through the cut. Such situation is shown by the left lower part of Fig. 12.8. Summing those gluons up results in a real breakthrough in the realization that the whole process can be described by a continuous semiclassical path, starting in the vacuum *under the barrier* and proceeding to a *turning point* and then to a *real (Minkowskian) evolution*. Whatever way the system is driven, it emerges from under

---

[5]Note that the coefficient between the current and filed is in this case T-even: so unlike the usual Ohmic current, this one is a *non-dissipative* one. This observation will lead to multiple applications of the CME in condense matter physics and perhaps even electronics.

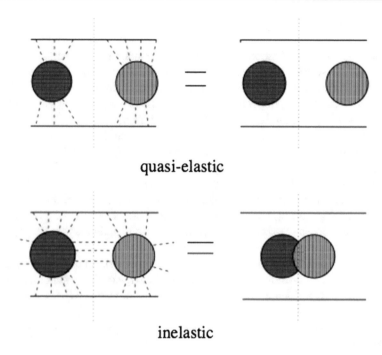

quasi-elastic

inelastic

Fig. 12.8 Schematic representation of the amplitude squared, with (without) gluon lines, is shown in the left (right) side of the figure. The dotted vertical line is the unitarity cut. The upper panel illustrates the quasi-elastic (at the parton level) amplitudes where only color is exchanged. The lower panel depicts inelastic processes in which some gluons cross the unitarity cut and some gluons are absorbed in the initial stage

the barrier via what we will call *a turning state*, familiar from WKB semiclassical method in quantum mechanics.

The turning states, released into the unitarity cut or Minkowski world, are the states we have already discussed. From there starts the real-time motion outside the barrier. Here the action is real and $|e^{iS}| = 1$. That means that whatever happens at this Minkowski stage has the probability 1 and cannot affect the total cross-section of the process: this part of the path is only needed for understanding of the properties of the final state.

So, how we describe inelastic collisions and what is produced? Let me summarize the main lesson of this section (and also of the section on instantaneous excitation in the previous chapter) in one sentence: if the particle under the barrier is hit, it jumps into the lowest state *at the barrier* with the same coordinate. The collision amplitude corresponds to the process described by "streamline" configurations, describing classical solution with an external force along the topological valley, exciting it to the sphaleron path.

12.3 Brief Summary

- Since the main application of sphaleron explosions is for cosmological electroweak phase transition, we start with some introductory facts about it. In short, in minimal standard model, it is a crossover, from symmetric to broken phases, occurring at temperature (12.6), which is nearly exactly three orders of magnitude higher than that of the QCD phase transition.
- Sphaleron explosions are seen in real-time lattice studies; their rate is known. These explosions freeze-out at another temperature $T_{fo} \approx 130\,\text{GeV}$.
- Near freeze-out the sphaleron explosions are out of equilibrium, which is one of Sakharov's conditions. Baryon number violation is there, so the question is what is the magnitude of the CP violation. Estimates in the literature are not yet precise enough to tell, whether known mechanism based on complexity of the CKM quark mass matrix is or is not sufficient to explain baryon asymmetry of the Universe.
- In hadronic and heavy ion collisions, one expect certain number of sphaleron explosions and therefore fluctuations in chiral charge of the system on event-by-event basis. Experiments looking for them are under way at RHIC collider.

References

Arnold, P.B., McLerran L.D.: Sphalerons, small fluctuations and Baryon number violation in electroweak theory. Phys. Rev. **D36**, 581 (1987)

Arnold, P.B., Son, D., Yaffe, L.G.: The Hot baryon violation rate is O (α-w**5 T**4). Phys. Rev. **D55**, 6264–6273 (1997). arXiv:hep-ph/9609481

Brauner, T., Taanila, O., Tranberg, A., Vuorinen, A.: Computing the temperature dependence of effective CP violation in the standard model. J. High Energy Phys. **11**, 076 (2012). arXiv:1208.5609

Burnier, Y., Laine, M., Shaposhnikov, M.: Baryon and lepton number violation rates across the electroweak crossover. J. Cosmol. Astropart. Phys. **0602**, 007 (2006). arXiv:hep-ph/0511246

Carson, L., Li, X., McLerran, L.D., Wang, R.-T.: Exact computation of the small fluctuation determinant around a sphaleron. Phys. Rev. **D42**, 2127–2143 (1990)

Crichigno, M.P., Flambaum, V.V., Kuchiev, M.Yu., Shuryak, E.: The W-Z-top bags. Phys. Rev. **D82**, 073018 (2010). arXiv:1006.0645

Dine, M., Kusenko, A.: The origin of the matter - antimatter asymmetry. Rev. Mod. Phys. **76**, 1 (2003). arXiv:hep-ph/0303065

D'Onofrio, M., Rummukainen, K., Tranberg, A.: Sphaleron rate in the minimal standard model. Phys. Rev. Lett. **113**(14), 141602 (2014)

Flambaum, V.V., Shuryak, E.: Possible role of the WZ-top-quark bags in baryogenesis. Phys. Rev. **D82**, 073019 (2010). arXiv:1006.0249

Fukushima, K., Kharzeev, D.E., Warringa, H.J.: The chiral magnetic effect. Phys. Rev. **D78**, 074033 (2008). arXiv:0808.3382

Garcia-Bellido, J., Garcia-Perez, M., Gonzalez-Arroyo, A.: Chern-Simons production during preheating in hybrid inflation models. Phys. Rev. **D69**, 023504 (2004). arXiv:hep-ph/0304285

Garcia-Recio, C., Salcedo, L.L.: CP violation in the effective action of the standard model. J. High Energy Phys. **07**, 015 (2009). arXiv:0903.5494

Hernandez, A., Konstandin, T., Schmidt, M.G.: Sizable CP violation in the bosonized standard model. Nucl. Phys. **B812**, 290–300 (2009). arXiv:0810.4092

Kalaydzhyan, T., Shuryak, E.: Gravity waves generated by sounds from big bang phase transitions. Phys. Rev. D **91**(8), 083502 (2015)

Kharzeev, D.E., McLerran, L.D., Warringa, H.J.: The effects of topological charge change in heavy ion collisions: 'Event by event P and CP violation'. Nucl. Phys. **A803**, 227–253 (2008). arXiv:0711.0950

Kharzeev, D., Shuryak, E., Zahed, I.: Baryogenesis and helical magnetogenesis from the electroweak transition of the minimal standard model. Phys. Rev. D **102**, 073003 (2019)

Mace, M., Schlichting, S., Venugopalan, R.: Off-equilibrium sphaleron transitions in the Glasma. Phys. Rev. **D93**(7), 074036 (2016). arXiv:1601.07342

Mattis, M.P.: The Riddle of high-energy baryon number violation. Phys. Rept. **214**, 159–221 (1992)

McLerran, L.D., Venugopalan, R.: Computing quark and gluon distribution functions for very large nuclei. Phys. Rev. **D49**, 2233–2241 (1994). arXiv:hep-ph/9309289

Moch, S., Ringwald, A., Schrempp, F.: Instantons in deep inelastic scattering: the Simplest process. Nucl. Phys. **B507**, 134–156 (1997). arXiv:hep-ph/9609445

Moore, G.D.: Fermion determinant and the sphaleron bound. Phys. Rev. **D53**, 5906–5917 (1996). arXiv:hep-ph/9508405

Nowak, M.A., Shuryak, E.V., Zahed, I.: Instanton induced inelastic collisions in QCD. Phys. Rev. **D64**, 034008 (2001). arXiv:hep-ph/0012232

Ostrovsky, D.M., Carter, G.W., Shuryak, E.V.: Forced tunneling and turning state explosion in pure Yang-Mills theory. Phys. Rev. **D66**, 036004 (2002). arXiv:hep-ph/0204224

Shuryak, E.: How quantum mechanics of the Yang-Mills fields may help us understand the RHIC puzzles. Nucl. Phys. **A715**, 289–298 (2003). arXiv:hep-ph/0205031

Shuryak, E., Zahed, I.: Prompt quark production by exploding sphalerons. Phys. Rev. **D67**, 014006 (2003a). arXiv:hep-ph/0206022

Shuryak, E., Zahed, I.: Semiclassical double pomeron production of glueballs and eta-prime. Phys. Rev. **D68**, 034001 (2003b). arXiv:hep-ph/0302231

Shuryak, E., Zahed, I.: CP violation during the electroweak sphaleron transitions (2016). arXiv:1610.05144

Tranberg, A., Smit, J.: Baryon asymmetry from electroweak tachyonic preheating. J. High Energy Phys. **11**, 016 (2003). arXiv:hep-ph/0310342

Chiral Matter

13

13.1 Examples of Chiral Matter

This terminology is rather recent, and we start with its explanation. By *chiral matter* we mean all forms of matter which:

(1) include massless (or near-massless) fermions;
(2) which interact with only vector/axial fields, classically preserving chirality;
(3) even including external fields and quantum anomalies (to be discussed below) the lifetime of fermion chirality is sufficiently long
(4) last but not least, a certain *imbalance* between the occupation of the left- and right-handed fermions is created initially, or even maintained indefinitely; so that

$$\langle \bar{\psi} \gamma_5 \psi \rangle \neq 0$$

In such setting quantum effects induced by the *chiral anomaly* can show their *macroscopic* manifestation. For extensive discussion of the history of the chiral effects, see, for example, Kharzeev (2014).[1]

Looking for examples, we naturally turn to QCD. It does have vector gauge fields and light quarks. However, as we discussed above, chiral symmetry is broken in its ground state (the QCD vacuum at $T = 0$) and, in fact, for any temperature below critical $T < T_c$. A nonzero *quark condensate* violates the chiral (left-right) symmetry present in the Lagrangian and creates quark effective mass, and therefore chirality relaxation time $\tau_5 \sim 1/M$ is only about half fm/c.

---

[1] For clarity, the subject of this section is not about violation of spatial P parity, but that of CP or T parity violation in some medium.

E. Shuryak, *Nonperturbative Topological Phenomena in QCD*
and Related Theories, Lecture Notes in Physics 977,
https://doi.org/10.1007/978-3-030-62990-8_13

However, at *supercritical* temperatures, $T > T_c$, one finds matter in quark-gluon plasma (QGP) phase, in which the chiral symmetry is *not* broken by quark condensate. So QGP is our first example of chiral matter. (Historically, it was also the first in which many effects to be discussed in this chapter were first considered.) Unfortunately, QGP is quite expensive—one needs to have a relativistic collider to create it—and so one may ask if there are other cheaper alternatives.

Chiral electron quasiparticles can also exist in the so-called *semimetals*. The terminology which needs to be explained at this point is as follows. As the reader surely knows, the *metals* have the chemical potential inside the allowed zone of the electron state, and thus there are Fermi spheres, with a nonzero surface area. Particle and hole excitations with small energies can exist near Fermi sphere surface, creating electric conductivity, etc. *Insulators* have the chemical potential located inside the forbidden energy zone, and thus there is no excitations with small energies. The metals have free electrons to move, by external fields, and the insulators have not.

Rather famous *graphene* made of carbon has approximately linear electron spectrum near certain momenta and can be approximated by analogs of massless fermions. However, it is a two-dimensional material, while the anomaly phenomenon (to be discussed below) exists in 1+3 dimensions: so *graphene* will not be discussed in this chapter.

There are other cases in between metals and insulators, called *semimetals*, which possess several points at which the valence and conduction bands touch. The so-called Dirac semimetals have the "Dirac point," shared by "left" and "right" fermions, near which linear relativistic-like dispersion relation is valid (see Fig. 13.1).

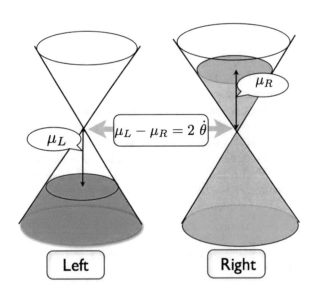

Fig. 13.1 Dirac cones of the left and right fermions. In the presence of the changing chiral charge, there is an asymmetry between the Fermi energies of left and right fermions $\mu_L - \mu_R = 2\mu_5$

There exist also "Weyl semimetals," for which two chiralities have two separate touching points. The quasiparticle modes near those points have the following Hamiltonian:

$$H = \pm v_F \vec{\sigma} \cdot \vec{k} \qquad (13.1)$$

where $\vec{\sigma}$ being Pauli spin matrices. It is of the kind originally suggested by Weyl for uncharged massless fermions. The first observation of CME in condense matter setting (Li et al. 2016) has been made using zirconium pentatelluride, $ZrTe_5$, a is three-dimensional Dirac semimetal.

13.2   Electrodynamics in a CP-Violating Matter

Chirality—the product of spin and momentum $(\vec{s}\,\vec{p})$—for massless particles, fermions, or bosons (e.g., photons) has two values, commonly referred to as "left-" and "right"-handed polarizations. It is odd under P parity transformation $\vec{x} \rightarrow -\vec{x}$ since momentum changes sign and spin does not. Thus matter in which there is left-right disbalance is P-odd (i.e., not the same as its mirror image).

To illustrate what we mean by "macroscopic" manifestations of chirality, let us recall some facts about vector and axial currents. Dirac equation for a free massive fermion conserves the vector current—the fermion number:

$$\partial_\mu \left(\bar{\psi} \gamma_\mu \psi \right) = 0 \qquad (13.2)$$

but not the axial current[2] (the difference between the number of the left and the right chirality components):

$$\partial_\mu \left(\bar{\psi} \gamma_\mu \gamma_5 \psi \right) = 2m \left(\bar{\psi} \gamma_5 \psi \right) \qquad (13.3)$$

because of the mass term which connects the left and right components of the fields in the Lagrangian.

In QGP one finds two light quarks, called up (u) and down d. Their masses are of the magnitude of few MeV, much smaller than any other relevant scales, and since they play no role, one can for simplicity consider them to be zero. If so, the r.h.s. of the equation above is also zero, and this means that axial charge is also conserved. A simpler way to understand the situation is to state that the number of left- and right-polarized quarks would be conserved. This is a classical statement of chiral symmetry.

---

[2]The spinor and its conjugate satisfy the Dirac equation. In vector case one gets $m - m = 0$ when acting to the left and right, but in axial case extra γ_5 change it to $m + m$. Recall also Dirac conjugation is defined as $\bar{\psi} = \psi^+ \gamma_0$.

However, on *quantum level* (including loop diagrams and/or nontrivial measure of fermionic path integral), some part of classical chiral symmetry (of the Lagrangian) does not survive: this is what we call *chiral anomaly*.

The physics of CP-violating matter can be conveniently discussed using some pseudoscalar effective field[3] $\theta(x)$. Our discussion of many interesting effects related to its existence we will discuss in the simplest framework of Abelian (or $U(1)$) gauge theory, called Maxwell-Chern-Simons electrodynamics. We will follow here Kharzeev (2010).

While the Maxwell theory historically emerged, term by term, from certain ingenious experiments, in modern textbooks, all of them are derived from a single *principle of gauge invariance*, requiring to change all derivatives of the charged fields by their covariant form:

$$\partial_\mu \rightarrow D_\mu = \partial_\mu - ieA_\mu \tag{13.4}$$

plus the statement that the gauge action can only be given by the only dimension 4 gauge invariant Lorentz scalar:

$$L_{Maxwell} = -\frac{1}{4}F^{\mu\nu}F_{\mu\nu} \tag{13.5}$$

A conventional media modify Maxwellian theory only slightly, renormalizing the coefficients of $(\vec{E})^2$, $(\vec{B})^2$ by certain coefficients, known as electric and magnetic permitivities ϵ and μ.

However, if the requirement of CP invariance is lifted, one may add another term $(\vec{E}\vec{B})$ of the same dimension. We will use it in the form

$$L_{MCS} = -\frac{1}{4}F^{\mu\nu}F_{\mu\nu} - A_\mu J^\mu - \frac{c}{4}\theta \tilde{F}^{\mu\nu}F_{\mu\nu} \tag{13.6}$$

where c is some coefficient and θ will be treated as a new field, which can be time- and space-dependent field. If one adds its kinetic energy, this ("axion") field θ can be promoted to a dynamical entity: but for our discussion, there is no need to do so. Therefore, one should think of θ as a matter-induced term in a Lagrangian.

Suppose first it is just a constant $\theta = const(t, \vec{x})$. If so, one finds that the last term does *not* change the equations of motion. It happens because the term we added is in fact a full divergence:

$$\tilde{F}^{\mu\nu}F_{\mu\nu} = \partial_\mu J^\mu_{CS} \tag{13.7}$$

---

[3]Extension of QCD including the so-called *axion* dynamics was proposed and extensively discussed in literature. While naturally appearing in string theory and holographic models, it is still hypothetical and is well beyond the scope of this book.

where the current is (the Abelian part of (6.1)) Chern-Simons current:

$$J^\mu_{CS} = \epsilon^{\mu\nu\rho\sigma} A_\nu F_{\rho\sigma} \tag{13.8}$$

The four-volume integral of the full divergence can be rewritten as three-dimensional integral of the current flux over the volume boundaries. If the boundary is a "large sphere," at which all fields are assumed to be zero, there is no boundary term.

If, however, $\theta(x)$ is time- and/or space-dependent, the derivative can be passed to it by integration by parts, so the last term in action can also be written as $+(c/4)(\partial_\mu\theta)J^\mu_{CS}$. To make the lessons more familiar, let us write the equations of motion in the nonrelativistic form, introducing the following notations for the vector and axial currents:

$$J_0 = \rho, \vec{J}, \quad M = \partial_0\theta, \quad \vec{P} = \vec{\nabla}\theta \tag{13.9}$$

Here they are

$$\vec{\nabla} \cdot \vec{E} = \rho + c\vec{P} \cdot \vec{B} \tag{13.10}$$

$$\vec{\nabla} \times \vec{E} + \frac{\partial \vec{B}}{\partial t} = 0 \tag{13.11}$$

$$\vec{\nabla} \cdot \vec{B} = 0 \tag{13.12}$$

$$\vec{\nabla} \times \vec{B} - \frac{\partial \vec{E}}{\partial t} = \vec{J} + c\left(M\vec{B} - \vec{P} \cdot \vec{E}\right) \tag{13.13}$$

One can see that both electric and magnetic fields get their sources—the r.h.s. of the first and last equations—modified. This leads to several new effects, two of them we will mention.

Witten effect appears for nonzero $\vec{P}$. For example, consider a spherical "defect," a region in which θ vanishes, while it is nonzero outside. Such a defect can contain a magnetic monopole. Without new terms, those would support magnetic field only, but at nonzero $\vec{P}$, the r.h.s. of the first equation sources the electric field as well! Thus a magnetic monopole obtains also an electric charge and becomes a *dyon*.

The electric charge separation in external magnetic field also appears for nonzero $\vec{P}$: the r.h.s. of the first equation may be zero when the two terms cancel each other.

If the $\theta(t)$ is *time*-dependent and therefore $M \neq 0$, one finds more unusual effects. We will mention just one of them: *chiral magnetic effect* (CME) which we will discuss more in the next section is one of them—it is a vector current along the magnetic field

$$\vec{J} = -cM\vec{B} \tag{13.14}$$

vanishing the r.h.s. of the last Maxwell-Chern-Simons equation. One can put $\vec{E} = 0$ and take constant $\vec{B}$, which vanishes all other terms.

13.3   Chiral Magnetic Effect (CME) and the Chiral Anomaly

We start this section with some general discussion of space-time symmetries and currents in the medium. It will explain the required conditions under which the CME *may* exist.

The expression we start with is the Ohmic current, induced by the electric field:

$$\vec{J} = \sigma_{Ohm}\vec{E} \tag{13.15}$$

If one watch this phenomenon in a mirror (which means performing the so-called P parity transformation $\vec{x} \to -\vec{x}$), both the l.h.s. and the r.h.s. vectors change sign, so the coefficient σ_{Ohm} remains unchanged.

Imagine now that one flips the sign of time $t \to -t$, performing the so-called T parity transformation. The current is related with velocity of the charge, and it changes sign. The electric field is only related with *positions* of the charges which created it: therefore it is unchanged. The conclusion is that $\sigma_{Ohm} \to -\sigma_{Ohm}$, which is not surprising since Ohmic current is dissipative, leading to increasing entropy of the media, and thus it should be also dissipative in an imagined world in which time goes backward.

In superconductors the expression for a current, proposed by London, is

$$\vec{J} = \sigma_{London}\vec{A} \tag{13.16}$$

where the inducing field is the vector potential. If one recalls that $\vec{E}$ contains $\partial_t \vec{A}$, it is clear that $\vec{A}$ must be T-odd. Therefore, σ_{London} should be the same in our world and in "backward time" Universe.[4] It follows that it cannot lead to dissipation and the entropy growth. And indeed, supercurrents are eternal and non-dissipative!

The CME is the vector current along the *magnetic* field:

$$\vec{J} = \sigma_{CME}\vec{B} \tag{13.17}$$

---

[4] According to Kharzeev, this argument has been made by V.I. Zakharov.

Transformation in a mirror of $\vec{J}$ is sign changing, while $\vec{B}$ remains unchanged. So σ_{CME} must be P-odd. So one may think about weak interaction effects, like with neutrinos, since they violate P parity by involving left-handed fermions only.

Under the T parity transformation, both $\vec{J} and \vec{B}$ change sign, as they are both related with velocities of the charges. Thus σ_{CME} must be T-even and therefore be non-dissipative(!), as the supercurrents are. This agrees with the idea that magnetic field, while exerting a force on a moving charge, does not make any work on it. So if one can find the conditions under which CME current can be produced, it will not dissipate. Unlike the supercurrent, it does not seem to require coherence and low temperatures.

The *chiral anomaly* not only leads to the existence of CME in a chiral matter but also provides the universal coefficient of the effect. In its operator form, the expression is

$$\vec{J} = \frac{e^2}{2\pi^2}\mu_5\vec{B} \qquad (13.18)$$

where μ_5 is the chemical potential for the chiral charge.[5]

Kharzeev and collaborators first suggested to use QGP—a good chiral matter— and superstrong magnetic field created by two positive colliding ions, directed normal to the collision plane. If vector CME current appears, it will create electric dipole along $\vec{B}$, which would be observed by a charge dependence of the elliptic flow of secondaries.

However, since strong interactions are CP-conserving, the chiral imbalance can only appear as a fluctuation. Indeed, in Chap. 12, we discussed sphaleron explosions inside the fireballs produced in heavy ion collisions. Their numbers fluctuate, on event-by-event basis, and so the total chirality of the fireball.

It means that we can only observe CP-even quadratic effect. Possible backgrounds of origins unrelated to CME, or even to $\vec{B}$, can influence observations. Therefore, strictly speaking, the observations of CME in heavy ion collisions are not yet completely clear: a special run at RHIC with two nuclei, having the same number of nucleons but different charge Z, has been performed but not yet analyzed, to at least separate $\vec{B}$-dependent effects from others.

It is easier to explain the CME experiment with the semimetal. In the absence of fields, there is of course no chiral imbalance, $\mu_L = \mu_R$, and there are equal numbers of left and right fermions. But when parallel electric and magnetic fields are applied,

---

[5] The first paper in which this formula was written was by Vilenkin. The setting was P-violating due to left-handed fermions—neutrinos—of weakly interacting sector, near a rotating black hole. The existence of this current in equilibrium was questioned by C.N. Yang, and in the next paper, Vilenkin showed that the equilibrium current is in fact zero. The resolution lies in the realization that the chiral matter is always metastable, not in equilibrium.

the change in the chiral density ρ_5 appears, as illustrated in Fig. 13.1 from Kharzeev (2010). The time evolution of the chiral imbalance can be written as

$$\frac{d\rho_5}{dt} = \frac{e^2}{4\pi^2}(\vec{E} \cdot \vec{B}) - \frac{\rho_5}{\tau_5} \tag{13.19}$$

where the first term in the r.h.s. again stems from the chiral anomaly and the second is the chiral density relaxation. The stationary condition is reached at late times, and then the l.h.s. is zero because the gain and loss terms in the r.h.s. cancel each other. So one get

$$\rho_5(t \to \infty) = \frac{e^2}{4\pi^2}(\vec{E} \cdot \vec{B})\tau_5 \tag{13.20}$$

The chemical potential $\mu_5 \sim (\vec{E} \cdot \vec{B})$ as well, and putting it into the expression (13.18) one would obtain the current $\vec{J} \sim \vec{B}(\vec{B}\vec{E})$. This is indeed what is observed (Li et al. 2016) in zirconium pentatelluride, $ZrTe_5$, and then in other materials. Potentially, the CME current may be used in electronics, providing non-dissipative currents at room temperatures.

13.4 Chiral Vortical Effect

Can a magnetic field $\vec{B}$ be substituted by another quantity, possessing a similar P, T parity, e.g., *vorticity* which we define in relativistic notations by

$$\omega_\mu = \left(\frac{1}{2}\right)\epsilon^{\mu\nu\lambda\rho}u_\nu\partial_\lambda u_\rho \tag{13.21}$$

with 4-velocity u_μ. The chiral vortical effect (CVE) introduced in Son and Surowka (2009), Kharzeev (2010) can be summarized by a relation:

$$j_\mu = \sigma_{CVE}\omega_\mu \tag{13.22}$$

with a vector current propagating along the vorticity. A similar relation—but of course with a different kinetic coefficient—can be written for the entropy current s_μ. Son and Surowka argued that entropy production must be positive $\partial_\mu s_\mu > 0$, but Kharzeev then argued that it should in fact be zero because of non-dissipative nature of the effect. These considerations lead to a specific expression for the coefficient σ_{CVE} in terms of matter EOS.

So far I am not aware of any specific applications of the CVE. In ultrarelativistic heavy ion collisions specifically, vorticity is not zero but is perhaps too small to be used for it.

13.5   The Chiral Waves

The CME expression (13.18) has an analog: interchanging vector current to axial one and axial chemical potential to the usual—vector—one μ, one also gets the following expression:

$$\vec{J}_A = \frac{e}{2\pi^2}\mu\vec{B} \qquad (13.23)$$

As argued in Burnier et al. (2011), combining the two together one finds new oscillation mode called the *chiral magnetic wave*. Indeed, divergence of the currents can be substituted by time derivatives of the corresponding densities, and two linear equations combined produce one equation of the second order.

So a heavy ion collision starting with certain baryon number density and μ leads to a quadrupole excitation in which the density and the chiral imbalance should oscillate into each other. In the previous chapter, we had shown how the density oscillations—the sound modes—were observed. A search for the chiral magnetic wave is in progress.

In early Universe chiral waves may perhaps lead to (at least locally) chiral magnetic fields: if it is true, it has a potential to contribute to Baryon asymmetry puzzle we discussed before.

13.6   Brief Summary

- Types of matter with chiral disbalance violate CP or T symmetry and possess unusual properties. Examples include quark-gluon plasma and certain semimetals.
- Introducing space- and time-dependent field $\theta(t, x)$ times $F_{\mu\nu}\tilde{F}^{\mu\nu}$ into Lagrangian, one gets modified Maxwell equations. Among arising new effects are Witten effect (e.g., turning monopoles into dyons), electric charge separation, and chiral magnetic effect (CME).
- CME should produce non-dissipative currents, as it was demonstrated in a number of materials, but not yet in QGP. Applications of it in electronics, and in particular quantum computers, potentially can significantly alleviate requirements for super-low temperatures.
- Rotation can substitute for magnetic field, creating chiral vortical effect (CVE).
- A combination of Ohmic and CME currents leads together to a new propagating mode, the chiral wave.

References

Burnier, Y., Kharzeev, D.E., Liao, J., Yee, H.-U.: Chiral magnetic wave at finite baryon density and the electric quadrupole moment of quark-gluon plasma in heavy ion collisions. Phys. Rev. Lett. **107**, 052303 (2011). arXiv:1103.1307

Kharzeev, D.E.: Topologically induced local P and CP violation in QCD x QED. Ann. Phys. **325**, 205–218 (2010). arXiv:0911.3715

Kharzeev, D.E.: The chiral magnetic effect and anomaly-induced transport. Prog. Part. Nucl. Phys. **75**, 133–151 (2014). arXiv:1312.3348

Li, Q., Kharzeev, D.E., Zhang, C., Huang, Y., Pletikosic, I., Fedorov, A.V., Zhong, R.D., Schneeloch, J.A., Gu, G.D., Valla, T.: Observation of the chiral magnetic effect in ZrTe5. Nature Phys. **12**, 550–554 (2016). arXiv:1412.6543

Son, D.T., Surowka, P.: Hydrodynamics with triangle anomalies. Phys. Rev. Lett. **103**, 191601 (2009)

Instanton-Dyons

14

14.1 The Polyakov Line and Confinement

14.1.1 Generalities

We have already introduced the *holonomy*, or Polyakov line, in Sect. 1.4.1, together with its relation to confinement. Here we remind the setting in a bit more details. The finite-temperature formulation of the periodic path integral defines the Euclidean time τ on a circle with circumference $\beta = \hbar/T$. A general mathematical construction allows closed loops around circles and toruses, known as holonomies. In the gauge theory, it generates a gauge invariant object known as the Polyakov line

$$P = Pexp \left(i \oint A_\mu^a T^a dx_\mu \right) \tag{14.1}$$

where $T^a = t^a/2$ is the color generator in a particular representation of the color group, namely, the fundamental one.

The temperature dependence of its VEV is plotted in Fig. 14.1. The left plot, from Kaczmarek et al. (2002), is for pure gauge theory. One can see that at high $T \ \frac{1}{N_c} < tr P >\to 1$: below we will call it "trivial holonomy" limit, because it corresponds to vanishing A_4. At $T < T_c$, the VEV is zero, which corresponds to strict confinement. As one can see in the plot, there is a finite jump, from the value of about 0.4 to 0, so the transition is of the first order.

The plot on the right side of Fig. 14.1, from Bazavov and Petreczky (2013), is for QCD with light quarks. In this case, the VEV is never strictly zero but decreases to rather small values gradually.

© The Author(s), under exclusive license to Springer Nature Switzerland AG 2021
E. Shuryak, *Nonperturbative Topological Phenomena in QCD and Related Theories*, Lecture Notes in Physics 977,
https://doi.org/10.1007/978-3-030-62990-8_14

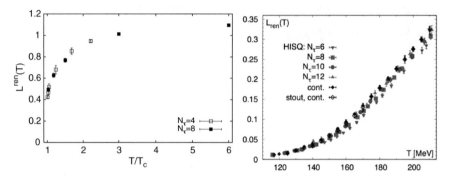

Fig. 14.1 The renormalized Polyakov line, in pure gauge SU(3) (left) and QCD with light quarks (right)

14.1.2 The Free Energy of the Static Quark on the Lattice

The Polyakov line is gauge invariant and thus a physical quantity, related to the free energy of the static quark:

$$\left\langle \frac{1}{N_c} tr\, P \right\rangle = exp(-F_Q/T) \tag{14.2}$$

Renormalized Polyakov loop was calculated in a wide range of temperatures by Bazavov et al. (2016). The resulting free energy of the static quark extrapolated to physical QCD is shown in Fig. 14.2.

In theories with light quark, there is level-crossing transition between the heavy quark $\bar{Q}Q$ state and the four-quark meson-meson state $\bar{Q}q\bar{q}Q$. It is also of course "dressed" with certain vacuum polarization around the static meson. The value of

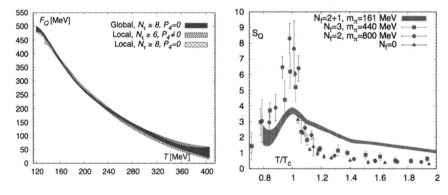

Fig. 14.2 Free energy and entropy of the static quark, extrapolated to physical QCD from different lattice measurements. The value of the quark mass is expressed via the pion mass: one can see that as it becomes smaller, the magnitude of the near-T_c peak is reduced

F_Q at the lowest T, ≈ 500 MeV, corresponds to the effective free energy of the extra light quark. In vacuum, at $T = 0$, it can be phenomenologically evaluated from the mass difference between heavy-light B meson and the b quark, $M_B - M_b$.

The so-called Polyakov loop susceptibility is defined by

$$\chi = (VT^3)\left(< |P|^2 > - < |P| >^2\right) \tag{14.3}$$

and it is also calculated in Bazavov et al. (2016). Using the gradient flow method, one can study how it changes as quantum fluctuations are reduced.

14.1.3   The Color Phases

Furthermore, it has temperature-dependent VEV $< P(T) >$ which is a unitary matrix. Its eigenvalues are complex number with modulus 1, so they can be written as an exponential with i times some phases, usually defined by

$$A_4 = 2\pi T \, diag(\mu_1, \mu_2, \ldots \mu_{N_c}) \tag{14.4}$$

Assuming they are ordered in magnitude $\mu_1 < \mu_2 \ldots$, and introducing one more $\mu_{N_c+1} = 1 + \mu_1$, one can proceed to their differences:

$$\nu_m = \mu_{m+1} - \mu_m \tag{14.5}$$

In Fig. 14.3 (left), we show examples of locations of the holonomy eigenvalues for the simplest case of the $SU(2)$ gauge group, for which most of the calculations are made, and a generic case with five colors. In the former case, $\mu_1 = -\mu_2 = \mu$, so there is only one parameter.

The VEV of the Polyakov line is

$$\left\langle \frac{1}{2} Tr P \right\rangle = \cos(\pi \nu) \tag{14.6}$$

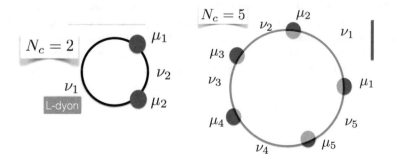

Fig. 14.3 The holonomy circle and the definition of the parameters μ_i and ν_i, for 2 and 5 colors

At high $T < P >\approx 1$, which means all $\mu_i \approx 0$. Only one $v_i \approx 1$ However in the temperature interval $(2...1)T_c$, it changes to zero (or small value in QCD with quarks). Accounting for this phenomenon leads Pisarski and collaborators to "semi-QGP" paradigm (Pisarski 2009) and construction of the so-called PNJL model. At $T < T_c$, in a confined phase, $< P >= 0$, which means that $v = 1/2$.

Generalization to $SU(N_c)$ follows the usual Cartan subalgebra of all diagonal (mutually commuting) generators and generalizes two dyons just described for the $SU(2)$ to N_c of them, $N_c - 1$ of M-type and one L-type. We will briefly introduce notations to be used below, starting from the HIggs VEV

$$A_4(\infty) = (2\pi T)diag(\mu_1 \ldots \mu_{Nc}); \quad \sum_i \mu_i = 0 \qquad (14.7)$$

where the latter condition follows from the required zero trace of A_4. Introducing $\mu_{Nc+1} = 1 + \mu_1$ and the differences

$$v_i = \mu_{i+1} - \mu_i; \quad \sum_i v_i = 1 \qquad (14.8)$$

These differences will determine the masses of the corresponding charged components and thus the core sizes, which are $\sim 1/(2\pi T v_i)$ for the i-th dyon.

It is also necessary to add the following. The so-called Cartan subalgebra of all diagonal (mutually commuting) generators for $SU(N_c)$ are made of $N_c - 1$ matrices of the type:

$$diag(1, -1, 0, \ldots 0), diag(0, 1, -1, 0, \ldots 0), \ldots, diag(0, 0, \ldots 1, -1)(14.9)$$

The corresponding $N_c - 1$ components of the gluon field remains massless, while the rest of them get nonzero mass from the term in the action $[A_4, A_\mu]^2$. For the case of two colors, there is only one "massless photon" A_μ^3 and two "massive gluons" A_μ^1, A_μ^2. For the physical case of three colors, there are two "massless photons" and six "massive gluons."

Since the instanton-dyons are basically SU(2) objects, they are made of the gauge fields with two colors. If one asks which ones, those are defined by two ends of the corresponding segment v_i with which it is identified with, namely, colors $i + 1$ and i. Two subsequent dyon types, say identified with segments v_i and v_{i+1}, have one color in common, namely, $i+1$, with charges plus and minus, respectively. The dyon types which are not subsequent on the circle have no common "photon charges" and thus cannot interact at large by Coulomb-like forces.

14.2 Semiclassical Instanton-Dyons

14.2.1 The Instanton-Dyon Field Configuration

In the chapter on monopoles, it was many times stated that QCD-like theories lack scalar fields. Some features depend heavily on that—in particular existence of chiral symmetries and physics related to them. But it also presents a number of difficulties to the physics of monopoles.

We already mentioned that one possible way to proceed is to use the fourth component of the gauge field A_4 as adjoint scalar. Of course, this is not a Lorentz invariant choice—but at nonzero T, the frame in which matter is at rest anyway. More importantly, if one attempts to analytically continue the theory to Minkowski formulation, A_4 gets imaginary and the construction loses its meaning.

It is perhaps worth repeating that unlike particle-monopoles discussed in the earlier chapter, those instanton-dyons are *not* particles in the ordinary sense. (The original name for instantons from Belavin et al. (1975) for any objects of the kind has been "pseudoparticles.") The distinction between the two types is described as follows: "particle" (or quasiparticle) dyons are described in the usual Minkowski world, and their mass squared is the sum of two positive terms, originating from $E^2 + B^2$, both limited from below by the so-called Bogomolny bounds, proportional to their integer electric and magnetic charges n_e, n_m, respectively.

The "pseudoparticle" dyons, like instantons, are self-dual (or antiself-dual) in Euclidean formulation. In Minkowski notations, this means that

$$\vec{E} = \pm i\,\vec{B} \qquad (14.10)$$

and thus negative electric energy cancels the magnetic one to $E^2 + B^2 = 0$. As we discussed above, they semiclassically describe vacuum transition between different gauge nonequivalent topological classical vacua, at zero classical energy.

Historically, their applications started "from the top," starting from the finite-T instantons, or calorons, generalized to the case of a nonzero holonomy, by Kraan and van Baal (1998). (More or less at the same time, Lee and Lu (1998) derived it from certain brain construction related to AdS/CFT.) Only after plotting the action distribution, it has been realized that instantons get split into N_c independent clusters—the instanton-monopoles or instanton-dyons. The technicality of the KvBLL solution is very interesting theoretically but rather involved: we return to it a bit later.

If the instanton constituents are very distant from each other, one can of course first study them as separate solitons and then try to do "bottom-up" superposition of them.

Before we discuss the solutions, let us mention their quantum numbers. For the simplest $SU(2)$ color group, there are four types of dyons: see Table 14.1. The charges and the mass (in units of $8\pi^2/e^2 T$) for four SU(2) dyons cover all four possibilities for the electric and magnetic charges:

Table 14.1 Electric charge, magnetic charge, and action (in units of the instanton action) for four types of the instanton-dyons of the $SU(2)$ gauge theory

Name	E	M	S/S_{inst}
M	+	+	v
$\bar{M}$	+	−	v
L	−	−	$\bar{v} = 1 - v$
$\bar{L}$	−	+	$\bar{v} = 1 - v$

The $M, \bar{M}$ dyons are the "ordinary" BPS dyons, which in the spherical "hedgehog" gauge is

$$A_4^a = \mp n_a v \Phi(vr) \tag{14.11}$$

$$A_i^a = \epsilon_{aij} n_j \frac{1 - R(vr)}{r}$$

where $n_a = r_a/r$ is the radial unit vector, the minus-plus is for self (antiself)-dual solutions, and the two functions are

$$\Phi(vr) = \left[\coth(vr) - \frac{1}{vr} \right] \rightarrow \mp n_a v \left[\left[1 - \frac{1}{vr} \right] + O(exp(-vr)) \right] \tag{14.12}$$

$$R(vr) = vr/\sinh(vr)$$

The corresponding magnetic field has the transverse and longitudinal structures, which contained derivatives of these functions, but can be rewritten without them as follows:

$$B_i^a = (\delta_{ai} - n_a n_i)[-v\Phi)vr)R(vr)/r] + n_a n_i [R^2(vr) - 1]/r^2 \tag{14.13}$$

While the former one exponentially decreases with the distance, the longitudinal has $1/r^2$ behavior indicating the nonzero magnetic charge. It of course matches the electric charge.

Construction of the $L, \bar{L}$ dyons starts from the same expressions, in which the following substitution

$$v \rightarrow \bar{v} = 2\pi T - v \tag{14.14}$$

is made. But then, it has a "wrong" asymptotics of the Higgs field A_4: this however can be remedied by the so-called twist. It is gauge transformation with the *time-dependent* matrix

$$U = exp\left[i2\pi T x^4 (\tau^3/2) \right] \tag{14.15}$$

the time derivative of which subtracts the unwanted $2\pi T$ from the Higgs asymptotics. Note that the color matrix τ_3 commutes with the diagonal (Abelian) part of the dyon field, leaving it as before. But the "core" of the L dyon is made of charged fields with colors 1,2: thus the core is time-dependent. So, the $L, \bar{L}$ dyons are not static solutions, and thus they do not exist in three-dimensional theory (and were not covered in the previous chapter). It can only be defined in the finite-T theory with the Matsubara time!

The actions of the dyons are $v_i(8\pi^2/e^2)$, so if all $self-dual$ ones are summed using $\sum_i v_i = 1$, the result is the instanton action. This statement, so far demonstrated for very distant (noninteracting) dyons, should in fact be true in general because self-duality relates the total action of the PBS dyons to their total charge.

I was often asked[1] how is it possible to have separate objects with a non-integer topological charge Q. The answer is they are independent in physical sense, not in mathematical one.

As any monopoles, they are interconnected by singular but invisible Dirac strings. This can be seen in two ways. One is explicit construction starting with well-separated monopoles, "combed" into a gauge in which the color direction of "Higgs" $< A_4 >$ is some fixed direction. SU(2) instanton is made of $L + M$ dyons, with a string connecting their centers.

The explicit solution, obtained for the single caloron in the Kraan-van Baal paper (Kraan and van Baal 1998), leads to this conclusion after plotting the action distribution. The solution itself can be described using the so-called prepotential scalar function.[2]

$$\psi(r) = (1/2)tr(A_N \ldots A_1) - \cos(2\pi r_4) \qquad (14.16)$$

where here and below the temperature and circumference of the Matsubara circle are temporarily put to $T = 1/\beta = 1$. The matrices need to be multiplied in the order written, and in the order corresponding to the magnitude of μ_i. The following 2×2 matrices A_m (not to be confused with the gauge potential A_μ) are defined by the product of two matrices

$$A_m = \begin{pmatrix} 1 & (\vec{y}_m - \vec{y}_{m+1})/r_m \\ 0 & r_{m+1}/r_m \end{pmatrix} \begin{pmatrix} c_m & s_m \\ s_m & c_m \end{pmatrix} \qquad (14.17)$$

with $c_m = \cosh(2\pi v_m r_m)$, $s_m = \sinh(2\pi v_m r_m)$. Here $r_m = |\vec{r} - \vec{y}_m|$ is the distance from the observation point to the m-th dyon center $\vec{y}_m$, and $\vec{y}_m - \vec{y}_{m+1}$ are

---

[1] Once by Polyakov himself.

[2] In order to explain how the gauge potential A_μ was constructed, one needs to understand Nahm construction, which I cannot describe better than done in Kraan-van Baal papers. Without that, the formulae for A_μ and $G_{\mu\nu}$ look like pure magic. I decided not to go into it, keeping only definition of ψ and the action distribution defined in terms of this function alone.

vector distances between dyons. Note further that the first matrix is "pure geometry" independent of holonomies, and the second is some hyperbolic rotation matrix by the angle including m-th holonomy and distance to m-th dyon.

The action density can be expressed in terms of prepotential in a surprisingly simple way

$$TrG_{\mu\nu}^2 = \partial^2\partial^2\log(\psi) \qquad (14.18)$$

with two four-dimensional Laplacians. Note that there are four derivatives because the matrices define the "prepotential" in terms of which the field potentials already have a derivative. Note also that the matrix A_m does not include time $\tau = r_4$, which is solely located in the cos function in the last term of ψ (thus the periodicity in the Matsubara box is explicit).

Exercise Calculating ψ for any holonomies and locations in Mathematica is very straightforward: do it and use the last expression for the action density. Do not look at the resulting huge expression but just plot its distribution, e.g., for parameters corresponding to Fig. 14.4.

The construction itself is based on ADHM multi-instanton construction and the so-called Nahm version of it, generalizing it to the monopoles. It is too technical to be presented here: see original paper (Kraan and van Baal 1998).

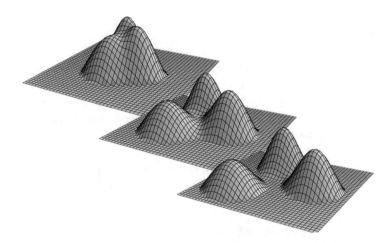

Fig. 14.4 Action densities for the SU(3) caloron at x0 = 0 in the plane defined by the centers of the three constituents for 1/T = 1.5, 3, and 4 (increasing temperature from top to bottom). We choose mass parameters $(\nu_1, \nu_2, \nu_3) = (0.4, 0.35, 0.25)$, implemented by $(\mu_1, \mu_2, \mu_3) = (-17/60, -2/60, 19/60)$. The constituents are located at $\vec{y}_1 = (-1, 1, 0)$, $\vec{y}_2 = (0, 1, 0)$, and $\vec{y}_3 = (1, -1, 0)$, in units of T . The profiles are given on equal logarithmic scales, cut-off at an action density below 1/e

14.3   Instanton-Dyon Interactions

14.3.1 Large-Distance Coulomb

Since the objects we now study have electric and magnetic Coulomb fields at large distances, $rv \gg 1$, one might naively expect that electric and magnetic Coulomb interaction are proportional to the usual products of the charges:

$$V_{naive}(r) = \frac{e_1 e_2 + m_1 m_2}{r} \qquad (14.19)$$

Yet this naturally looking expression turned out to be *wrong*!

Let me start with a counterexample showing that it *must* be wrong. We know that two self-dual solitons—M and L in particular—cannot classically interact at all, since the total action is simply given by the total topological charge, which cannot depend on the distance between them r.

And indeed, plugging solutions into the action and performing the large-distance expansion, one finds another—funny looking but correct—expression

$$V(r) \sim \left(\frac{-e_1 e_2 + m_1 m_2}{r} \right) \qquad (14.20)$$

with a minus sign in the electric term. Extra correction to the naive formulae above comes from the commutator of A_4 with other components of the gauge field. So, in the L-M case, with charges from table above, there is no potential.

This expression for $\bar{M}M$ produces *double* attraction, since their magnetic charges are opposite and electric are the same. The $\bar{L}M$ case has the opposite; thus it generates a *double* repulsion.

14.3.2 The Dyon-Antidyon Classical Interaction

14.3.2.1 Combing the Hedgehogs

Superposition of the dyons at nonzero A_4 is nontrivial since it should match not only in magnitude but also in its direction in the color space. This is achieved by the following four-step procedure:

1. "combing," or going to a gauge in which the "Higgs field" $A_4 = v$ of a dyon at large distances is the same in all directions and for all objects
2. performing a time-dependent gauge transformation which removes v
3. superimposing the dyons in this gauge
4. making one more time-dependent gauge transformation, re-introducing v back

The Details of the Combing

The hedgehog ansatz is invariant with respect to gauge transformations of the form

$$U = e^{i\hat{r}\cdot\tau f(r)} \,. \tag{14.21}$$

However to superpose the dyon solutions, we need to first have the "Higgs" go to a constant value at infinity, which does not depend on the direction. This is accomplished by doing gauge transformation which converts $\hat{r} \cdot \tau$ into one direction in the color space. Since the "Higgs" field is associated with the zero component of the gauge field A_0, applying the time-independent gauge transformation simply rotates the color direction of the A_0 field. Further, the self-dual and antiself-dual sector have asymptotic values of the "Higgs" A_0 differ by a sign (see Diakonov (2009)), so two matrices have to be used to gauge comb the self-dual and the antiself-dual dyon. The gauge transformations are given by

$$S_+ = e^{-i\phi\tau_3/2} e^{i\theta\tau_2/2} e^{i\phi\tau_3/2} \tag{14.22}$$

$$S_- = e^{-i\phi\tau_3/2} e^{-i(\pi-\theta)\tau_2/2} e^{i\phi\tau_3/2} \,. \tag{14.23}$$

We calculate the fields of a dyon and antidyon in these fields. If the self-duality (antiself-duality) equations are satisfied, then the solution to the fields are given by

$$A_0^a = \mp v\Phi(vr)\hat{r}^a \,, \tag{14.24}$$

$$A_i^a = \epsilon_{aij} n_j \frac{1-R(vr)}{r} \,, \tag{14.25}$$

where

$$\Phi(x) = \coth x - \frac{1}{x} \,, \tag{14.26}$$

$$R(x) = \frac{x}{\sinh x} \,. \tag{14.27}$$

We work in cylindrical coordinates, in which this ansatz becomes

$$A_r = 0 \,, \tag{14.28}$$

$$A_\theta = \frac{R(vr)-1}{r} \frac{\vec{\phi}\cdot\vec{\tau}}{2} \,, \tag{14.29}$$

$$A_\phi = \frac{1-R(vr)}{r} \frac{\vec{\theta}\cdot\vec{\tau}}{2} \,, \tag{14.30}$$

The upper sign in all equation corresponds to self-dual solutions. This means that if we want to superpose the self-dual dyon and its antiself-dual counterpart, we first have to make sure that A_0 goes to a common constant value. We accomplish this with the matrices (14.22), namely, the matrices $S_\pm$ will take $\vec{r} \cdot \vec{\tau} \rightarrow \pm\tau^3$. The choice of the matrices which accomplish this is, of course, not unique. We always have a choice of residual $U(1)$ symmetry $U = \exp(i\varphi\tau^3/2)$.

Proceeding with the calculation, we obtain that the holonomy of both the self-dual and antiself-dual dyon after this gauge transformation is

$$A_0 = v \frac{\tau^3}{2} \, , \tag{14.31}$$

and their spatial components are given as

$$A_r = 0 \, , \tag{14.32}$$

$$A_\theta = \frac{R(vr)}{r} \frac{\vec{\hat{\phi}} \cdot \vec{\tau}}{2} \, , \tag{14.33}$$

$$A_\phi = \frac{R(vr)}{r} \frac{\vec{\hat{\rho}} \cdot \vec{\tau}}{2} - \frac{1}{r} \cot \frac{\theta}{2} \frac{\tau^3}{2}, \tag{14.34}$$

for the self-dual dyon and

$$A_r = 0 \, , \tag{14.35}$$

$$A_\theta = \frac{R(vr)}{r} \frac{\vec{\hat{\phi}} \cdot \vec{\tau}}{2} \, , \tag{14.36}$$

$$A_\phi = -\frac{R(vr)}{r} \frac{\vec{\hat{\rho}} \cdot \vec{\tau}}{2} + \frac{1}{r} \tan \frac{\theta}{2} \frac{\tau^3}{2}, \tag{14.37}$$

However this gauge transformation, apart from possessing a singular Dirac string in the third color direction, has also multivalued color $1, 2$. The reason for this is multivaluedness of the gauge transformation $S_{+,-}$ for the lines $\theta = \pi, 0$ respectively.

We suggest a different gauge transformation in which the Dirac string splits in $\pm z$ direction, but the fields remain single-valued. What's more, since we want to consider the superposition of dyon and anti-dyon, the Dirac strings will cancel everywhere except in between two objects like in Fig. 14.5.

The gauge transformation we need is given by

$$U_+ = e^{i\theta \, \tau_2/2} e^{i\phi \, \tau_3/2}$$

$$U_- = e^{i(\pi-\theta) \, \tau_2/2} e^{i\phi \, \tau_3/2}$$

Fig. 14.5 $D\bar{D}$ with Dirac flux canceling outside, and adding inside

The gauge field then looks as follows

$$A_i = \begin{cases} A_r = 0 \,, \\ A_\theta = \frac{R(vr)}{2r} \tau_2 \\ A_\phi = -\frac{R(vr)}{2r} \tau_1 + \frac{1}{2r} \cot\theta \ \tau_3 \end{cases} \tag{14.38}$$

where $R(z) = z/\sinh(z)$.

However, superposing the two solutions is not as trivial as one may think. Surely the Abelian components will superpose properly, and one may think that the contribution of the core is negligible, as the core has exponentially small influence. However the string singularity makes any contribution of the core large, and one must proceed with caution.

We first examine a single monopole and compute its magnetic field in spherical coordinates. The metric is given by the line element:

$$ds^2 = dr^2 + r^2 d\theta^2 + r^2 \sin^2\theta d\phi^2 \tag{14.39}$$

The radial component of the magnetic field B_r can then be calculated as[3]

$$B^r = \frac{F_{\theta\phi}}{\sqrt{g}} \,. \tag{14.40}$$

where we calculate[4]

$$\vec{F}_{\theta\phi} = \partial_\theta(\vec{A}_\phi r \sin\theta) - \partial_\phi(\vec{A}_\theta r) + \vec{A}_\theta \times \vec{A}_\phi r^2 \sin\theta \,, \tag{14.41}$$

Taking into account just the Abelian part, we obtain exactly what we expect. Namely, the $\sin\theta$ term above turns $\cot\theta$ into $\cos\theta$ which, upon differentiation, becomes $\sin\theta$. Then dividing by $\sqrt{g} = r^2 \sin\theta$, we obtain $B^{r,a} = \delta^{3,a} 1/r^2$, where a is the color index. The commutator of the core of A_θ and A_ϕ will correct the filed so that it is not divergent in the center. The more interesting cancelation is one of the string and the core. In fact we will see that they conspire to cancel the contribution of the string. The relevant term in A_ϕ is the core. Then

$$\partial_\theta(A_\phi r \sin\theta) = -\partial_\theta(R \sin\theta \tau_1/2) + \cdots = -R\cos\theta \,, \tag{14.42}$$

---

[3] The magnetic field B is defined as

$$B^i = \frac{1}{2\sqrt{g}} \epsilon^{ijk} F_{jk}$$

where the $\sqrt{g}$ is put because ϵ^{ijk} is a tensor density.

[4] the factors of $r \sin\theta$ and r are inserted because A_ϕ and A_θ are *neither* contravariant nor covariant components of the vector field A_i, but $A_\phi r \sin\theta$ and $A_\theta r$ are covariant components.

where dots indicate the Abelian part which we argued contributes to the expected field of the monopole. The commutator term becomes

$$(\vec{A}_\theta \times \vec{A}_\phi r^2 \sin\theta) \cdot \frac{\vec{\tau}}{2} = R\cos\theta \qquad (14.43)$$

which cancels the term in the derivative part of field strength tensor.

However, now we consider the superposition of the dyon and antidyon. We will have then that the fields are that of dyon and the Abelian part of the antidyon, as the core contribution can be neglected. Therefore, apart from the cancelation which we described before, there will be a term of the form

$$-\partial_\theta R \pm \frac{r}{r_2} R\cot\theta_2 = 0 . \qquad (14.44)$$

where we assumed now that R is a function of both θ and r. The sign depends on whether the superposed field is a monopole or antimonopole, as well as how it is oriented, i.e., whether the south poles of the monopole-(anti)monopole system are facing each other or the south pole of the one facing the south pole of the other. We will see that the sign we want is actually a relative minus sign between the two terms.

The equation above is just a simple, separable, differential equation. All that we have to do is express r_2, θ_2 in terms of the spherical coordinates r, θ. The relation is the following:

$$r_2 = \sqrt{r^2 + d^2 + 2rd\cos\theta} \qquad (14.45)$$

The actual equation is

$$\partial_\theta \ln R = \frac{r}{r_2} \tan\frac{\theta_2}{2} , \qquad (14.46)$$

and the solution is

$$-\frac{2(1 - \xi_2 + \xi^2)\cot\theta}{\xi\xi_2} + \frac{2(\xi_2 - \cos 2\theta - 1)}{\xi_2 \sin\theta} - \frac{2(1 + \xi)}{\xi} E\left(\frac{\theta}{2}, \frac{4\xi}{(1 + \xi)^2}\right)$$

$$+\frac{2}{\xi(1 + \xi)} F\left(\frac{\theta}{2}, \frac{4\xi}{(1 + \xi)^2}\right) \qquad (14.47)$$

We start by writing the fields in cylindrical coordinates. Then we may simply superpose the two fields. The Dyon becomes

$$A_\rho = A_\theta \cos\theta = \frac{R(vr)z}{\rho^2+z^2}\,, \tag{14.48}$$

$$A_\phi = \frac{R(vr)}{\sqrt{\rho^2+z^2}}\frac{\vec{\rho}\cdot\vec{\tau}}{2} - \frac{1}{r}\frac{\sqrt{\rho^2+z^2}+z}{\sqrt{\rho^2+z^2}-z}\frac{\tau^3}{2}\,;\,, \tag{14.49}$$

$$A_z = -A_\theta \sin\theta = -\frac{R(vr)\rho}{\rho^2+z^2}\frac{\vec{\phi}\cdot\vec{\tau}}{2}\,, \tag{14.50}$$

14.3.2.2 Following the Gradient Flow Down the Streamline

As it is well known, a "combed" monopole or dyon must possess the Dirac string, a singular gauge artifact propagating one unit of magnetic flux from infinity to the dyon center. By selecting appropriate gauge, one can direct the Dirac string to have arbitrary direction. Superimposing into a sum of two dyons with different directions of the Dirac string, one gets *non-equivalent* configurations: the interference of singular and regular terms make the Dirac strings no longer invisible or pure gauge artifact. (However, this is cured during the gradient flow process, as we will discuss below.)

Two obvious extreme selections for the Dirac strings are (a) a "minimally connected dipole" when it goes along the line connecting two dyon centers and (b) a "maximally disconnected" pair, in which the Dirac strings go into the centers from two opposite directions (see Fig. 14.6). Under the gradient flow, the former is supposed to reach magnetically trivial configuration, while the latter must relax to a pure gauge Dirac string-like state passing the flux through the system, from minus to plus infinity. The former case seems to be simpler and more natural to use: but our experience has shown the opposite, that (b) generates smaller artifacts since the Dirac strings interfere less with the gradient flow changes between the two objects. So we will use case (b) as our starting configurations below.

An important role in what follows is a color current:

$$j_\mu^a = -\frac{\delta S}{\delta A_\mu^a}|_{A=A_{\text{ansatz}}} = (D_v^{ab}G_{v\mu}^b)|_{A=A_{\text{ansatz}}} \neq 0. \tag{14.51}$$

The current vanishes for extrema (solutions of the YM equation, such as a single dyon), but it is nonzero for dyon-antidyon configurations which we study. It has

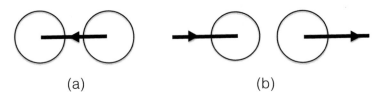

(a)   (b)

Fig. 14.6 Two extreme positions for the Dirac strings, for the $M\bar{M}$ pair

the meaning of the force in the functional space showing the direction toward a reduction of the action.

The study of dyon-antidyon streamline configurations we will follow[5] is due to Larsen and Shuryak (2016a). The gradient flow is a process in a computer time τ; thus the current would be the driving force. In the paper we follow in this section, a dyon-antidyon pair was put on the lattice.[6]

The gradient flow process was found to proceed via the following stages:[7]

(i) *near initiation*: starting from relatively arbitrary ansatz, one finds rapid disappearance of artifacts and convergence toward the streamline set

(ii) following the *streamline itself*: the action decrease at this stage is small and steady. The dyons basically approach each other, with relatively small deformations: thus the concept of an interaction potential between them makes sense at this stage

(iii) a *metastable state* at the streamline's end: the action remains constant; evolution is very slow and consists of internal deformation of the dyons rather than further approach

(iv) *rapid collapse* into the perturbative fields plus some (pure gauge) remnants

We will detail properties of these stages below, for now restricting to general comments. One is the existence of the stage (iii) which has not been anticipated on general grounds. Since all configurations corresponding to it have the same action, one can perhaps lump all of them into a new class of states, corresponding to the same dyon-antidyon distance. Unlike the instanton-dyons themselves, such states have not yet been identified on the lattice.

Our other comment is the action value even at the end of the streamline is not that far from the sum of the two dyon masses. In other words, the classical interaction potential happens to be rather small numerically, a welcoming feature for statistical mechanics simulations.

It should also be noted that while the topological and magnetic charges of M and $\bar{M}$ are opposite, their electric charges are the same. Thus total value of it is 2, rather than zero. That is why they cannot annihilate each other, as instanton and anti-instanton do.

Last but not least, we do observe the *universality* of the streamline. As expected, independent on the initial dyon separation, we found that gradient flow

---

[5]We later learned that stable monopole-antimonopole configuration and the potential leading to it has also been calculated in Shnir (2005a).

[6]I was asked if it is possible to put it on the standard lattice with periodic boundary conditions. The answer is negative: the $M\bar{M}$ (and any other interesting dyon pairs) always have one of the charges—electric or magnetic—uncompensated. Furthermore, for arbitrary holonomy v, the pair also does not have an integer topological charge. So, the lattices we used had *no* periodicity conditions at all.

[7]Let me supply more colloquial names for these stages: running downhill to the stream, following the streamline, finding a lake, and then a waterfall.

proceeds through essentially the same set of configurations at stages (ii)–(iv). Thus one-parameter characterization of those is possible. A parameter we found most practical in this work is simply its *lifetime*—duration in our computer time τ needed for a particular configuration to reach a final collapse. (Of course, for statistical mechanics applications, one better map that into some collective coordinate, such as the dyon separation, whatever way it can be defined).

We now show the results for a M and $\bar{M}$ dyon separated by a distance (in natural units $1/v$) of the order 0 to 10 along the z-axis which is cooled using gradient flow. The action of an individual dyon on the lattice was found to be 11.94, 5% lower than the analytic value 4π. This gives the action of 23.88 for two well-separated dyons. Any action lower than this therefore is ascribed to an attractive interaction between the dyons.

The simulations of the streamline for different configurations start out with a slightly higher action around that of two individual dyons; this then converges to an almost stable configuration, the Streamline. If the separation is bigger than 4, the two cores of the dyons are seen to move toward each other, and their action smoothly decreases. We show four important computer times in Fig. 14.7, where the action density is plotted along the z-axis (along the dyon separation) at start configuration with a separation of 10 between the dyon and antidyon. A typical action history is shown in Fig. 14.8, for the initial separation values from 0 to 10.

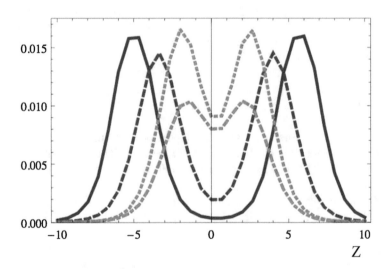

Fig. 14.7 Action density along the z axis in natural units for a separation $|r_M - r_{\bar{M}}|v = 10$ between the centers of the two dyons. The configuration with the maximums furthest from each other is the start configuration. After 3000 iterations, it has moved further toward the center. At 12,000 iterations, the configuration has reached the metastable configuration with a separation between the maximums of around 4. At 13,700, the configuration has collapsed around halfway and will continue to shrink until the action is 0. Times are as shown in Fig. 14.8

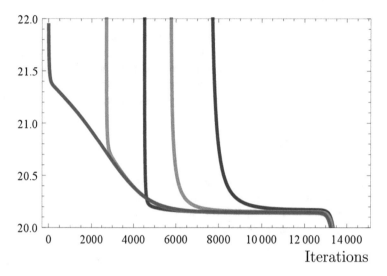

Fig. 14.8 Action for $v = 1$ as a function of computer time (in units of iterations of all links) for a separation $|r_M - r_{\bar{M}}|v = 0, 2.5, 5, 7.5, 10$ between the M and $\bar{M}$ dyon from right to left in the graph. The action of two well-separated dyons is 23.88

14.4 The Partition Function in One Loop

14.4.1 Electric Screening

The basic physics of the electric screening can be explained most simply following the original derivation by one of us (Shuryak 1978) in the Coulomb gauge. If some electrically charged object, with nonzero A_4, is immersed into a finite-temperature QCD, gluons and quarks of the heat bath are scattered on it. The simplest diagram comes from the quartic term in the gauge Lagrangian, $e^2 A_m^2 A_4^2$ which couples the heat bath gluons directly to square of A_4, but there are also other diagrams contributing to the forward scattering amplitude. The result was the expression for the QCD Debye mass:

$$M_D^2 = e^2 T^2 (1 + N_f/6) \tag{14.52}$$

In 1976, when QCD was only 3 years old, the main finding was its positivity, which ensured *screening* of a charge, as opposed to antiscreening by vacuum loops: thus the "plasma" name.

The next relevant paper is Pisarski and Yaffe (1980) (PY) who found in the one-loop action of the calorons (the finite-T instantons) the famous PY term $\sim \rho^2 M_D^2/e^2$, where ρ is the instanton radius. It also comes from the forward scattering amplitude of the thermal plasma quanta on the A_4^2 of the instanton. This term is important because it ensures that at $T > T_c$ (only in this case there are thermal gluons!), the semiclassical expression provides finite instanton density.

Going forward to calorons at nonzero holonomy, the one-loop effective action has been computed by Diakonov, Gromov, Petrov, and Slizovskiy (DGPS) (Diakonov et al. 2004). The caloron is now a superposition of the M and L dyons, separated by distance r_{ML}, and the basic expression from which the effect came looks as follows

$$\int d^3r \left(\frac{1}{r_L} - \frac{1}{r_M} \right)^2 = 4\pi r_{ML} + \dots \tag{14.53}$$

where r_L, r_M are distances from the dyon centers. It thus creates an effect resembling linear confinement, with a force independent on the separation. At zero holonomy, this result matches the PY answer because the relation between the instanton size and ML separation r_{ML} is

$$\rho^2 = \frac{r_{ML}}{\pi T} \tag{14.54}$$

The interaction is given by $\int d^r A_4^2$. One obvious consequence would be that if one would generalize the DGPS derivation to theory with fermions, they will simply get extra factor $1 + N_f/6$ as in (14.52). Another one, worth mentioning, is that for pairs $L\bar{L}$ or $M\bar{M}$ with the *same* electric charge, there will be plus in the integral above, and thus the effect becomes repulsive.

The electric screening effect ensures LM "binding" into finite-size instantons, into an object with a size $r_{ML} \sim e^2 T/M_D^2$. (Note that the coupling e cancels here; it is because the nonperturbative fields are always $\sim 1/e$.)

Although asymptotically at $N_f \to \infty$ this size is $O(1/N_f)$, the coefficient 1/6 in (14.52) makes it less important for "interesting" $N_f = 0...10$. We will later see that the direct fermionic interaction discussed in the preceding section binds $L\bar{L}$ pairs stronger than LM interaction.

Since we will be discussing charge-zero clusters consisting of all four dyons, let us give an example of the potential electric screening in this case. For simplicity we will only discuss $L\bar{L}$ at the same point and M, $\bar{M}$, and $L\bar{L}$ to be on one line. The integral

$$\int d^3r \left(\frac{2}{r} - \frac{1}{r_M} - \frac{1}{r_{\bar{M}}} \right)^2 \tag{14.55}$$

leads to a potential for M shown in Fig. 14.9. As one can see, like for DGPS case, it consists of linear segments but is now deformed away from the companion dyon. (Note, that it is not due to their Coulomb repulsion, which is also there but will be discussed in the next subsection.)

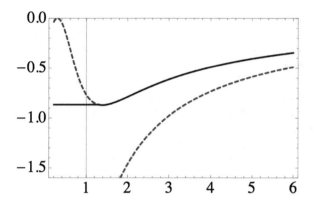

Fig. 14.9 Black solid line is the $\bar{M}M$ dimensionless potential (action); red dash line is the asymptotic Coulomb term; blue dashed line is a sketch of possible "core" shape

14.4.2 The One-Loop Measure, Perturbative Coulomb Corrections and the "Core"

Another important result of the DGPS paper (Diakonov et al. 2004), see also the influential Diakonov's lectures (Diakonov 2009), was the derivation of the forces decreasing with the distance also coming from the one-loop gluonic determinant.[8]
  Combined with Manton's result for identical dyons, it was generalized to the arbitrary number of self-dual L, M dyons. The resulting volume element in space of collective variables is expressed elegantly $\sqrt{g} = det[G]$ via a determinant of the so-called Diakonov matrix $\hat{G}$, defined by

$$\hat{G} = \left[\delta_{mn}\delta_{ij}\left(4\pi v_m - 2\sum_{k\neq i} \frac{1}{T|x_{i,m} - x_{k,m}|} + 2\sum_{k} \frac{1}{T|x_{i,m} - x_{k,p\neq m}|} \right) \right. \tag{14.56}$$

$$\left. + 2\delta_{mn} \frac{1}{T|x_{i,m} - x_{j,n}|} - 2\delta_{m\neq n} \frac{1}{T|x_{i,m} - x_{j,n}|} \right],$$

Here $x_{i,m}$ denote the position of the i-th dyon of type m. This form is an interpolation of the exact metric between a M and L dyon, true at any distance, with the metric of the two dyons of the same type at large distances. We introduce a cutoff on the separation via $r \to \sqrt{r^2 + cutoff^2}$, such that for one pair of dyons of the same type, the diagonal goes to 0 for $v = 0.5$, instead of minus infinity.

---

[8]More precisely, what was calculated and presented below is the volume element of the "moduli space" of all multi-dyon configurations. The geometry, the metric tensor, and the volume element of space of all classical multi-monopole solutions have been studied in mathematical literature, especially by Attiya, Hitchin et al. which started from the 1970s.

Note that while the matrix itself has only 1/r terms, but after the effective one-loop action $\log(det(G))$ is computed, one gets all powers of 1/r, involving nontrivial many-body interactions.

In subsequent works (see lectures Diakonov (2009) for summary and references), Diakonov had used this interaction in the L, M sector to perform many-body calculation. He argued that it can also lead to some integrable model, which can be solved and even results in the holonomy potential leading to confinement.

We think those conclusions were a bit premature, because it somehow assumes that the self-dual L, M sector and antiself-dual one $\bar{L}, \bar{M}$ are invisible to each other and do not interact at all. Not only do we not see why this should be the case, we have found that dyon-antidyon interactions are important and in theory with many fermions even dominant.

Another issue is that for many configurations, $det(G) < 0$, which obviously makes no sense for a volume element. As some eigenvalue of G approaches zero, it means such configuration approaches zero measure in the space of solutions. What it means is that only those with $det(G) > 0$ should be included in the partition function. This is relatively easy to do in numerical studies, but was not taken care of in the original studies.

14.5 Fermionic Zero Modes

14.5.1 How Quark Zero Modes Are Shared Between the Dyons

The instanton has unit topological charge $Q = 1$ and, according to Attiyah-Singer index theorem, has one fermionic zero mode. In the chapter about instantons, we have discussed 't Hooft's effective Lagrangian arising due to it. Now we discuss nonzero holonomy environment at $T \neq 0, < P > \neq 1$, in which the instanton is split into N_c constituents. The question is how the zero mode is distributed among these constituents.

As was determined by van Baal and collaborators, fermionic zero mode "hops" from one type of dyon to the next at certain critical values of the holonomy. Periodicity condition along the Matsubara circle can be defined with some arbitrary angles ψ_f for quarks with the flavor f. The resulting rule is it belongs to the dyon corresponding to the segment of the holonomy circle ν_i to which the periodicity phase belongs: $\mu_i < \psi_f < \mu_{i+1}$.

In physical QCD, all quarks are fermions, and therefore $\psi_f = \pi$ for all f. This case is schematically shown by blue dots in Fig. 14.10 (left): all fermions fall on the same segment of the circle, and therefore only one, of N_c dyons, has zero modes and interacts with quarks.

But one can introduce other arrangements of these phases. In particular, for $N_c = N_f$, the opposite extreme is the so-called $Z(N_c)$ QCD, proposed in Kouno et al. (2013), who put them symmetrically around the circle (see Fig. 14.10 (right)). In this case, the instanton-dyon framework becomes very symmetric: each dyon interacts with "its own" quark flavor.

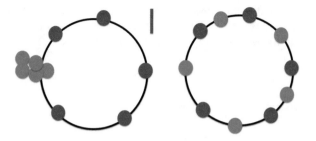

Fig. 14.10 Schematic explanation of the difference between the usual QCD (left) and the $Z(N_c)$ QCD (right)

14.5.2 The Zero Mode for the Fundamental Fermion

We start with the Dirac equation

$$\not{D}\psi = 0 , \tag{14.57}$$

and look for a normalizable solution within the hedgehog ansatz:

$$A_i^a = \epsilon_{aji}\mathcal{A}\hat{r}^j , \tag{14.58}$$

$$A_0^a = \mathcal{H}\hat{r}^a . \tag{14.59}$$

Since the Dirac operator is chiral, we may write the fermion in terms of upper and lower components ψ_L and ψ_R. We do the calculation for the lower component, namely, ψ_R. The Dirac equation then reads

$$- (\sigma^\mu)_{\alpha\beta}(D_\mu)_{AB}(\psi_R)_\beta^B , \tag{14.60}$$

where we explicitly wrote the Dirac indices α, β and color indices A, B. Now we ansatz (see Shnir)

$$\psi_\alpha^A = \alpha(r)\epsilon_{A\alpha} + \beta(r)[(\vec{\hat{r}} \cdot \vec{\sigma})\epsilon]_{A\alpha} . \tag{14.61}$$

We may choose to consider the matrix

$$\eta_{A\alpha} = -\psi_\beta^A \epsilon_{\beta\alpha} \tag{14.62}$$

in which case

$$\eta = \alpha(r)\vec{1} + \beta(r)\vec{\hat{r}} \cdot \vec{\sigma} \tag{14.63}$$

The rule of acting with a color and the spin sigma matrices on this object is such that we multiply by a color matrix τ from the left, and if we multiply by a spin matrix σ, then we multiply from the right and put a minus sign, i.e.,

$$\sigma \psi = \eta \epsilon \sigma^T = -\eta \sigma \epsilon \ . \tag{14.64}$$

If we wish to construct fermion density, we see that

$$\psi^{*A}_{\alpha} \psi^A_{\alpha} = \text{Tr } (\eta^\dagger \eta) \tag{14.65}$$

We now plug the ansatz into (14.60) and obtain the following two equations

$$\alpha'(r) + \tfrac{\mathcal{H}+2\mathcal{A}}{2}\alpha + \tfrac{z}{\beta}\beta = 0 \ , \tag{14.66}$$

$$\beta'(r) + \left(\tfrac{\mathcal{H}-2\mathcal{A}}{2} + \tfrac{2}{r}\right)\beta + \tfrac{z}{\beta}\alpha = 0 \ . \tag{14.67}$$

where we have assumed $\psi_R \propto e^{izt/\beta}$, i.e., that the Fermion has arbitrary periodicity condition in the imaginary time direction.

Let us look at the asymptotic behavior of dyons, i.e., when $\mathcal{H}(r \to \infty) = v$ and $\mathcal{A}(r \to \infty) = 0$, then

$$\alpha'(r) + \tfrac{v}{2}\alpha(r) + \tfrac{z}{\beta}\beta = 0 \ , \tag{14.68}$$

$$\beta'(r) + \tfrac{v}{2}\beta(r) - \tfrac{z}{\beta}\alpha = 0 \ . \tag{14.69}$$

This equation is easily solvable by taking the substitution $\alpha_\pm = \alpha \pm \beta$ where we get

$$\alpha_\pm = e^{-\left(\frac{v}{2} \pm \frac{z}{\beta}\right)} \ . \tag{14.70}$$

In order for the solution to be normalizable, we must have that both $\alpha_\pm$ vanish at infinity. This is only possible if $|z| < |v|\beta/2$.

We now proceed to numerical solution of the differential equation. The plots are shown in Fig. 14.11.

We may rewrite this equation in a two-component form as

$$\frac{d}{dr}\vec{\alpha} = -\vec{M}\vec{\alpha} \ , \tag{14.71}$$

where

We can write the formal solution as a path-ordered exponent:

$$\vec{\alpha} = \mathcal{P}\exp\left(-\int_{r_0}^r \vec{M}\, dr\right)\vec{\alpha}(r_0) \tag{14.72}$$

$\alpha_1(r)$ – solid, $\alpha_2(r)$ – dashed

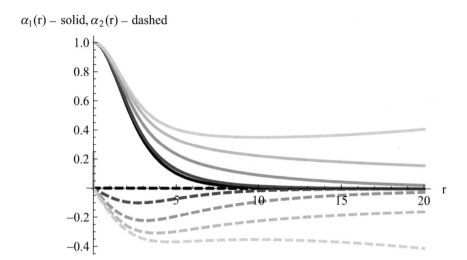

Fig. 14.11 Plot shows profile of zeromode components $\alpha_{1,2}$, for four different values of $z = 0, 0.2v/\beta, 0.4v/\beta, 0.5v/\beta, 0.55v/\beta$. Note that the zero mode delocalizes at $z = 0.5v/\beta$

The matrix $\vec{M}$ can be written as follows:

$$\vec{M}(r) = \left(\frac{\mathcal{H}}{2} + \frac{1}{r}\right)\vec{1} + \frac{z}{\beta}\sigma_1 + \left(\mathcal{A} - \frac{1}{r}\right)\sigma_3 \qquad (14.73)$$

The factor proportional to the unit matrix commutes with everything else and may be factored out. On the other hand,

$$\mathcal{A} - 1/r = -v/\sinh(vr) \sim e^{-vr}$$

is small, and we may neglect it except at the origin. So for $r \gg 1/v$, we may integrate the exponent, as it is proportional to σ_3 only, and then expand the result. Since

$$\int\left(\frac{\mathcal{H}}{2} + \frac{1}{r}\right) = \frac{1}{2}\ln\left[r\sinh(rv)\right] ,$$

we have that

$$\vec{\alpha}(r) = \frac{e^{-\frac{v}{2}(r-r_0)}}{\sqrt{r/r_0}} \times \left[\cosh\left(\frac{z(r-r_0)}{\beta}\right) + \sigma_1 \sinh\left(\frac{z(r-r_0)}{\beta}\right)\right]\vec{\alpha}(r_0) \quad (14.74)$$

Now we will solve the equation exactly. To do this, we separate the matrix $M(r)$ as

$$M(r) = M_0(r) + M_1(r) , \qquad (14.75)$$

where

$$M_0(r) \;\;= \left(\tfrac{\mathcal{H}}{2} + \tfrac{1}{r}\right)\vec{1} = , \qquad (14.76)$$

$$M_1(r) = \tfrac{z}{\beta}\sigma_1 + \left(\mathcal{A} - \tfrac{1}{r}\right)\sigma_3 \qquad (14.77)$$

$$= \tfrac{z}{\beta}\sigma_1 - \tfrac{v}{\sinh(vr)}\sigma_3 . \qquad (14.78)$$

The solution can then be written as $\vec{\alpha} = \exp(-\int_0^r M_0(r)dr)\vec{\chi}$, or

$$\vec{\alpha} = \frac{1}{\sqrt{r \sinh rv}} \qquad (14.79)$$

with the differential equation for $\vec{\chi}$ reading

$$\frac{d}{dr}\vec{\chi} = -M_1(r)\vec{\chi} , \qquad (14.80)$$

i.e.

$$\chi_1'(r) = \tfrac{v}{\sinh(vr)}\chi_1(r) - \tfrac{z}{\beta}\chi_2(r) , \qquad (14.81)$$

$$\chi_2'(r) = -\tfrac{v}{\sinh(vr)}\chi_2(r) - \tfrac{z}{\beta}\chi_1(r) , \qquad (14.82)$$

we may take a change of variables $\xi = rv$. Then the equation reads

$$\chi_1'(\xi) = \tfrac{1}{\sinh(\xi)}\chi_1(r) - \varsigma\chi_2(r) , \qquad (14.83)$$

$$\chi_2'(\xi) = -\tfrac{1}{\sinh(\xi)}\chi_2(r) - \varsigma\chi_1(r) , \qquad (14.84)$$

where we labeled $\varsigma = x/(v\beta)$. We now eliminate ξ_2 and obtain the second-order differential equation:

$$-\frac{d^2}{d\xi^2}\chi_1 - \frac{1}{2\cosh^2\tfrac{\xi}{2}}\chi_1 = -\varsigma^2\chi_1 . \qquad (14.85)$$

A general solution with arbitrary constants $c_{1,2}$ is

$$\chi_1(\xi) = c_1\left(-2\varsigma + \tanh\frac{\xi}{2}\right)e^{\varsigma\xi} + c_2\left(2\varsigma + \tanh\frac{\xi}{2}\right)e^{-\varsigma\xi}. \qquad (14.86)$$

Using the first-order equation, we can write χ_2 as

$$\chi_2(\xi) = c_1\left(2\varsigma - \coth\frac{\xi}{2}\right)e^{\varsigma\xi} + c_2\left(2\varsigma + \coth\frac{\xi}{2}\right)e^{-\varsigma\xi} \qquad (14.87)$$

The function $\chi_2(\xi)$ is divergent when $\xi \to 0$, except if $c_1 = c_2$, in which case $\xi_2(0) = 0$. Therefore $c_2 = c_1$. The constant c_1 can be determined by overall normalization. The solution then becomes

$$\chi_1(\xi) = 2c_1\left(-2\varsigma\sinh(\xi\varsigma) + \tanh\frac{\xi}{2}\cosh(\xi\varsigma)\right) \qquad (14.88)$$

$$\chi_2(\xi) = 2c_1\left(2\varsigma\cosh(\xi\varsigma) - \coth\frac{\xi}{2}\sinh(\xi\varsigma)\right) \qquad (14.89)$$

Finally we obtain

$$\alpha_{1,2} = \frac{\sqrt{v}}{\sqrt{\xi\sinh\xi}}\chi_{1,2}\,. \qquad (14.90)$$

$\sqrt{v}$ can be absorbed into constant c_1, and our final expression is

$$\alpha_{1,2} = \frac{\chi_{1,2}}{\sqrt{\xi\sinh\xi}}\,. \qquad (14.91)$$

with functions $\chi_{1,2}$ given by (14.88), $\xi = vr$, $\varsigma = z/(v\beta)$. Note that the value of $\alpha_1(\xi \to 0)$ is given by

$$c1(1 - 4\varsigma^2)\,, \qquad (14.92)$$

and the solution is completely regular at $r = 0$.

14.5.2.1 Elements of Quark "Hopping Matrix"
We calculate the fermionic transition amplitude between the dyons, which is the operator $\not{D}$ with the proper background gauge field, expressed in the basis of the fermion zero modes of all dyons and antidyons. The transition element will look like

$$T_{D\bar{D}} = \int d^4x\, \psi_{\bar{D}}^\dagger \not{D} \psi_D\,, \qquad (14.93)$$

where $\psi_{D,\bar{D}}$ are dyonic and antidyonic zero modes. We assume a superposition ansatz of dyons and have that

$$\not{D} = \not{\partial} - i\not{A}_D - i\not{A}_{\bar{D}} \qquad (14.94)$$

then since $(\partial\!\!\!/ - i A\!\!\!/_{D,\bar{D}})\psi_{D,\bar{D}} = 0$

$$T_{D\bar{D}} = \int d^4x\ \psi_{\bar{D}}^{\dagger}\partial\!\!\!/\,\psi_D ,\tag{14.95}$$

In our solution, we use $\eta_{D,\bar{D}}$, and the transition element becomes

$$- \operatorname{Tr}\,(\eta_{\bar{D}}^{\dagger}U_{\bar{D}}^{\dagger}\partial_{\mu}(U_D\eta_D)\sigma^{\mu})\tag{14.96}$$

where $U_{D,\bar{D}}$ are gauge combing matrices for the dyon and antidyon, respectively (remember that zero modes were found in a hedgehog gauge). Explicitly we use the form of the matrices $U_D = U_-(\theta, \phi)$ and $U_{\bar{D}} = U_-(\vartheta, \varphi)$, where θ, ϕ, and ϑ, φ are spherical angles in the center of the dyon and antidyon, respectively. These matrices have the following properties

$$U_D\hat{\theta}\cdot\vec{\sigma}U_D^{\dagger} = \sigma^1 \quad U_D\hat{\phi}\cdot\vec{\sigma}U_D^{\dagger} = \sigma^2\tag{14.97}$$
$$U_D\hat{r}\cdot\vec{\sigma}U_D^{\dagger} = \sigma^3 ,$$

$$U_{\bar{D}}\hat{\vartheta}\cdot\vec{\sigma}U_{\bar{D}}^{\dagger} = -\sigma^1 \quad U_{\bar{D}}\hat{\varphi}\cdot\vec{\sigma}U_{\bar{D}}^{\dagger} = \sigma^2\tag{14.98}$$
$$U_{\bar{D}}\hat{s}\cdot\vec{\sigma}U_{\bar{D}}^{\dagger} = -\sigma^3 ,$$

where $\hat{r}, \hat{\theta}, \hat{\phi}$, and $\hat{s}, \hat{\vartheta}, \hat{\varphi}$ are spherical unit coordinate vectors around dyon and antidyon, respectively.

We first consider the spatial index $\mu = i$. We may use the matrices $U_{D,\bar{D}}$ to transform $\eta_{D,\bar{D}} = \alpha_{D,\bar{D}} + \sigma\cdot\hat{r}_{D,\bar{D}}\beta_{D,\bar{D}}$ into an object $\zeta_{D,\bar{D}} = \alpha_{D,\bar{D}}\pm\beta_{D,\bar{D}}\sigma_3$, i.e.,

$$\zeta_{D,\bar{D}} = U_{D,\bar{D}}\eta_{D,\bar{D}}U_{D,\bar{D}}^{\dagger}$$

It can be shown that the transition element is given by the expression

$$\operatorname{Tr}\,\Big\{U_{\bar{D}}U_D^{\dagger}\zeta_{\bar{D}}\partial_r\zeta_D\sigma_3\tag{14.99}$$

$$+\zeta_{\bar{D}}^{\dagger}\zeta_D\Big[\frac{1}{r}(\partial_\theta U_D)U_D^{\dagger}\sigma_1 + \frac{1}{r\sin\theta}(\partial_\phi U_D)U_D^{\dagger}\sigma_2\Big]\Big\}\tag{14.100}$$

Since

$$(\partial_\theta U_D)U_D = \frac{i\sigma_2}{2} ,\tag{14.101}$$

$$(\partial_\phi U_D)U_D = \tfrac{i}{2}(-\sigma_1\sin\theta + \sigma_3\cos\theta) ,\tag{14.102}$$

After straightforward manipulations, we obtain that the spatial part of the expression for the transition element is given by

$$\text{Tr} \left\{ \sigma_3 U_{\bar{D}} U_D^{\dagger} \left[\zeta_{\bar{D}}^{\dagger} \partial_r \zeta_D \right. \right. \tag{14.103}$$

$$\left. \left. + \frac{1}{r} \zeta_{\bar{D}}^{\dagger} \zeta_D \left(1 - \frac{i\sigma_2}{2} \cot\theta \right) \right] \right\} \tag{14.104}$$

The matrix $U_{\bar{d}} U_d^{\dagger}$ can be calculated as

$$U_{\bar{d}} U_d^{\dagger} \tag{14.105}$$

$$= \left(\sin \frac{\vartheta + \theta}{2} + i\sigma_2 \cos \frac{\vartheta + \theta}{2} \right) \cos \frac{\Delta\phi}{2}$$

$$+ \left(\sin \frac{\vartheta - \theta}{2} + i\sigma_2 \cos \frac{\vartheta - \theta}{2} \right) i\sigma_3 \sin \frac{\Delta\phi}{2}$$

We write this expression as

$$U_{\bar{d}} U_d^{\dagger} = (a + i\sigma_2 b) + (c + i\sigma_2 d)i\sigma_3 . \tag{14.106}$$

where

$$a = \sin \frac{\vartheta+\theta}{2} \cos \frac{\Delta\phi}{2} \tag{14.107}$$

$$b = \cos \frac{\vartheta+\theta}{2} \cos \frac{\Delta\phi}{2} \tag{14.108}$$

$$c = \sin \frac{\vartheta-\theta}{2} \sin \frac{\Delta\phi}{2} \tag{14.109}$$

$$d = \cos \frac{\vartheta-\theta}{2} \sin \frac{\Delta\phi}{2} \tag{14.110}$$

14.5.3 Fermionic Zero Mode for a Set of Self-Dual Dyons

As a special example, we consider the so-called instanton-anti-instanton molecule. Each of the instantons has one zero mode (for fundamental fermions), and there is only one amplitude of fermion exchange T_{IA} , so the contribution of zero modes to the fermionic determinant is in this case simply

$$det \not{D} = - |T_{IA}|^2 \tag{14.111}$$

which corresponds to the fact that instantons exchange one quark and one antiquark (per flavor).

At nonzero holonomy, the instanton is described by N_c dyons. However simple generalization of the determinant construction to dyons would be wrong, as dyons possess only a fraction of the topological charge. In the instanton-anti-instanton setting, there is still only one left-handed and one right-handed fermionic zero mode. Those are however located in "lumps" near each dyon, with some holonomy-dependent coefficients normalized as below

$$\psi_{L,R}(x) = \sum c_i^{L,R}(\mu)\psi_i(x); \quad \sum |c_i^{L,R}|^2 = 1 \tag{14.112}$$

where for L and R, the sum runs only over dyons of particular self-duality. It is then obvious that the fermionic determinant is

$$|det\, \slashed{D}| = |< R|\slashed{D}|L >|^2 = \left| \sum_{i,j} c_i^{*,R} c_j^L < i|\slashed{D}|j > \right|^2 \tag{14.113}$$

where $< i|\slashed{D}|j >$ is an amplitude of hopping between a dyon and antidyon. As usual, if the dyons do not strongly overlap in space-time, such hopping amplitude is described by some coupling constants a_i, obtained by standard "cutting the tail" procedure for modes, and the free fermionic propagator

$$< i|\slashed{D}|j >= a_i^* a_j S(x_i - x_j) \tag{14.114}$$

Note that at nonzero, holonomy fermions are massive; thus at zero $T = 1/\beta$ $S \sim exp(-m_f r_{ij})/r_{ij}^3$ and at nonzero T, it depends differently on temporal and spatial distance between the dyons. We also remind that different color fermions have different masses, although in the eigenframe of the holonomy, the mass matrix (and thus S) is diagonal in color. (Of course, fermions moving in the background color field of the gauge solitons do not conserve color, but they do when they move in "empty" space in between them.)

Explicit expressions for c_i, a_i can be found from expression for zero modes worked out in reference.

It is now straightforward to generalize this to configurations which have Q instantons and Q anti-instantons, still with total topological charge zero: there are Q left-handed and Q right-handed zero modes, and the fermionic determinant in the partition function can be approximated by the determinant of the "hopping" matrix $T_{ij}.S$.

14.6   Instanton-Dyons on the Lattice Are Seen via Their Fermionic Zero Modes

When we discuss topology on the lattice, we only once mentioned the instanton-dyons, in the section on constrained Cooling, with the Polyakov line preserved (Langfeld and Ilgenfritz 2011): while the total topological charge of the lattice was

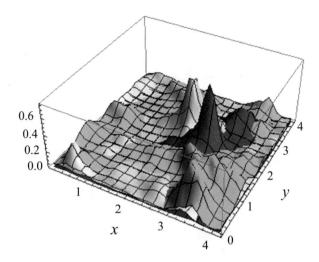

Fig. 14.12 Density $\rho(x, y)$ of the zero mode of conf. 2660 at $T = T_c$. $\phi = \pi$(red), $\phi = \pi/3$(blue), $\phi = -\pi/3$(green). Peak height has been scaled to be similar to that of $\phi = \pi$

always integer, the clusters observed had smaller topological charge (and the same actions, as they were self-dual or antiself-dual).

Using fermionic method allows to get better understanding of these objects, since changing the periodicity phases, one can see *all* types of the dyons separately. While "cooling" still distorts the configurations, hunting for lowest (or even zero) Dirac eigenvalues allows one to get the dyons *as they are* in the gauge ensemble. One of the early studies of the kind was done by Gattringer (2003) who used quarks with modified periodicity phases as a tool to locate all kinds of instanton-dyons. Further studies along these lines have been continued by Mueller-Presussker, Ilgenfritz, and collaborators (see e.g. Bornyakov et al. (2016, 2017)).

We however will jump to recent work by Larsen, Sharma, and myself (Larsen et al. 2018), which shows the underlying instanton-dyons in lattice QCD with utmost clarity, due to application of the "overlap" fermions possessing exact chiral symmetry and thus exact index theorems. Out of configurations at $T = 1, 1.08T_c$ of QCD with realistic quark masses, we selected those which have $|Q_{top}| = 1$, thus with one exactly zero mode. By varying the periodicity phase, we can identify location of all three type of dyons: see e.g. Fig. 14.12 in which three dyons are well separated. We also see configurations of strongly overlapping ones.

Not only we see that semiclassical formulae for zero modes well describe the lattice measurements when the dyons are far from each other, they also work well in the case of partial or even complete overlap. In Fig. 14.13, the profiles for single instanton is compared with that of overlapping dyons at the appropriate holonomy: the latter is closer to the data. We have analyzed many cases, in which the semiclassical expression from Kraan and van Baal is in agreement with the data much better than expected.

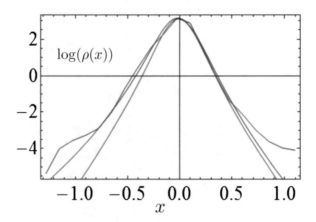

Fig. 14.13 $\log(\rho(x))$ of the zero mode of conf. 2960 at $\phi = \pi$ (black) and the log of the analytic formula for Polyakov loop $P = 0.4$ and $P = 1$ though the maximum. $T = 1.08 T_c$. Red peak only has been scaled to fit in height, while blue peak uses the found normalization. The positions of the other dyons are (blue) (0.13,0.1,0.0) and (0.1, −0.1, 0.0) and (red) (0.14,0.0,0.0) and (−0.14, 0.0, 0.0)

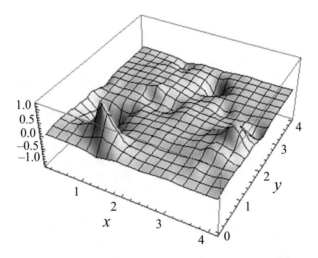

Fig. 14.14 Chiral density $\rho_5(x, y)$ of the first near-zero mode of conf. 2660 at $\phi = \pi$. $T = T_c$

Finally, the lowest nonzero modes show collectivized zero modes (see example in Fig. 14.14). Furthermore, near T_c we can even see that one type of dyons shows collectivized zero mode and thus a nonzero chiral condensate, while in the same configuration, the lowest nonzero modes of the other dyon types are still localized (with chiral symmetry unbroken).

14.7   Brief Summary

- Euclidean finite-temperature formulation of gauge theories allows introduction of important gauge invariant observable, the Polyakov line (14.1). It is called "holonomy" in mathematics, as an integral of connection over noncontractible cycle. Log of its average has a meaning of free energy of external pointlike static quark. Therefore, zero VEV indicates confinement. This happens in theories which do and do not possess center symmetry.
- Lattice studies specified the temperature dependence of $\langle P \rangle (T)$.
- In presence of $\langle P \rangle \neq 0$, all classical solutions should be modified, as the field A_0 is nonzero at infinity. Modification of instanton solution revealed that it becomes a (nonlinear) superposition of N_c objects called $instanton - dyons$. They are (anti)self-dual monopoles and therefore connected by Dirac strings. While those are unobservable gauge artifacts, they negate topological theorems and allow instanton-dyons to have $non - integer$ topological charges (see Fig. 14.4).
- As P is unitary matrix $N_c \times N_c$, its eigenvalues can be expressed as N_c phases, or points on a circle. Actions and topological charges of instanton-dyons are proportional to the *length of arcs* between the phase. In QCD with $N_c = 3$, there are three types of such dyons and three types of antidyons.
- Classically dyon-dyon interaction is zero, as total action is directly proportional to the topological charge Q. Dyon-antidyon interaction was studied by gradient flow method. Quantum interaction can be put into an elegant form known as Diakonov determinant (14.57).
- Dirac equation in the field of instanton-dyons possesses zero modes. However, only one of the species, out of N_c, has it *normalizable* and therefore physical. Which one depends on periodicity phase required from fermion fields around the thermal circle, see Fig. 14.10 and discussion nearby.
- Defining those phase differently for each quark flavors allows interesting "soft deformation" of QCD. While it does not affect gluons and even (full) instantons, it changes allocation of fermionic zero modes.
- Quark zero modes near T_c of the lattice configurations show convincing evidences that under these conditions, the topological objects are indeed the instanton-dyons. Even in the case of strongly overlapping clusters, the shape of zero modes from lattice matches well the analytic description based on Dirac eigenmodes for dyons, without any thermal gluons and quarks.

References

Bazavov, A., Brambilla, N., Ding, H.T., Petreczky, P., Schadler, H.P., Vairo, A., Weber, J.H.: Polyakov loop in 2+1 flavor QCD from low to high temperatures. Phys. Rev. D **93**(11), 114502 (2016). arXiv:1603.06637

Bazavov, A., Petreczky, P.: Polyakov loop in 2+1 flavor QCD. Phys. Rev. D **87**(9), 094505 (2013). arXiv:1301.3943.

Belavin, A.A., Polyakov, A.M., Schwartz, A.S., Tyupkin, Yu.S.: Pseudoparticle solutions of the Yang-Mills equations. Phys. Lett. B **59**, 85–87 (1975) [350 (1975)]

Bornyakov, V.G., Ilgenfritz, E.M., Martemyanov, B.V., Muller-Preussker, M.: Dyons near the transition temperature in lattice QCD. Phys. Rev. D **93**(7), 074508 (2016). arXiv:1512.03217

Bornyakov, V.G., Boyda, D.L., Goy, V.A., Ilgenfritz, E.M., Martemyanov, B.V., Molochkov, A.V., Nakamura, A., Nikolaev, A.A., Zakharov, V.I.: Dyons and Roberge -Weiss transition in lattice QCD. EPJ Web Conf. **137**, 03002. arXiv:1611.07789

Diakonov, D.: Topology and confinement. Nucl. Phys. Proc. Suppl. **195**, 5–45 (2009). arXiv:0906.2456

Diakonov, D., Gromov, N., Petrov, V., Slizovskiy, S.: Quantum weights of dyons and of instantons with nontrivial holonomy. Phys. Rev. D **70**, 036003 (2004). arXiv:hep-th/0404042

Gattringer, C.: Calorons, instantons and constituent monopoles in SU(3) lattice gauge theory. Phys. Rev. D **67**, 034507 (2003). arXiv:hep-lat/0210001

Kaczmarek, O., Karsch, F., Petreczky, P., Zantow, F.: Heavy quark anti-quark free energy and the renormalized Polyakov loop. Phys. Lett. B **543**, 41–47 (2002). arXiv:hep-lat/0207002

Kouno, H., Makiyama, T., Sasaki, T., Sakai, Y., Yahiro, M.: Confinement and $\mathbb{Z}_3$ symmetry in three-flavor QCD. J. Phys. G **40**, 095003 (2013). arXiv:1301.4013

Kraan, T.C., van Baal, P.: Monopole constituents inside SU(n) calorons. Phys. Lett. B **435**, 389–395 (1998). arXiv:hep-th/9806034

Langfeld, K., Ilgenfritz, E.-M.: Confinement from semiclassical gluon fields in SU(2) gauge theory. Nucl. Phys. B **848**, 33–61 (2011). arXiv:1012.1214.

Larsen, R., Shuryak, E.: Classical interactions of the instanton-dyons with antidyons. Nucl. Phys. A **950**, 110–128 (2016a). arXiv:1408.6563

Larsen, R.N., Sharma, S., Shuryak, E.: The topological objects near the chiral crossover transition in QCD **794**, 14–18 (2019)

Lee, K.-M., Lu, C.-H.: SU(2) calorons and magnetic monopoles. Phys. Rev. D **58**, 025011 (1998). arXiv:hep-th/9802108

Pisarski, R.D.: Towards a theory of the semi-Quark Gluon Plasma. Nucl. Phys. Proc. Suppl. **195**, 157–198 (2009)

Pisarski, R.D., Yaffe, L.G.: The density of instantons at finite temperature. Phys. Lett. **97B**, 110–112 (1980)

Shnir, Y.: Electromagnetic interaction in the system of multimonopoles and vortex rings. Phys. Rev. D **72**, 055016 (2005a)

Shuryak, E.V.: Theory of hadronic plasma. Sov. Phys. J. Exp. Theor. Phys. **47**, 212–219 (1978) [Zh. Eksp. Teor. Fiz.**74**, 408 (1978)]

Instanton-Dyon Ensembles

15

We will study three approaches to instanton-dyon ensembles: (i) the "parametrically dilute" ones; (ii) then very dense, in which case one can perhaps use the mean field methods; and (iii) statistical simulations which in principle can work at any density. The last section of the chapter will be related with the so-called flavor holonomies, or complex chemical potentials, used as some diagnostic tool.

15.1 Deformed QCD and Dilute Ensembles with Confinement

15.1.1 Perturbative Holonomy Potential and Deformed QCD

In the previous chapter, we had discussed holonomy field

$$A_4 = O(1/g) = const(x) \neq 0 \tag{15.1}$$

as some classical constant background. At this level, it obviously does not lead to any energy, since the corresponding fields $G_{\mu\nu} = 0$.

The next semiclassical approximation follows when one considers some quantum field $a_\mu = O(g^0)$ interacting with it. Since the holonomy is assumed to be diagonal in the color space, the commutator $[a_m, A_4]$ is nonzero for non-diagonal quantum gluons, which are "Higgsed" to become massive, like in Georgi-Glashow model. Quarks (fundamental fermions) interact with the holonomy via the $A_4 \bar{q} \gamma_4 q$ term in the Dirac Lagrangian, also getting a mass. Adjoint fermions, to be briefly discussed below, have color indices like those of the gauge fields.[1]

---

[1] Similarly to gluons, only the adjoint quark species corresponding to the non-diagonal color generators become massive. This is in contrast to the usual fundamental quarks, which are all affected by the Polyakov "Higgsing."

© The Author(s), under exclusive license to Springer Nature Switzerland AG 2021
E. Shuryak, *Nonperturbative Topological Phenomena in QCD and Related Theories*, Lecture Notes in Physics 977,
https://doi.org/10.1007/978-3-030-62990-8_15

All of this leads to certain positive free energy, first calculated in the (appendix of) review by Gross, Pisarski, and Yaffe (1981). In our current notation, for adjoint bosonic Majorana N_a fermions, it has the form

$$V_{GPY} = (1 \mp N_a)\frac{2\pi^2 T^4}{3}\sum_{i,j}(\mu_i - \mu_j)^2(\mu_i - \mu_j - 1)^2 \qquad (15.2)$$

where minus sign stands for periodic and plus for antiperiodic boundary conditions. In the $SU(2)$ case, it simplifies to

$$V_{SU(2)} = (1 \mp N_a)\frac{4\pi^2 T^4}{3}v^2(1 - v)^2 \qquad (15.3)$$

In pure gauge theory, $N_a = 0$, the potential is positive and has minima at two symmetric points $v = 0, \bar{v} = 0$, corresponding to the *trivial* holonomy and its copy. In these limits, one dyon is massless and another has the action of the instanton.

Normal fermions are *antiperiodic* on a circle,[2] so the potential is the sum of two positive terms, as expected from the mass argument given above.

However one may consider a theory in which the periodicity angle has any value one wishes.[3] Following (Unsal 2008), let us for now select *bosonic* spinor field with *periodic* boundary conditions. This leads to the following observations:

(i) Standard finite-T periodicity conditions are different for fermions and bosons; thus supersymmetry is violated at $T \neq 0$. If however one forces the same periodicity conditions, it is preserved. The $N_a = 1$ case corresponds to $\mathcal{N} = 1$ supersymmetric theory, and correspondingly the potential vanishes as gluon and gluino contributions cancel each other.

(ii) At $N_a > 1$ (no supersymmetry), the sign is flipped, and the minimum of the potential corresponds to $v = \bar{v} = 1/2$—the confining value. In the latter case— one of the "deformed QCD" versions—there is confinement both at small and large circle β; thus there is no deconfinement phase transition in this setting.

Although we will not discuss them, let us mention another—simpler—versions of the "deformed QCD," also discussed in the literature. One may simply add to the QCD action some artificial potential, depending on the Polyakov line, $V_{deform}(P)$, pushing the minimum, from the trivial $P = 1$, $A_4 = 0$ point to the confining value.

---

[2]Recall that spinors get half rotation angle of the vectors, so if the latter rotates by 2π, spinor rotates by π.

[3]For example, it can be interpreted as some external Abelian gauge field flux put through the circle in extra dimension. We will return to this idea several times below, as it provides an excellent diagnostic tool to test our understanding of topological phenomena in gauge theories.

15.1.2 The Instanton-Dyons in $N_a = 1$ QCD (or $\mathcal{N} = 1$ SYM)

The pair gluon-gluino (adjoint Majorana fermion field) constitutes the shortest supersymmetric multiplet, so the theory we are going to discuss in this section is mostly known as $\mathcal{N} = 1$ super-Yang-Mills. Before we turn to our main subject—properties of this theory compactified to $R^3 \times S^1$ with a small circle and periodic boundary conditions—let me briefly describe what we know about this theory. It is very much QCD-like, and if compactification is "thermal" (fermions are antiperiodic), it also has deconfinement and chiral restoration phase transitions. According to lattice simulations (Bergner et al. 2014), those two transitions happen at about the same critical temperature T_c.

One difference with QCD is that the number of zero modes for adjoint fermion is $2N_c$, so for $N_c = 2$, 't Hooft effective vertex for an instanton has four zero modes. When chiral symmetry is broken, $< \lambda\lambda > \neq 0$, its *sign* remains undetermined: so unlike QCD, this theory has remaining Z_2 symmetry and two equivalent vacua. As a result, this theory has domain walls, or kinks.

From this point on, we follow the work of Davis, Hollowood, and Khose (1999), the first serious application of the the instanton-dyons. The setting of the paper is supersymmetric Yang-Mills theory (SYM) with a single supersymmetry $\mathcal{N} = 1$, defined on $R^3 \times S^1$: it corresponds to $N_a = 1$ and *periodic* compactification, as described in the previous subsection.

The importance of this application is in the fact that it has resolved the so-called *gluino condensate puzzle*. Two methods to evaluate the value of the gluino condensate have two different answers, namely,

$$< tr\lambda^2 >_{WCI} = 16\pi^2 \Lambda_{PV}^3 \tag{15.4}$$

$$< tr\lambda^2 >_{SCI} = 16\pi^2 \Lambda_{PV}^3 \frac{2}{[(N_c - 1)!(3N_c - 1)]^{1/N_c}} \tag{15.5}$$

The abbreviations here stand for strong coupling instanton (SCI) and weak coupling instanton (WCI) approaches. We will only review the former one.

Now back to the setting. Gluino λ is the super-partner of a gluon, and $\mathcal{N} = 1$ means that there is only one type of the gluino. It is real adjoint field with spin 1/2, so there are two fermionic states. With two gluonic polarizations, it completes the simplest SUSY multiplet. The selection of periodicity condition $\alpha = 0$ (periodic) for gluino preserves the supersymmetry, which removes the GPY potential.

The circle S^1, if small in length $\beta \ll 1/\Lambda$, ensures weak coupling (like high-T). If the circle is large, $\beta \to \infty$, the theory is strongly coupled, like in the low-T QCD. The main difference between the two theories is that—unlike QCD—in this setting for the $\mathcal{N} = 1$ SYM, there are *neither deconfinement nor chiral restoration phase transitions*, at any β!

Discussion in the previous subsection had prepared the reader to the conclusion that the holonomy can be confining, at any β, since the main obstacle—the GPY potential—is in this case absent. Now we need to understand why at small β, one may still have a broken chiral symmetry. Since it is a crucially important point, let us for the moment interrupt the discussion of instanton-dyons and return to the discussion of the issue in historical order, starting with the instantons.

Adjoint color gluinoes, unlike the fundamentally charged quarks, have not one but $2N_c$ zero modes per unit topological charge Q. For the simplest gauge group we discuss, $SU(2)$, it is four (instead of two) fermionic sources in the 't Hooft effective vertex per topological charge. Therefore, unlike the $N_f = 1$ QCD in which this vertex has the structure $\bar{q}q$, in the $N_a = 1$ theory, it is instead $\sim \lambda^4$, with four gluino lines. This is similar to $N_f = 2$ QCD: but in this case, we know that the chiral symmetry is spontaneously broken only at sufficiently low $T < T_c$ (large β), not at all T.

Here is an outline of the instanton-based calculation of the condensate. The condensate has only two gluino fields: so in the SCI calculation, one did averaging of the *square* of the condensate $< tr\lambda^2(x)tr\lambda^2(y) >$ with a single instanton amplitude and then argues that this function $f(x - y)$ cannot depend on the distance and thus is the same when $|x - y| \rightarrow \infty$, and one can apply the so-called cluster decomposition

$$< tr\lambda^2(x)tr\lambda^2(y) > \rightarrow < tr\lambda^2(x) > < tr\lambda^2(y) > \qquad (15.6)$$

and get the SCI answer for the condensate mentioned above.

The alternative calculation (Davies et al. 1999; Hollowood et al. 2000) was revolutionary in that they had realized that in this setting, the semiclassical objects which need to be used are not instantons but *instanton-dyons*. The reason for it is that, even in weak coupling small circle setting, the holonomy is not trivial but the *confining* one. In general, $N_c\,\mu_i$ has "homogeneous distribution" on the circle:

$$< A_4 > = -\left(\frac{i\pi}{\beta}\right)diag\left(\frac{N_c - 1}{N_c}, \frac{N_c - 3}{N_c}, ... - \frac{N_c - 1}{N_c}\right) \qquad (15.7)$$

For $N_c = 2$, there are just two holonomy values, as usual.

With such holonomy setting and two colors, the fermionic zero modes are spread equally between M and L dyons. Since $M+L$=instanton, the total number of modes is still 4, thus it is 2 per dyon. So, the situation is *not* like in QCD with $N_f = 2$ and quartic vertex but rather like in the $N_f = 1$ theory and the quadratic vertex! That is why there is no chiral restoration transition, and the condensate $< tr\lambda^2 >$ can be calculated directly from dilute gas of the dyons.

The result of the explicit calculation is[4] based on the dyon measure in the form

$$dn_M = M_{PV}^3 e^{-S_M} d^3x \left(\frac{g^2 S_M}{2\pi}\right)^{3/2} d\phi \left(\frac{g^2 S_M}{2\pi v^2}\right)^{1/2} \frac{d\xi^2}{2g^2 S_M} \qquad (15.8)$$

---

[4]The original notations used in this paper are BPS monopole for M dyon and KK monopole for L dyon.

where three-dimensional x is the center coordinate, $\phi \in [0, 2\pi]$ is the fourth collective coordinate corresponding to dyon color rotation around the holonomy direction, and ξ are Grassmannian fermionic coordinates for zero modes. The condensate is calculated in a standard diagram using zero modes

$$< \lambda_\alpha(y)\lambda_\beta(y) >= \int dn\lambda_\alpha(y - x)\lambda_\beta(y - x) \qquad (15.9)$$

and additional simplification at large distances

$$\lambda_\alpha \approx 8\pi S_\alpha^\rho(x)\xi_\rho, \quad S(x) = \frac{\gamma_\mu x^\mu}{16\pi^2|x|^2} \qquad (15.10)$$

where S is the massless fermion propagator at zero Matsubara frequency (time integrated). The result is

$$< tr\lambda^2 >_M =< tr\lambda^2 >_L = (16\pi^2)\frac{M_{PV}^3}{2}\exp\left(-\frac{4\pi^2}{g^2}\right) = 16\pi^2\Lambda_{PV}^3 \qquad (15.11)$$

and it agrees exactly with the WCI value but is different from the SCI one.

The lesson is the vacuum of (bosonically compactified) $\mathcal{N} = 1$ SYM in weak coupling regime is a dilute gas of the *independent* instanton-dyons, *not* a gas of instantons.

15.1.3 QCD(adj) with $N_a > 1$ at Very Small Circle: Dilute Molecular (or "bion") Ensembles

Study of the QCD (adj) $N_a > 1$ compactified to parametrically small circle was due to Unsal (2009). The setting is the same as in the previous subsection, namely, the periodicity is "bosonic," reversing the sign of the quark contribution to the GPY vacuum energy. As we had discussed previously, for $N_a = 1$, this cancels the GPY potential, but for $N_a > 1$, the sign of the GPY potential is reversed, and its minimum corresponds to the confining holonomy.

The first point to focus on is the distinction between the R^3 space in the case of Polyakov's confinement and the $R^3 \times S^1$ setting, with a small circle, we consider now. Both setting have time-independent monopoles we call the M-type, but in the latter case, there exists also the "time-twisted", KK, or L-type monopole as well.

The next question is what happens with chiral symmetries in such setting, when $N_a > 1$. In general the effective 't Hooft Lagrangian per dyon is $\sim \lambda^{2N_a}$. For example, for $N_a = 2$, it is the 4-fermion vertex similar to that of the NJL model. Since the effective NJL coupling—the density of the dyons—is exponentially small at weak coupling setting (small circle), there is no spontaneously broken chiral symmetries. The density of the individual dyons is thus zero!

Note that this is the same phenomenon which we discussed in the chapter on instantons in the usual QCD with massless quarks. Since instantons become effective vertices of the type $\sim \lambda^{2N_f}$, for $N_f > 1$, and since there is no chiral symmetry at high T, in QGP, the density of the individual instantons vanishes. There remain however clusters with the topological charge zero, particularly the instanton-anti-instanton "molecules." The ensemble in this case is a "molecular gas," made of soliton-antisoliton pairs bound by quark exchanges (Ilgenfritz and Shuryak 1989).

In the confining setting on $R^3 \times S^1$ for QCD(adj), there are dyon-antidyon pairs bound by fermion exchanges: Unsal (2009) call these binary objects "bions." Introducing deviation from confining holonomy as ϕ and magnetic holonomy[5] σ, one can write down amplitudes for all four types of dyons (of the SU(2) gauge group)

$$M = BPS \sim e^{-\phi+i\sigma-S_0/2}(\psi\psi)^{N_a}; \quad \bar{M} = B\bar{P}S \sim e^{-\phi-i\sigma-S_0/2}(\bar{\psi}\bar{\psi})^{N_a}$$

$$L = KK \sim e^{+\phi-i\sigma-S_0/2}(\psi\psi)^{N_a}; \quad \bar{L} = \bar{K}K \sim e^{+\phi+i\sigma-S_0/2}(\bar{\psi}\bar{\psi})^{N_a}$$

where we use both Unsal' and our notations for the instanton-monopoles (instanton-dyons). The instanton is LM pair, so in the combined amplitude—the LM product of individual amplitudes—all prefactors cancel out except the instanton action e^{-S_0}, with $2N_a N_c$-fermion operator.

Let us now, still following (Unsal 2009), form all possible dyon-antidyon[6] pairs. The fermions propagate between the pairs, for any N_a, so we do not write them anymore (although they of course lead to extra factors in actual expression of the "molecules"):

$$\bar{M}M = B\bar{P}SBPS \sim e^{-S_0-2\phi}, \quad \bar{L}L = \bar{K}KKK \sim e^{-S_0+2\phi},$$

$$\bar{L}M = \bar{K}KBPS \sim e^{-S_0-2i\sigma}, \quad \bar{M}L = B\bar{P}SKK \sim e^{-S_0-2i\sigma}$$

The main idea of Unsal was to focus on the second row, the bions which are *twice* magnetically charged. Since those have nonzero (but still exponentially small) density and nonzero magnetic charge, they will screen the magnetic charge precisely as a single monopole does in the Polyakov confinement. So, these magnetically charged bions do enforce the confinement.

The next important point from this work is that the effective action due to all four types of bions should lead, if the $N_a = 1$ case, to the effective action of the type

$$L_{eff} = \left(\frac{1}{2}\right)\partial\phi^2 + \left(\frac{1}{2}\right)\partial\sigma^2 + ae^{-S_0}(\cos(2\sigma) - \cosh(2\phi)) \qquad (15.12)$$

---

[5]Note that in the Polyakov confinement subsection it was called χ.

[6]Only for zero topological charge objects, one can correctly couple zero modes to each other, so that there remains no zero modes left.

based on supersymmetry arguments. The first comment is the minimum of the potential requires $\sigma = \phi = 0$, or it is the confining one. The second—surprising— observation follows from the expansion of this Lagrangian to $O(\sigma^2, \phi^2)$ terms: both holonomies, σ, ϕ, should have in this theory *the same* screening masses! (As we will see below, in non-supersymmetric theories, those masses are always different and have very different T-dependence.) The question is: how can it be understood microscopically? Earlier in this chapter, we discussed long-distance classical binary interactions and concluded that $\bar{M}M$, $\bar{L}L$ channels are attractive while $\bar{M}L$, $\bar{L}M$ are repulsive. In the latter case, the repulsion can be overcome by fermion exchanges, so the integral over the inter-dyon distance is converging both at small and at large distances. However in the former case, *both* bosonic and fermionic interactions are attractive; the integral is thus converging only due to the "core." Why are both integrals the same? I cannot answer it is one of the miracles induced by supersymmetry.

15.1.4 QCD(adj) with $N_a = 2$ and Periodic Compactification on the Lattice

In this subsection, we continue to discuss the same setting, a theory on $R^3 \times S^1$ with periodic compactification, but add one more adjoint (Majorana) gluino. Of course, this theory is not supersymmetric. For its general discussion, see (Myers and Ogilvie 2008).

There were lattice studies of this theory (Cossu and D'Elia 2009), with variable circumference of the circle β, called in this paper L_c.

The lattice simulations had found *four distinct phases*, which we subsequently briefly describe:

(i) At large L_c (low "temperature"), one finds the usual confining phase, with the Polyakov VEV $< P >= 0$ and symmetric distribution of its eigenvalue, consistent with unbroken center symmetry.

(ii) As the L_c gets shorter, one observes the deconfinement transition, in which $< P > \neq 0$ and its eigenvalues distributed along one of the center elements, breaking the center symmetry.

(iii) As the L_c gets even shorter, in some finite interval of L_c, there exists *another* deconfined phase, in which eigenvalues distributed along a direction *opposite* (making angle π)to that in the usual one (ii). This phase was predicted by (Myers and Ogilvie 2008).

(iv) At very small L_c (high "temperature"), the theory returns to center-symmetric confined phase. This is consistent with the arguments made above, based on the GPY potential (15.2). The authors called it "re-confined" phase.

Figure 15.1(upper) from (Cossu and D'Elia 2009) shows Polyakov line eigenvalues as a function of the circle circumference L_c. The most important consequence of this study is that it contradicts to the conjectured "volume independence": contrary to naive interpolation, two confined phases are *not* smoothly connected. Since it is

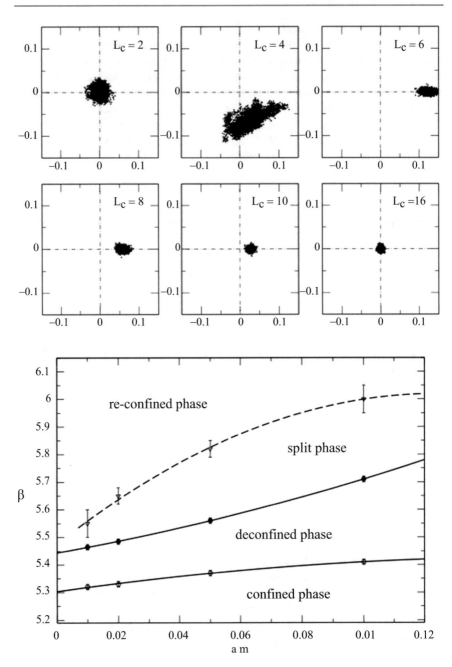

Fig. 15.1 (upper) Scatter plot of the Polyakov line eignevalues, as a function of the size of the compactified dimension L_c: four subsequent phases are seen. (lower) The phase diagram on the plot quark mass am and lattice gauge coupling $\beta = 6/g^2$. Extrapolation of the phase boundaries to the chiral limit for split phase is not obvious

a lattice work, with a nonzero gluino masses and finite lattice spacings, one may ask to what extent their zero limits have been reliably reached. Figure 15.1(lower) shows, that in the $m \rightarrow 0$ limit, the phase (iii) *may* disappear, as such option is within the numerical accuracy, but the other deconfined phase (ii) seems definitely to be there, also in the chiral limit.

What about the fate of the chiral symmetry breaking through all these transitions? Remarkably (but in agreement with other studies of the QCD(adj) with thermal compactification), it was observed in (Cossu and D'Elia 2009) that $< \bar{\psi}\psi > \neq 0$ in the whole region of L_c studied. So, massless fermions keep their nonzero "constituent quark mass" throughout: the two (or maybe one) deconfined phases are then a plasma of "constituent quarks."

The magnitude of the condensate however decreases by about an order of magnitude from "vacuum" (large L_c) value, suggesting that chiral symmetry is going to be eventually restored, just at very small L_c not included in these particular set of simulations (Cossu and D'Elia 2009). Eventually, the $N_a = 2$ theory with small circle should of course have zero gluino condensate, as was argued in the previous section.

15.2   Dense Dyon Plasma in the Mean Field Approximation

The main idea of the mean field approximation is that a particle interacts simultaneously with many.[7] One may think that it can be used in cases when the ensemble of the dyons is dense enough, producing strong screening, which effectively reduces the pair-wise correlations. Anyway, using some average mean field is the simplest approach, in which one can get analytical evaluation of the observables.

In this section, we will follow a series of papers by Liu, Zahed, and myself using the mean field approximation. The first paper of the series, (Liu et al. 2015a), had established the approximation in the technical sense. Here there is no place to present technical details of these works, and we just summarize few important points and the results.

The main result is that dense enough dyon ensemble does overcome the GPY potential and shift the minimum of the free energy to the confining value, $v = 1/2$, for the SU(2) gauge theory considered. The key expression for the partition function is put into the form

$$\ln Z / V_3 = -V_{ideal} - \frac{1}{2} \int \frac{d^3 p}{(2\pi)^3} \ln \left| 1 - \frac{V^2(p)}{16} \frac{p^8 M_D^4}{(p^2 + M_D^2)^4} \right|$$

(15.13)

---

[7]Note that this condition is necessary but in general not sufficient. For example, the approximation is valid in the perturbative Debye theory of plasma, e.g., for the monopole plasma discussed by Polyakov. It is not valid in strongly coupled plasmas, which may be strongly correlated with liquids or even solids, producing mean fields very different from being space-independent.

where V_{ideal} describes free energy of the noninteracting dyons, and $V(p)$ is the Fourier transform of the dyon-antidyon potential (classical or including one loop). The very presence of dyons, with electric charges, generates a Debye electric screening mass $M_D = \sqrt{2n_D/T}$ related to the dyon density n_D. When this density is large, M is large. From expression of the partition function, it follows that the effect of the potential (inside the logarithm) gets reduced (screened out), which in principle justifies the mean field method. Its specific applicability limits can be derived from requirement that the second term inside the log is *less* than one.[8]

Figure 15.2 illustrates one of the results of this work, the temperature dependence of the electric and magnetic screening masses, in comparison to what has been derived from numerical simulations of the $SU(2)$ gauge theory. Note that the electric screening mass[9] —shown by the closed circles—has a drop downward, as T is reduced below the deconfinement transition, while the magnetic screening is expected to get larger than the electric one there.

The next work of the series (Liu et al. 2015b) applies MFA to the $N_c = 2$ color theory with $N_f = 2$ light quark flavors. At high density, the minimum of the free energy still corresponds to the confining ensemble with $\nu = 1/2$. The gap equation for the effective quark mass (proportional the quark condensate) of (Liu et al. 2015b) is in the usual form

$$\int \frac{d^3 p}{(2\pi)^3} \frac{M_{eff}^2(p)}{p^2 + M_{eff}^2(p)} = n_L \qquad (15.14)$$

where the r.h.s. is the density of the dyon type possessing the fermion zero mode, namely, the L-dyons. The equation is actually for the parameter λ in the effective mass $M_{eff}(p) = \lambda p T(p)$, in which $T(p)$ being the Fourier transform of the

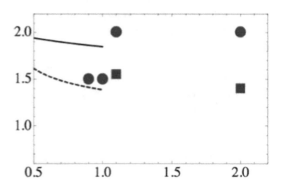

Fig. 15.2 The electric M_E/T (dashed line) and magnetic M_M/T (solid line) screening masses versus T/T_c. The points are the lattice data for $SU(2)$ gauge theory shown for comparison; (blue) circles are electric; (red) squares are magnetic

---

[8]Otherwise the argument is negative, and logarithm gives an imaginary part, signaling appearance of an instability.

[9]For clarity: while the calculation includes only dyons but not gluons, it does include the one-loop GPY potential. Its derivative over the holonomy value is the perturbative one-loop electric Debye mass, due to gluons.

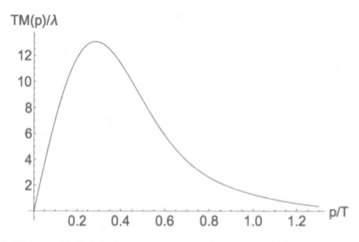

Fig. 15.3 The momentum-dependent constituent quark mass $TM(p)/\lambda$ versus momentum in units of temperature p/T

"hopping matrix element" calculated using the fermionic zero mode. Momentum dependence of $M(p)/\lambda$ is universal and is shown in Fig. 15.3. In practice, the best way to solve the gap equation is to calculate the momentum integral in its l.h.s. numerically and then parameterize the dependence on parameter λ.

A generalization of the mean field treatment to arbitrary number of colors and flavors in (Liu et al. 2015b) shows that this gap equation has nonzero solutions for the quark condensate only if

$$N_f < 2N_c \tag{15.15}$$

So, the critical number of flavors is $N_f = 6$ for $N_c = 3$. The lattice simulation indeed shows weakening of chiral symmetry violation effects with increasing N_f, but specific results on the end of chiral symmetry breaking are so far rather incomplete: for $N_c = 3$, we know that in the $N_f = 4$ case, the chiral symmetry is broken, the case $N_f = 8$ is not yet completely decided, and $N_f = 12$ seems to be already in the conformal window.

Another important generalization—for quarks in the adjoint representation— is made in a separate paper (Liu et al. 2016a). The number of fermionic zero modes increases, and they are more complicated. In the symmetric dense phase, both M and L dyons have two zero modes. But the actual difficulty is not some longer expressions but the fact that one of them has rather singular behavior—gets delocalized—exactly at the confining value of the holonomy, $v = 1/2$. Therefore, in the case of adjoint quarks, the approach toward the confining phase needs some special care. In the case of $N_c = 2$, $N_a = 1$, the deconfinement and chiral restoration happen at about the same temperature, in agreement with lattice result we discussed above for this theory.

15.3   Statistical Simulations of the Instanton-Dyon Ensembles

15.3.1   Holonomy Potential and Deconfinement in Pure Gauge Theory

The first direct simulation of the instanton-dyon ensemble with dynamical fermions has been made by (Faccioli and Shuryak 2013). The general setting follows the example of the "instanton liquid"; it included the determinant of the so-called "hopping matrix," a part of the Dirac operator in the quasizero-mode sector. It has been done for $SU(2)$ color group, and the number of fermions flavors $N_f = 1, 2, 4$. Except in the last case, chiral symmetry breaking has been clearly observed, for dense enough dyon ensemble.

The second one, by Larsen and myself (Larsen and Shuryak 2015), uses direct numerical simulation of the instanton-dyon ensemble, both in the high-T dilute and low-T dense regime. The holonomy potential as a function of all parameters of the model is determined and minimized.

In Fig. 15.4(left) from this work, we show the dependence of the total free energy on holonomy value, for different ensemble densities. This important plot[10] had shown, for the first time in direct simulations, that at high density of the dyons, their back reaction does generate confinement! Indeed, the minimum of the holonomy potentials shifts to $\nu = 1/2$, the confining value for $SU(2)$ ($\cos(\pi \nu) = 0$). The self-consistent parameters of the ensemble, minimizing the free energy, are determined for each density.

As the action parameter S is growing, corresponding to the growing temperature, the dyon-symmetric phase goes into an asymmetric phase, in which the densities of M and L dyons are not the same (see Fig. 15.4(right)). The l dyon has larger action due to time-dependent "twist" and thus smaller density.

The next simulations of the instanton-dyon ensemble for $SU(2)$ gauge group have been done by Lopez-Ruiz, Y. Jiang and J. Liao (2018). In Fig. 15.5 from this work, we show the shapes of the holonomy potential $V(\nu)$ near the critical point and the fit to the average value of the Polyakov line, fitted to the expected second-order behavior with indices of the three-dimensional Ising model:

$$\beta \approx 0.3265, \quad \omega \approx 0.84 \tag{15.16}$$

Within the statistical accuracy of the calculation, the expected second-order behavior is indeed observed.

The free energy potential between static quark and antiquark from (Lopez-Ruiz et al. 2018) is shown in Fig. 15.6. Note first that the potential is nearly temperature-independent below T_c (the first two sets) but rapidly decreases above T_c. While the

---

[10]Later analysis improved statistical accuracy of the data points: we nevertheless show here the first plot in which the confinement had came out of the simulations.

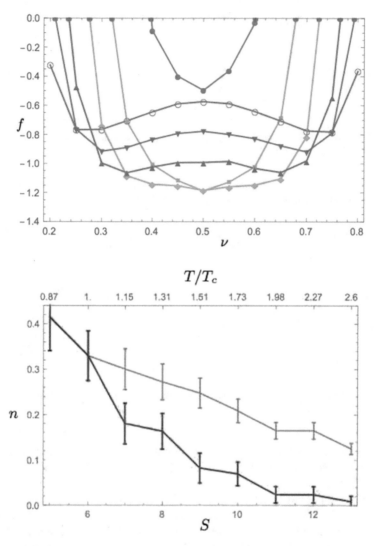

Fig. 15.4 (left) Free energy density f as a function of holonomy v at $S = 6$, $M_D = 2$ and $N_M = N_L = 16$. The different curves corresponds to different densities. $filled\ circle$ $n = 0.53$, $filled\ square\ n = 0.37$, $filled\ diamond\ n = 0.27$, $filled\ triangle\ n = 0.20$, $filled\ triangledown\ n = 0.15$, $circle\ n = 0.12$. (right) Density n (of an individual kind of dyons) as a function of action S (lower scale) which is related to T/T_c (upper scale) for M dyons(higher line) and L dyons (lower line)

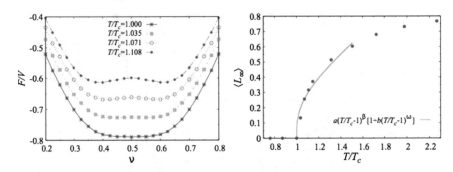

Fig. 15.5 (left) Free energy density of the dyon ensemble near Tc ($S = 7$); (right) Fit of the average value of the Polyakov line to the expected 2-nd order critical point

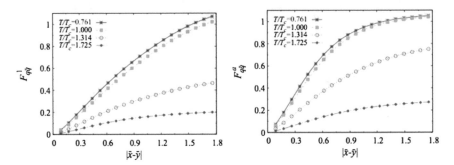

Fig. 15.6 The free singlet energy potential, for static charges in fundamental (left) and adjoint (right) color representations

fundamental quarks show nearly linear confining potential, the adjoint one shows screening above certain distance. This is as expected for pure gauge theory, without quarks.[11]

Another important issue addressed in (Lopez-Ruiz et al. 2018) is the area law for the spatial Wilson loop. The corresponding measurements are reported, the existence of the spatial tension is demonstrated, but its temperature dependence is not yet compared to the available lattice data.

15.3.2 Instanton-Dyon Ensemble and Chiral Symmetry Breaking

The issue of chiral symmetry breaking using numerical simulations was addressed by (Larsen and Shuryak 2016b). Including the fermionic determinant in "hopping" approximation, we calculated the spectrum of the lowest Dirac eigenvalues.

---

[11] Note that the simulation includes dyons but not gluons, which however are also integer charged and can screen the adjoint charge.

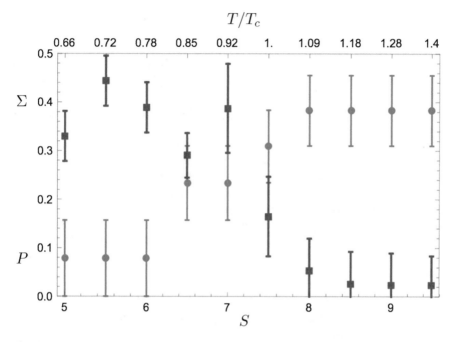

Fig. 15.7 The Polyakov loop P (blue circles) and the chiral condensate Σ (red squares) as a function of action $S = 8\pi^2/g^2$ or temperature T/T_c. Σ is scaled by 0.2

Extracting the quark condensate is complicated, as usual, by finite-size effects. Using two sizes of the system, with 64 and 128 dyons, we identify the finite-size effects in the eigenvalue distribution and extrapolate to infinite-size system. The location of the chiral transition temperature is defined by both extrapolation of the quark condensate, from below, and the so-called "gaps" in the Dirac spectra, from above.

We do indeed observe, for $SU(2)$ gauge theory with two flavors of light fundamental quarks, both the confinement-deconfinement transition and chiral symmetry breaking, as the density of dyons goes up at lower temperature (see Fig. 15.7). Determination of the transition point by vanishing of $< P >$ or $< \bar{\psi}\psi >$ is difficult for technical reasons. Since both transitions appear to be in this case just a smooth crossover ones, it is by now a well-established procedure to define the transition points via maxima of corresponding susceptibilities. Those should correspond to *inflection points* (change of curvature) on the plots to be shown. Looking from this perspective at Fig. 15.7, one would locate the inflection points of both curves, for $< P >$ or $< \bar{\psi}\psi >$, at the same location, namely, $S = 7$–7.5. Thus, within the accuracy our simulations have, we conclude that both phase transitions happen at the same conditions.

15.4   QCD with Flavor-Dependent Quark Periodicity Phases

For applications, such as heavy ion collisions, one needs to know properties of the QCD matter not only as a function of the temperature T but quark (baryon number, isospin, etc.) densities as well. Unfortunately, Euclidean partition function at nonzero quark chemical potentials μ_f contains complex factor $e^{i\mu_i/T}$ which cannot be interpreted as probability: so standard Monte Carlo simulation algorithms cannot be used.

One can however introduce *imaginary chemical potentials*, proceed with calculations, and then extrapolate, in the μ^2 plot, across zero. This is done by several lattice groups, but we will not discuss those results here.

The reason is we are not really focused on the physical problems with chemical potential here but on the use of the imaginary chemical potentials, or "flavor holonomies" as they are also-called, as some diagnostic tool. Observing how QCD ensembles react on their magnitude, on the lattice, and in the models, we hope to gain better understanding of the underlying mechanisms of the QCD phase transitions.

15.4.1 Imaginary Chemical Potentials and Roberge-Weiss Transitions

The phase diagram of QCD-like theories with imaginary quark chemical potential has been discussed in the fundamental paper by Roberge and Weiss (1986). Imaginary chemical potential denoted by $\theta = i\mu/T$ appears in the periodicity condition over the Matsubara circle:

$$\psi(x, \beta) = e^{i\theta}\psi(x, 0) \tag{15.17}$$

As we already mentioned, for real θ, the sign problem is absent, and standard Monte Carlo algorithms can be applied to simulate lattice QCD. QCD possesses rich phase structure at nonzero θ, details of which depend on the number of flavors N_f and the quark mass (masses) m_f. Later in the section, we will make it even richer, by considering different phases for different quark flavors θ_f.

Since $\theta = i\mu/T$ is an angle, it is obvious that the QCD partition function Z is a periodic function of it, with the period 2π. However, as noted in (Roberge and Weiss 1986), the period is actually smaller $2\pi/N_c$, and there is the so-called Roberge-Weiss symmetry

$$Z(T, \theta) = Z\left(T, \theta + \frac{2\pi}{N_c}\right) \tag{15.18}$$

because of N_c branches of the gluonic GPY potential.

The main point is that the imaginary chemical potential θ simply shifts holonomy values

$$2\pi \mu_j \rightarrow 2\pi \mu_j + \theta \qquad (15.19)$$

in the quark term (B.16) of the GPY one-loop effective potential. In pure gauge theory at sufficiently large T, there is spontaneous breaking of the center symmetry, and one of the N_c branches is selected. For example, $SU(3)$ theory has one real and two complex conjugated branches. Recall that in QCD, the quark term is *not* Center-symmetric; thus the free energy is tilted and the real branch is the preferred one: the mean $< P >$ slowly moves as a function of T along the real axes, as described above.

If there is a nonzero θ, it effectively rotates the quark part of the potential and at certain values, $\theta_k = (2k + 1)\pi/3$ in this theory, the free energies of different brunches cross. As a result, there appear kinks—the first-order phase transitions—at such values. These points are crossings of the N_c branches of the effective potential, as shown in Fig. 15.8: at any θ, the physical branch is the lowest one.

Since these arguments are derived using the one-loop GPY potential, they of course are valid only at high T, or weak coupling. In reality, the RW transition exists at $T > T_{RW}$, where T_{RW} are the critical endpoints of the Roberge-Weiss first-order transitions.

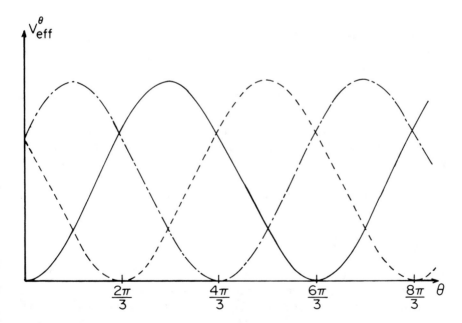

Fig. 15.8 Effective GPY potential as a function of the imaginary chemical potential θ, for three colors $N_c = 3$

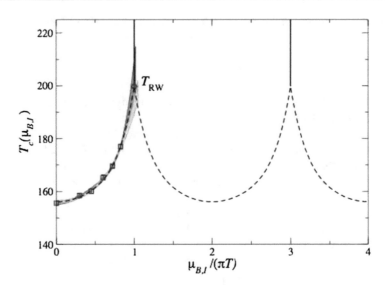

Fig. 15.9 Phase diagram of QCD in the presence of an imaginary baryon chemical potential obtained from numerical simulations on Nt = 8 lattices alone. Bands denote fits to polynomials in μ_B^2: the orange (longer) band is obtained using terms up to order μ_B^4, the violet (shorter) one using up to μ_B^6 terms

At sufficiently low T, there is the confinement phase, with $< P >= 0$, and the cusp disappears. First-order transitions must end at some critical points. Where exactly it happens can be calculated on the lattice. Lattice investigation (Bonati et al. 2016b) has located it at

$$T_{RW} = 1.34(7)T_c = 208(5) \text{ MeV} \qquad (15.20)$$

Does the pseudocritical line really get to the RW endpoint, as suggested by early studies on the subject? Or two pseudocritical lines meat each other, and then go vertically to the critical point? Figure 15.9 from (Bonati et al. 2016b) illustrates current state of lattice answer to this question. Inside the accuracy of the current data, the answer to this question seems to select the former option.

15.4.2 $Z(N_c)$ QCD

The main idea of this deformation of $N_c = N_f$ QCD (Kouno et al. 2013) is a "democratic" distribution of flavor holonomies, putting those in between the subsequent holonomy phases μ_i (see Fig. 14.10(right)). The framework in which it has been suggested is the PNJL model.

In the theory of the instanton-dyons, $Z(N_c)$ QCD has a very simple meaning: *each dyon type gets a zero mode of one quark flavor.*

The $Z(N_c)$ QCD has been studied in the mean field framework (Liu et al. 2016b), by statistical simulations (Larsen and Shuryak 2016c) and also by lattice simulations (Misumi et al. 2016). In the dilute limit, it also has been studied by (Cherman et al. 2016).

The first two papers consider the $N_c = N_f = 2$ version of the theory, while the last one focuses on the $N_c = N_f = 3$. In the former case, the set of phases are $\psi_f = 0, \pi$, so one quark is a boson and one is a fermion. In the latter, $\psi_f = \pi/3, \pi, -\pi/3$.

All these works find deconfinement transition to strengthen significantly, compared to QCD with the same N_c, N_f in which it is a very smooth crossover. While in (Liu et al. 2016b) the $< P >$ reaches zero smoothly, a la second-order transition, the simulations (Larsen and Shuryak 2016c) and lattice (Misumi et al. 2016) both see clear jump in its value which indicated strong first-order transition. The red squares at Fig. 15.10(left) from (Larsen and Shuryak 2016c) are comparing the behavior of the mean Polyakov line in Z_2 and ordinary QCD. The parameter S used as measure of the dyon density is the "instanton action," related with the temperature by

$$S = \left(\frac{11N_c}{3} - \frac{2N_f}{3} \right) \log(\frac{T}{\Lambda}) \tag{15.21}$$

The dyons share it as $S_M = vS$, $S_L = \bar{v}S$. So, larger S at the r.h.s. of the figure corresponds to high T and thus to more dilute ensemble, since densities contain $\exp(-S_i)$.

All three studies see nonzero chiral condensates in the studied region of densities: perhaps no chiral restoration happens at all. The values for the condensate are shown in Fig. 15.10(right) from (Larsen and Shuryak 2016c).

The simulation (Liu et al. 2016b) demonstrates that the spectrum of the Dirac eigenvalues has a very specific "triangular" shape, characteristic of a single-flavor QCD. This explains why the $Z(N_c)$ QCD has much larger condensate than ordinary QCD, at the same dyon density, and also why there is no tendency to restoration. As expected, all works see different condensates, $< \bar{u}u > \neq < \bar{d}d >$ but with difference smaller than one could expect from the difference in the dyon density.

In (Cherman et al. 2016), the authors study the periodic compactification to a small circle, with $Z(N_c)$ QCD flavor holonomies. The main statement is that even in the limit of very small circle—exponentially small dyon density—in this setting, the chiral symmetry remains spontaneously broken but in a very specific way. There are however only $N_c - 1$ massless pions, not $N_f^2 - 1$ as usual, equal to the number of Cartan subalgebra generators.

15.4.3 Roberge-Weiss Transitions and Instanton-Dyons

In Sect. 15.4.1, we introduced the Roberge-Weiss transitions and discussed some lattice studies of them. Let us remind the reader that imaginary chemical potential

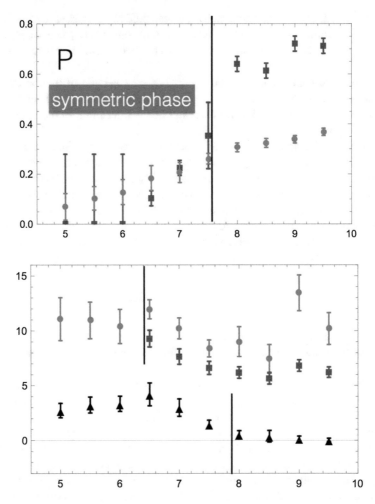

Fig. 15.10 (left) The mean Polyakov line P versus the density parameter S. Red squares are for $Z_2 QCD$, while blue circles are for the usual QCD, both with $N_c = N_f = 2$. (right) The quark condensate versus the density parameter S. Black triangles correspond to the usual QCD: and they display chiral symmetry restoration. Blue and red points are for two flavor condensates of the $Z_2 QCD$: to the left of vertical line, there is a "symmetric phase" in which both types of dyons and condensates are the same

$\theta = i\mu/T$ is a phase, or flavor holonomy, which effectively rotates quarks on the holonomy circle.

Here we will discuss the role of $\theta \neq 0$ in the theory of the instanton-dyons. As discussed in the previous chapter and also previous subsection, the fermionic zero modes "jump" from one kind of instanton-dyons to another, when the flavor and color holonomies coincide. Therefore, instanton-dyons should play a very important role in the Roberge-Weiss transitions.

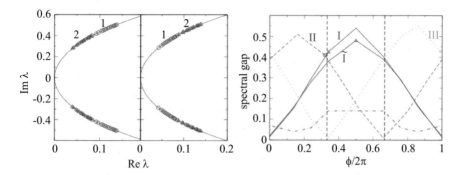

Fig. 15.11 (left) Spectra of the overlap Dirac operator for two configurations. As the RW transition is crossed, the eigenvalues corresponding to different dyon types exchange places. (right) Spectral gap of the overlap Dirac operator as function of the periodicity angle ϕ for four thermal configurations (two red solid curves, one dashed blue curve, and one green dotted curve) generated at $T = 1.35 T_c$

The first lattice study of this phenomenon has been performed in (Bornyakov et al. 2017). Its main result is demonstration of dramatic changes in the Dirac eigenvalue spectra as the lines of the Roberge-Weiss transitions are crossed.

The authors use variable periodicity angle ϕ as a diagnostic tool, allowing to monitor eigenvalues for each kind of dyons separately. The left figure Fig. 15.11 demonstrates eigenvalue spectra for two configurations on both sides of RW transition: the dyon spectra are clearly interchanged. The right figure shows how the spectral gap (for configurations above T_c) depends on the variable periodicity angle ϕ.

All results are exactly as anticipated, based on the instanton-dyon mechanism of chiral symmetry breaking.

15.5 Brief Summary

- The ordinary perturbation theory around constant holonomy field $A_0 = const$ in pure gauge theory has minimum corresponding to "trivial holonomy" or $\langle P \rangle = 1$. In this case, L dyon has the action of the instanton, $8\pi^2/g^2$, and others are massless. Theories with adjoint fermions, e.g., supersymmetric ones, have this potential canceled, with any value of $\langle P \rangle$ possible. Instanton-dyons were first studied on this context. It allowed to resolve a puzzle concerning two conflicting derivation of the value of quark condensate in $\mathcal{N} = 1$ SYM. That historically was the first indication that instanton-dyon ensemble is correct, not the ensemble of instantons.
- In supersymmetric theories, it was also possible to study dyon-antidyon molecules (or "bions") and deduce their back reaction to total free energy (see (15.12)).

- Dense plasma of dyons was studied in the mean field approximation. In a number of papers, one was able to study deconfinement transition as well as chiral symmetry restoration.
- Numerical statistical mechanics of the dyon ensemble also was subject of several papers, so far only for the simplest gauge theory with $SU(2)$ color. It was shown that deconfinement and chiral transition coincide in ordinary QCD.
- And yet, if different periodicities of quark flavors are imposed, these two transitions can be significantly modified. A particular setting, called $Z_N QCD$, is then $N_c = N_f$, and N_f periodicity phases are selected in such a way that all species of dyons have *one* quark zero mode each. In this theory, the deconfinement transition strengthened to become the first order, while chiral symmetry moved to infinity!
- All these observations confirm the view that instanton-dyons are indeed the objects, in terms of which one can understand all phase transitions, in QCD and in its deformed versions.

References

Bergner, G., Giudice, P., Munster, G., Piemonte, S., Sandbrink, D.: First studies of the phase diagram of N=1 supersymmetric Yang-Mills theory. PoS **LATTICE2014**, 262 (2014). arXiv:1501.02746

Bonati, C., D'Elia, M., Mariti, M., Mesiti, M., Negro, F., Sanfilippo, F.: Roberge-Weiss endpoint at the physical point of $N_f = 2 + 1$ QCD. Phys. Rev. **D93**(7), 074504 (2016b). arXiv:1602.01426

Bornyakov, V.G., Boyda, D.L., Goy, V.A., Ilgenfritz, E.M., Martemyanov, B.V., Molochkov, A.V., Nakamura, A., Nikolaev, A.A., Zakharov, V.I.: Dyons and Roberge - Weiss transition in lattice QCD. EPJ Web Conf. **137**, 03002 (2017). arXiv:1611.07789

Cherman, A., Schaefer, T., Unsal, M.: Chiral Lagrangian from duality and monopole operators in compactified QCD. Phys. Rev. Lett. **117**(8), 081601 (2016). arXiv:1604.06108

Cossu, G., D'Elia, M.: Finite size phase transitions in QCD with adjoint fermions. JHEP **07**, 048 (2009). arXiv:0904.1353

Davies, N.M., Hollowood, T.J., Khoze, V.V., Mattis, M.P.: Gluino condensate and magnetic monopoles in supersymmetric gluodynamics. Nucl. Phys. **B559**, 123–142 (1999). arXiv:hep-th/9905015

Faccioli, P., Shuryak, E.: QCD topology at finite temperature: statistical mechanics of self-dual dyons. Phys. Rev. **D87**(7), 074009 (2013). arXiv:1301.2523

Gross, D.J., Pisarski, R.D., Yaffe, L.G.: QCD and instantons at finite temperature. Rev. Mod. Phys. **53**, 43 (1981)

Hollowood, T.J., Khoze, V.V., Lee, W.-J., Mattis, M.P.: Breakdown of cluster decomposition in instanton calculations of the gluino condensate. Nucl. Phys. **B570**, 241–266 (2000) arXiv:hep-th/9904116.

Ilgenfritz, E.-M., Shuryak, E.V.: Chiral symmetry restoration at finite temperature in the instanton liquid. Nucl. Phys. **B319**, 511–520 (1989)

Kouno, H., Makiyama, T., Sasaki, T., Sakai, Y., Yahiro, M.: Confinement and $\mathbb{Z}_3$ symmetry in three-flavor QCD. J. Phys. **G40**, 095003 (2013). arXiv:1301.4013

Larsen, R., Shuryak, E.: Interacting ensemble of the instanton-dyons and the deconfinement phase transition in the SU(2) gauge theory. Phys. Rev. **D92**(9), 094022 (2015) arXiv:1504.03341

Larsen, R., Shuryak, E.: Instanton-dyon ensemble with two dynamical quarks: the chiral symmetry breaking. Phys. Rev. **D93**(5), 054029 (2016b). arXiv:1511.02237

Larsen, R., Shuryak, E.: Instanton-dyon ensembles with quarks with modified boundary conditions. Phys. Rev. **D94**(9), 094009 (2016c). arXiv:1605.07474

Liu, Y., Shuryak, E., Zahed, I.: Confining dyon-antidyon coulomb liquid model. I. Phys. Rev. **D92**(8), 085006 (2015a). arXiv:1503.03058

Liu, Y., Shuryak, E., Zahed, I.: Light quarks in the screened dyon-antidyon Coulomb liquid model. II. Phys. Rev. **D92**(8), 085007 (2015b). arXiv:1503.09148

Liu, Y., Shuryak, E., Zahed, I.: Light adjoint quarks in the instanton-dyon liquid model IV. Phys. Rev. **D94**(10), 105012 (2016a). arXiv:1605.07584

Liu, Y., Shuryak, E., Zahed, I.: The instanton-dyon liquid model V: twisted light quarks. Phys. Rev. **D94**(10), 105013 (2016b). arXiv:1606.02996

Lopez-Ruiz, M.A., Jiang, Y., Liao, J.: Confinement, holonomy and correlated instanton-dyon ensemble I: SU(2) Yang-Mills theory. Phys. Rev. **D97**(5), 054026 (2018). arXiv:1611.02539

Misumi, T., Iritani, T., Itou, E.: Finite-temperature phase transition of $N_f = 3$ QCD with exact center symmetry. PoS **LATTICE2015**, 152 (2016). arXiv:1510.07227

Myers, J.C., Ogilvie, M.C.: New phases of SU(3) and SU(4) at finite temperature. Phys. Rev. **D77**, 125030 (2008). arXiv:0707.1869

Roberge, A., Weiss, N.: Gauge theories with imaginary chemical potential and the phases of QCD. Nucl. Phys. **B275**, 734–745 (1986)

Unsal, M.: Abelian duality, confinement, and chiral symmetry breaking in QCD(adj). Phys. Rev. Lett. **100**, 032005 (2008). arXiv:0708.1772

Unsal, M.: Magnetic bion condensation: A new mechanism of confinement and mass gap in four dimensions. Phys. Rev. **D80**, 065001 (2009). arXiv:0709.3269

The Poisson Duality Between the Particle-Monopole and the Semiclassical (Instanton) Descriptions

16

We discussed a number of applications of the monopoles. In particular, the confinement is indeed a Bose-Einstein condensation of monopoles, at $T < T_c$. There are multiple applications to physics of quark-gluon plasma and heavy ion collisions. Basically, QGP is a *"dual plasma"* made of electrically charged quasiparticles, quarks, and gluons, and magnetically charged monopoles. This view lead to explanation of various observed phenomena. The key to all of it is the notion that monopoles can be treated as quasiparticles and can be used in calculations involving both Euclidean (thermodynamics) and Minkowski (kinetics) times as needed. However, the 't Hooft-Polyakov solution requires an adjoint scalar (Higgs) with a nonzero VEV. This is the case in the Georgi-Glashow model and in other theories with an adjoint scalar field, notably in theories with extended supersymmetry $\mathcal{N} = 2, 4$. Yet it is *not* so in QCD-like theories without scalars, and thus one cannot use this solution.

In the chapter devoted to instanton-dyons, the 't Hooft-Polyakov solution is used with the time component of the gauge field A_4 as an adjoint scalar. The semiclassical theory built on them obviously can only be used in the Euclidean time formulation: an analytic continuation of A_4 to Minkowski time includes an imaginary field which makes no sense. So, the instanton-dyons cannot be used as quasiparticles. And yet, the presence of magnetic charge of the instanton-dyons does suggest that they should *somehow* be related to particle-monopoles.

A gradual understanding of this statement began some time ago but remained rather unnoticed by the larger community. One reason for that was the setting in which it was shown, which was based on extended supersymmetry. Only in these cases was one able to derive reliably *both* partition functions—in terms of monopoles and instanton-dyons—and show them to be equal (Dorey and Parnachev 2001; Poppitz and Unsal 2011; Poppitz et al. 2012). Furthermore, they were not summed up to an analytic answer but shown instead to be related by the so-called Poisson duality.

© The Author(s), under exclusive license to Springer Nature Switzerland AG 2021
E. Shuryak, *Nonperturbative Topological Phenomena in QCD and Related Theories*, Lecture Notes in Physics 977,
https://doi.org/10.1007/978-3-030-62990-8_16

16.1 The Rotator

Another classic example, which displays features important for physics to be discussed in this book, is a rotating object, which we will call the *rotator* or *the top*. What is special in this case is that the coordinates describing its location are *angles*, which are always defined with some natural periodicity conditions. Definition of the path integrals in such cases requires important additional features.[1]

The key questions and solutions can be explained following Schulman (1968) using the simplest $SO(2)$ top, a particle moving on a circle. Its location is defined by the angle $\alpha \in [0, 2\pi]$, and its (initial) action contains only the kinetic term

$$S = \oint dt \, \frac{\Lambda}{2} \dot{\alpha}^2 \qquad (16.1)$$

with $\Lambda = mR^2$, the corresponding moment of inertia for rotation.

All possible paths are naturally split into topological homotopy classes, defined by their *winding number*. The paths belonging to different classes cannot be continuously deformed to each other. Therefore a fundamental question arises: *how should one normalize those disjoint path integrals over classes of paths?* Clearly, there is no natural way to define their relative normalization, or rather their relative *phase*.

Following Aharonov and Bohm (1959), one may enhance physical interpretation of this setting. Suppose our particle has an Abelian electric charge, and certain device (existing in extra dimensions invisible to the rotator) creates a nonzero magnetic field flux $\Phi \neq 0$ through the circle. Stokes theorem relates it to the circulation of the gauge field:

$$\oint d\alpha \, A_\alpha = \int \vec{B} d\vec{S}$$

While $A_\mu(x)$ is gauge-dependent, its circulation (called holonomy) is gauge invariant, since it is related to the field flux.[2]

The extra phase is thus physical. Furthermore, it propagates into the energy spectra and the partition function. One can write it in a Hamiltonian way, as the sum over states with the angular momentum m at temperature T

$$Z_1 = \sum_{m=-\infty}^{\infty} \exp\left(-\frac{m^2}{2\Lambda T} + im\omega \right), \qquad (16.2)$$

---

[1] The Feynman-Hibbs book does not have its discussion and contains only a comment that the authors cannot describe, say, an electron with spin $1/2$ and that it was a "serious limitation" of the approach.

[2] For non-Abelian case, there is no Stokes theorem, but gauge invariance of all closed paths is still true: it follows from direct calculation of gauge transformation of path-ordered exponents.

where ω is the holonomy phase, which is so far arbitrary.

Although physical, the effect is invisible at the classical level. This can be seen from the inclusion of the additional term in the action $\sim (\omega/2\pi) \int d\tau \dot{\alpha}$ which would "explain" the holonomy phase. This term in Lagrangian however is a full derivative, $\dot{\alpha}$, so the action depends on the endpoints of the paths only and is insensitive to its smooth deformations. It therefore generates no contribution to classical equations of motion, thus failing to "exert any force" on the particle in classical sense. In summary, an appearance of the holonomy phase is a nontrivial quantum effect, not coming from the classical equation of motion.

Now one can also use Lagrangian approach, looking for paths periodic in Euclidean time on the Matsubara circle. Classes of paths which make a different number n of rotations around the original circle can be defined as "straight" classical periodic paths

$$\alpha_n(\tau) = 2\pi n \frac{\tau}{\beta}, \tag{16.3}$$

plus small fluctuations around them. Carrying out a Gaussian integral over them leads to the following partition function:

$$Z_2 = \sum_{n=-\infty}^{\infty} \sqrt{2\pi \Lambda T} \exp\left(-\frac{T\Lambda}{2}(2\pi n - \omega)^2\right). \tag{16.4}$$

The key point here is that these quantum numbers, m used for Z_1 and n for Z_2, are very different in nature. The dependence on the temperature is different. Also, for Z_1 each term of the sum is periodic in ω, while for Z_2, this property is also true but recovered only after summation over n.

In spite of such differences, both expressions are in fact the same! In this toy model, it is possible to do the sums numerically and plot the results. Furthermore, one can also derive the analytic expressions, expressible in terms of the elliptic theta function of the third kind

$$Z_1 = Z_2 = \theta_3\left(-\frac{\omega}{2}, \exp\left(-\frac{1}{2\Lambda T}\right)\right), \tag{16.5}$$

which is plotted in Fig. 16.1 for few values of the temperature T.

In order to prove that one may use the Jacobi identity,

$$\theta_3(z, t) = (-it)^{-1/2} e^{z^2/i\pi t} \theta_3(z/t, -1/t)$$

As emphasized by our recent work (Ramamurti et al. 2018), one can observe that two statistical sums are related by the Poisson summation formula, in a form

$$\sum_{n=-\infty}^{\infty} f(\omega + nP) = \sum_{l=-\infty}^{\infty} \frac{1}{P} \tilde{f}\left(\frac{l}{P}\right) e^{i2\pi l\omega/P}, \tag{16.6}$$

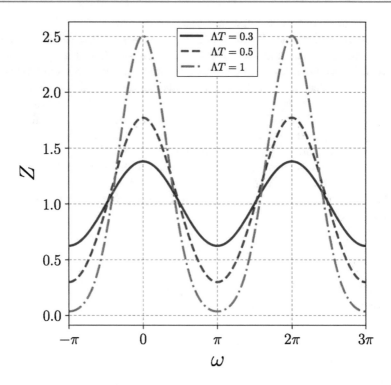

Fig. 16.1 The partition function Z of the rotator as a function of the external Aharonov-Bohm phase ω (two periods are shown to emphasize its periodicity). The (blue) solid, (red) dashed, and (green) dash-dotted curves are for $\Lambda T = 0.3, 0.5, 1$

where $f(x)$ is some function, $\tilde{f}$ is its Fourier transform, and P is the period of both sums as a function of the "phase" ω. In this particular example, the function is Gaussian, with Fourier transform being a periodic Gaussian: but we will later encounter examples of the Poisson duality with other functions as well.

The generalization to path integrals defined on other groups can proceed similarly. Schulman (1968) in particularly was interested in the rotation over the $SO(3)$ group, a manifold with three Euler angles. Instead of infinitely many topological classes of paths, in this case, there are two classes. The arbitrariness reduces to the relative sign between them: case plus leads to bosons and minus to fermions. The interested reader should consult literature on path integrals over Lie groups: for our purposes, the simplest $SO(2)$ (circle) case would be sufficient.

In summary, the rotator serves as an example of path integral on manifolds which have topologically distinct classes of paths. Lesson number one is that their ambiguous relative normalization allows to recognize "hidden quantum phase" ω. Lesson number two is that this is the simplest example in which the Hamiltonian and Lagrangian ways to calculate the statistical sum lead to differently looking, although "Poisson dual," results.

16.2 Monopoles Versus Instantons in Extended Supersymmetry

The setting of these studies are in weak coupling $g \ll 1$ and compactification to $R^3 \times S^1$. In the $\mathcal{N} = 4$ theory, the charge does not run, and g is simply an input parameter. In the $\mathcal{N} = 2$ theory, however, the coupling does run, and one needs to select the circumference of the circle β to be small enough such that the corresponding frequencies $\sim 2\pi/\beta$ are large enough to ensure weak coupling. Compactification of one coordinate to the circle is needed to introduce "holonomies," gauge invariant integrals over the circle $\oint dx_\mu A^\mu$, and $\oint dx_\mu C^\mu$ of the electric and magnetic potentials, respectively. Their values can have nonzero expectation values, which can be viewed as external parameters given by Aharonov-Bohm fluxes through the circle induced by fields in extra dimensions. These holonomies will play important role in what follows. Dorey et al. (2001) call these external parameters ω and σ, respectively. Finally, in order to make the discussion simpler, one assumes the minimal non-Abelian color group $SU(N_c)$ with the number of colors $N_c = 2$. This group has only one single diagonal generator τ^3, breaking the color group $SU(2) \to U(1)$.

The theories with extended supersymmetry $\mathcal{N} = 2, 4$ have one and six adjoint scalar fields, respectively. Recall that these two theories also have, respectively, two and four fermions, so that the balance between bosonic and fermionic degrees of freedom is perfect. For simplicity, all vacuum expectation values (VEV) of the scalars, as well as both holonomies, are assumed to be in this diagonal direction, so the scalar VEVs and ω and σ are single-valued parameters without indices. In the general group $SU(N_c)$, the number of diagonal directions is the Abelian subgroup, and thus the number of parameters is $N_c - 1$.

The particle-monopole mass is

$$M = \left(\frac{4\pi}{g^2}\right)\phi. \tag{16.7}$$

We will only discuss the $\mathcal{N} = 4$ case, following Dorey and collaborators (2001). Six scalars and two holonomies can be combined to vacua parameterized by eight scalars, extended by supersymmetry to eight chiral supermultiplets. These eight fermions have zero modes, describing their binding to monopoles. We will, however, not discuss any of those in detail.

The $SU(2)$ monopole has four collective coordinates, three of which are related with translational symmetry and location in space, while the fourth is rotation around the τ^3 color direction:

$$\hat{\Omega} = \exp(i\alpha\hat{\tau}^3/2). \tag{16.8}$$

Note that such rotation leaves unchanged the presumed VEVs of the Higgses and holonomies, as well as the Abelian $A_\mu^3 \sim 1/r$ tails of the monopole solution. Nevertheless, these rotations are meaningful because they do rotate the monopole core—made up of non-Abelian A^1_μ, A^2_μ fields—nontrivially. It is this rotation in the angle α that makes the monopole problem similar to a quantum rotator. As was explained by Julia and Zee (1975), the corresponding integer angular momentum is nothing but the electric charge of the rotating monopole, denoted by q.

Now that we understand the monopoles and their rotated states, one can define the partition function at certain temperature, which (anticipating the next sections) we will call $T \equiv 1/\beta$,

$$Z_{mono} = \sum_{k=1}^{\infty} \sum_{q=-\infty}^{\infty} \left(\frac{\beta}{g^2}\right)^8 \frac{k^{11/2}}{\beta^{3/2} M^{5/2}}$$
$$\times \exp\left(ik\sigma - iq\omega - \beta kM - \frac{\beta \phi^2 q^2}{2kM} \right),$$

(16.9)

where k is the magnetic charge of the monopole. The derivation can be found in the original paper, and we only comment that the temperature $T = 1/\beta$ in the exponent only appears twice, in the denominators of the mass and the rotation terms, as expected. The two other terms in the exponent, $\exp(ik\sigma - iq\omega)$, are the only places where holonomies appear, as the phases picked up by magnetic and electric charges over the circle.

Now we derive an alternative four-dimensional version of the theory, in which we will look at gauge field configurations in all coordinates including the compactified "time coordinate" τ. These objects are versions of instantons, split by a nonzero holonomy into instanton constituents. Since these gauge field configurations need to be periodic on the circle, and this condition can be satisfied by paths adding arbitrary number n of rotations, their actions are

$$S^n_{mono} = \left(\frac{4\pi}{g^2}\right)\left(\beta^2 |\phi|^2 + |\omega - 2\pi n|^2\right)^{\frac{1}{2}},$$

(16.10)

including the contribution from the scalar VEV ϕ, the electric holonomy ω, and the winding number of the path n. In the absence of the holonomies, the first term would be M/T as one would expect.

The partition function then takes the form (Dorey 2001)

$$Z_{inst} = \sum_{k=1}^{\infty} \sum_{n=-\infty}^{\infty} \left(\frac{\beta}{g^2}\right)^9 \frac{k^6}{(\beta M)^3}$$
$$\times \exp\left(ik\sigma - \beta kM - \frac{kM}{2\phi^2 \beta}(\omega - 2\pi n)^2 \right),$$

(16.11)

where $M = (4\pi\phi/g^2)$, the BPS monopole mass without holonomies; thus the second term in the exponent is interpreted as just the Boltzmann factor. The "temperature" appears in the unusual place in the last term (like for the rotator toy model). The actions of the instantons are large at high-T (small circumference β); the semiclassical instanton theory works best at high-T.

The Poisson duality relation between these two partition functions, Eqs. (16.9) and (16.11), was originally pointed out by Dorey and collaborators (2001). In this book, following (Ramamurti et al. 2018), it was explained earlier using the toy model of a *quantum rotator*. In fact the Poisson duality relation between two sums is in this case exactly the same.

16.3   Monopole-Instanton Duality in QCD

The authors of (Ramamurti et al. 2018) went further, performing the Poisson duality transformation over the semiclassical sum over twisted instanton-dyons. The resulting expression for the semiclassical partition function is

$$Z_{inst} = \sum_n e^{-\left(\frac{4\pi}{g_0^2}\right)|2\pi n - \omega|} \tag{16.12}$$

It is periodic in the holonomy, as it should be. Note that, unlike in Eq. (16.11), it has a modulus rather than a square of the corresponding expression in the exponent. This is due to the fact that the sizes of L_n and their masses are all defined by the same combination $|2\pi n - \omega|T$ and therefore the moment of inertia $\Lambda \sim 1/|2\pi n\beta - \nu|$.

Using the general Poisson relation, Eq. (16.6), the Fourier transform of the corresponding function appearing in the sum in Eq. (16.12) reads

$$F\left(e^{-A|x|}\right) \equiv \int_{-\infty}^{\infty} dx\, e^{i2\pi\nu x - A|x|}$$
$$= \frac{2A}{A^2 + (2\pi\nu)^2}, \tag{16.13}$$

and therefore the monopole partition function is

$$Z_{mono} \sim \sum_{q=-\infty}^{\infty} e^{iq\omega - S(q)}, \tag{16.14}$$

where

$$S(q) = \log\left(\left(\frac{4\pi}{g_0^2}\right)^2 + q^2\right)$$

$$\approx 2\log\left(\frac{4\pi}{g_0^2}\right) + q^2\left(\frac{g_0^2}{4\pi}\right)^2 + \ldots, \qquad (16.15)$$

where the last equality is for $q \ll 4\pi/g_0^2$.

The resulting partition function can be interpreted as being generated by *moving and rotating monopoles*. The results are a bit surprising. First, the action of a monopole, although still formally large in weak coupling, is only a logarithm of the semiclassical parameter; these monopoles are therefore quite light. Second is the issue of monopole rotation. The very presence of an object that admits rotational states implies that the monopole core is not spherically symmetric. The Poisson-rewritten partition function has demonstrated that the rotating monopoles are *not* the rigid rotators, because their action, Eq. (16.15), depends on the angular momentum q and is quadratic only for small values of q. The slow (logarithmic) increase of the action with q implies that the dyons are in fact shrinking with increased rotation. In the moment of inertia, this shrinkage is more important than the growth in the mass, as the size appears quadratically. As strange as it sounds, it reflects on the corresponding behavior of the instanton-dyons L_n with the increasing n.

Although such rotations are well known in principle as Julia-Zee dyons with *real* electric charge (unlike that of the instanton-dyons, which only exist in the Euclidean world) and studied in theories with extended supersymmetries, to our knowledge, the existence of multiple rotational states of monopoles has not yet been explored in monopole-based phenomenology. In particular, one may wonder how the existence of multiple rotational states affects their Bose condensation at $T < T_c$, the basic mechanism behind the deconfinement transition. The electric charges of the rotating monopoles should, therefore, also contribute to the jet quenching parameter $\hat{q}$ and the viscosity, which was not yet included in literature. The monopole partition function with action per monopole (16.15) is in good correspondence with lattice data on monopole density shown in Fig. 16.2.

16.4   Brief Summary

- The appearance of what gets called "Poisson duality" was demonstrated using the simplest problem, a particle rotating on a circle at finite temperatures. "Hamiltonian" approach is to find all energy levels and then do statistical sum with appropriate Boltzmann factors. "Lagrangian" approach is to consider classical paths winding around Euclidean time. Two different sums obtained, when calculated, turned out to give identical answer for partition function.

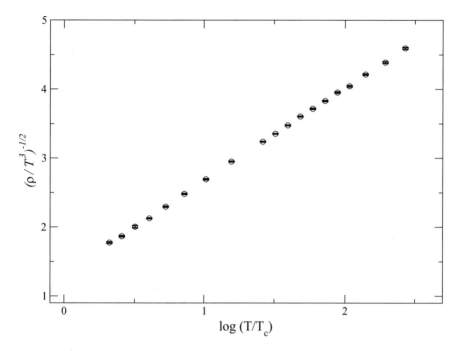

Fig. 16.2 The normalized monopole density in $SU(2)$ gauge theory in power $-1/2$, $(\rho/T^3)^{-1/2}$ versus $\log(T/T_c)$ shows an apparent linear dependence

- In $\mathcal{N} = 4$ theory, one is able to do two similar sums, with monopole motion and with the instanton-dyons. The total summation is not reached, but two sums were related by famous Poisson formula relating two series (16.6).
- Recall that moving monopoles produce electric fields, and rotating monopoles may become dyons. The instanton-dyons therefore have the meaning of some extrema of monopole paths with rotation.
- Assuming that this is the case in QCD, in which the semiclassical expression for instanton-dyon partition sum can be calculated, one can perform its Poisson transformation. As a result, one can derive the monopole mass and see that it is not $\sim 1/g^2$, as it would be for any classical solution, but only $\log(1/g^2)$. This result is in good agreement with lattice data on monopole density, e.g., those shown in Fig. 2.6.
- Monopoles and instanton-dyons describe the same physics and produce the same partition function. There are just different ways to describe it. Depending on the problem, one or the other should be used. Adding the contributions would be double counting, a sin in theoretical physics.
- There is however important difference: instanton-dyons are classical solutions of equation of motion, with full semiclassical theory behind it. Monopoles in theories like QCD (without adjoint scalars) are not such solutions, and no semiclassical theory can be built on them.

References

Aharonov, Y., Bohm, D.: Significance of electromagnetic potentials in the quantum theory. Phys. Rev. **115**, 485–491 (1959). [95 (1959)]

Dorey, N.: Instantons, compactification and S-duality in N=4 SUSY Yang-Mills theory. 1. JHEP **04**, 008 (2001). arXiv:hep-th/0010115

Dorey, N., Parnachev, A.: Instantons, compactification and S duality in N=4 SUSY Yang-Mills theory. 2. JHEP **08**, 059 (2001). arXiv:hep-th/0011202

Julia, B., Zee, A.: Poles with both magnetic and electric charges in nonabelian gauge theory. Phys. Rev. **D11**, 2227–2232 (1975)

Poppitz, E., Schafer, T., Unsal, M.: Continuity, deconfinement, and (super) Yang-Mills theory. JHEP **10**, 115 (2012). arXiv:1205.0290

Poppitz, E., Unsal, M.: Seiberg-Witten and 'Polyakov-like' magnetic bion confinements are continuously connected. JHEP **07**, 082 (2011). arXiv:1105.3969

Ramamurti, A., Shuryak, E., Zahed, I.: Are there monopoles in the quark-gluon plasma? (2018). arXiv:1802.10509

Schulman, L.: A path integral for spin. Phys. Rev. **176**, 1558–1569 (1968)

The QCD Flux Tubes

17

17.1 History

The story of QCD flux tubes started in the 1960s, prior to the discovery of QCD, with two important *hints*.

At the time, experimental discoveries of multiple hadronic states were the main occupation of high-energy physics, and discovery and tests of flavor $SU(3)$ symmetry were the focus. It became obvious that mesons and baryons cannot be "elementary particles," as they were expected to be earlier. Indeed their number starts growing exponentially!

It was pointed out by (Hagedorn 1965) that if such rapid growth of the density of states is indeed there, it should impede growth of temperature. If it is indeed exponential, $\rho(m) \sim \exp(m/T_H)$, it would lead to the "ultimate temperature limit" of hadronic matter, because the partition function

$$Z = \int \frac{d^3p}{(2\pi)^3} dm\rho(m)e^{(-\sqrt{p^2+m^2})/T} \sim \int e^{\left[m\left(\frac{1}{T_H}-\frac{1}{T}\right)\right]}dm \qquad (17.1)$$

becomes divergent at $T \to T_H$.

Another line of studies were related to hadron-hadron scattering. A phenomenological breakthrough was discovery that hadron seems to belong to certain *Regge trajectories*, like quantum-mechanical bound states in some nonrelativistic potentials. It means that there exist some formulae for angular momentum as a function of energy, producing energy levels when the values are integer. In relativistic notations, such expression is written as $l = \alpha(m^2)$. Furthermore, it was shown in many papers[1] starting with (Chew and Frautschi 1962) that the trajectories for mesons and

---

[1]For a recent review on Regge trajectories of light and heavy mesons, see (Sonnenschein and Weissman 2014).

© The Author(s), under exclusive license to Springer Nature Switzerland AG 2021
E. Shuryak, *Nonperturbative Topological Phenomena in QCD
and Related Theories*, Lecture Notes in Physics 977,
https://doi.org/10.1007/978-3-030-62990-8_17

baryons are approximately linear and can be approximated by only two constants:

$$\alpha(t) \approx \alpha(0) + \alpha'(0)t \tag{17.2}$$

the dimensionless "intercept" $\alpha(0)$ and the "slope" $\alpha'(0)$. The expression works not only for positive $t = m^2 > 0$ but also for $t < 0$ in scattering. For high energies $s \gg |t|$, the cross sections have the following form

$$\frac{d\sigma(s, t)}{dt} \sim s^{\alpha(t)-1} \tag{17.3}$$

in good agreement with the data. The largest $\alpha(t)$ belongs to the so-called "leading" trajectory called the *Pomeron*, named after Pomeranchuk. Its $\alpha(0) \approx 1.08$ is above one, and therefore all total cross sections grow with s, although in a rather slow pace.

Perhaps the most influential (and the most beautiful) paper of that period was (Veneziano 1968) which constructed expression for the amplitude, based on linearity of the trajectories and possessing a marvelous *duality* property:[2] it can be derived either as sum of s-channel resonances or t-channel Regge exchanges.

In due time, it was realized that straight Regge trajectories and Veneziano amplitude indicated that the object under consideration is basically a *rotating string*. This development has led to important historic point, the birth[3] of the *string theory*.

Of course, the QCD strings are not pointlike but some complicated gluonic finite-size objects, with certain properties and structure we are going to discuss in this chapter. This was not tolerated by purists among string theorists, and a bifurcation happened, namely, most of them proceeded to study theories of some idealized fundamental pointlike strings, not intimidated by the serious obstacles. (One of them was that one needed to quit our four-dimensional space-time and go into much larger number of dimensions, D=26.) Fortunately, many years later, with the advent of AdS/CFT duality, it became possible to reunite the two theories together, describing the QCD strings as *holograms* of the pointlike fundamental strings in higher dimensions. We will return to this point in Sect. 17.7.

Dramatic events of the 1970s included not only the discovery of QCD but also experimental discoveries of heavy c, b quarks and quarkonia states. It soon became apparent that the potential needed to explain them was linear, $V(r) \sim r$, and thus the QCD strings got another name, the *confining flux tubes*. By the end of the 1970s, numerical studies of the non-Abelian fields on the lattice have developed to the

---

[2]Perhaps the first time such notion was explicitly demonstrated, the same answer followed from two entirely different and seemingly unrelated derivations. Before this work, the phenomenologists were inclined to sum these contribution together, but Veneziano formula elucidated that it was a double counting.

[3]As a birthmark, proving the connection, note that modern string theorists still call the string scale α'. I doubt many remember what prime stands for.

point that it was possible (Creutz 1980) to relate the Yang-Mills Lagrangian and asymptotic freedom to the string tension.

17.2   The Confining Flux Tubes on the Lattice vs the "Dual Superconductor" Model

Permanent confinement of color-electric charges (or "confinement," for short) is the most famous nonperturbative feature of the gauge theories. The Lagrangian of QCD-like gauge theories is similar to that of QED, with massless photons substituted by massless gluons and massive electrons by (very light, or even massless) quarks. There are multiple definitions of the term itself, e.g., the statement that no object with a color charge can appear in physical spectrum.[4]

There is perhaps no need to remind the reader the general setting of the lattice gauge theory, or any technical details about it. Most physicists trust that the limit of vanishing lattice spacing $a \rightarrow 0$ is taken correctly, since, starting from the pioneering work (Creutz 1980), it was many times demonstrated that the string tensions (and other relevant quantities) do scale in correspondence with the correct renormalization group prescription and are thus physical.

What was observed on the lattice, for pure gauge theories, was that the electric flux from a color charge is not distributed radially outward, as in electrodynamics, but instead, being expelled from the QCD vacuum, is confined into a *flux tube* between the charges. At large distances, the leading contribution to the static and heavy quark-antiquark potential $V_0(r)$ in pure Yang-Mills theory is the famous linear potential

$$V_0(r) = \sigma_T r \tag{17.4}$$

with σ_T as the fundamental string tension, the energy per length. Its numerical value (in QCD with physical quarks) is

$$\sigma_T \approx (420 \, \text{MeV})^2 \approx 1 \ \text{GeV}/fm \tag{17.5}$$

which also served[5] as the *definition* of absolute units in any confining theory.

Figure 17.1 (displaying the result of lattice simulations summarized in the review (Bali 1998)) shows distribution of the electric field (left) along the flux tube and

---

[4]There is a pending million-dollar prize offered for a mathematical proof that pure gauge theory has a finite mass gap. Physicists are already sure that it is the case, beyond any reasonable doubt. Billions of high-energy collisions of hadrons and nuclei (we already briefly discussed above) observed produced large number of secondaries, and none of them ever was a quark or a gluon. The formal limits on that are so small that there is no sense to even mention them.

[5]Recall that in QCD with light quarks, this behavior is only valid till some distance due to screening by light quarks in the form of two heavy-light mesons. So now lattice units are usually set via location of a point at which the potential times the distance take some prescribed value.

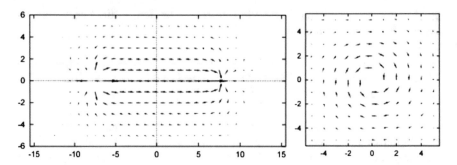

Fig. 17.1 Lattice data on distribution of the electric field strength (left) and the magnetic current (right), for two static quark-antiquark external sources. The profile of the electric field is shown by squares, where lines just fit

magnetic current (right) in a transverse plane. So, in numerical simulations of the gauge fields, not only the flux tube with the longitudinal electric field is clearly seen but also the stabilizing "coil" around it. Its physical origin we will discuss later in the chapter, in Sect. 17.4.

The "dual superconductor" idea has been mentioned many times above, and need not be repeated here. The specific relation between the QCD flux tubes and their dual flux tubes in superconductors has been pointed out in (Nielsen and Olesen 1973). At this point, one would like to test whether the duality relation between them is really quantitative. More specifically, let us test whether the shape of the confining flux tubes can indeed be described by the same generic effective model, the Ginzburg-Landau theory[6] as the magnetic flux tubes in superconductors.

There is no need to describe here the Ginzburg-Landau theory in detail: it is enough to say that its expression for the effective free energy (analog of the action) includes Abelian QED gauge field and a charged scalar described by complex field ϕ, representing condensate of Cooper pairs.[7]

The key Maxwell equation we will focus on is[8]

$$\vec{\nabla} \times \vec{B} = \vec{j}$$

---

[6]Let me remind that when Ginzburg-Landau paper was written, the physical nature of electric object which makes the condensate was also unknown: they argue for the form of effective action on general grounds.

[7]There was an instructive story about the charge of ϕ. The GL paper was written well before the microscopic BCS theory of superconductivity. Ginzburg initially put some "effective charge" e_{eff}, but Landau objected, saying that if the charge be dependent on matter parameters, like temperature, it would spoil gauge invariance of electrodynamics, and so they put e, the electron charge. After BCS, it became clear that the charge must indeed be fixed, but not to one but $2e$.

[8]A clarification: we are looking for static solutions only, so time derivative of $\vec{E}$ in it is omitted.

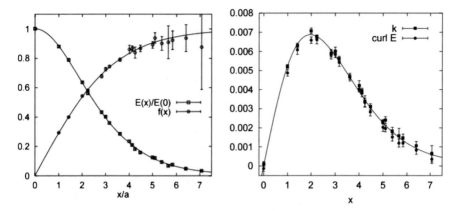

Fig. 17.2 (left) Transverse profile of the electric field and the condensate. (right) The transverse distribution of the magnetic current. The points in both are lattice data, and the lines are fits using Abrikosov's flux tube solution of the Ginzburg-Landau equations

It tells us that Abrikosov flux tube solution, with nonzero magnetic field B inside the tube and zero outside, needs a "coil" with current, confining the field inside. The current is the gradient of the scalar's phase. In the dual case we discuss, one should substitute

$$\vec{B} \rightarrow \vec{E}, \quad \vec{j} \rightarrow \vec{j}_{magnetic}$$

and ϕ representing the (magnetically charged) monopole condensate.

The curl of E shown in Fig. 17.2(right) coincides well with the separately measured magnetic current k from monopole motion. So, the second equation—basically dual Maxwell equation—is indeed satisfied. With some accuracy, also the first equation is satisfied, and $f(x)$ shown in Fig. 17.2(left) is the radial profile of the condensate observed on the lattice. Note, in particular, that like in the superconductors, the "dual Higgs" scalar field vanishes at the center. The resulting parameters for two basic lengths, in "physical units" obtained fixing lattice scale to physical σ, are

$$\lambda = 0.15 \pm 0.02 \, fm, \quad \xi = 0.251 \pm 0.032 \, fm, \kappa = \frac{\lambda}{\xi} = .59 \pm 0.14 < \frac{1}{\sqrt{2}}$$
$$(17.6)$$

Recall that the so-called Ginzburg-Landau ratio κ of them is *smaller* than the critical value (shown at the end of the previous equation). This implies that the QCD vacuum (we live in) is the *dual superconductor of type I*[9]

---

[9]Those who are not convinced by not-too-impressive accuracy of this numerical statement may wonder if there are more direct manifestation of it. We will return to the issue of flux tube interaction in section on multi-string systems.

Rather extensive calculations of the static potential, at large and also finite r, were performed in the framework of the "dual Higgs model" (see review (Baker et al. 1991)). We will not reproduce here the results but just provide few comments:

Comment 1:   In this and subsequent works, not only the static classical linear potential is described but also the velocity-dependent relativistic corrections, ultimately rather successfully compared to phenomenological relativistic terms derived from the quarkonia spectra. Later Baker and collaborators had also calculated the Regge trajectories (see (Baker and Steinke 2002)).

Comment 2:   Generally speaking, the "dual Higgs models" constitute quite interesting examples of "an effective magnetic theory" approach. Not only the scalar fields in them are magnetically charged—describing the BEC of monopoles—but the gauge field itself is also treated using the dual potential rather than the usual A_μ. A specific form of the model is motivated and defined in (Baker et al. 1991).

As an example of further progress in lattice technology, let me mention (Yanagihara et al. 2018) in which the flux tube has been studied from the point of view of the underlying stress tensor. The QCD operator of the stress tensor is of course well known, but for many years, its direct evaluation has been blocked by very large statistical noise. In the paper under consideration, the authors used the *gradient flow* smoothening procedure, appended by additional extrapolation back to zero value of the gradient flow time. The authors have demonstrated that the procedure is consistent with lattice studies of the equation of state, for example, their $\langle T^{00}(T)\rangle$ agrees with the energy density, $\langle T^{11}(T)\rangle$ agrees with pressure, etc.

Needless to say, in order to study the properties of the flux tube, one needs sufficient statistical precision to subtract these mean values, present everywhere. In Fig. 17.3 from this work, one can see transverse distribution of the diagonal components of the stress tensor. Note that the signs of them, as shown, are selected in such a way that in pure electric field, all four would be the same: this apparently

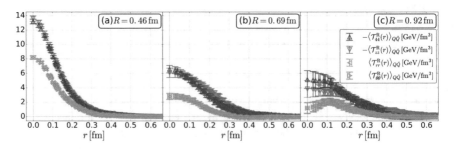

Fig. 17.3 Mid-plane distribution of various components of the stress tensor, in cylindrical coordinates. Three pictures differ by the value of R, the distance between static quarks

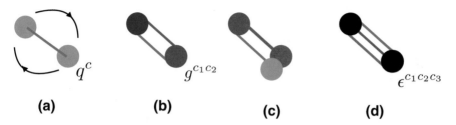

Fig. 17.4 Various types of Reggeons: (**a**) rotating $\bar{q}q$ pair, held by a string; (**b**) rotating gluon pair (positive C-parity) held by two strings; (**c**) three gluon (negative C-parity) objects held by three strings; (**d**) rotating baryon junctions held by N_c (=3) strings

is *not* the case. Yet two transverse pressures, the rr and $\theta\theta$ components, seem to be always the same.

One other comment is that the energy density at the center of the flux tube, as read from Figs. 17.3b and c, is about 6 GeV/fm^3. This is very large value. In particular, it is about two orders of magnitude larger (!) than the value suggested by the early MIT bag model of the 1970s, which tried to describe flux tube compressed by some "bag pressure." In those days, the magnitude of nonperturbative effect were grossly underestimated.

17.3   Regge Trajectories and Rotating Strings

Static potentials are not the only place where one can infer existence of the (fundamental) flux tubes. Another impressive confirmation comes from hadronic spectroscopy.

Quarkonia—the nonrelativistic bound state of heavy quark-antiquarks—are indeed well described by the sum of Coulomb and linear potential. Hadrons made of light quarks also show a very spectacular confirmation to the idea that mesons are basically quark-antiquarks connected by a flux tube, and (at least some) baryons can be approximated by similar quark-diquark systems, also connected by a flux tubes, see Fig. 17.4 and discussion below of several cases shown.

Quantization of such system with a string predicts that such hadrons should appear in the form of Regge trajectories. As it was noticed in the 1960s, excited states of light mesons and baryons are indeed located on near-linear trajectories, in the total angular momentum J—squared mass M^2. Let me not present historic Chew-Frautschi plot but go into relatively recent Fig. 17.5 from (Sonnenschein and Weissman 2014), in which there are many states and also lines indicating the model we will be discussing.[10] For future reference, let me note that all vector trajectories at $M^2 \to 0$ go to $\alpha(0) \approx 0.5$.

---

[10]For completeness, let me mention that slope is universal $\alpha' = .884\,\text{GeV}^2$, and effective quark masses (those including the chiral symmetry breaking, not the ones in QCD Lagrangian) are 60, 220, 1500 MeV, for light, strange, and charm quark, respectively.

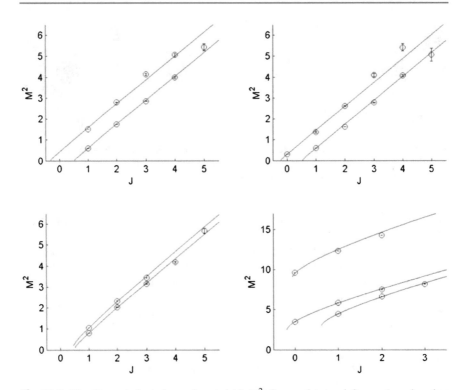

Fig. 17.5 Nine Regge trajectories on inverted (J, M^2) Regge plot: top left π, ρ (pseudoscalars and vectors with isospin $I = 1$), top right η, ω (pseudoscalars and vectors with isospin $I = 0$), bottom left K^*, ϕ (vectors with one or two strange quarks), bottom right D, D_s^*, J/Ψ (with one or two charmed quarks)

Furthermore, even the hadrons *without* quarks—the glueballs—can be described in terms of rotating *closed* strings, forming another set of Regge trajectories. The lattice data on spectroscopy of pure gauge theories have, in my view, produced rather significant support to this statement. Since it is much less known and will be needed in connection to Pomerons, we will discuss later in this chapter, let us see them, following (Kharzeev et al. 2018). The corresponding plot for the masses of glueball with positive charge C parity, taken from the lattice study (Meyer 2004), is shown in Fig. 17.6.

A comment: Naive approach to rotating closed string states suggests that the string tension should be doubled (two strings rather than one), or that the slope of C-even glueball Regge trajectories should be *half* of that in mesons. Yet, as seen from these plots, the mesonic $\alpha' = 0.88\,\mathrm{GeV}^{-2}$ and the glueball slope (calculated from $J = 0$ and $J = 3$) is $\alpha'_{C=1} = 0.36\,\mathrm{GeV}^{-2}$, so this ratio is 2.4 rather than 2. What it implies is that the two strings must not be independent but interacting with each other. For $C = -1$ plot, one again finds that in three pairs, the slope is the same, $\alpha'_{C=-1} = 0.33\,\mathrm{GeV}^{-2}$, not far from the other glueballs. Does it mean that,

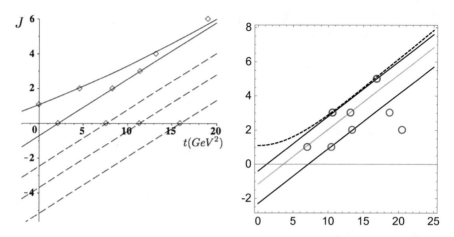

Fig. 17.6 Glueballs with positive charge parity $C = 1$ (left) and negative $C = -1$ (right) on Regge plots, their angular momentum J versus their squared mass $M^2(\text{GeV})^2$. The two upper (blue) points and lines are for the negative *spatial* parity $P = -1$ glueballs, and the lower (red) ones are for the $P = +1$. The lines are the (hypothetical) Regge trajectories

like baryons, the 3-gluon states are not Y-shaped, but 1+2 gluon type? Perhaps. To my knowledge, nobody had worked it out. Also I am not aware of any calculations of the Odderon slope.

QCD has one more mysterious colored gluonic object, the *baryonic junction*. In case of N_c colors, it connects together N_c flux tubes with all colors. Its algebraic structure is antisymmetric $\epsilon^{c_1, c_2 \cdots c_{N_c}}$.

One place where it can appear is in the so-called Y-shape baryons, with three strings joint at the junction at the center. Rotation of it should then lead to Regge trajectories with slope 1/3 of the usual, as there are three strings involved. However, Regge trajectories and other theoretical studies had convincingly shown that at least nucleon-like baryons are *not* of this type, having quark-diquark structure instead and standard single slope, the same as mesons have.

Another use[11] of baryonic junctions appeared in hadronic (e.g., pp) collisions in which "stopped baryons" are observed far from the beam rapidity, for example, near the center of mass energy (where colliders have best detection capabilities). Effective Regge diagram for this process must include "the junction Reggeon" (nobody proposed any name for it so far) with (unknown to me) intercept and the slope close to 1/3 of the usual one. I am not aware of any resonance or state attributed to this trajectory.

In summary, confining (fundamental) flux tubes have been seen and studied on the lattice, and they also have strong support via hadronic spectroscopy and reactions. Their properties are in agreement with predictions based on the dual Higgs models.

---

[11]This paragraph is based on (Kharzeev 1996) and recent private communications.

17.4 Flux Tubes and Finite Temperatures: The Role of Monopoles

In this section, we extend our discussion of the flux tubes to QCD *at finite temperatures*. A general expectation—based on analogy to superconductors—is that they exist in the confining phase $T < T_c$ and disappear above it. However, as we will see shortly, the situation turned out to be posessing unexpected and rather peculiar features, not present in the case of superconductors.

At finite temperatures, the natural quantity to calculate, for the observed flux tubes between static charges, is the *free energy*. It can be written as

$$F(r, T) = V(r, T) - T S(r, T), \quad S(r, T) = \frac{\partial F(r, T)}{\partial T} \tag{17.7}$$

where $S(r, T)$ is the *entropy* associated with the pair of static quarks. Since it can be calculated from the free energy itself, as indicated in the r.h.s., one can subtract it and plot also the *potential energy* $V(r)$. The derivatives over r of both potentials— the force—are what we call the *string tension*.

The lattice calculations have shown that in certain range of r, the tension is constant (the potential is approximately linear in r). We do not show this but proceed directly to the temperature dependence of the two resulting tensions, shown in Fig. 17.7(left) (based on lattice calculations by the Bielefeld-BNL group, see (Kaczmarek and Zantow 2005) and earlier works mentioned there).

The tension of the free energy shows the expected behavior: $\sigma_F(T)$ vanishes as $T \to T_c$. But the tension of the potential energy $\sigma_V(T)$ shows drastically different behavior, with large *maximum* at T_c, and nonzero value above it. This unexpected

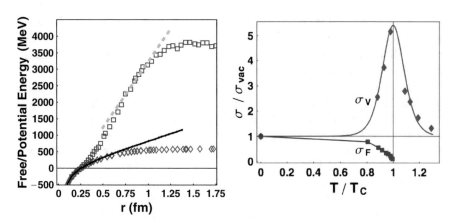

Fig. 17.7 Left: Free (red rhombs) energy $F(r)$ and potential (blue squares) energy $V(r)$, at T_c, compared to the zero temperature potential (black line). Right: Effective string tension for the free and the internal energy

behavior was hidden in $\sigma_F(T)$, studied in many previous works, because in it a large energy and a large entropy cancel each other.

The explanation to this effect has been proposed by (Liao and Shuryak 2010), which we will here follow. But before we follow this particular explanation, related with monopoles, let us make some general comments:

Comment 1:   A large entropy implies exponentially large $\exp\left[S(T, r)\right]$ number of states, associated with static quark pair. Furthermore, the nonzero tension—derivative over r—means that such states are not concentrated at the string's end but are also distributed along the string. What can the physical origin of those states be?

Comment 2:   A nonzero tension $\sigma_V(r)$ at $T > T_c$ implies the existence of flux tubes *above* T_c.

Comment 3:   The *free* energy, by its nature, corresponds to physical conditions of complete thermal equilibrium, which can only be reached at long time. If the time is limited—for example, if color dipole is only created for a finite time, or in other situations with moving (nonstatic) charges—there would be deviations from equilibrium, and therefore cancellations between energy and entropy may be only partial.

Comment 4:   Therefore, the effective potentials for quarkonia, which have nonrelativistic but still nonstatic heavy quarks, should be somewhat intermediate between $F(r)$ and $V(r)$. Liao and Shuryak (2010) discussed a setting in which heavy quark and antiquark slowly move away from each other with some velocity v and argue that the entropy production can be calculated using Landau-Zener theory of level crossings, used originally for description of bi-atomic molecules. This theory describes how the resulting population of both crossed levels depends on v.

Now let us proceed to the dynamical explanation proposed by (Liao and Shuryak 2010). Its main point is that the "dual superconductor" picture is not sufficient: one should also recognize the existence of *uncondensed* (or "normal") component[12] of the monopole density.

Liao and myself argued (Liao and Shuryak 2010) that electric flux tubes can be (mechanically) stable even without the "dual superconductor," or BEC of monopoles. Indeed presence of magnetic flux tubes in various plasmas is well known.[13] The difference between such flux tubes, at $T > T_c$, and the one due

---

[12]Note that in the BCS superconductors, there are no "uncondensed" Cooper pairs.

[13]In fact, in a good telescope, one can directly see hundreds of them in solar corona! Bunches of flux tubes compose the black spots on the Sun made famous by observations by Galileo, who used them to discover solar rotation. The Pope of the time interpreted black spots on the Sun as allegory criticizing him personally and initiated trials ended in Galileo home arrests.

to "dual superconductor" at $T < T_c$ is that in the latter case, the "coil" includes non-dissipative supercurrent, making them permanently stable, while the latter ones have Ohmic losses and are therefore metastable.

Using elliptic coordinates, Liao and Shuryak (2010) had derived a solution for the electric field in the monopole plasma, even for finite distance between the quark charges, reproducing the potential from Coulomb-like behavior at small distances to long flux tubes at large. We will not give any details here and only note that because at high-T the monopole density drops rapidly at $T > 1.5T_c$, and thus the metastable flux tubes do not exist there.

Finally, let me add some comments about more recent lattice study of the flux tubes at finite temperature (Cea et al. 2018). Using certain smoothening procedure, the shape of the longitudinal electric field as a function of *transverse* coordinate is measured, for a number of temperatures, for pure gauge $SU(3)$ theory, and for QCD with realistic quark masses. In Fig. 17.8, we show some of their results. Note that this theory has the first-order transition, seen as a jump in the field strength. And yet, the overall flux tube shape persists, in the pictures of the electric field, approximately till $T = 1.5T_c$. These observations support the idea that the flux tubes do exist in the QGP, in spite of absence of the "the dual superconductor" there.

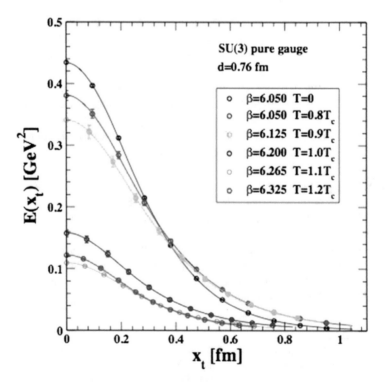

Fig. 17.8 The longitudinal electric field as a function of *transverse* coordinate is measured, for a number of temperatures, for pure gauge $SU(3)$ theory, from (Cea et al. 2018)

17.5 Effective String Theory (EST) Versus Precise Lattice Data

This section starts with a pedagogical introduction, introducing classical string solutions and explaining elements of string quantization, and then jumps to a brief review of the current status of EST, in connection with empirical and lattice data.

A particle moving in D space-time dimensions can be described by its path $X^\mu(\tau)$ with proper time τ and $\mu = 1..D$. Similarly, a string moving in D space-time dimensions is described by coordinates $X^\mu(\tau, \sigma)$ with two *internal* coordinates, time-like τ and space-like σ. The simplest geometrical Nambu-Goto action is simply the area of the corresponding "membrane" times the tension

$$S_{NG} = -\sigma_T \int d\tau d\sigma \sqrt{-h}, \quad h_{\alpha\beta} = \eta_{\mu\nu} \partial_\alpha X^\mu \partial_\beta X^\nu \tag{17.8}$$

where $\eta_{\mu\nu}$ is called the external metric, assumed for now just flat Minkowski metric, and $h_{\alpha\beta}$ in which $\alpha, \beta = 1, 2$ is the *internal* metric of worldsheet. The area element includes the 2*2 determinant of it, $h = det(h_{\alpha\beta})$. We will also use a dot for derivative in τ and a prime for derivative in σ, e.g. $\dot{X}^\mu, X'^\mu$. If the endpoints are massive, the following "particle" term is added to the action:

$$S_{ends} = -m \int d\tau \sqrt{-\dot{X}^2} \tag{17.9}$$

(For simplicity we assume both masses to be the same m.)

Classical equation of motion for the string is

$$\partial_\alpha \left(\sqrt{-h} h^{\alpha\beta} \partial_\beta X^\mu \right) = 0 \tag{17.10}$$

and the boundary conditions should be

$$\sigma_T \sqrt{-h} \partial^\sigma X^\mu \pm m \partial_\tau \left(\frac{\dot{X}^\mu}{\sqrt{-\dot{X}^2}} \right) = 0 \tag{17.11}$$

It may look complicated, but we will not study complicated string dynamics. The most straightforward rotating string configuration is given by

$$X^0 = \tau, \ X^1 = \sigma \cos(\omega\tau), \ X^2 = \sigma \sin(\omega\tau) \tag{17.12}$$

which solves the string equation of motion. The boundary condition takes the form

$$\sigma_T \sqrt{1 - \omega^2 l^2} = \frac{m\omega^2 l}{\sqrt{1 - \omega^2 l^2}} \tag{17.13}$$

which has the obvious meaning, l is half of string length. Using standard Noether procedure to calculate the energy and angular momentum, and substituting this solution into them, one gets

$$E = \frac{2m}{\sqrt{1 - \omega^2 l^2}} + \sigma_T \int_{-l}^{l} \frac{d\sigma}{\sqrt{1 - \omega^2 \sigma^2}} \tag{17.14}$$

$$J = \frac{2m\omega l^2}{\sqrt{1 - \omega^2 l^2}} + \sigma_T \omega \int_{-l}^{l} \frac{d\sigma \sigma^2}{\sqrt{1 - \omega^2 \sigma^2}} \tag{17.15}$$

Using simpler notation $v \equiv \omega l$, the velocity of the string ends, and performing the integrals, one gets the following rather intuitive results

$$E = \frac{2m}{\sqrt{1 - v^2}} + 2\sigma_T l \frac{arcsin(v)}{v} \tag{17.16}$$

$$J = \frac{2mvl}{\sqrt{1 - v^2}} + \sigma_T l^2 \left(\frac{arcsin(v) - v\sqrt{1 - v^2}}{v^2} \right) \tag{17.17}$$

where the length $2l$ can be substituted from the boundary condition $\sigma_T l = mv^2/(1 - v^2)$, and one has the resulting Regge trajectory, in a parametric form. How well it describes the mesons one can see in (Sonnenschein and Weissman 2014), let us just mention the case of light quarks, small m and v close to 1. Expanding the function in the r.h.s., one has then the linear Regge trajectory with corrections

$$J = \alpha' E^2 \left(1 - \frac{8\sqrt{\pi}}{3} (\frac{m}{E})^{3/2} + ... \right) \tag{17.18}$$

where we also use the standard relation between the string tension σ_T and the Regge slope

$$\alpha' \equiv \frac{1}{2\pi \sigma_T} \tag{17.19}$$

Quantization of the problem must include not only quantum motion of masses but also that of the string, and it is not simple at all.[14] People obviously did first the case in which string ends are fixed. For the Nambu-Goto action, one can solve this problem and obtain string energy including quantum string vibrations (Arvis 1983)

$$E(r) = \sigma_T r \sqrt{1 - \frac{\pi}{6} \frac{1}{\sigma_T r^2}} \qquad (17.20)$$

which appends the classical linear potential by a quantum factor close to one at large r but generating certain expansion in powers of $1/\sigma_T r^2$.

For Regge trajectories, transition from classical to quantum results is more involved in general, but for massless endpoints, it can be done by the following *additive* substitution

$$J = \alpha' E^2 \quad \rightarrow \quad J + n - a = \alpha' E^2$$

where n is the quantum number for radial excitations and a is the "quantum addition to the intercept."[15] Note how elegantly two quantum numbers for orbital and radial excitations—J and n—appear together. In a Chew-Frautschi plot $J(M^2)$, various integer values of n generate "daughter" trajectories, which are simply shifted downward from the "parents" by one or more units. We will not discuss them but just mention that those predict correctly certain observed mesons and baryons as well.

As a parting comment, let us note that masses at the string ends can be viewed holographically, as just extra piece of a string, reaching in the fifth dimension to the "flavor brane." For a review on holography-inspired stringy hadrons, see (Sonnenschein 2017).

Now we have completed the pedagogical part introducing stringy potentials and Regge trajectories in their simplest form. Now we will address much more difficult questions related with real-life QCD strings. Those are complicated extended objects, and one has no general reasons to assume that they simply follow Nambu-Goto geometric action. Twisting of a string may cause extra energy: therefore higher-order terms may appear in the effective string action. Let us briefly summarize what is known about them at this time.

---

[14]String quantization is a complicated topic going well beyond this course. Unless one deals with the so-called critical dimension of space-time $D = 26$ for bosonic string, certain anomalies appear. Their cancellation is possible via complicated addition to string Lagrangian. To my knowledge, it is not important issue for stationary string but appears, e.g., for rotating one, dealt with by Sonnenschein and collaborators. For stringy Pomeron solution to be discussed below, it is not yet resolved, to my knowledge.

[15]In (Hellerman and Swanson 2015), it was shown, using very general assumptions, that for massless endpoints, $a = 1$.

Long strings are described uniquely by the expanded form of the Nambu-Goto action:

$$S = -\sigma_T \int_M d^2x \, (1 + \partial_\alpha X^i \partial^\alpha X_i) \tag{17.21}$$

The integration is over the worldvolume of the string M with embedded coordinates X^i in D-dimensions. The first contribution is the area of the worldsheet, and the second contribution captures the fluctuations of the worldsheet in leading order in the derivatives.

Since the QCD string is extended and therefore not fundamental, its description in terms of an action is "effective" in the generic sense, organized in increasing derivative contributions each with new coefficients. These contributions are generically split into bulk M and boundary ∂M terms. The former add pairs of derivatives to the Polyakov action. The first of such contribution in the gauge (fixed as in (17.21)) was proposed by Polyakov (1986)

$$+ \frac{1}{\kappa} \int d^2x \, \left(\ddot{X}^\mu \ddot{X}_\mu + 2\dot{X}'^\mu \dot{X}'_\mu + X''^\mu X''_\mu \right) \tag{17.22}$$

which is seen to be conformal with the dimensionless extrinsic curvature. Higher derivative contributions are restricted by Lorentz (rotational in Euclidean time) symmetry. The boundary contributions are also restricted by symmetry. The leading contribution is a constant μ, plus higher derivatives. We will only consider the so-called b_2 contribution with specifically

$$S_b = \int_{\partial M} d^2x \, \left(\mu + b_2 \, (\partial_0 \partial_1 X^i)^2 \right) \tag{17.23}$$

All the terms in (17.21–17.23) contribute to the static potential (17.4). The first contribution stems from the string vibrations as described in the quadratic term (17.21):

$$\sigma_T r \left(1 + \frac{V_0}{\sigma_T r^2} \right) \tag{17.24}$$

It is Luscher universal term with $V_0 = -\pi/12$ in four dimensions. Using string dualities, Luscher and Weisz (2004) have shown that also the next term is *universal* (true for any string action):

$$\sigma_T r \left(1 + \frac{V_0}{\sigma_T r^2} - \frac{1}{8} \left(\frac{V_0}{\sigma_T r^2} \right)^2 \right) \tag{17.25}$$

Note that these contributions are the two terms of quantum string contributions resummed (Arvis 1983) mentioned above, so they also follow from the Nambu-Goto action, as of course they should. But the next terms can be modified. For further discussion of the static $Q\bar{Q}$ potential stemming from the EST, we refer to (Aharony and Klinghoffer 2010).

Summarizing, quantum and boundary corrections to the potential, at large r to order $1/r^4$, have the form

$$V(r) \approx \sigma_T r - \mu - \frac{\pi D_\perp}{24r} - \frac{\pi^2}{2\sigma r^3}\left(\frac{D_\perp}{24}\right)^2 + \frac{\tilde{b}_2}{r^4} + \dots \tag{17.26}$$

The third and fourth contributions in (17.26) are Luscher and Luscher-Weisz universal terms in arbitrary dimensions, both reproduced by expanding Arvis potential; see (Petrov and Ryutin 2015) for a related discussion of the role of Luscher terms in the Pomeron structure. The last contribution is induced by the derivative-dependent string boundary contribution (17.23).

Comment 0:   Even if the string ends are constant in external space, they still may depend on the two-dimensional coordinates on the worldvolume of a vibrating string.

Comment 1:   The number of transverse dimensions $D_\perp = D - 2$ is 2, if string vibrations occur in the usual $D = 4$ space-time. However, in holography $D = 5$, $D_\perp = 3$. An extra vibration can be physically viewed as radial string excitation, as we will discuss in Sect. 17.7.

Comment 2:   The μ term receives both perturbative and nonperturbative contributions. The former are UV sensitive and in dimensional regularization renormalize to zero, as we assume throughout. The latter are not accounted for in the conformal Nambu-Goto string but arise from the extrinsic curvature term (17.23) in the form (Hidaka and Pisarski 2009; Qian and Zahed 2015).

$$\frac{D_\perp}{4}\sqrt{\sigma\kappa} \to \mu \tag{17.27}$$

Note that this contribution amounts to a negative boundary mass term in (17.23). While it vanishes for $D = 4$ space-time dimensions, but is finite for $D_\perp > 2$ in the holographic AdS/QCD approach.

We will not discuss the extensive holographic studies of the EST and related potential (Aharony and Klinghoffer 2010) but proceed to lattice simulations of the heavy-quark potential. These studies have now reached a high degree of precision, shedding light on the relevance and limitation of the string description. In a recent investigation by Brandt (2017), considerable accuracy was obtained for the potential at zero temperature and for pure gauge $SU(2)$ and $SU(3)$ theories. As can be seen from Fig. 3 in (Brandt 2017), the inter-quark potential is described to an accuracy of one-per-mille, clearly showing that both Luscher's universal terms, $1/r$, $1/r^3$, are correctly reproduced by the numerical simulations. Indeed, for $r/r_0 > 1.5$ (or $r >$

0.75 fm for Sommer's parameter $r_0 = 0.5$ fm), these two contributions describe the potential extremely well.

Expanding further to order $1/r^5$, or keeping the complete square root in Arvis potential, would not improve the agreement with the lattice potential, since the measured potential turns up and opposite to this expansion. Brandt lattice simulations (Brandt 2017) have convincingly demonstrated that the next correction is of order $1/r^4$ with the opposite sign. The extracted contribution fixes the b_2 coefficient in (17.26) as

$$\tilde{b}_2 = -\frac{\pi^3 D_\perp}{60} b_2 \tag{17.28}$$

with the numerically fitted values

$$b_2^{SU(2)} \sigma_T^{3/2} = -0.0257(3)(38)(17)(3)$$

$$b_2^{SU(3)} \sigma_T^{3/2} = -0.0187(2)(13)(4)(2) \tag{17.29}$$

(for the details and explanation regarding the procedure and meaning of the errors, we refer to (Brandt 2017)). Note that the overall contribution of this term to the potential is positive.

In summary, according to modern lattice studies, at $r \approx r_0 = 0.5$ fm, the static potential contains a wiggle, visible however only with a good magnifying glass since its relative magnitude is 10^{-3}. Above this point, EST describes the potential accurately, with four terms of the expansion defined.

Applications of QCD strings in general (and EST in particular) include not only (i) the static potential (the Wilson loop) but, via certain duality transformation, also two more important applications:

(ii) the correlator of two Polyakov lines at finite temperatures;
(iii) the "stringy Pomeron," or the tube-like stringy instanton describing amplitude of the elastic hadron-hadron scattering at high energies.

Basically, all three applications stem from description of a rectangular piece of stringy membrane. Therefore any progress in understanding of (i) thus induces some progress in (ii) and (iii) as well. We will briefly describe those for (iii), following (Kharzeev et al. 2018), in the next section.

17.6 The Stringy Pomeron

Let us start this section with a general motivation, explaining why I decided to go into the subject of hadron scattering and Pomerons, in spite of its apparent complexity.

One motivation is that hadronic scattering amplitudes depend on properties of QCD strings *exponentially*. At large-impact parameter b (exceeding the r.m.s. hadronic sizes), the stringy configuration produced must be *virtual*, instanton-like, so that the amplitude has a tunneling form $\sim \exp(-S_{cl}) \sim \exp(-b^2)$. It is indeed confirmed by the experimental data at large b. Furthermore, the experimentally observed peripheral collisions reach[16] b as large as $2\,fm$. So, in collisions one has access to string lengths much *larger* than what we observe in quarkonia or Reggeons. The second motivation is that, put in the exponent, even relatively small quantum corrections can be seen more clearly.

One of the specific important issues of the field is whether one would be able to locate a transition between the perturbative regime at small b and "stringy one" at large b. We have shown above that in static potentials, the transition is now detected: similar study is badly needed as a function of t or b.

The Pomeron can be defined as the *non-positive* t (zero or negative near-zero mass squared) object located at the leading (highest $\alpha(t)$) Regge trajectory. It has the vacuum quantum number, which means no charge or flavor[17] is transferred from one beam to another. The universal behavior of all hadronic elastic amplitudes at large $s \sim s^{\alpha(t)-1}$. In Fig. 17.6, we already presented current data on the glueball spectroscopy and located this trajectory, containing $J = 2, 4, 6$ lowest mass states. Phenomenologically, the Pomeron *intercept* $\alpha(t = 0) \approx 1.08$.[18]

The Regge calculus, with Reggeon exchange diagrams, has been created phenomenologically in the 1960s, mostly by Pomeranchuk, Gribov, and Veneziano. With the development of pQCD, it has been derived from re-summation of ladder diagrams, describing multiple productions of gluons. The so-called BFKL Pomeron (Kuraev et al. 1976) collects collinear logarithms and produces power of s known as the BFKL Pomeron intercept:

$$\alpha^{\text{BFKL}} = 1 + \frac{g^2 N_c}{\pi^2} \ln 2 \tag{17.30}$$

The Pomeron slope, as dimensional quantity, cannot of course occur in pQCD.

There is ongoing debate about the experimental observation of the Odderon, the C-odd exchange which would make a difference between the pp and $\bar{p}p$ elastic amplitudes, and the pQCD predictions for its trajectory. It is supposed to be calculated from the bound state of three Reggeized gluons: some studies put its intercept at $\alpha(0) < 1$, some exactly at $\alpha(0) = 1$. I am not aware of any calculations for stringy Odderon.

---

[16]At very small scattering angles or t, electromagnetic Coulomb forces dominate the strong interactions.

[17]The Regge trajectory with, e.g., the pion has isospin, and thus can be studied via isospin transfer reactions like $pn \to np$.

[18]High sensitivity to the Pomeron parameters can be illustrated by the fact that that this small deviation from 1 is the reason why all cross section slowly grows with s. This "small effect" is in fact responsible for about doubled NN cross section, between the collision energies used in the 1960s and today.

Let us now introduce the *semiclassical stringy* Pomeron. It originates from the paper (Basar et al. 2012) and thus will be called the BKYZ Pomeron. Using the instanton method and stringy Lagrangian, these authors had calculated the forward scattering amplitude between two small dipoles relativistically moving relative to each other.

In order to explain the stringy Pomeron, let me first take a detour and consider related classic problem *of the e + e− pair production in constant electric field*. It is widely known as the Schwinger process, as he solved it in detail in the 1950s. However we will not discuss neither the Schwinger paper, nor even earlier Heisenberg-Euler paper, but much earlier semiclassical work (Sauter 1931) from 1931 (well before anyone else).

The EOM of a charge relativistically moving in constant electric field is a classic problem which everybody had encounter in E/M classes. Writing it in the form

$$\frac{dp}{dt} = \frac{d}{dt}\left(\frac{v}{\sqrt{1-v^2}}\right) = \frac{eE}{m} \equiv a \qquad (17.31)$$

one finds the solution

$$v(t) = \frac{at}{\sqrt{1-a^2t^2}}, \quad x(t) = \frac{1}{a}\left(\sqrt{1+a^2t^2} - 1\right) \qquad (17.32)$$

(check small and large time limits).

Transformation into Euclidean time $\tau = it$ of the trajectory yields

$$x_E(\tau) = \frac{1}{a}\left(\sqrt{1-a^2\tau^2} - 1\right) \qquad (17.33)$$

and between $\tau = -1/a$ and $\tau = 1/a$, it describes the Euclidean path in shape of the semicircle. This should not surprise us: in the Euclidean world, time is no different from other coordinates, and electric field G_{01} is no different from the magnetic ones, so in the 0-1 plane, the paths are circles, like they are in all other planes.

The physical meaning of the semicircle is as follows: it describes tunneling through the "mass gap" in the spectrum of states; there are no states between $E = -\sqrt{p^2 + m^2}$ and $E = \sqrt{p^2 + m^2}$ with real momentum p, but on the Euclidean path, we found the momentum is imaginary.

Calculating the action $S = \int(-mds - eExdt)$, one gets the Euclidean version of it for the semicircle $S_E = \pi m^2 / 2eE$. The semiclassical probability (square of the amplitude) of the pair production is then

$$P \sim e^{-2S_E} \sim e^{-\pi m^2/eE} \qquad (17.34)$$

In QCD problem, we want to solve two colliding dipoles, *each* having a flux tube in between two charges. If *each* of them produces $\bar{q}q$ pair, since quarks are under constant tension force, so the problem is analogous to that just considered.

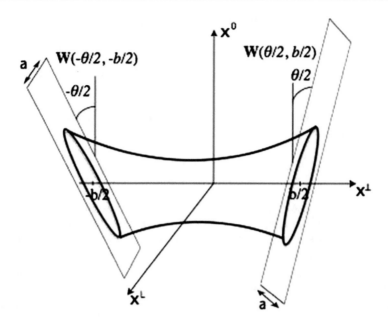

Fig. 17.9 The dipole-dipole scattering due to closed string exchange. The impact parameter **b** is the transverse distance between two colliding dipoles. Reclined by angles $\pm\theta/2$, dipole paths will collide after the result is transferred to Minkowski kinematics via $\theta \to iy$ where y is relative rapidity

The probability to create two quark pairs (and split each dipole into *two*) would correspond to trajectories of quarks making *two* circles, on the worldsheet of each dipole.

Now, imagine that, instead of production of massive quarks, we would like to think of purely stringy process, in which two "circular holes" on the worldsheets get connected by some "tube-like" configuration connecting the worldsheets of two dipoles. The setting is sketched in Fig. 17.9. Such stringy object with minimal Euclidean action (area times the tension) is the "stringy instanton" describing tunneling between two colliding dipole worldsheets. The semiclassical probability of it to happen in the forward scattering amplitude will give us the "stringy Pomeron."

Some introduction to the formal setting of BKYZ paper is perhaps needed. The starting expression is

$$\frac{i}{2s}T(\theta, q) = \int d^2b e^{i(\vec{q}_\perp \vec{b})} \langle (W(-\theta/2, -b/2) - 1)(W(\theta/2, b/2) - 1) \rangle$$

$$(17.35)$$

where $q_\perp$ is the momentum transfer, b is impact parameter, and W is the Wilson loop for a dipole:

$$W(\theta, b) = \frac{1}{N_c} Tr \left(Pexp(ig \oint dx A) \right)$$

For clarity, the authors start in perturbation theory and calculate this amplitude due to two-gluon exchange

$$T(\theta, b) \approx \frac{N_c^2 - 1}{32\pi^2 N_c^2} \left(\frac{ga}{b} \right)^4 cotan^2(\theta)$$

with a being the dipole size. Minkowski analytic continuation is done via $\theta \rightarrow iy$ where y is relative rapidity of the dipoles. Note that the scattering profile (its b-dependence) is power-like and can be obtained just by dimensional argument: pQCD, lacking any dimensionful quantities, cannot give anything else.

The effective string theory has a parameter $\sigma_T b^2$, which, as we will see, will appear in the action and then in the scattering profile.

The classical solution itself is obtained with simplified Polyakov form of the action:

$$S = \frac{\sigma_T}{2} \int_0^T d\tau \int_0^1 d\sigma (\dot{X}^\mu \dot{X}_\mu + X'^\mu X'_\mu)$$

The length of the tube is obviously the impact parameter between the two dipoles (protons) b, assumed to be large. The circumference of the tube is $\beta = 2\pi b/\chi$ where the quantity in denominator represents the collision energy $\chi = \log(s/s_0)$, and $s = (p_1 + p_2)^2$ is the Mandelstam invariant related to the collision energy. χ can also be viewed as the rapidity difference between the two colliding beams.

The product $b\beta$ is the tube area, which, times the string tension σ_T, gives the action of stringy instanton (presumed large for semiclassical setting to be valid).

One can map the two problems—static potential and the Pomeron—to each other, as discussed in the Appendix of (Shuryak and Zahed 2018). It is done via some duality relation, by exchanging time and space. One also needs to add another mirror image of a potential, to match the boundary conditions, which explains appearance of factor 2 below. In this case, the two partition functions of the string and its excitations become identical. The explicit transformation is

$$2b \leftrightarrow \frac{\hbar}{T}, \quad \beta \leftrightarrow 2r \tag{17.36}$$

Assuming the correspondence between the potential and the Pomeron is exact, we can map the potential (17.26) onto the Pomeron scattering amplitude in impact parameter space as

$$\mathcal{A}(\beta, b) \approx 2is \, \mathbf{K} \approx 2is \, e^{-S(\beta, b)} \tag{17.37}$$

with

$$S(\beta, b) = \sigma_T \beta b - 2\mu b - \frac{\pi D_\perp}{6} \frac{b}{\beta} - \frac{8\pi^2}{\sigma} \frac{b}{\beta^3} \left(\frac{D_\perp}{24} \right)^2 - \frac{2^5 b \tilde{b}_2}{\beta^4} \qquad (17.38)$$

with $\sigma = \sigma_T/2$ and $2\pi\sigma_T = 1/\alpha'_R$. Now, we can recall the parameters of the "tube" and set $\beta = 2\pi b/\chi$, $\chi = \ln(s/s_0)$.

Following this substitution, one observes that the leading and subleading terms have very different roles and energy dependence. The leading two contributions

$$e^{\chi \frac{D_\perp}{12} - \frac{b^2}{4\chi\alpha'_P}} \qquad (17.39)$$

give the Pomeron form of the amplitude, with the intercept value

$$\alpha(0) - 1 = \frac{D_\perp}{12}$$

The Gaussian dependence on b is consequence of the famous "Gribov diffusion," derived originally in perturbative setting, due to random emission of gluons in ladder diagrams. In fact strings also follow the same "diffusive law"

$$b^2 \sim \chi = \ln \left(\frac{s}{s_0} \right) \qquad (17.40)$$

which exists equally for perturbative gluons and strings.

Furthermore, one should recognize that the stringy Pomeron approach exists in two versions, the flat space and the holographic ones. In the former case, the space has two flat transverse directions $D_\perp = 2$, while in the latter, the string also propagates in the third and curved dimension. Since Gribov diffusion also takes place along this coordinate, identified with the "scale" of the incoming dipoles, the expressions we will use are a bit modified from the standard expressions. One such effect, derived for the BKYZ Pomeron, is the modification of the Pomeron intercept due to extra dimension:

$$\frac{D_\perp}{12} \to \frac{D_\perp}{12} \left(1 - \frac{3(D_\perp - 1)^2}{2D_\perp \sqrt{\lambda}} \right) \qquad (17.41)$$

Here $D_\perp = 3$ and $\lambda = g^2 N_c$ is the 't Hooft coupling, assumed to be large. In the range of $\lambda = 20$–40, (17.41) is in the range 0.14–0.18. For the numerical analyses to follow, we will use for the Pomeron intercept the value $\alpha_P(0) - 1 = \Delta_P = 0.18$. (This happens to be not far from the flat space value of $\frac{1}{6} = 0.166$.)

Experimentally, the Pomeron scattering amplitude exhibits both a real and imaginary part. The real part can in fact be measured at two locations: (i) at small $t \approx 0$, by observing the interference with the electromagnetic scattering induced by a photon exchange and (ii) at the location of the diffractive node t_{node} where the imaginary part vanishes and the subleading real part gets visible. For the interference measurement, the results are expressed in terms of the so-called ρ parameter:

$$\rho = \frac{\text{Re}(\mathcal{A}(s, t = 0))}{\text{Im}(\mathcal{A}(s, t = 0))} \tag{17.42}$$

The TOTEM data (Antchev et al. 2016) give

$$\rho(\sqrt{s} = 8\,\text{TeV}) = 0.12 \pm 0.03, \quad \rho(\sqrt{s} = 13\,\text{TeV}) = 0.098 \pm 0.01 \tag{17.43}$$

The textbook description of the Regge scattering amplitudes relates the ρ-parameter with the signature factor which for small t is captured by the phase factor $e^{i\pi\Delta_{\mathbf{P}}}$. It is small if $\Delta_{\mathbf{P}}$ is small, in agreement with data.

A real part of the amplitude may appear in Reggeon calculus because the Pomeron can be exchanged both in s and cross-channel u. In the Euclidean calculation of the scattering amplitude, one needs to include two contributions, with both the Euclidean angle θ and $\pi + \theta$, representing the u channel. The resulting amplitude, after analytic continuation to Minkowski space and the Fourier transform from the impact parameter to momentum transfer $\sqrt{-t}$, has the form $s^{\alpha(t)} + u^{\alpha(t)}$.

The main information we have about the profile of the scattering amplitude can be summarized as follows. One observable is the total cross section, with few data points shown in the upper plot of Fig. 17.10. Another important parameter is the

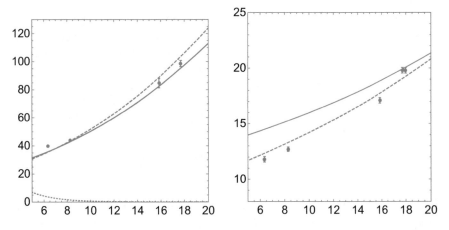

Fig. 17.10 The upper plot shows σ_{tot} in mb versus the log of the collision energy $\chi = \ln(s/s_0)$. The left-side (low energy) data points at $\sqrt{s} = 27, 63\,\text{GeV}$ are from the old ISR measurements, and the three right-side points, for $\sqrt{s} = 2.76, 7, 8\,\text{TeV}$, are from the TOTEM measurements. The dotted line in the lower plot indicates the contribution of the Reggeons other than the Pomeron (from the PDG fit.) The right plot shows the elastic slope B (GeV^{-2}). The curves are for model profiles discussed in our paper

so-called slope of the elastic scattering amplitude B:

$$B(t = 0) = \left(-\frac{d\ln\sigma_e}{d|t|}\right)_0 = \frac{1}{2}\langle b^2\rangle \tag{17.44}$$

The corresponding data are shown in the right plot of Fig. 17.10. Both grow with the collision energy due to the effective growth of the proton size induced by Gribov diffusion process.

Our last comment is that there is a serious problem with the description via Pomerons of the pp collisions at LHC energies: at small enough $b < b_{bd}$, the protons basically are black discs, with probability of scattering very close to unity. Obviously any structure in the amplitude inside the black disc is unobservable. One can model it with multi-Pomeron expressions and unitarization of the amplitude, but inherently there is no accuracy at small b. The only information remaining is the large-b or small-t slope we discussed above. To go around this difficulty, one can hope to get data on γp collisions from the future electron-ion collider: the photon coupling to Pomeron is small, and no multi-Pomeron processes will be needed.

17.7 Interaction of QCD Strings: Lattice, AdS/QCD, and Experiments

In the "dual superconductor" approach, the electric flux tubes are treated as dual to Abrikosov's solution for magnetic flux tubes in semiconductors. Depending on the ratio of the two lengths of the problem, associated with the gauge field and "Higgs" field masses, their interaction at large distances can be attractive or repulsive. For superconductors, this generates type-I and type-II superconductors. We already discussed above that QCD vacuum is of the type-I, which means that QCD strings should attract each other at large distances.

Obviously, one can arrange lattice configuration with four static charges and two strings and study their mutual interaction. And indeed, for pure gauge theories, their mutual attraction has been confirmed. However, the situation is different for pure gauge theories and QCD with the light quarks. In the former case, the lightest hadron is the scalar glueball, with a mass of about $m_{0^{++}} \approx 1.5\,\text{GeV}$. Therefore the interactions can only be very short-range $\sim \exp(-m_{0^{++}}/r)$.

In the real-world QCD, the lightest mesons are pions, of mass $m_\pi = 0.138\,\text{GeV}$ and its scalar chiral partner σ meson with a mass of about $m_\sigma \approx 0.5\,\text{GeV}$. Therefore much longer-range string-string interactions are possible. The pion is isovector, and cannot be emitted by pure glue state, so we are left with sigma.[19]

---

[19]Analogy to NN nuclear forces suggests that one needs to include the isoscalar vector ω meson as well, as its repulsive force nearly cancels the attractive sigma term. In the application discussed below, we have all kind of string pairs, string-string and string-antistrings, with equal probability, so one may think in this case the sign-changing vector exchange averages out to zero. Yet in the

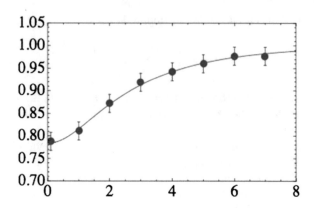

Fig. 17.11 The normalized chiral condensate perturbed by a flux tube, as a function of the coordinate transverse to the tube. The lattice data are from (Iritani et al. 2014); the curve is a fit with the sigma meson 2d propagator described in the text

In order to understand how QCD vacuum is modified around a string, one can perform lattice studies, measuring VEVs of various operators around it. For sigma meson, the operator is the isoscalar scalar quark density $\langle\sigma(x)\rangle = \langle\bar{q}q(x)\rangle$. In Fig. 17.11, the (normalized to vacuum) value of this VEV is shown as a function of transverse distance from the string (in lattice units), from (Iritani et al. 2014). Note that all deviations from 1—the vacuum value of the quark condensate—is indeed small, as it is expected to be suppressed by $1/N_c^2$. Even at the string center, the suppression effect is only about 1/5 or so.

The curve at this plot is from (Kalaydzhyan and Shuryak 2014); it is a fit to the lattice data by the expression

$$\frac{\langle\bar{q}q(x_\perp)W\rangle}{\langle W\rangle} = 1 - CK_0(m_\sigma\tilde{x}_\perp) \tag{17.45}$$

where the regulated transverse distance is defined by $\tilde{x}_\perp^2 = x_\perp^2 + s_{string}^2$. The K_0 is Bessel function, corresponding to massive scalar propagator in $d - 1 = 2$ spatial dimensions normal to the string. The fit parameters used are $C = .26, s_{string} = .176\,fm, m_\sigma = 600\,\text{MeV}$. In the string-string interaction, the following dimensionless parameter enters

$$g_{N\sigma T} = \frac{\langle\sigma\rangle^2 C^2}{4\sigma_T} \ll 1 \tag{17.46}$$

case of a Pomeron or glueball Reggeons, or the baryon junction Reggeon, we have strong-antistring or three strings, respectively. Perhaps in this case, the omega meson exchanges need to be included.

which is of the order of few percents. Thus one finds string ensembles subject to scalar effective description.[20]

The final topic in this section, devoted to flux tube interactions, is its experimental aspect. In fact producing a single string is hardly possible: the color field flux needs to be returned. We have already mentioned that because the Pomeron can be viewed either as an exchange of a closed string or production of two strings connecting the colliding hadrons: thus the minimal number of produced strings is *two*. But there are occasions in which many more strings are produced.[21]

Completing the subject of QCD string-string interactions, let us briefly discuss what such interactions imply for multi-string systems. Collective interaction of an ensemble of strings was studied by (Kalaydzhyan and Shuryak 2014) using the sigma exchange in 1+3 d space time and in holographic setting later. In both of them, the strings were assumed to be stretched in the same longitudinal direction, as it is the case in not-too-early time in high-energy collisions. We call configurations with many parallel strings a "spaghetti" state.

So, classical string dynamics is restricted to motion the transverse $d - 1 = 2$ or 3 space, in which strings are just points. The simulations are molecular dynamics (MD), or simply solving classical equation of motion of the strings. In Fig. 17.12, we show an example of snapshots at subsequent time: as one can see, the central part of the multi-string system undergoes clustering resembling the gravitational collapse. For each configuration, one can calculate the value of the quark condensate, modified according to collective influence of all strings: typically it rapidly develops regions in which such suppression is about complete. What it means is that there is chiral symmetry restoration at the center of the system, or that multi-string systems rapidly create a QGP fireball.

---

[20]Another theoretical approach in which string interactions can be studied is based on *holographic* models, originating from the AdS/CFT correspondence, of the conformal $\mathcal{N} = 4$ supersymmetric gluodynamics to string theory in $AdS_5 \times S^5$ ten-dimensional space-time. We put some elementary introduction to it in Appendix. Since these models start with string theories with ten-dimensional superstrings, the strings are natural pointlike objects "in the bulk." Their ends lead naturally to fundamental charges on the boundary—that is, in the four-dimensional manifold where the gauge theory (and ourselves) is located. While the string shape in curved space is not so simple, its total energy (static potential) is $V(r) \sim 1/r$. Indeed, it is obvious by dimension, because conformal theories lack any dimensional parameters. The original AdS/CFT correspondence has been generalized to some "bottom-up" holographic models, collectively known as AdS/QCD: for a review, see (Gursoy and Kiritsis 2008).

[21]For example, when a proton flies through a diameter of a heavy nucleus, the mean number of protons it interacts with at LHC energies is about $n_A \sigma_{NN}(2R_A) \approx 16$. If so, one needs to deal with at (minimum) 32 strings in such "central pA events." This number of course returns to two strings or a single Pomeron for very peripheral pA collisions. The open question is: at which impact parameters one has to describe the system as a set of strings, and at which all strings get "collectivized" into a common QGP fireball?

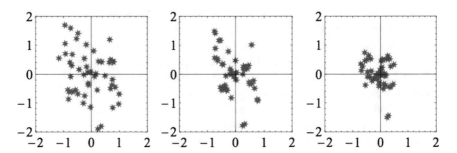

Fig. 17.12 Snapshots of multi-string configuration evolution in time, at $t = 0.1, 0.5, 1.\ fm/c$. The string locations (in fm) in the transverse plane are shown by stars; they are assumed to be all stretched in the same longitudinal direction (not shown)

17.8   String Balls

Historically, studies of the *self-interacting* string balls started in the framework of fundamental string theory in critical dimensions (26 for bosonic strings). The theoretical questions discussed were related to the understanding of the transition from the free strings, via *string balls*, to black holes.

The main question relevant for this transition is very simple. As we discussed in the previous section, the interaction between QCD strings is weak, and the same weak coupling regime is believed to hold for fundamental strings.[22]   So, for a short string, the self-interaction is negligible. A very large string (or in fact any large object), if described by gravity which grows with mass more than any other interaction, is subject to gravitational collapse. Therefore, gravity becomes a dominant force, and sufficiently massive strings should be black holes of the classical gravity. These two limits are obvious. The main idea is that *at some intermediate mass range*, the gravity and other forces can be balanced, producing stable gravitationally bound object.[23]

Let us start with free strings: a "random walk" process, of M/M_s steps, where $M_s \sim 1/\sqrt{\alpha'}$ is the typical mass of a straight string segment. If so, the string entropy scales as the number of segments:

$$S_{ball} \sim M/M_s \qquad (17.47)$$

The Schwarzschild radius of a black hole in d spatial dimensions is

$$R_{BH} \sim (M)^{\frac{1}{(d-2)}} \qquad (17.48)$$

---

[22]We remind that massless modes of closed strings include gravitons; therefore, it is a candidate for the theory of quantum gravity.

[23]For example, stars and (gaseous) planets exist due to balance between thermal pressure and gravity.

and the Bekenstein entropy

$$S_{BH} \sim Area \sim M^{\frac{d-1}{d-2}} \qquad (17.49)$$

Thus, the equality $S_{ball} = S_{BH}$ can only be reached at some special critical mass M_c. When this happens, the Hawking temperature of the black hole is exactly the string Hagedorn value T_H, and the radius is at the string scale. So, at least at such value of the mass, a near-critical string ball can be identified—at least thermodynamically—with a black hole.

However, in order to understand how exactly this state is reached, one should first address the following puzzle. Considering a free string ball (described by the Polyakov's near-critical random walk), one would estimate its radius to be

$$\frac{R_{ball,r.w.}}{l_s} \sim \sqrt{M} \qquad (17.50)$$

for any dimension d. This answer does not fit the Schwarzschild radius R_{BH} given above (17.48).

The important element missing is the self-interaction of the string ball: perhaps, Susskind was the first who pointed it out. A more quantitative study (Horowitz and Polchinski 1998) had used the mean field approach, and then (Damour and Veneziano 2000) completed the argument, by using the correction to the ball's mass due to the self-interaction. Their reasoning can be nicely summarized by the following schematic expression for the entropy of a self-interacting string ball of radius R and mass M,

$$S(M, R) \sim M \left(1 - \frac{1}{R^2}\right)\left(1 - \frac{R^2}{M^2}\right)\left(1 + \frac{g^2 M}{R^{d-2}}\right) \qquad (17.51)$$

where all numerical constants are for brevity suppressed and all dimensional quantities are in string units given by its tension. The coupling g in the last bracket is the string self-coupling constant to be much discussed below. For a very weak coupling, the last term in the last bracket can be ignored, and the entropy maximum will be given by the first two terms; this brings us back to the random walk string ball. However, even for a very small g, the importance of the last term depends not on g but on $g^2 M$. So, very massive balls can be influenced by a very weak gravity (what, indeed, happens with planets and stars). If the last term is large compared to 1, the self-interacting string balls become much smaller in size and eventually fit the Schwarzschild radius.

Let us now switch back to QCD strings. In the preceding section, their long-range interaction has been ascribed to σ meson exchanges. We also have demonstrated there that sufficiently dense multi-string states can collapse (in this case, into a QGP fireball). Now we wonder if this attractive force can be balanced by entropy, leading to some stable configurations at some intermediate parameters of the problem.

This is the numerical model we use to study the string balls with self-interaction. While we discuss the details of the setting below in this section, let us emphasize on the onset its main physics prerequisites, namely, that the ball surface should be approximately near the Hagedorn temperature, making the string fluctuate widely outward. The string-string interaction established in the previous section is in the vacuum, $T = 0$, while the string balls we are discussing are expected to be produced at $T \approx T_c$. Therefore, the effective σ meson mass is expected to be reduced, in fact to zero in QCD with strictly massless quarks.

Following a bit Wilson's strong coupling expansion, we place the strings on links of a $(d = 3)$-dimensional lattice. Strings are assumed to be in contact with a heat bath, and a partition function includes all possible string configurations.

Intersections of the strings are not included because of the repulsive interaction at small distances. Even for the Abelian fields, which add up simply as vectors, the action is quadratic in fields (no commutators), and intersections are energetically not favorable. An exception (in the lattice geometry) is the case of exactly oppositely directed fluxes, when a part of the string should basically disappear. We had not included this complication believing that the total entropy and energy of the string ball will not be affected much.

Instead of using boxes (with or without periodic boundary conditions) as is customary in the lattice gauge theory and many other statistical applications, we opted for an infinite space (no box). Instead the temperature T is *space dependent*. We think it better corresponds to the experimental situation. Furthermore, the string ball surface is automatically near criticality and thus strongly fluctuating; this aspect will be important for our application of initial deformations below.

The "physical units" in gluodynamics, as in lattice tradition, are set by putting the string tension to its value in the real world: $\sigma_T = (0.42\,\text{GeV})^2$. Numerical lattice simulations have shown that gluodynamics with $N_c > 2$ has a first-order deconfinement phase transition, with $T_c/\sqrt{\sigma_T}$ very weakly dependent on N_c (for review, see, e.g., (Teper 2009)). Numerically, the critical temperature of the gluodynamics is $T_c \approx 270\,\text{MeV}$.

It has been further shown that the effective string tension of the *free* energy $\sigma_F(T)$ decreases with T; a point where it vanishes is known as the Hagedorn point. Since this point is above T_c, some attempts have been made (Bringoltz and Teper 2006) to get closer to it by "superheating" the hadronic phase, yet some amount of extrapolation is still needed. The resulting value was found to be

$$\frac{T_H}{T_c} = 1.11 \qquad (17.52)$$

The nature of the lattice model we use is very different from that of the lattice gauge theory (LGT). First of all, we do not want to study quantum strings and generate two-dimensional surfaces in the Matsubara $R^d S^1$ space, restricting ourselves to the thermodynamics of strings in d spatial dimensions.

The lattice spacing a in LGT is a technical cutoff, which at the end of the calculation is expected to be extrapolated to zero, reaching the so-called continuum limit. In our case, a is a physical parameter characterizing QCD strings: its value is selected from the requirement that it determines the correct density of states. Since we postulate that the string can go to any of $2d - 1$ directions from each point (going backward on itself is prohibited), we have $(2d - 1)^{L/a}$ possible strings of length L. Our partition function is given by

$$Z \sim \int dL \exp\left[\frac{L}{a} \ln(2d - 1) - \frac{\sigma_T L}{T}\right], \tag{17.53}$$

and hence the Hagedorn divergence happens at

$$T_H = \frac{\sigma_T a}{\ln(2d - 1)}. \tag{17.54}$$

Setting $T_H = 0.30\,\text{GeV}$, according to the lattice data mentioned above and the string tension, we fix the three-dimensional spacing to be

$$a_3 = 2.73\,\text{GeV}^{-1} \approx 0.54\,fm. \tag{17.55}$$

It is, therefore, a much more coarse lattice, compared to the ones usually used in LGT.

If no external charges are involved, the excitations are closed strings. At low T, one may expect to excite only the smallest ones. With the "no self-crossing" rule, we apply that would be an elementary plaquette with four links. Its mass,

$$E_{plaquette} = 4\sigma_T a \approx 1.9\,\text{GeV}, \tag{17.56}$$

is amusingly in the ballpark of the lowest glueball masses of QCD. (For completeness, the lowest "meson" is one link or mass 0.5 GeV, and the lowest "baryon" is three links—1.5 GeV of string energy—plus that of the "baryon junction.")

At temperatures below and not close to T_H, one finds extremely dilute $\mathcal{O}(e^{-10})$ gas of glueballs, or straight initial strings we put in. Only close to T_H do multiple string states get excited; the strings rapidly grow and start occupying a larger and larger fraction of the available space.

Before we show the results of the simulation, let us discuss the opposite "dense" limit of our model. We do not allow strings to overlap; the minimal distance between them is one link length, or again about $0.5\,fm$. Is it large enough for the string to be considered well separated? We think so, as it is about three times the string radius.

The most compact (volume-filling or Hamiltonian) string wrapping visits each site of the lattice. If the string is closed, then the number of occupied links is the same as the number of occupied sites. Since in $d = 3$ each site is shared among eight neighboring cubes, there is effectively only one occupied link per unit cube, and this wrapping produces the maximal energy density,

$$\frac{\epsilon_{max}}{T_c^4} = \frac{\sigma_T a}{a^3 T_c^4} \approx 4.4 \tag{17.57}$$

(we normalized it to a power of T_c, the highest temperature of the hadronic phase). It is instructive to compare it to the energy density of the gluonic plasma, for which we use the free Stefan-Boltzmann value

$$\frac{\epsilon_{gluons}}{T^4} = (N_c^2 - 1)\frac{\pi^2}{15} \approx 5.26 \qquad (17.58)$$

and conclude that our model's maximal energy density is comparable to the physical maximal energy density of the mixed phase we would like to study.

The algorithm consists of a sequence of updates for the each string segment, such that the configuration gradually approaches equilibrium. The spatial distribution over all three coordinates is close to the Gaussian one, as is exemplified in the upper figure. Yet it is not just a Gaussian ensemble of random points, as the points constitute extended objects—strings.

In Fig. 17.13 (left figure), we show the calculated relation between the average string length L and its energy E. Each point is a run of about 10^4 iterations of the entire string updates after equilibration. While at small coupling E and L are simply proportional to each other, like for noninteracting strings described above, this behavior changes abruptly. As the negative self-interaction energy becomes important, the total energy E of the ball becomes *decreasing* with the string length L. In Fig. 17.13 (right figure), we show more details of this behavior: this plot demonstrates how total energy E depends on the coupling value g_N. We find a jump at the critical coupling (for this setting) g_N^{cl}, which in a simulation looks like a first-order transition, with double-maxima distributions in the energy and length. As is seen from the figure, the precise value of the coupling somewhat depends on the system size. At this coupling, the jump in energy is always about a factor 3, and the jump in string length (or entropy) is even larger.

In this way, we observe a new regime for our system, which we will call the "entropy-rich self-balanced string balls." For a given fixed mass M, we thus find that string balls may belong to two very distinct classes: (i) small near-random balls

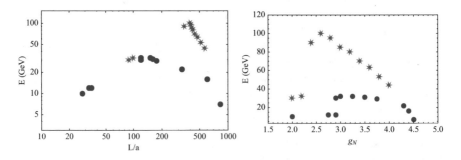

Fig. 17.13 Left plot: the mean energy of the cluster $E(g_N)$ [GeV] vs the mean length of the string $L(g_N)/a$. Lower plot: the mean energy of the cluster $E(g_N)$ [GeV] vs the "Newton coupling" g_N [GeV^{-2}]. Points show the results of the simulations in setting $T_0 = 1$ GeV and size of the ball $s_T = 1.5a, 2a$, for circles and stars, respectively

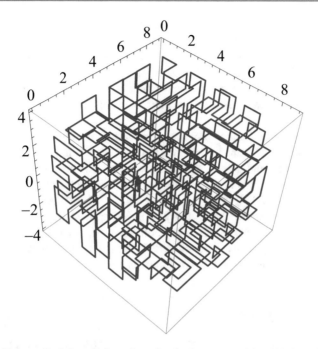

Fig. 17.14 (Color online) A typical configuration in the entropy-rich self-balanced string balls ensemble. Simulation parameters: $T_0 = 1\,\text{GeV}$, $s_T = 1.5a$, $g_N = 4.4\,\text{GeV}^{-2}$

and (ii) large ones in which the string can be very long but balances its tension by a comparable collective attraction. Discovery of this second regime is the main result of this paper.

An example of a corresponding configuration is shown in Fig. 17.14. Note that, in spite of a very large string length $L/a \sim 700$, the total energy is only $E \approx 17\,\text{GeV}$, as a result of the balancing between the mass and self-interaction. Note furthermore that that configurations are very asymmetric: one string is excited much more than the other, since the longer string has many more states than the shorter one. The same feature has been noticed on the lattice as well: typically, one very long string forms a large cluster, dominating over a few small clusters. Note further that nearly all space inside the ball with $T > T_H$ is occupied. High entropy corresponds to a (astronomically) large number of shapes this string may have.

Finally, there exists the second critical coupling, which is found to be $g_N^{c2} \approx 4.5\,\text{GeV}^{-2}$, above which balancing the energy becomes impossible and simulations show immediate collapse of the system, in which the energy quickly falls to large negative values, clearly of no physical meaning.

Finally, admitting that the QCD string balls still remain a theoretical construction, let us discuss whether the QCD string balls can still be produced in experiments.

We already discussed, in the preceding section, that "tube" geometry of the surface naturally leads to a periodic coordinate and thermal description: the circumference of the tube is identified with the Matsubara time $\tau = 1/T$, inverse to the effective string temperature. At certain values of the impact parameter b, this temperature corresponds to the Hagedorn value; the effective tension of the string decreases, and its high excitations become possible. As a result, as one can expect (and, indeed, sees it directly in the observed elastic scattering profile), the scattering amplitude for such b exceeds the value interpolated by a Pomeron string expression from large b.

There are two explanations proposed in literature for this rapid increase of the scattering profile at certain b. The "mainstream" one is that Pomeron amplitude becomes too large and needs "unitarization," or shadowing by certain multi-Pomeron amplitudes. It is also possible that the Hagedorn transition suggested in (Shuryak and Zahed 2014) is at play, so that it is the mixed phase with long strings.

Whatever is the interpretation, the way to experimentally proceed is to study *double diffraction*, or Pomeron-Pomeron collisions. As discussed in Refs for a long time, we have already seen a production of scalar and tensor glueballs. Studying in detail the created system with few-GeV mass is a way to go.

17.9   Brief Summary

- QCD flux tubes were studied by analogy with Abrikosov flux tubes in super-conductors of the second kind. The difference is that the latter carry *magnetic* flux and are surrounded by a "coil" made of rotating *electrons*, while the former carry *electric* flux and are surrounded by rotating *monopoles*. Both are solutions of equation of motion stemming from Ginzburg-Landau Lagrangian.
- The picture of the field and magnetic currents was observed on the lattice (see Fig. 17.1). Ginzburg-Landau deception of lattice data is in fact remarkably accurate (see Fig. 17.2).
- We discuss Regge trajectories corresponding to rotating charges connected by flux tubes.
- A separate issue is whether one can form flux tube-like objects at $T > T_c$, when there is no supercurrent of monopoles. Those were suggested by monopole models and are perhaps observed on the lattice as well.
- Various attempts to construct effective theory of flux tubes are discussed. The simplest version is Nambu-Goto action, containing total area of the string worldvolume, but one can and perhaps should add terms with derivatives. We discuss precision lattice data of static potential, aiming on testing these models.
- String exchange leads to Pomeron-like scattering amplitude for high-energy hadronic collisions. The shape of classical string configuration is explained in Fig. 17.9.
- Flux tube interaction, studied on the lattice, is found to be weakly attractive and described by σ meson exchange. We discussed studies of multi-string "spaghetti" configurations, produced in heavy-ion collisions. There is a transition between a dilute regime, in which flux tubed retain individuality, and dense regime, in which a common QGP fireball is formed.

- Flux tube may hypothetically also form "string balls," in between expanding separate strings and imploding dense systems.

References

Aharony, O., Klinghoffer, N.: Corrections to Nambu-Goto energy levels from the effective string action. JHEP **12**, 058 (2010). arXiv:1008.2648

Antchev, G. et al.: Measurement of elastic pp scattering at $\sqrt{s} = 8\,TeV$ in the Coulomb-nuclear interference region: determination of the ρ-parameter and the total cross-section. Eur. Phys. J. **C76**(12), 661 (2016). arXiv:1610.00603

Arvis, J.F.: The exact $q\bar{q}$ potential in Nambu string theory. Phys. Lett. **127B**, 106–108 (1983)

Baker, M., Ball, J.S., Zachariasen, F.: Dual QCD: A review. Phys. Rept. **209**, 73–127 (1991)

Baker, M., Steinke, R.: Semiclassical quantization of effective string theory and Regge trajectories. Phys. Rev. **D65**, 094042 (2002). arXiv:hep-th/0201169

Bali, G.S.: The mechanism of quark confinement. In: Quark Confinement and the Hadron Spectrum III. Proceedings, 3rd International Conference, Newport News, USA, June 7–12, 1998, pp. 17–36 (1998). arXiv:hep-ph/9809351

Basar, G., Kharzeev, D.E., Yee, H.-U., Zahed, I.: Holographic Pomeron and the Schwinger mechanism. Phys. Rev. **D85**, 105005 (2012). arXiv:1202.0831

Brandt, B.B.: Spectrum of the open QCD flux tube and its effective string description I: 3d static potential in SU(N = 2, 3). JHEP **07**, 008 (2017). arXiv:1705.03828

Bringoltz, B., Teper, M.: In search of a Hagedorn transition in SU(N) lattice gauge theories at large-N. Phys. Rev. **D73**, 014517 (2006). arXiv:hep-lat/0508021

Cea, P., Cosmai, L., Cuteri, F., Papa, A.: QCD flux tubes across the deconfinement phase transition. EPJ Web Conf. **175**, 12006 (2018). arXiv:1710.01963

Chew, G.F., Frautschi, S.C.: Regge trajectories and the principle of maximum strength for strong interactions. Phys. Rev. Lett. **8**, 41–44 (1962)

Creutz, M.: Asymptotic freedom scales. Phys. Rev. Lett. **45**, 313 (1980)

Damour, T., Veneziano, G.: Selfgravitating fundamental strings and black holes. Nucl. Phys. **B568**, 93–119 (2000). arXiv:hep-th/9907030

Gursoy, U., Kiritsis, E.: Exploring improved holographic theories for QCD: Part I. JHEP **02**, 032 (2008). arXiv:0707.1324

Hagedorn, R.: Statistical thermodynamics of strong interactions at high-energies. Nuovo Cim. Suppl. **3**, 147–186 (1965)

Hellerman, S., Swanson, I.: String theory of the Regge intercept. Phys. Rev. Lett. **114**(11), 111601 (2015). arXiv:1312.0999

Hidaka, Y., Pisarski, R.D.: Zero point energy of renormalized Wilson loops. Phys. Rev. **D80**, 074504 (2009). arXiv:0907.4609

Horowitz, G.T., Polchinski, J.: Selfgravitating fundamental strings. Phys. Rev. **D57**, 2557–2563 (1998). arXiv:hep-th/9707170

Iritani, T., Cossu, G., Hashimoto, S.: Analysis of topological structure of the QCD vacuum with overlap-Dirac operator eigenmode. PoS **LATTICE2013**, 376 (2014). arXiv:1311.0218

Kaczmarek, O., Zantow, F.: Static quark anti-quark interactions at zero and finite temperature QCD. II. Quark anti-quark internal energy and entropy (2005). arXiv:hep-lat/0506019

Kalaydzhyan, T., Shuryak, E.: Collective interaction of QCD strings and early stages of high multiplicity pA collisions. Phys. Rev. **C90**(1), 014901 (2014). arXiv:1404.1888

Kharzeev, D.: Can gluons trace baryon number? Phys. Lett. **B378**, 238–246 (1996). arXiv:nucl-th/9602027

Kharzeev, D., Shuryak, E., Zahed, I.: Higher order string effects and the properties of the Pomeron. Phys. Rev. **D97**(1), 016008 (2018). arXiv:1709.04007

Kuraev, E.A., Lipatov, L.N., Fadin, V.S.: Multi-Reggeon processes in the Yang-Mills theory. Sov. Phys. JETP **44**, 443–450 (1976). [Zh. Eksp. Teor. Fiz. **71**, 840 (1976)]

Liao, J., Shuryak, E.: Static $\bar{Q}Q$ potentials and the magnetic component of QCD plasma near T_c. Phys. Rev. **D82**, 094007 (2010). arXiv:0804.4890

Luscher, M., Weisz, P.: String excitation energies in SU(N) gauge theories beyond the free-string approximation. JHEP **07**, 014 (2004). arXiv:hep-th/0406205

Meyer, H.B.: Glueball regge trajectories. Ph.D. thesis, Oxford U (2004). arXiv:hep-lat/0508002

Nielsen, H.B., Olesen, P.: Vortex line models for dual strings. Nucl. Phys. **B61**, 45–61 (1973). [302 (1973)]

Petrov, V.A., Ryutin, R.A.: High-energy scattering versus static QCD strings. Mod. Phys. Lett. **A30**(18), 1550081 (2015). arXiv:1409.8425

Polyakov, A.M.: Fine structure of strings. Nucl. Phys. **B268**, 406–412 (1986)

Qian, Y., Zahed, I.: A stringy (holographic) Pomeron with extrinsic curvature. Phys. Rev. **D92**(8), 085012 (2015). arXiv:1410.1092

Sauter, F.: Uber das Verhalten eines Elektrons im homogenen elektrischen Feld nach der relativistischen Theorie Diracs. Z. Phys. **69**, 742–764 (1931)

Shuryak, E., Zahed, I.: New regimes of the stringy (holographic) Pomeron and high-multiplicity pp and pA collisions. Phys. Rev. **D89**(9), 094001 (2014). arXiv:1311.0836

Shuryak, E., Zahed, I.: Regimes of the Pomeron and its intrinsic entropy. Ann. Phys. **396**, 1–17 (2018). arXiv:1707.01885

Sonnenschein, J.: Holography inspired stringy hadrons. Prog. Part. Nucl. Phys. **92**, 1–49 (2017). arXiv:1602.00704

Sonnenschein, J., Weissman, D.: Rotating strings confronting PDG mesons. JHEP **08**, 013 (2014). arXiv:1402.5603

Teper, M.: Large N and confining flux tubes as strings - a view from the lattice. Acta Phys. Polon. **B40**, 3249–3320 (2009). arXiv:0912.3339

Veneziano, G.: Construction of a crossing - symmetric, Regge behaved amplitude for linearly rising trajectories. Nuovo Cim. **A57**, 190–197 (1968)

Yanagihara, R., Iritani, T., Kitazawa, M., Asakawa, M., Hatsuda, T.: Distribution of Stress Tensor around Static Quark–Anti-Quark from Yang-Mills Gradient Flow (2018). arXiv:1803.05656

Holographic Gauge-Gravity Duality

18

In the QCD-related setting, the first example of the duality is that between the *weak-coupling* description in UV, in which the fields are quarks and gluons, and the *chiral effective Lagrangians* describing IR properties in terms of light mesons.

More attention in these lectures was devoted to *electric-magnetic duality*, also related with the RG evolution from the UV to IR momentum scales. It goes from the "electric" description in terms of quarks and gluons to "magnetic" one in terms of monopoles and dyons. It is important that magnetic coupling is inverse of electric one, $1/g$: therefore in the inconvenient *strong coupling limit*, one better switch to magnetic description, which is in this case weak coupling.

The dualities we are going to discuss in this and the next chapters are a bit different. They are "holographic"; the UV and IR ends are in this case connected by an extra coordinate (called z or $u = 1/z$). The Lagrangians and equations of motion are written for fields living "in the bulk," between the two limits. The RG flow becomes just their dynamics in extra dimensions. After the problem in the bulk is solved, the physical predictions are extracted "holographically," by projecting (via certain procedures) the "bulk" results "to the boundary," where the QCD-like gauge theories are located.

Like for electric-magnetic duality, if the "boundary gauge theory" is in strong coupling, the bulk one is weakly coupled. So the problem can better be solved in the bulk, using standard perturbative tools.

As this projection goes, pointlike fundamental objects in the bulk become somehow "blurred" on the UV boundary. A single point source located in a bulk, at distance ρ from the boundary, generates an *instanton* on the boundary, of some size ρ. A pointlike fundamental string in the bulk corresponds to a finite-width QCD string (we also called the flux tube). Any object falling into IR direction, in which gravity force acts, appears expanding in its size on the boundary: this is how one describes "fireball explosions" of high-energy collisions.

© The Author(s), under exclusive license to Springer Nature Switzerland AG 2021   429
E. Shuryak, *Nonperturbative Topological Phenomena in QCD
and Related Theories*, Lecture Notes in Physics 977,
https://doi.org/10.1007/978-3-030-62990-8_18

18.1   D-branes

D-branes are topological solitons of the string theory: but since we are not expecting/requiring its knowledge, we will not discuss their structure in depth. (The reader interested in their historical origins and structure should consult, e.g., the TASI lecture of Polchinski (Rodrigues and Wotzasek 2006) or the original papers mentioned there.) Our aim in this chapter is to explain the geometric properties of these objects and their connections to our main object of interest, the gauge theories.

There is no place here to present string theory systematically, but we still need to remind its most elementary points. *Strings* are extended object with one spatial extension, moving in time. Their dynamics is described by $D = d + 1$ dimensional *external* coordinates X^μ, $\mu = 0..d$ of some manifold M, all being the functions of *two* internal coordinates σ_a, $a = 0, 1$, usually called τ, σ, with the time-like and space-like metric signatures, respectively. The resulting two-dimensional surface is a world history of the string propagation. Obviously the string *action* should be invariant under re-parameterization of both sets of coordinates. Classic example is the Nambu-Goto action

$$S = \frac{1}{2\pi\alpha'} \int_M d^2\sigma \, \partial_a X^\mu \partial_a X_\mu \qquad (18.1)$$

where the (dimensional) coefficient for historical reasons is written like this, with the constant α'. (It has prime—derivative—because it has been first introduced as the "slope of the Regge trajectories.") For QCD strings, it defined their tension, and also for fundamental string theory, its value defines the basic scale of the theory. The bosonic strings are consistent[1] in $d = 26$, while superstrings (with fermions living on them) can be formulated in $d = 10$ or 11.

The D_p branes are generalization of the concept of a string, an object with $p > 1$ spatial coordinates. The letter D in the name came from the Dirichlet boundary condition $X^\mu = const$ on the manifold boundary ∂M (as opposed to Neumann boundary condition for the derivative at the boundary). Brane world history is $p + 1$ dimensional manifold, as they propagate in time. Its classical action has the same Nambu-Goto form, except the coefficients, the brane tensions, are of course different.

(Examples include $p = 0$ case, a particle, and $p = 1$, a fundamental string. Sometimes one can even find some abused notation D_{-1}, which refers to an "event," a point in space and time.)

For reasons soon to become obvious, a choice most popular for applications is the D_3 branes. Including time, this one has $p + 1 = 4$ internal coordinates. Since this number is less than the space-time of the superstring theory, $D = 10$, there

---

[1] Anomalies appearing during string quantization cancels in these dimensions. We will not consider string quantization: but we discussed some of its consequences in chapter on the QCD strings.

are six extra dimensions in which such branes appear as points and can move and interact.

As for any soliton, symmetries induce the zero modes of the branes, such as simple shifts of the object as a whole. When collective coordinates corresponding to shifts (and other zero modes) are gently modulated (coordinate dependence has small gradients), these near-symmetries generate massless Goldstone modes. One can classify all of them and calculate the corresponding effective Lagrangian. It turns out that spin of these modes can be 2,1,0. Therefore interactions induced by their exchange between branes appear to be gravity-like,[2] vector-like, and scalar-like.

18.2   Brane Perturbations Induce Effective Gauge Theories

Suppose one finds a p-dimensional membrane solution for the D-branes, with $p + 1 < d$. Without knowing its internal structure, one can in general determine its lowest excitations. The issue is no different from any other membrane, e.g., that of a drum. One may start from a flat membrane with $p + 1$ coordinates inside it, x^μ, $\mu = 0..p$, and then deform it to

$$y^\mu = x^\mu + A^\mu(x) \tag{18.2}$$

The displacements A_μ are of two kinds:

(1) one with $\mu > p + 1$ *external* to the brane,
(2) and another with $\mu = 0..p$ *internal* to the brane.

From the point of view of a p-dimensional observer *living on the brane*, the first type corresponds to some scalar fields, while the second are vector fields, as the index direction can be recognized.

The corresponding effective action of the deformed membrane is known as the *Born-Infeld action*. Its form is determined from geometric principles of re-parameterization invariance. This action is also proportional to the area, times some undetermined coefficient, the tension T_p:

$$S_p = -T_p \int d^{p+1} x \, e^{-\phi} \sqrt{det\,(G_{ab} + B_{ab} + 2\pi\alpha' F_{ab})} \tag{18.3}$$

The differential geometry tells us that the square root of the determinant of the internal metric G_{ab}, $a, b = 0..p$ is required here, in order to get the invariant volume element, and the dilaton ϕ appears from this determinant as well.

---

[2]We will not discuss "theory of everything" based on superstrings, aiming to find a theory unifying gravity with the standard model.

The term F_{ab}, central to our discussion, is the (Abelian) field strength corresponding to vector displacement field A_a (internal part). If the gauge field coefficient is considered small, the square root can be expanded to second order, resulting in the Lagrangian $\sim (F_{ab})^2$. It then looks like effective $U(1)$ theory, called "induced electrodynamics." One can understand its appearance as follows. Suppose two dimensions X^1, X^2 do belong to the brane and there is one nonzero field component F_{12}, so that one can set the displacement to be $A_2 = X^1 F_{12}$. What it means is certain linear stretching of a brane along two directions by amount proportional to X^1, such that the volume element will increase, now containing a factor $dX^1 \sqrt{1 + (2\pi\alpha' F_{12})^2}$. This is precisely a correction which a nonzero $F_{12} = -F_{21}$ in the Born-Infeld determinant will produce in such setting. (The factor $2\pi\alpha'$ is just a matter of field normalization.)

The term B_{ab} is related with the magnetic charge density of the brane: let me not specify its role and structure at this point, as it is not that important for what follows.

Let us now make a more complicated example with N p-branes placed together, at the same place. Open strings then obtain indices at both ends, $i, j = 1..N$. The displacements can be described by the matrix-valued vector fields A_{ij}^μ. The central consequence of this construction is that the corresponding Born-Infeld action will then contain the correct *non-Abelian* field strength $(F^{\mu\nu})^2$, with the commutator term as prescribed by Yang and Mills. This means that the vibrations of N p-branes are described by some effective non-Abelian gauge theory.

This is the reason why *the string theory can be related with the physical gauge theories*. How important this observation is depends on the opinion, which vary vastly across the theory community.

The *maximalists* in the string theory community think that the $U(1) \times SU(2) \times SU(3)$ gauge groups of the standard model are effective theories in this very sense and that we do live on the D_3 brane and hope that one day we will be able to find out how exactly the compactification of unwanted six extra dimensions will explain "everything," or even discover these extra dimensions experimentally.

The *minimalists* (to which most people outside the string community belong) take this fact as a mathematical statement, providing very useful theoretical tools to *model* the problem one wants to solve.

Whatever people's hopes may be, the fact remains that we have no good ideas about compactification scale of the extra dimensions. Experimentally, it can be from something like TeV to the inverse Plank mass or $1/10^{19}$ GeV. Thus, even if we do live on a brane, we may never be able to find that out, given obvious experimental limitations.

But nothing stops us from using the theoretical models which originated from this construction, as a mathematical tool. Indeed, as we will see below, one can invent certain brane constructions whose vibration spectrum is capable of reproducing the QCD spectrum of hadrons. Furthermore, one can calculate effective chiral

Lagrangians, thermal and even transport properties of strongly coupled quark-gluon plasma.[3]

18.3 Brane Constructions

18.3.1 A Stack of D_3 Branes

The main construction leading to AdS/CFT is a stack of N_c "coincident D_3 branes," which is located at the same spatial point in six external coordinates. A brane is a location at which string can end. The (nonexcited) string has a mass equal to its length times the tension. So, if the length is zero, we have $N_c \times N_c$ massless vector fields: thus this construction has just enough vector fields to generate $SU(N_c)$ effective gauge theory.

As another example, one can consider "non-coincident D_3 branes," separated in external coordinates by some distances $R_{ij}, i, j = 1..N_c$. The strings between them would have masses proportional to these distances. In this way, one can generate models with "Higgsing," like Georgi-Glashow model of weak interactions.

Let us just point out that the extra coordinate can be compactified to a circle. This will generate strings with the lengths proportional to N_c lengths of the segments. This model will split gluons into massless and massive ones, modeling the adjoint Higgsing happening when the Polyakov line has a nonzero VEV. Recall that precisely in this setting, we came to understand instanton-dyons, or specifically why in this case the instanton (a complete D_1 wrapping of the whole circle) can be seen as consisting out of N_c massive segments, the N_c types of the (self-dual) instanton-dyons. We will not discuss this idea in full: see the original paper (Lee and Lu 1998).

18.3.2 The Seiberg-Witten Curve from the Branes

One more beautiful brane construction[4] (Witten 1997) explains how the celebrated curve for $\mathcal{N} = 2$ SYM and related theories can be interpreted (and calculated) directly from the equilibrium (minimal action) brane shape.

It includes 5-branes extended in coordinates (0..5), located at $x^7 = x^8 = x^9 = 0$ and *some* x^6 values to be called α_i. Like in a previous subsection, those are

---

[3]There are of course multiple applications of the gauge-string correspondence outside QCD, e.g., for strongly coupled systems in condense matter physics, which we do not discuss.

[4]This subsection can be skipped in first reading.

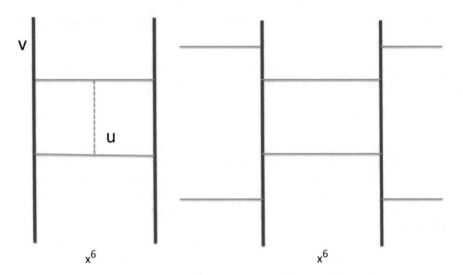

Fig. 18.1 The vertical thick lines are 5-branes; this direction corresponds to complexified coordinate v. The horizontal lines (along x^6) are 4-branes, we show two of them which correspond to SU(2) color. Left picture has adjoint matter, and the forces on the 5-branes are unbalanced: thus the branes are logarithmically bent at large v. Right picture has four external lines depicting four flavors of fundamentally charged matter: those balanced the forces on 5-branes and make its position at large v balanced, which thus produces a conformal theory with fixed coupling. Thin dashed line on the left connecting 4-branes will correspond to adjoint matter particle (or monopole): its length is associated with the Higgs VEV a (or dual a_D)

connected by some lower dimensional branes, now of the dimension 4.[5] They have infinite extension in coordinates $(0,1,2,3)$ and finite in x^6, namely, between the positions of the $D5$ $\alpha_i < x^6 < \alpha_{i+1}$. It is depicted in the Fig. 18.1 in which x^6 is the horizontal direction and the complexified coordinate $v = x^4 + ix^5$ runs vertically. (One may think of x^6 to be also complexified together with the 11th coordinate x^{10} into

$$t = exp\left[-\left(x^6 + ix^{10}\right)/R\right] \tag{18.4}$$

where R is the radius of a circle to which the latter can be compactified. (This variable is then single-valued.)

The horizontal branes would pull the vertical ones, and if the tensions are not balanced (as in the left picture), they would displace the vertical branes. Furthermore, they would create certain "dimples" on the $D5$. At large distance (in

---

[5] Type IIB string theory has odd-dimensional branes which naturally generate gauge fields/theories, and so far we dealt with them only. Dimension 4 branes lead to multi-index forms, and to get a gauge field out of it is done by another procedure related with nonzero holonomy for those forms: see the paper.

$|v|$) from such dimple, the curve $x^6(v)$ must satisfy the Laplace equation

$$\nabla^2 x^6(v) = 0 \tag{18.5}$$

which in two dimensions (one complex v) produce solution

$$x^6 \sim log(v) \tag{18.6}$$

which is not constant even at large v. Adding or subtracting those changes the coefficient of the $log(v)$: this is a change of a perturbative beta function.

If tensions are balanced (like the right picture with four flavors of fundamental matter), a conformal theory emerges. (In fact this paper leads to many new conformal models.) The positions of those horizontal branes in v correspond to masses of all those flavors.

The four-dimensional world and also the four-dimensional gauge theory appear in the worldvolume of horizontal 6-branes: and their mutual "Higgsing" depends in the (vertical) distance in v (see the dashed line) and the log corresponds to perturbative running of that coupling.

Lower-dimensional ones (4-branes) can be considered as becoming narrow tubes made of 5-branes. Furthermore, simple plots in the figure are for rigid strings, represented by the straight lines. In fact they are flexible and will bend to get to a minimal energy shape. (Its zero Laplacian is obtained, as usual, by complexification and then using function depending on z but not its conjugate z^*.)

Let us imagine the shape of the 5-branes is given by some equation. Since there are two of them, it should be quadratic

$$A(v)t^2 + B(v)t + C(v) = 0 \tag{18.7}$$

where the power of v in the coefficient is k, the number of horizontal branes (e.g., two in our left figure but six on the right one). Zero t corresponds to infinite x^6: so roots of $C(v)$ are right-going branes. A root of $A(v)$ is $t = \infty$ or $x^6 = -\infty$, so those count lines going to the left.

Suppose we don't want any flavors of fundamental matter: then A, C can be constants, rescaled to the equation

$$t^2 + B(v)t + 1 = 0 \tag{18.8}$$

which can be further rescaled by $\tilde{t} = t + B/2$ into a form

$$\tilde{t}^2 = \frac{1}{4}B(v)^2 - 1 \tag{18.9}$$

with B being degree-k polynomial, which is a known curve for SU(k) SYM. Adding fundamental matter is then making A, C non-constant.

In summary, this brane construction leads to explicit *derivation* of the Seiberg-Witten curve, while they had to guess it in their original paper.

18.4 Brane Interactions

Closed strings form certain massless bosonic excitations, described by fields in the $d = 10$ dimensional external string space. Those are the graviton $g_{\mu\nu}$, dilaton ϕ, antisymmetric tensor $B^{\mu\nu}$, and vector field A_μ. The corresponding classical (tree level) effective action is

$$S = \frac{1}{2\kappa_0^2} \int d^{10}x e^{-2\phi} \left(R + 4(\nabla\phi)^2 - \frac{1}{12} H_{\mu\nu\kappa} H^{\mu\nu\kappa} \right) - \frac{c}{4} e^{-\phi} F_{\mu\nu} F^{\mu\nu}$$

(18.10)

The constants κ_0, c as well as loop corrections of higher orders in α' can in principle be calculated. Here R is the scalar curvature defined via the metric tensor (Einstein-Hilbert action of general relativity), and H, F are the field strength tensors for B, A fields. Supersymmetric extension of this action makes full classical ten-dimensional supergravity (SUGRA).

Any objects existing in the theory interact at large distances via an exchange of those massless fields. D-branes are such objects, and in fact they do have certain mass and charge densities (in respect to ϕ, A, B fields).

Let us now be more specific and discuss the fields induced by the branes around them in the bulk. As they have no extension in some coordinates (e.g., just mentioned $x_4...x_9$ coordinates for D_3 brane), their fields are that of a point objects: in general relativity, this naturally implies that they are black hole-like in such coordinates. As they are extended in other coordinates, the proper name of such objects is *black branes*. To find out their properties in classical (no loops or quantum fluctuations) approximation, one has to solve coupled Einstein-Maxwell-Scalar equations, looking for static spherically symmetric solutions. This is no more difficult than to find the Schwarzschild solution for the usual black hole, which we remind has a metric tensor in spherical coordinates

$$ds^2 = g_{\mu\nu}dx^\mu dx^\nu = -(1 - r_h/r)dt^2 + \frac{dr^2}{(1 - r_h/r)} + r^2 d\Omega^2 \qquad (18.11)$$

and the horizon radius (in full units) is $r_h = 2G_N M/c^2$, containing the mass M and Newton constant G_N. The horizon—zero of g_{00}—is the "event horizon": a distant observer cannot see beyond it.[6] Similar expressions, with appropriate powers of the distance can be derived for p-branes.

Note that at large distance $r \gg r_h$, deviations from flat metrics are a small correction—there the nonrelativistic potential description of Newtonian gravity is possible. Similarly far from a brane, the fields are just given by Newton+Coulomb+scalar formulae in corresponding dimensions, and it is not hard to figure out what are their mutual interactions. It is important that these densities are related in a nontrivial way. Like the BPS magnetic monopoles we had studied above, the D-branes are BPS objects. It means attractive forces (Newtonian due to gravity as well as to the dilaton exchange) and repulsive Coulomb forces due to vector exchanges cancel each other, so that two parallel D-branes do not interact. (The D and anti-D are not BPS.) As a result, a "brane engineer" may consider a number of parallel branes to be put at some random points: and they will stay there. A large ($N_c \to \infty$) number of branes put into the same point combine their mass and thus create strong gravity field, justifying the use of classical Einstein-Maxwell equations. Open string states, which keep their ends on some branes i and j, lead to effective gauge theory on the brane with the $U(N_c)$ group.

(The BPS objects still have quantum interaction and can form bound states. In particular electric particles (quarks, gluons) as well as monopoles and dyons can form positronium-like series of Coulombic bound states. Explicit examples are, e.g., well studied in the context of $\mathcal{N} = 2$ SYM, the Seiberg-Witten theory. Generalization of all of that to multidimensional SUGRA had also been studied: in this case of course, gravity is added to Coulombic electric-magnetic forces as well as (massless) Yukawa ones. Tuning the parameters, one may make gravity to become even dominant: the corresponding bound states have been called "galaxies" in such limits.)

When the black hole has a nonzero vector charge, the Gauss' law insists on constant flux through sphere of any radius; thus there are nonzero fields at large r, and Schwarzschild solution gets modified into a "charged black hole." For asymptotically flat spaces, there is the famous statement, much emphasized by Wheeler: there are "no scalar hairs" of black holes, as there is no Gauss theorem to protect them. This happens to be not true in general, and for spaces which are AdS-like, thus scalars should *not* be left over.[7]

The solution for D_3 brane happens to be the so-called *extremal* (six-dimensional) charged black hole, which has the lowest possible mass for a given charge. When the mass decreases to the extreme value, the horizon shrinks to nothing, and thus it is especially simple. Spherical symmetry in six dimensions allows one to separate the five angles (making the five-dimensional sphere S_5) from

---

[6]Note that both metric elements g_{00}, g_{rr} change sign at the horizon, so in a way, the time and space exchange places. It indicates that the Schwarzschild-like metrics cannot be used beyond horizon. Some other versions of metric however can do it: see GR textbooks.

[7]In fact there is a whole recent direction based on solution with a "scalar atmosphere" around black branes, providing gravity dual to superconductors on the boundary.

the radius r (in six dimensions orthogonal to "our world" space $x_1, x - 2, x - 3$) and write the resulting ten-dimensional metrics as follows:

$$ds^2 = \frac{-dt^2 + dx_1^2 + dx_2^2 + dx_3^2}{\sqrt{1 + L^4/r^4}} + \sqrt{1 + L^4/r^4}\left(dr^2 + r^2 d\Omega_5^2\right) \qquad (18.12)$$

Again, at large r, all corrections are small, and we have the asymptotically flat space there.

18.5   AdS/CFT Correspondence

AdS/CFT correspondence (Maldacena 1999) is a *duality* between a specific conformal field theory (CFT, $\mathcal{N} = 4$ super-Yang-Mills theory in four dimensions), and the (ten-dimensional) superstring theory in anti-de-Sitter (AdS) metrics.

Step one toward the AdS_5 space starts from the black hole metric (18.12) which can be in certain region be simplified. In fact we will only need the metric in the "near-horizon region," at $r << L$, when 1 in both roots can be ignored. If so, in the last term, two r^2 cancel out, and the five-dimensional sphere element gets constant coefficient and thus gets decoupled from five other coordinates. Thus quantum numbers or motion in S^5 becomes kind of internal quantum numbers like flavor in QCD, and it will be mostly ignored from now on. What is left is a very simple five-dimensional metric known as anti-de-Sitter metric. Using a new coordinate $z = L^2/r$, we get it into the "standard AdS_5 form" used below:

$$ds^2 = \frac{-dt^2 + dx_1^2 + dx_2^2 + dx_3^2 + dz^2}{z^2} \qquad (18.13)$$

Note that z counts the distance from "the AdS boundary" $z = 0$. This metric has no scale and is *not* asymptotically flat even at the boundary. Performing dilatation on these five coordinates, we find that the metric remains invariant: in fact one can do any conformal transformation. It is this metric which is AdS in the AdS/CFT correspondence, and string theory in this background is "holographically gravity dual" to some conformal gauge theory at the boundary.

So far nothing unusual happened: all formulae came straight from string and general relativity textbooks. A truly remarkable theoretical discovery is the so-called holography: the exact duality (one-to-one correspondence) between the five-dimensional "bulk" effective theory in AdS_5 to 4-dim "boundary" ($r \rightarrow \infty$) gauge theory. There is a dictionary relating any observable in the gauge theory to another one in string theory: the duality implies that all answers are the same in both formulations. We will see below how it works "by examples."

The last step, which makes it useful, is the Maldacena relations between the gauge coupling, the AdS radius L, and the string tension α' (which comes from the total mass of the brane set):

$$L^4 = g^2 N_c (\alpha')^2 = \lambda (\alpha')^2 \qquad (18.14)$$

It tells us that large gauge coupling $\lambda \equiv g^2 N_c >> 1$ corresponds to large AdS radius (in string units), and one can use classical (rather than quantum) gravity. At the same time, the string and gravity couplings $g_s \sim g^2$ may remain small: so one may do perturbative calculations in the bulk!

At this point, many readers are probably very confused by the new fifth dimension of space. One way to think of it is as just another mathematical trick, somewhat analogous to more familiar introduction of the complex variables.[8] However, there is a perfectly physical meaning of the fifth coordinate. One hint is provided by the fact that distance along it $\int_a^b dl = \int_a^b dz/z = log(b/a)$ is the logarithm of the ratio. Thus its meaning is the "scale," the argument of the renormalization group. If one takes a bulk object and moves it into larger z, its hologram at the boundary ($z = 0$) grows in size: this direction thus corresponds to the infrared direction. The running coupling constant would thus be treated as a z-dependent field called "dilaton." Indeed, there are theories with gravity dual, in which this field (related to the coupling) does "run" in z: unfortunately, known examples do not (yet?) include QCD! In spite of that, there are efforts to build its gravity dual "bottom-up," introducing weak coupling at ultraviolet (small z) (Shuryak 2007) and confinement in infrared (large z) (Karch et al. 2006b; Gursoy and Kiritsis 2008) by certain modification of the rules.

The simplified AdS_5 metric above is very simple; it apparently has no scale and has a conformal symmetry matching that of the effective boundary theory which is $\mathcal{N} = 4$ SYM. That observation was the basis for Maldacena AdS/CFT correspondence. Let me start with our first example of the "AdS/CFT at work," related with the strong-coupling version of the Coulomb law. The setting—to be called "the Maldacena dipole"—shown in Fig. 18.2 includes two static charges (heavy fundamental quarks) separated by the distance L.

At weak coupling—the usual QED—we think of one charge creating the electric potential in which the other is placed, leading to the usual Coulomb law which in our notation is

$$V(L) = -\frac{g^2}{4\pi} \frac{1}{L} \qquad (18.15)$$

In the $\mathcal{N} = 4$ theory at *weak* coupling, the only difference is that one can exchange massless scalars on top of gluons. It is always attractive and for two

---

[8] Suppose an experimentalist measured some complicated cross section which is approximately a sum of Breit-Wigner resonances. His friend phenomenologist may be able to write the answer as an analytic function with certain pole singularities in the complex energy plane, which will help for fitting and for evaluating integrals. Even better, their other friend theorist cleverly developed a "bulk theory," deriving the pole positions from some interaction laws on the complex plane.

Fig. 18.2 Setting of the
Maldacena dipole: two
charges at the boundary
(black dots) are connected by
the string (shown by solid
curve) pending under gravity
toward the AdS center
$z \to \infty$. Classical graviton
propagator (the dashed line)
should be used to calculate
the "hologram"—the stress
tensor at the observation point
y. The string is the gravity
source; the point A has to be
integrated over

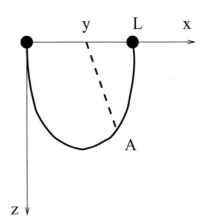

heavy quarks leads to cancellation of the force, with doubling for quark-antiquark
(we discuss now). The QED coupling g^2 changes to 't Hooft coupling $\lambda = g^2 N_c$
proportional to the number of colors N_c.

Now we turn to the AdS/CFT for the $\mathcal{N} = 4$ theory at *strong* coupling
(Maldacena 1998). The electric flux in the bulk forms a singular object—the
fundamental or F string (shown by the solid curve in Fig. 18.2)—which bends from
the boundary $z = 0$ due to gravity force into the fifth dimension, like in the famous
catenary (chain) problem.[9] The calculation thus follows from Nambu-Goto action
for the string, whose general form is

$$S = \frac{1}{2\pi\alpha'} \int d\sigma d\tau \sqrt{det\left(G_{MN}\partial_\alpha X^M \partial_\beta X^N\right)} \tag{18.16}$$

where determinant is in indices $\alpha, \beta = 0, 1$ numerating internal coordinates σ, τ
parameterize the string world line $X^M(\tau, \sigma)$. Here M, N are space-time indices in
the whole space (ten dimensions reduced to five dimensions in AdS/CFT). G_{MN} is
the space metric and det stands for 2*2 matrix with all α, β. In the AdS_5 metric,
we need the components $-G_{00} = G_{11} = G_{55} = 1/z^2$, and we can think of σ, τ as
our coordinates x, t: the string is then described by only one function $z(t, x)$, and
its action is reduced to

$$S \sim \int dt dx \frac{1}{z^2} \sqrt{1 + (\partial z/\partial x)^2 - (\partial z/\partial t)^2} \tag{18.17}$$

---

[9]Another—more Einsteinian—way to explain it is to note that this is simply the shortest string
possible: it is not straight because the space is curved. It is the same reason why the shortest path
from New York to London does not look straight on the map.

We will use this action for "falling strings" below and now proceed to further simplifications for static string, for which there is no time derivative and the function is $z(x)$. Maldacena uses $u(x) = 1/z(x)$, and thus the Lagrangian becomes $L = \sqrt{(u_{,x})^2 + u^4}$ with comma meaning the x-derivative. One more simplification comes from the fact that x does not appear in it: thus an "energy" is conserved

$$H = p\dot{q} - L = \frac{\partial L}{\partial u_{,x}} u_{,x} - L = E = const \tag{18.18}$$

which reduces the EOM from second-order equation to just $(u_{,x})^2 = u^4(u^4/E^2 - 1)$ which can finally be directly integrated to

$$x(u) = \int_{u_m}^{u} \frac{du'}{(u')^2\sqrt{(u')^4/E^2 - 1)}} \tag{18.19}$$

The minimum position of the string u_m is related to E by the relation following from this formula at $x = L/2$. Plugging the solution back into action and removing divergence (which is independent of L) give finally the total string energy, which is the celebrated *new Coulomb law at strong coupling*:

$$V(L) = -\frac{4\pi^2}{\Gamma(1/4)^4} \frac{\sqrt{\lambda}}{L} \tag{18.20}$$

The inverse power of the distance $1/L$ is in fact the only one possible by dimension, as the conformal theory has no scales of its own. What is remarkable is the $\sqrt{\lambda}$ appearing here instead of just λ, which we find in the Coulomb law in the weak coupling. (The numerical coefficient in the first bracket is 0.228, to be compared to the result from a diagrammatic re-summation below.)

What is the reason for this modification? For pedagogical reasons, let me start with two "naive but reasonable guesses," both to be shown to be wrong later in this section:

(1) One idea is that strongly coupled vacuum acts like some kind of a space-independent dielectric constant, $\epsilon \sim 1/\sqrt{\lambda}$ which is reducing the effect of the Coulomb field, similarly at all points.
(2) Perhaps such dielectric constant has nonlinear effects, and thus is not the same at different points: but the fields created by static dipole are still just the electric field $\vec{E}$.

As we get glimpse of some first results from AdS/CFT, we see that they are quite different from those in weak coupling. One would like to understand them better, both from the bulk (gravity) side and from the *gauge theory* side.

But before we discuss this problem, we perhaps need to recall how this problem is solved for the familiar Coulomb forces in weak coupling. Since we consider a static (time-independent) problem, one can rotate time t into its Euclidean version $\tau = it$, which simplifies calculation of the perturbative diagrams. A static charge generates a Wilson line operator

$$W = Pexp\left(ig \int d\tau A_0\right) \tag{18.21}$$

and the negative charge produces a conjugated line: the two are separated by the distance L. The exponents can be expanded in powers of the coupling g: the lowest order is the first one. The correlator of the two A fields is the propagator. Its endpoints resign at the two lines, thus a double integral running over both lines:

$$V(L)(T_{max}) \sim -g^2 \int_0^{T_{max}} \frac{d\tau_1 d\tau_2}{L^2 + (\tau_1 - \tau_2)^2} \sim -\frac{g^2 T_{max}}{L} \tag{18.22}$$

where large regulator time T_{max} is introduced. The denominator—the square of the Euclidean distance between the gluon emission and absorption points—comes from the Euclidean propagator in flat $3 + 1$ space-time. Note that the propagation time of a virtual exchange gluon is of the order of the distance $(\tau_1 - \tau_2) \sim L$: this cancels one power of L, and thus one gets the usual Coulomb potential $1/L$.

It turns out the first is relatively easy. For example, both in the total energy and energy density, we get $\sqrt{\lambda}$ because this factor is in front of the Nambu-Goto Lagrangian (in proper units). The reason the field decays as y^7 is extremely natural: in the AdS_5 space-time, the field of a static object (a single time integral over the five-dimensional propagator, analogous to Coulomb $1/r$ in flat 3d) has that very power of distance. In fact it is

$$P_s = \frac{15}{4\pi} \frac{z^2}{(z^2 + r^2)^{\frac{7}{2}}} \tag{18.23}$$

with z being the fifth coordinate of the source and r the three distances between it and the observation point. Thus it is likely to be a bulk gravity/scalr exchanges rather than the vector fields. (Note that gravity/scalar forces have no cancellations between opposite charges, unlike the vector ones.)

In order to understand the same results from the gauge side, we will need a bit of pedagogical introduction: the resolution will be given by the idea of *short color correlation time* by Shuryak and Zahed (2004a). In QCD, with its running coupling, higher-order effects modify the zeroth order Coulomb field/potential.

Higher-order diagrams include self-coupling of gluons/scalars and multiple interactions with the charges. A famous simplification proposed by 't Hooft is the large number of color limits in which only *planar* diagrams should be considered. People suggested that as g grows, those diagrams are becoming "fishnets" with smaller and smaller holes, converging to a "membrane" or string worldline: but although this idea was fueling decades of studies trying to cast gauge theory into stringy form, it has not strictly speaking succeeded. It may still be true: just nobody was smart enough to sum up all planar diagrams.[10]

If one does not want to give up on re-summation idea, one may consider a subset of those—the *ladders*—which can be summed up. Semenoff and Zarembo (2002) have done that: let us look what have they found. The first point is that in order that each rung of the ladder contributes a factor N_c, emission time ordering should be strictly enforced, on each charge; let us call these time moments $s_1 > s_2 > s_3...$ and $t_1 > t_2 > t_3....$ Ladder diagrams must connect s_1 to t_1, etc.; otherwise it is nonplanar and subleading diagram. Thus the main difference from the Abelian theory comes from the dynamics of the color vector. The (re-summed) Bethe-Salpeter kernel $\Gamma(s, t)$, describing the evolution from time zero to times s, t at two lines, satisfies the following integral equation:

$$\Gamma(\mathcal{S}, \mathcal{T}) = 1 + \frac{\lambda}{4\pi^2} \int_0^{\mathcal{S}} ds \int_0^{\mathcal{T}} dt \frac{1}{(s-t)^2 + L^2} \Gamma(s, t) \tag{18.24}$$

If this equation is solved, one gets re-summation of all the ladder diagram. The kernel obviously satisfies the boundary condition $\Gamma(\mathcal{S}, 0) = \Gamma(0, \mathcal{T}) = 1$. If the equation is solved, the ladder-generated potential is

$$V_{\text{lad}}(L) = - \lim_{\mathcal{T} \to +\infty} \frac{1}{\mathcal{T}} \Gamma(\mathcal{T}, \mathcal{T}) , \tag{18.25}$$

In weak coupling, $\Gamma \approx 1$ and the integral on the rhs is easily taken, resulting in the usual Coulomb law. For solving it at any coupling, it is convenient to switch to the differential equation

$$\frac{\partial^2 \Gamma}{\partial \mathcal{S} \partial \mathcal{T}} = \frac{\lambda/4\pi^2}{(\mathcal{S} - \mathcal{T})^2 + L^2} \Gamma(\mathcal{S}, \mathcal{T}) . \tag{18.26}$$

and change variables to $x = (\mathcal{S} - \mathcal{T})/L$ and $y = (\mathcal{S} + \mathcal{T})/L$ through

$$\Gamma(x, y) = \sum_m \mathbf{C}_m \gamma_m(x) e^{\omega_m y/2} \tag{18.27}$$

with the corresponding boundary condition $\Gamma(x, |x|) = 1$. The dependence of the kernel Γ on the relative times x follows from the differential equation:

$$\left(-\frac{d^2}{dx^2} - \frac{\lambda/4\pi^2}{x^2 + 1} \right) \gamma_m(x) = -\frac{\omega_m^2}{4} \gamma^m(x) \tag{18.28}$$

---

[10]Well, AdS/CFT is kind of a solution, actually, but it is doing it indirectly.

For large λ, the dominant part of the potential in (18.28) is from *small* relative times x resulting into a harmonic equation (Semenoff and Zarembo 2002)

At large times $\mathcal{T}$, the kernel is dominated by the lowest harmonic mode. For large times $\mathcal{S} \approx \mathcal{T}$ that is small x and large y

$$\Gamma(x, y) \approx \mathbf{C}_0\, e^{-\sqrt{\lambda}x^2/4\pi}\, e^{\sqrt{\lambda}y/2\pi} . \tag{18.29}$$

From (18.25) it follows that in the strong coupling limit, the ladder-generated potential is

$$V_{\text{lad}}(L) = -\frac{\sqrt{\lambda}/\pi}{L} \tag{18.30}$$

which has *the same parametric form* as the one derived from the AdS/CFT correspondence (18.20) except for the overall coefficient. Note that the difference is not so large, since $1/\pi = 0.318$ is larger than the exact value 0.228 by about 1/3. Why did it happen that the potential is reduced relative to the Coulomb law by $1/\sqrt{\lambda}$? It is because the relative time between gluon emissions is no longer $\sim L$, as in the Abelian case, but reduced to parametrically small time of relative color coherence $\tau_c \sim 1/L\lambda^{1/2}$. Thus we learned an important lesson: in the strong coupling regime, even the static charges communicate with each other via high-frequency gluons and scalars, propagating (in Euclidean formulation!) with a super-luminal velocity $v \approx \lambda^{1/2} \gg 1$.

This idea—although it seemed to be too bizarre to be true—we will see below to explain some of the AdS/CFT results. Klebanov et al. (2006) have pointed out that the reason the stress tensor around the dipole is different from perturbative one by a factor $\sim (L/r)/\sqrt{\lambda}$ is actually explained by limited relative emission time by color coherence time. I am sure possible usage of this idea does not end here.

Before we leave the subject of Maldacena dipole, one more interesting question is what happens if one of the charges makes a small accelerated motion near its original position. Will there be a radiation? In a somewhat different setting than used above, the answer was provided by Mikhailov (2003). He studied perturbations of the string and found that the radiated energy is described by familiar classical Liénard formula

$$\Delta E = A \int_{-\infty}^{\infty} \frac{\ddot{x}^2 - \left[\dot{x} \times \ddot{x}\right]^2}{\left(1 - \dot{x}^2\right)^3} dt \tag{18.31}$$

in which QED weak coupling constant $A = \frac{2}{3}e^2$ is substituted by CFT strong coupling $A = \frac{\sqrt{\lambda}}{2\pi}$.

18.6 Holography at Work

The first topic we will focus on is the *bulk-boundary relation*. Let us recall the discussed setting: there is the time X^0, plus p "internal" brane coordinates, plus $d - p$ "external" coordinates in which the brane can move.

Imagine some creatures which live inside the internal coordinates and are not aware of the external ones. We of course do live on the surface of the Earth, with $p = 2$, but we can see up and down and are aware of stars above, so at this stage, one should imagine, e.g., some blind insects which cannot jump or dig in, with only these two coordinates imprinted in their branes. Suppose these creatures are intelligent enough to develop their theory of the world, obviously $p + 1$ dimensional, describing what happens in it. Say consequences of earthquakes or fallen meteorites, while originated externally, create displacements of the surface which can be described by certain scalar and vector fields as we described above. Perhaps our creatures can send surface vibrations themselves and communicate by those as we use sounds. Perhaps they can even travel and understand that Earth is topologically a sphere and even develop a theory about existence of the radial "extra dimension," unifying their previously separate scalar and vector fields into common vector displacement fields in $3 + 1$ dimensions.

(It may be that our four dimensions are a brane in larger dimensions, and events we observe like our Big Bang have in fact simple external causes. There are many fascinating papers about that, but this text is not a place to follow this, at this point, purely speculative possibility.)

The second statement is truly unusual; it deals with the "holography" between the near-horizon region of the bulk and the observers at the boundary of the black holes. It historically started with the discussions of the information paradox of classical gravity, which states that information can fall into a black hole and get lost for the outside observers.[11]

Hawking famously found that black holes should emit thermal radiation, with T given by inverse total mass M. Bekenstein argued that since in thermodynamics $dM = T dS$, one can evaluate the *universal amount of entropy* produced when you increase the black hole mass by dM.[12] The black hole entropy scales as its area, so it is expected to be stored in the vicinity of the horizon. These developments lead to an idea that some *holographic*[13] relation should exist between the dynamics/history observed far from black hole in d dimensional space and its "compactified form" (by strong gravity) in the $d - 1$ dimensional region near horizon.

---

[11] Loss of information, which is the same as entropy growth, is of course rather standard phenomenon in statistical mechanics. Everyone knows that if you throw a book into a bonfire, the information contained in it is lost. Same may happen if you drop a book into a black hole. However, from quantum mechanical point of view, the pure state should remain pure state, so the information should be somehow recoverable. We will not discuss huge literature on this "information paradox."

[12] So, dropping your Ph.D. thesis weighing 1 kg, or an equivalent 1 kg stone, produces the same amount of entropy!

[13] A hologram keeps all 3-d information in a 2d plane.

The holographic AdS/CFT duality is in fact a precise correspondence between the two, namely, that *certain* boundary and near-horizon theories can be *dual* to each other, in the sense of complete/exact equivalence. The boundary effective theory is not only applicable to describe something like long-wavelength sounds, as is always the case, but it is in fact an exact copy of the bulk quantum theory. AdS/CFT correspondence is believed to be one of such examples: many tests have been performed and they all passed well, yet we don't have a formal proof of it.

18.6.1 A Hologram of a Point Space-time Source: An *Instanton*!

Instead of discussing deep issues related to holography, let me demonstrate few examples of how one uses it in applications. Our first example would be an image of a point space-time event—sometimes called the D_{-1} brane because its word history has dimension $p + 1 = 0$—seen on the boundary (see Fig. 18.3a). Since a point cannot produce vector or tensor field (it lacks directions for the indices), the only light bulk field it can emit are the two scalars, the axion and the dilaton. It thus immediately follows that the holographic image would be lacking any energy and momentum, since only $g_{\mu\nu}$ can induce the stress tensor $T_{\mu\nu}$, but there would be nonzero values of the operators G^2 and $G\tilde{G}$. If the object has the same couplings to both, the images may be also equal (perhaps up to a sign).

In order to calculate this simplest diagram, one needs bulk-to-boundary propagators in AdS_5 which we discuss in Appendix. In the usual coordinates, the metric is

$$ds^2 = \frac{dX^\mu dX_\mu + dz^2}{z^2} \tag{18.32}$$

with $\mu = 0..3$ and convolution done with the usual Minkowski metric, with minus one for the zeroth component. Feynman (the analytically continued Euclidean) propagator is expected to be just the function of the invariant distance $d(X, X')$ between the two points, which in this space is

$$cosh(d(X, X')) = 1 + \frac{z^2 + z'^2 + \sum_{i=0}^{p}(X^p - X'^p)^2}{zz'} \tag{18.33}$$

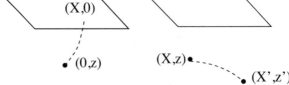

Fig. 18.3 (a) Setting of the bulk point event (black dots) at distance z from the boundary at the boundary, or (b) two such events (black dots)

in fact the propagator is just a power of the combination in the r.h.s. above, namely,

$$D(X, X') \sim \frac{z^4 z'^4}{\left[X^\mu X_\mu + z^2 + z'^2 \right]^4} \qquad (18.34)$$

The "holography" is a projection from the (d+1 dimensional) bulk to (d-dimensional) boundary. It is done in two steps: (1) one should take one point to the boundary, $z' \to 0$; (2) the power of z' of the amplitude corresponding to the dimensionality of the boundary operator in question should be "amputated."

Example 1 Both operators G^2 and $G\tilde{G}$ have mass dimension 4; thus one should amputate exactly the 4 powers of z' in the propagator. Thus we recover the distributions of these quantities at the boundary:

$$G^2(X), G\tilde{G}(X) \sim \frac{z^4}{\left[X^\mu X_\mu + z^2 \right]^4} \qquad (18.35)$$

So, it now becomes clear that the object found on the boundary, as *an image of the bulk point event*, is nothing else but the *gauge field instanton*! The fifth coordinate of the bulk point z is nothing else but the instanton radius ρ, confirming the idea that it is a "scale" variable.

Note further that a bulk point even cannot be couple to gravitons, as a point lacks needed indices to make a tensor field. In the example at hand, it implies that gauge instantons must have *zero stress tensor* $T^{\mu\nu} = 0$ at any point, which is also true and nontrivial.[14]

Relations between the gauge theory instantons in $\mathcal{N} = 4$ SYM and the AdS_5 space have been noticed independently and simultaneously with the AdS/CFT correspondence, while studying the semiclassical (weak coupling) instantons in Dorey et al. (1999), see also a review (Dorey et al. 2002). Basically, the instanton moduli space includes the AdS_5 space, if $\rho = z$ is the fifth coordinate. For example, the volume element of the AdS_5 metric (18.13) is

$$\sqrt{det(g)} = \frac{1}{\rho^5} \qquad (18.36)$$

which is a factor well familiar from the semiclassical instanton measure. What this agreement tells us is that instantons of the $\mathcal{N} = 4$ SYM populate the internal AdS_5 space *homogeneously*.

The same logic then implies that the *interaction between instanton-anti-instanton pair* is given by the diagram Fig. 18.3b in which one should use the same bulk-

---

[14]Note that the fact that instantons describe classical tunneling paths only requires zero stress globally, not locally everywhere.

to-bulk scalar propagator in AdS_5, which is the function of "geodesic distance" between two points in the bulk. One indeed obtains this nontrivial result, now in much simpler (and "geometrically motivated") way than we did in the instanton chapter.

Let us complete this example with one important comment: the "holography" just described cannot give us boundary gauge field A_μ of the instanton (or in fact anything having a color index): we only get colorless operators $G^2(X)$, $G\tilde{G}(X)$ and a statement about $T_{\mu\nu} = 0$ which are coupled to bulk massless fields.

18.6.2 A Hologram of the Maldacena Dipole

In the leading order in weak coupling, the only field is electric, and the field of a dipole is just a linear superposition of the two Coulomb fields of the charges:

$$E_m(y) = \left(\frac{g}{4\pi}\right)\left(\frac{y_m - (L/2)e_m}{|y_m - (L/2)e_m|^3} - \frac{y_m + (L/2)e_m}{|y_m + (L/2)e_m|^3}\right)$$

Because of the cancellation between these two terms, at large distances from the dipole, the field decays as $E \sim L/y^3$. The corresponding energy density (and other components of the stress tensor) is thus of the order $T_{00} \sim E^2 \sim g^2 L^2/y^6$.

In the strong coupling, one has to calculate the hologram of the "pending string" construction explained above. The stress tensor of matter $< T_{\mu\nu}(y) >$ at any point y on the boundary comes from the gravitational field of that string, calculated in the bulk and then approximated near the boundary. Lin and Shuryak (2007) have found it: the solution is rather technical to be presented here in full, and even the resulting stress tensor expressions are rather long. Let me just comment that there are some general requirements which should be satisfied, used to verify explicitly that no mistake in the calculation was made. The stress tensor should be traceless $T_\mu^\mu = 0$ because the boundary theory is conformal, and it should have zero covariant divergence $T_{\mu\nu}^{;\nu}$ because there are no sources away from the charges themselves. Let me just illustrate one point, using the leading large-distance term, at the $y >> L$

$$T_{00} = \sqrt{\lambda}L^3\left(\frac{C_1 y_1^2 + C_2 y^2}{|y|^9}\right) f(\theta) \tag{18.37}$$

where C_1, C_2 are numerical constants whose values can be looked up in the paper and $f(\theta)$ is the angular distribution. It is shown by solid line in Fig. 18.4, to be compared to that in weak coupling (the dashed line).

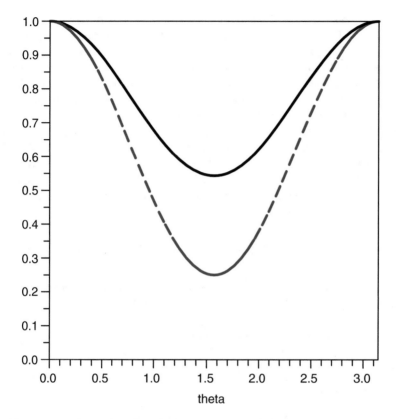

Fig. 18.4 Angular distribution of the far field energy versus the polar angle ($cos(\theta) = y_1/|y|$). Solid black line is the AdS/CFT result, compared to the perturbative dipole energy $(3cos^2\theta + 1)/4$ (the dashed blue line), both normalized at zero angle

18.6.3 A Hologram of a Particle Falling in the Bulk

The example in this subsection can be put in a setting quite different in scope from AdS/CFT correspondence. Now we ask the reader to imagine that instead of using it as a mathematical tool (similar in spirit to extending functions to complex plane), we for a moment assume it to be the true picture of the physical universe. Specifically, imagine that we indeed live on the D_3 brane, where all gauge fields of the standard model are defined, but there are more dimensions of space and gravity act "in bulk."

What would be our perception (the hologram) of a *bulk particle*—so to say a bulk meteorite—flying by? Of course any particle[15] must fly on some geodesic, as a stone thrown into a nonzero gravity field. For example, at $t < 0$, it can fly

---

[15]The example before was a point even in the bulk or D_{-1} brane. A particle has a path; it is a D_0 brane, no space extension but with time-dependent path.

toward the boundary in z direction, then stop at $t = 0$ at some distance from it, and then fall back to the AdS center. Since "stone" has a nonzero stress tensor $T_{\mu\nu}^{stone} = m\dot{X}^\mu \dot{X}^\nu$, it can be coupled to the bulk metric perturbations $\delta g_{\mu\nu}$ and, via the gravitational propagator, produce a holographic stress tensor on the boundary. The hologram can again be obtained by finding the bulk $\delta g_{\mu\nu}(z, X)$, expanding it at small z, and reading the coefficient of the fourth power of z, as we did above. The result is a spherically symmetric implosion and then explosion.

However, this time we will use a different method, using this occasion to learn a bit more about geometry of the AdS_5 space. We will follow Friess et al. (2007) which uses the so-called global coordinates. Recall that, e.g., a D-dimensional sphere of radius R can be described by its imbedding into the $D + 1$ space, in which it is defined by the equation $(X^1)^2 + (X^D)^2 = R^2$. Similarly, the AdS_5 space can be described by the equation

$$-\left(X^{-1}\right)^2 - \left(X^0\right)^2 + \left(X^1\right)^2 + \left(X^2\right)^2 + \left(X^3\right)^2 + \left(X^4\right)^2 = -L^2 \qquad (18.38)$$

in the six-dimensional space $X^{-1}..X^4$. (Note, that there are *two* negative signs!) The relation to the coordinates, the usual ones for hyperboloids, is

$$X^{-1} = \sqrt{L^2 + \rho^2}\cos\left(\frac{\tau}{L}\right), \quad X^0 = \sqrt{L^2 + \rho^2}\sin\left(\frac{\tau}{L}\right), \quad X^i = \rho\Omega^i \qquad (18.39)$$

The last term contains the coordinates of the three-dimensional unit sphere, with standard line element:

$$d\Omega^2 = d\chi^2 + \sin^2\chi \left(d\theta^2 + \sin^2\theta d\phi^2\right) \qquad (18.40)$$

The point is there exist another set, known as Poincare coordinates, defined by $(i = 1..3)$

$$X^{-1} = \frac{z}{2}\left(1 + \frac{L^2 + \vec{x}^2 - t^2}{z^2}\right), \quad X^0 = L\frac{t}{z}, \qquad (18.41)$$

$$X^i = L\frac{x^i}{z}, \quad X^4 = \frac{z}{2}\left(-1 + \frac{L^2 - \vec{x}^2 + t^2}{z^2}\right)$$

One can eliminate global coordinates and get the relation between these two sets, $\tau, \rho, \chi, \theta, \phi$, and $t, \vec{x}$ (left as an exercise).

Understanding of the construction can be started from the boundary, the large $\rho \gg L$ limit in which all $X^A \sim \rho$. Here τ just runs a circle and Ω is a three-dimensional sphere with constant spatial curvature. This world is known as "Einstein's static universe." The Poincare coordinates at large z are just Minkowskian.

The boundary relation is relatively simpler to do; it is

$$\frac{t}{L} = \frac{sin(\tau/L)}{cos(\tau/L) + cos(\chi)}, \frac{\vec{x}}{L} = \frac{sin(\chi)}{cos(\tau/L) + cos(\chi)}\Omega^i \qquad (18.42)$$

This is a map between the curved Einsteinian and flat Minkowskian four-dimensional universes.

Let us now switch to the relation between the differentials of both set of coordinates and work out their correspondence and the metrics. After going through it, one finds the following metric tensor

$$ds^2 = -dt^2 + \left(dx^i\right)^2 = W^2\left(-d\tau^2 + L^2 d\Omega^2\right) \qquad (18.43)$$

where the so-called *conformal factor* is

$$W^2 = \frac{1}{(cos(\tau/L) + cos(\chi))^2} = \frac{t^2}{L^2} + \frac{1}{4}\left(1 + \frac{r^2}{L^2} - \frac{t^2}{L^2}\right)^2 \qquad (18.44)$$

The next step is upgrading the AdS to the so-called Global AdS-Schwarzschild black hole, GAdSBH, a charged five-dimensional black hole which we will need below anyway. Its line element corresponds to the metric

$$ds^2 = -f d\tau^2 + \frac{d\rho^2}{f} + \rho^2 d\Omega^2 \quad f = 1 - \frac{\rho_0^2}{\rho^2} + \frac{\rho^2}{L^2} \qquad (18.45)$$

where τ, ρ, Ω have the same meaning as "standard" coordinates above. This metric is a solution to the Einstein equation with a particular cosmological term

$$R_{ab} + \frac{4}{L^2}g_{ab} = 0 \qquad (18.46)$$

and ρ_0, L are two parameters related to the mass and charge of the black hole. Here, for reference, are such relations including the Bekenstein entropy and the Hawking temperature

$$M = \frac{3\pi^2\rho_0^2}{\kappa^2}, \quad S = \frac{4\pi^3\rho_h^3}{\kappa^2}, \quad T = \frac{\rho_h}{\pi L^2}\left(1 + \frac{L^2}{2\rho_h^2}\right) \qquad (18.47)$$

where the horizon radius, the upper root of $g_{00} = f = 0$, is

$$\rho_h/L = \sqrt{\sqrt{\left(1 + 4\rho_0^2/L^2 - 1\right)}/2} \qquad (18.48)$$

κ is the five-dimensional Newton constant, which is related to L and the brane number N via

$$\frac{L^3}{\kappa^2} = \left(\frac{N}{2\pi}\right)^2 \qquad (18.49)$$

Note that if $N \gg 1$ is large, then the space parameter L is large compared to the five-dimensional Plank scale, and thus gravity is classical.

This black hole is to play a role of a "meteorite" flying in the bulk away from our "brane world." In principle, one can consider its mass, and thus ρ_h to be small parameters, and expand in it to the first order: but it is not really needed. In the way we just described it, it is a static BH sitting at the origin $\rho = 0$: but this is just because standard coordinates used describe everything "from the BH point of view". "Our coordinates," the same Poincare ones as defined above, are related with the standard ones by the *same* coordinate transformation. Since those are moving (time-dependent) relative to the BH, in such coordinates, the BH does fly.

The elegant solution we describe calculates $T_{\mu\nu}$ using *static* black hole metric in global AdS, in standard coordinates, and then just transforms this tensor into the Poincare coordinates. The result (written in the usual Minkowski spherical coordinates t, r, θ, ϕ) takes relatively simple form

$$T_{\mu\nu} \sim \begin{pmatrix} 3L^4 + 4t^2r^2/W^2 & -2tr(L^2 + t^2 + r^2)/W^2 & 0 & 0 \\ -2tr(L^2 + t^2 + r^2)/W^2 & L^4 + 4t^2r^2/W^2 & 0 & 0 \\ 0 & 0 & r^2L^4 & 0 \\ 0 & 0 & 0 & r^2L^4 \sin^2(\theta) \end{pmatrix}$$

where L is a parameter and W is the conformal factor (18.44) of metric transformation given above. It is instructive to check that this stress tensor is traceless and conserved,

$$T_{\mu\nu}^{;\nu} = 0, \qquad (18.50)$$

so there is no source for it in the Minkowski world. It is so to say spherical tsunami,[16] an implosion at $t < 0$ and explosion at $t > 0$, coming and going to infinity at $t = \pm\infty$.

---

[16]Note, however, that this phenomenon has nothing to do with hydrodynamics, as this tensor *cannot* be written in the form of a moving fluid.

18.6.4 "Holographic e^+e^- Collisions" Show no Signs of Jets!

Here we present results of calculation (Lin and Shuryak 2008b,a) modeling what happens when the two ends of the open string, "charges" located on the boundary, move away from each other with certain velocities[17] $\pm v$.

This setting corresponds to a process in which some neutral object (highly virtual photon or Z boson) decays into quark-antiquark pair. In QCD, due to asymptotic freedom, at large total mass $M \gg \Lambda_{QCD}$, the corresponding coupling $g^2(M)$ is weak, and quarks produce a sequence of rather distant gluons along their way. This is indeed what is observed as two-jet events experimentally.

The question we now address is: what happens in strong coupling, when $g^2(M)$ is large and gluons interact strongly, between quarks and themselves? In particular, would there be any visible jets along the direction of motion of the charges?

In order to answer it, one needs to solve first equation of motion for the string, simultaneously falling into the AdS space due to gravity and expanding between the receding charges. Then, one needs to calculate the effective stress tensor source the falling string provides. And finally, using propagators in the AdS space, one can calculate the induced stress tensor perturbation on the boundary. All these steps are highly technical, and details should be looked at in the original papers. In Fig. 18.5, we show as an example the momentum density distribution on the boundary (charges move relativistically in the horizontal direction). The answer to the question raised above is now clear: at strong coupling, there are no jets. There is some kind of collective explosion. It however cannot be a hydrodynamical flow of some ideal fluid: we see this because stress tensor is very anisotropic even in the rest frame of each element.

18.7 Thermal AdS/CFT and Strongly Coupled Plasma

Let me briefly remind few more facts from the black hole toolbox. Studies of how quantum field theory can sit in a background of a classical black hole metrics have resulted in two major discoveries: the *Hawking radiation* and the *Bekenstein entropy*, related to horizon radius and area, respectively. Hawking radiation makes black holes in asymptotically flat space unstable: it heats the universe till the black hole disappears. Putting black hole into a finite box also does not help: it is generically thermodynamically unstable and gets smaller and hotter till it finally burns out. Only in appropriate curved spaces, black branes can be in thermal equilibrium with their "universe," filled with radiation at some finite temperature T. As shown by Witten, all one has to do to get its metric is to consider non-extreme (excited extreme) black brane solution, which has a horizon.

---

[17]If $v = 0$, it is static Maldacena dipole. If v is small, it complements modified Coulomb law by the modified Ampere law. These case are discussed in the paper, but in this subsection, we only consider ultrarelativistic v.

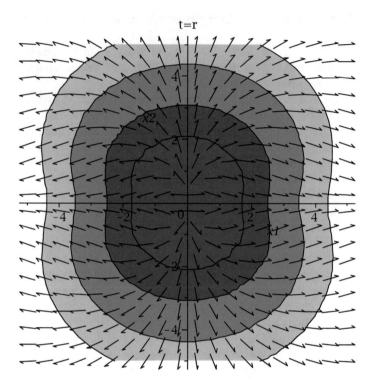

Fig. 18.5 Contour plot of momentum density T^{0m}. The direction of the momentum density is indicated by normalized arrows

Let me now make a logical jump, and instead of just giving the metric for finite-T solution, let me first provide an "intuitive picture" for non-experts, explaining the finite-temperature Witten's settings in which most[18] pertinent calculations are done. The three-dimensional space boundary $z = 0$ is flat (Minkowskian) and corresponds to "our world," where the gauge theory lives. In the bulk, there is a black hole metric with horizon at $z = z_h$. The b.h. center is located at $z = \infty$, but all $z > z_h$ are in fact irrelevant as they are not observable from the boundary. Studies of finite-T conformal plasma by AdS/CFT famously started exactly by evaluation of the Bekenstein entropy (Gubser et al. 1998), $S = A/4$ via calculating the horizon *area A*.

Now comes the promised "intuitive picture": this setting can be seen as a *swimming pool*, with the gauge theory (and us, to be referred below as "distant observers") living on its surface, at zero depth $z = 0$, enjoying the desired

---

[18]The exception is heavy-quark diffusion constant which needs more complicated settings, with a Kruskal metric connecting a world to an anti-world through the black hole.

temperature T. In order to achieve that, the pool's bottom looks *infinitely* hot for observers which are sitting at some fixed z close to its coordinate z_h: thus diving to such depth is not recommended. Strong gravity takes care and stabilizes this setting thermodynamically: recall that time units, as well as those of energy and temperature, are subject to "warping" with g_{00} component of the metric, which vanishes at z_h.

When astronomers found evidences for black holes and accretion into those, the physics of black hole became a regular part of physics since lots of problem have to be solved. Here important step forward was the so-called *membrane paradigm* developed by many people and best formulated by Thorne et al. (1986), known also under the name of "stretched horizons." Its main idea is to imagine that there is a physical membrane at some small distance ϵ away from the horizon and that it has properties exactly such that all the equations (Maxwell's, Einstein's, etc.) would have the same solutions outside it as *without* a membrane but with a continuation through horizon. For example, a charged black hole would have a membrane with a nonzero charge density to terminate the electric field lines. The fact that Poynting vector at the membrane must be pointed inward means some (time-odd!) relation between $\vec{E}$ and $\vec{B}$: this is achieved by giving the membrane finite *conductivity*, which in turn leads to Ohmic losses, heat, and entropy generation in it. Furthermore, as shown by Damour back in the 1980s, displacements of the membrane and relations for gravitational analog of the Poynting vector lead to nonzero *viscosity* of the membrane, and its effective low-frequency theory takes the form of Navier-Stokes hydrodynamics. For a bit more modern derivations of the effective action and field theory point of view, look at Parikh and Wilczek (1998). As we will see below in this chapter, all of those ideas have resurfaced now in AdS context, generating new energy of young string theorists who now pushed the "hydrodynamics of the horizon membrane" well beyond the Navier-Stokes to a regular construction of systematic derivative expansions to any needed order.

18.8   Brief Summary

- D-branes are multidimensional solitons of string theory. Their displacements are described by vector fields A^μ. If μ in external to the brane, those components are perceived as scalar in internal space. The action for those fields has geometric Born-Infeld form (18.3).
- A stack of N_c of D_3 coincident branes generate $U(N_c)$ gauge theory. For large $N_c \to \infty$, fields around them can be described by classical supergravity, with tensor graviton, with certain scalars and fermions. The masses and charges are such that Newton, Coulomb, and scalar-induced forces cancel each other, so in this approximation, branes do not interact. The nature of this cancellation is the same as BPS cancellation of interaction between monopoles or instantons, which we discussed in the corresponding chapters.
- Like a point mass in GR creates a black hole solution, a brane stack (pointlike in six external dimensions) creates similar solution for the metric. In certain

limit (Maldacena 1999), it may be represented as simple anti-de-Sitter (ADS) conformal metric. "Holography" is a correspondence between $\mathcal{N} = 4$ superconformal gauge theory on its boundary with string theory in AdS metric. The central point is that the former is *strongly coupled*, while the latter "bulk theory" is coupled weakly and can be treated by usual perturbation theory.

- One of the first problems solved is the strong coupling version of the Coulomb law (18.20) with a *square root* of product of charges. It is based on finding the shape of a string ending on two charges and pending in gravity field in the bulk. We discuss approximate ladder diagram re-summation in gauge theory leading to the same dependence.

- A hologram is image of any object in the bulk on the boundary. The simplest case is a pointlike defect at distance ρ from the boundary. Its hologram turned out to be an *instanton* of the gauge theory! The complicated interaction law of instanton and anti-instanton of sizes ρ_I, $\rho_{\bar{I}}$ at distance R turned out to be just a geodesic distance between two points in AdS space.

- We discuss hologram of pair of charges, connected by a string, static and moving away from each other. We derive "strong coupling Ampere law" and demonstrate that hologram of two charges moving away from each other is *not* a pair of jets (see Fig. 18.5).

- While AdS metric is that of "extreme" black hole with zero horizon size, one may wander what would be a hologram of *non-extreme* black brane in the bulk. It turns out to be thermal ensemble of gauge fields. Since it is in strong coupling, it is exactly what is needed to study strongly coupled QGP phase. In literature there are extensive studies of hydrodynamics and kinetics of it, with shock waves and collisions, but those applications go beyond this book.

References

Dorey, N., Hollowood, T.J., Khoze, V.V., Mattis, M.P., Vandoren, S.: Multi-instanton calculus and the AdS/CFT correspondence in $N = 4$ superconformal field theory. Nucl. Phys. **B552**, 88–168 (1999)

Dorey, N., Hollowood, T.J., Khoze, V.V., Mattis, M.P.: The Calculus of many instantons. Phys. Rep. **371**, 231–459 (2002)

Friess, J.J., Gubser, S.S., Michalogiorgakis, G., Pufu, S.S.: Expanding plasmas and quasinormal modes of anti-de Sitter black holes. JHEP **04**, 080 (2007)

Gubser, S.S., Klebanov, I.R., Tseytlin, A.A.: Coupling constant dependence in the thermodynamics of $N = 4$ supersymmetric Yang-Mills theory. Nucl. Phys. **B534**, 202–222 (1998)

Gursoy, U., Kiritsis, E.: Exploring improved holographic theories for QCD: Part I. JHEP **02**, 032 (2008). 0707.1324

Karch, A., Son, D.T., Katz, E., Stephanov, M.A.: Linear confinement and AdS/QCD. In: Proceedings of the 7th Workshop Continuous Advances in QCD (QCD 2006), Minneapolis, USA, 11–14 May 2006, pp. 96–102 (2006b)

Klebanov, I.R., Maldacena, J.M., Thorn, III, C.B.: Dynamics of flux tubes in large N gauge theories. JHEP **04**, 024 (2006)

Lee, K.-M., Lu, C.-h.: SU(2) calorons and magnetic monopoles. Phys. Rev. **D58**, 025011 (1998). hep-th/9802108

Lin, S., Shuryak, E.: Stress tensor of static dipoles in strongly coupled $N = 4$ gauge theory. Phys. Rev. **D76**, 085014 (2007)

Lin, S., Shuryak, E.: Toward the AdS/CFT gravity dual for high energy collisions. 2. The stress tensor on the boundary. Phys. Rev. **D77**, 085014 (2008a)

Lin, S., Shuryak, E.: Toward the AdS/CFT gravity dual for high energy collisions: I. Falling into the AdS. Phys. Rev. **D77**, 085013 (2008b). hep-ph/0610168

Maldacena, J.M.: Wilson loops in large N field theories. Phys. Rev. Lett. **80**, 4859–4862 (1998)

Maldacena, J.M.: The Large N limit of superconformal field theories and supergravity. Int. J. Theor. Phys. **38**, 1113–1133 (1999). [Adv. Theor. Math. Phys. **2**, 231 (1998)]

Mikhailov, A.: Nonlinear Waves in AdS/CFT Correspondence (2003)

Parikh, M., Wilczek, F.: An Action for black hole membranes. Phys. Rev. **D58**, 064011 (1998)

Rodrigues, D.C., Wotzasek, C.: 3D and 4D noncommutative electromagnetic duality and the role of the slowly varying fields limit. PoS **IC2006**, 048 (2006)

Semenoff, G.W., Zarembo, K.: Wilson loops in SYM theory: From weak to strong coupling. Nucl. Phys. Proc. Suppl. **108**, 106–112 (2002). [106(2002)]

Shuryak, E.: A 'Domain Wall' Scenario for the AdS/QCD (2007)

Shuryak, E., Zahed, I.: Understanding the strong coupling limit of the $N = 4$ supersymmetric Yang-Mills at finite temperature. Phys. Rev. **D69**, 046005 (2004a)

Thorne, K.S., Price, R.H., Macdonald, D.A. (eds.) Black Holes: The Membrane Paradigm (1986)

Witten, E.: Solutions of four-dimensional field theories via M theory. Nucl. Phys. **B500**, 3–42 (1997). [452(1997)]

Holographic QCD

19

As we emphasized it few times already, there is *no* known holographic dual for QCD. And yet, people tried to build it, so to say, by "bottom-up" approach. It does not have the same status as AdS/CFT correspondence discussed above but obviously a set of made-up models, which however can be rather instructive. We start with two particular models, to be generalized later.

19.1 Witten and Sakai-Sugimoto Models

The way from the holographic dual of the $\mathcal{N} = 4$ (SYM) theory to the usual Yang-Mills gluodynamics (YM) was suggesting in Witten (1998b). He emphasized that this step included the high-temperature limit of YM theory, implying breaking of supersymmetry, yet preserving holographic correspondence.

In short, the idea is to start with D_4 branes (rather than D_3 ones used for ADS/CFT). The "unwanted" direction of space x^4 is compactified on a circle with circumference $\beta_4 \equiv 2\pi/M_{KK}$, the M_{KK} being the so-called Kaluza-Klein mass. The metric of the Witten background is

$$ds^2 = \left(\frac{u}{R}\right)^{3/2}\left(dx^\mu dx_\mu + f(u)dx_4^2\right) + \left(\frac{R}{u}\right)^{3/2}\frac{du^2}{f(u)} + R^{3/2}u^{1/2}d\Omega_4^2$$

(19.1)

where $f(u) = 1 - u_0^3/u^3$ is the warping factor[1] depending on the holographic coordinate u, defined for $u > u_0$. There is also a dilaton field and 4-form Ramon-Ramon field in this background, which we would not go into. Here $\mu = 0, 1, 2, 3$ are

---

[1]It is completely analogous to Schwarzschild solution of the usual black hole, except the power of distance in the Newtonian potential is not $1/r$, as in three dimensions, but $1/r^3$ in five dimensions.

© The Author(s), under exclusive license to Springer Nature Switzerland AG 2021
E. Shuryak, *Nonperturbative Topological Phenomena in QCD and Related Theories*, Lecture Notes in Physics 977,
https://doi.org/10.1007/978-3-030-62990-8_19

Minkowski coordinates, and the extra spatial coordinate is x_4. The key feature here is that the radius of x_4 vanishes at u_0: the geometry thus has the so-called "cigar" form with its tip at u_0.

The Euclidean form of the finite-T theory corresponds to the time, $\tau = ix_0$, also compactified to the Matsubara time, with circumference $\beta = 1/T$. Thus there are two competing phases of this theory: a cigar in one cyclic variable and "tube" in another. The lowest free energy corresponds to the winning phase. This phase transition Witten identified with the deconfinement phase transition in thermal YM theory. The deconfinement temperature scale is therefore related with M_{KK}, defined above.

Generalization of this model to nonzero θ angle can be done by turning on nonzero 1-form potential C_1 along the x_4 circle, Aharonov-Bohm style, so that there is a nontrivial flux of the field $F_2 = dC_1$ through the cigar.

In order to get a QCD-like model, one needs to complement Witten's construction for YM theory by inclusion of the light quarks. This can be done by incorporating some N_f copies of extra branes possessing fundamental fermions. Since the number of flavors N_f is assumed to be small compared to N_c, they are treated in "probe approximation," which is neglecting their role in overall metric formation.

The model we will now discuss is Sakai-Sugimoto model (Sakai and Sugimoto 2005), based on N_f sets of $D8$ and $\bar{D}8$ branes located at the opposite sides of the Witten's cigar. In the upper part of Fig. 19.1 from this paper, we show extensions of the $D4$ and $D8$ branes. The left lower sketch shows how they are connected, with the circle being in x_4 coordinate which $D8$ lack. If the geometry has two independent $D8$ branes, corresponding to left and right quarks, the chiral symmetry is unbroken.

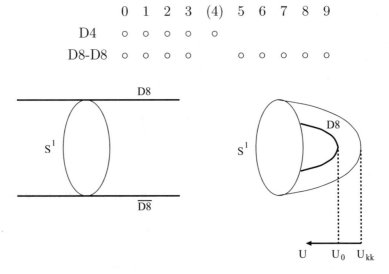

	0	1	2	3	(4)	5	6	7	8	9
D4	o	o	o	o	o					
D8-D8	o	o	o	o		o	o	o	o	o

Fig. 19.1 Upper: extensions of the coordinates of D_4 and D_8 branes. Lower, two phases, with chiral symmetry and with its breaking

It gets however broken if the geometry is as shown in the lower-right corner, in which there is a single $D8$ brane turning back after it reached some point U_0.

In the geometry thus constructed, one can move further, studying small perturbation of this brane construction. The displacement of the $D8$ is described by nine-dimensional gauge field A_M. Those with $M = 0, 1, 2, 3$ would be Lorentz 4-vector, while others scalars. The action of perturbation has a general form

$$S = -T \int d^9 x e^{-\phi} \sqrt{-det\left(g_{MN} + 2\pi\alpha' F_{mn}\right)} \qquad (19.2)$$

where T is tension, and F_{MN} is the field strength corresponding to A_M. Assuming the field is small, one can expand the square roots in powers of the field. In quadratic order in F_{MN}, it describes masses of (pseudo)vector and (pseudo)scalar mesons. We would not go into details but just comment that mass ratios between those can be calculated from the model, and their relation to experiment is reasonable. Let me emphasize that in fact we have much more than predictions of the meson masses here: we have a complete form of *effective chiral Lagrangian*, a dream of Weinberg, Leutwyler, etc., which in QCD is so hard to build! Note further that the relation between vector and scalar mesons fulfills also another dream, that of *hidden gauge symmetry* related with the flavor $SU(N_f)$ group!

Since in these lectures we devoted a lot of attention to chiral anomaly, in its relation to topology and also to chiral effects in general, let us explain how (Sakai and Sugimoto 2005) had approached these issue holographically. They added the following topological action term

$$S_{topo} = \frac{N_c}{24\pi^2} \int_{D8} \omega_5(A) \qquad (19.3)$$

where N_c comes from integral over F_4 charge, of the RR 4-form. The Chern-Simons form in five dimensions x_0, x_1, x_2, x_3, z is (with all indices and 5d epsilon symbols suppressed)

$$\omega_5(A) = \text{Tr}\left(AF^2 - \frac{1}{2}A^3 F + \frac{1}{10}A^5\right) \qquad (19.4)$$

Recall that the CS form is not by itself gauge invariant, but its change is, because[2]

$$d\omega_5 = \text{Tr}\left(F^3\right)$$

---

[2]This expression is analogous to $d\omega_3 = \text{Tr}(F^2)$ relation we used relating three-dimensional to four-dimensional gauge topology. Note again, that this expression includes epsilon symbols, in six and four dimensions.

Therefore gauge invariant action, reproducing QCD chiral anomaly, can be defined as the difference between the boundary values at two ends of $D8$, the left minus the right-handed one.

We end our discussion of Witten-Sakai-Sugimoto settings by introducing the θ dependence of the vacuum energy and calculate it. For details, we refer to Bigazzi et al. (2019) and only note that always $E_{vac} \sim \cos(\theta)$, as in the dilute instanton gas.

To make this connection more clearly, we would like to mention the topological susceptibilities obtained in these two settings. In $N = 4$ or SYM theory, they have, at finite-temperature setting

$$\chi_{SYM}(T) = \frac{15}{128}\pi^{3/2}\sqrt{N_c}T^4 exp\left(-\frac{8\pi^2}{g_{YM}^2}\right) \qquad (19.5)$$

and the exponent indeed corresponds to the instanton action. The preexponent however has strange square root of N_c and no coupling: both features I cannot explain in simple terms.

In WYM (Witten-Sakai-Sugimoto) high-T phase in the topological susceptibility (deduced from the θ- dependence of the vacuum energy $\sim \cos(\theta)$), the result is

$$\chi_{WYM}(T) = \frac{3285\pi^{3/2}}{42}\left(\frac{4\pi}{3}\right)^4\sqrt{\frac{N_c}{\lambda}}T^4 exp\left(-\frac{8\pi^2}{g_{YM}^2}\right) \qquad (19.6)$$

In this case, N_c and $\lambda \equiv g_{YM}^2 N_c$ dependence is such that they cancel dependence on N_c and only the YM coupling remains!

In summary, holographic model of the theta vacua/axions produces predictions for the topological susceptibilities which look very much like the semiclassical instanton density in exponent but with very different preexponential factors. This is hardly surprising by itself, since in weak and strong coupling, the fluctuating fields (making the bosonic and fermionic determinants) are quite different. Yet some "microscopic" derivation of expressions given above would be much appreciated.

19.2   Using Gauge Instantons as Baryonic Solitons

Let me start admitting that we had not covered in this book the *skyrmions*, modeling baryons as classical solitons made of the pion field.

This approach is natural in the limit $N_c \to \infty$ (in which baryon mass is large, $O(N_c)$) and the chiral limit $m_q \to 0$ (in which pions are massless). The so-called Wess-Zumino-Witten term in the chiral Lagrangian provides the topological definition of the baryon number which therefore is conserved.

Skyrmions naturally predict large-distance forces between nucleons to be pion exchanges. However we know from nuclear physics that phenomenology suggests nuclear forces to be described rather in terms of scalar σ and vector ρ, ω mesons, producing attractive and repulsive parts of the forces. Note furthermore that attraction and repulsion obviously are well tuned, nearly canceling each other in sum.

The WSS model just described provides a very curious realization of "mesonic soliton" idea, in the setting explained in the previous section. As shown by Hata et al. (2007), the classical solution to four-dimensional Euclidean YM equations, used at the lectures at the top of the course describing tunneling in the topological (Chern-Simons) landscape, can now be interpreted in four-dimensional spatial dimensions of the WSS model as a soliton.

Its "scalar part" corresponds to the Skyrmion pion cloud, while the vector field A_M made $U(N_f)$ vector mesons, ρ, ω, representing the repulsive baryon cores (missing from the original Skyrme model). Since instanton is a solution to the nonlinear YM equations, it means that interactions between all types of mesons are taken into account semiclassically.

The excited states of instanton-baryons, with say other meson attached, were used as models for other hadrons, e.g., pentaquarks.

The study of high density of instantons, turned to be a four-dimensional crystalline lattice, has been first constructed in Shuryak and Verbaarschot (1990),; it was the first example of "time crystal" later emphasized by Wilczek. The same crystalline lattice in the holography setting was later reused as a model for dense baryonic matter. If the density is high, instantons can be further split into its constituents with fractional topological charge (see Rho et al. 2010).

19.3   Confining Holographic Models with "Walls" in the Infrared

In spite of strong coupling regime, the AdS/CFT setting is scale invariant, with Coulomb-like forces between charges. (Only the coefficients of $1/r$ changes at strong coupling, as we already discussed.) In QCD-like holographic models, one needs to somehow *generate confinement* and linear potential between charges.

The first crude idea aims at it was introduced in Polchinski and Strassler (2002) who suggested to simply cut off the most infrared part of the AdS_5 space, by "brute force." Indeed, if there is no space for the bending string to fall into, it would reach "the bottom of space" and thus generate a linear—confining—potential.[3]

A generic idea can be formulated as follows: hadrons would correspond to normalizable modes of the fields in the AdS space, modified by some *confinement-related cutoff*. Let us show how it works, following Karch et al. (2006a). Suppose the metric takes the form

$$ds^2 = exp(2A(z)) \left(dz^2 + \eta_{\mu\nu} dx^\mu dx^\nu \right) \tag{19.7}$$

---

[3] The reader should be warned that this idea cannot be really true, as it also limits D_1 magnetic strings and thus confines the magnetic monopoles. We are not expected to do so in QCD!

where $\eta_{\mu\nu} = diag(-1, 1, 1, 1)$ is the flat Minkowski metric. Another function, the dilaton profile, is in the action:

$$S = \int d^5x \sqrt{g} e^{-\Phi(x)} \mathcal{L} \qquad (19.8)$$

The model has a general minimal Lagrangian containing the scalar field X and left/right vector fields A_L, A_R, both matrices in terms of some flavor group $SU(N_f)$.

$$\mathcal{L} = |DX|^2 + 3X^2 - \frac{1}{2g_5^2}\left(F_L^2 + F_R^2\right) \qquad (19.9)$$

where $D_\mu X = \partial_\mu X - i A_{L\mu} X + i A_{R\mu} X$ and $F_{L,R}$ are the fields strength made of A_L, A_R. We will be using the gauge $A_z = 0$. Vector/axial fields are $V, A = (A_L \pm A_R)/2$ as usual. They satisfy the eigenvalue equation $(B(z) = \Phi(z) - A(z))$

$$\partial_z e^{-B} \partial_z v_n(z) + m_4^2(n) e^{-B} v_n(z) = 0 \qquad (19.10)$$

where we already look for a four-dimensional plain wave solution and substituted the 4-momentum squared by the mass squared. Standard substitution $v_n = e^{B/2}\psi_n$ transfers it to Schreodinger-like form:

$$\psi_n'' + \left(\frac{(B')^2}{4} - \frac{B''}{2}\right)\psi_n = 0 \qquad (19.11)$$

It was argued that the best way to cut the space off is by a Gaussian factor in the metric $B \sim z^2$ at large z. One motivation (Karch et al. 2006a) is that bulk fields generate in this case quite perfect Regge trajectories, as we will see shortly. Another motivation (Shuryak 2006) is that the instanton size distribution in QCD (unlike in the CFT) does show a Gaussian cutoff, and we now know that the instanton size ρ is in holography just the fifth coordinate of its pointlike source.

For the preferred choice of the functions,

$$B = z^2 + \log(z), \quad \frac{(B')^2}{4} - \frac{B''}{2} = z^2 + \frac{3}{4z^2} \qquad (19.12)$$

there is the following analytical solution

$$m_n^2 = 4(n+1), \quad v_n = z^2 L_n^1(z^2)\sqrt{\frac{2n}{(1+n)}} \qquad (19.13)$$

where L_n^m are associated Laguerre polynomials. Note the equidistant spectrum of the vector meson masses m^2, which is phenomenologically correct. Furthermore, introducing spin S fields, as five-dimensional symmetric tensors of rank S (Karch et al. 2006a) one sees that the equation of motion is modified to

$$B = z^2 + (2S - 1)\ln(z) \qquad (19.14)$$

and the spectrum to

$$m_{n,S}^2 = 4(n + S) \qquad (19.15)$$

which is the (phenomenologically correct) linear Regge trajectory.

The model can include explicit and spontaneous chiral symmetry breaking (Erlich et al. 2005). By including the scalar X with the boundary conditions $X = Mz/2 + \Sigma z^3/2 + \ldots z \to 0$, where M is the (flavor matrix) of quark masses and Σ is the quark condensate, Karch et al. also got reasonable parameters of the scalar and axial mesons. Schafer (2008) has demonstrated that the whole Euclidean vector and axial correlators $V(\tau)$, $A(\tau)$ built out of these modes reproduce experimental data within 10–20% accuracy, at any Euclidean time τ. (These results are extremely similar to what was obtained by Schafer and myself in much more sophisticated instanton model (Schafer and Shuryak 2001a). The remaining 10% deviations, at least at small distances, are clearly just neglected perturbative correction $1 + \alpha_s/\pi + \ldots$.)

We will however argue that vector/axial channels are in fact *exceptions* rather than the rule, and for generic operators/fields, one cannot hope to use this simplest model. Coupling of the boundary sources to fields propagating in the (modified-AdS) bulk does indeed fit to the general philosophy of holography. Yet I do not see any justification to why all those spin-S bulk fields may be massless in the five-dimensional sense. String theory can only supply the massless states up to gravitons ($S = 2$): and even for strict AdS/CFT, generic bulk fields have nonzero (and large at large spins and strong coupling) anomalous dimensions. In the form of the five-dimensional mass, those would appear in the effective Schreodinger potential as follows:

$$V(z) = z^2 + 2(S - 1) + \frac{S^2 + M_5^2 - 1/4}{z^2} \qquad (19.16)$$

(Only the vector fields can be exempted, on the basis of vector current conservation.)

We follow notations of Karch et al. (2006a), except that we have added extra five-dimensional mass term. Standard substitution $\phi = e^{B/2}\psi$ transforms this into a Schreodinger-like equation without first-order derivatives

$$-\psi'' + V(z)\psi = m_4^2\psi \qquad (19.17)$$

with

$$V(z) = z^2 + 2(S - 1) + \frac{S^2 + M_5^2 - 1/4}{z^2} \tag{19.18}$$

KKSS needed only bound state wave functions, given in terms of Laguerre polynomials. Quadratic IR potential was tuned by Karch et al. (2006a) to reproduce nice Regge trajectories: for absent bulk mass $M_5 = 0$, they are linear and $m_4^2 = 4(n + S)$.

The absolute units we will use also follow from KKSS notations and can be fixed by calculating some physical mass. Rho meson is an example of the *protected* state, associated with conserved vector current: it has with $M_5 = 0$ and $S = 1$: a solution without nodes ($n = 0$) gives $m_\rho^2 = 4$. Using it as an input, we fix our unit of length as

$$length\,unit = 2/m_\rho = 0.51\,fm \tag{19.19}$$

The expected position of the domain wall at large N_c is

$$z_{dw} \approx 0.4\,fm = 0.777(length\,unit) \tag{19.20}$$

Thus *most of the wave function is in fact located in the strongly coupled domain*, and modifications due to weakly coupled domain at $z < z_{dw}$ turn out to be very small, as far as the four-dimensional spectroscopy goes. However hard processes will be sensitive to this small tail of the wave function, and their relative normalization is crucial for what follows.

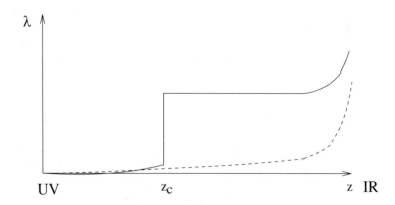

Fig. 19.2 Schematic dependence of the 't Hooft coupling λ on the holographic fifth coordinate z

19.4   The "Domain Wall" in the Ultraviolet?

The second crude idea tries to deal with an unpleasant fact: the QCD coupling runs into weak in the UV, and so the AdS/CFT-like description cannot possibly be valid there. I have proposed (Shuryak 2006, 2007) to include this feature by introducing the so-called *UV wall* beyond which the space is still AdS but all five-dimensional masses of the fields return to their noninteracting dimensions see Fig. 19.2.

Where should the UV wall be placed? What is its physical nature? The guiding idea is again related with the instantons. We had discussed large-N_c limit of the instanton ensemble and have concluded that their size distribution becomes a delta function, at some limiting instanton size ρ_*.

What can be calculated in this model? Example includes hard exclusive processes and transition from strong coupling at large distances to weak coupling at small one. A jump in power should be seen, as a consequence of the change from large anomalous dimensions in strong coupling to "naive" dimensions for noninteracting fields.

Specific example is the simplest object of the kind, which is the pion form factor. Unfortunately experiment has not yet seen how transition to old perturbative prediction takes place, as it is very hard to measure it at large Q. All we know is that the observed $Q^2 F_\pi(Q)$ at $Q \sim 1 - 2\,\text{GeV}$ is few times large than the asymptotic value, so some decrease and then leveling at new level should happen. As far as I know, nobody has seen it on the lattice either. The magnitude of the pion form factor at $Q \sim 1 - 2\,\text{GeV}$ induced by instantons has been calculated in Faccioli et al. (2004), see also our Sect. 10.6. So the transition to the perturbative regime has not been seen in this approach as well. More or less similar situation is with the nucleon form factor and many other exclusive reactions.

Another hard reaction involving the pion is pion diffractive dissociation into two jets

$$\pi \rightarrow jet(k_1) + jet(k_2) \tag{19.21}$$

first theoretically discussed by Frankfurt et al. (1993) and studied experimentally by the Fermilab experiment E791. It seems to be showing a transition from the nonperturbative to the perturbative regime we are looking for.

19.5   Improved Holographic QCD

"Realistic bottom-up" holographic models with the dilaton/axion potentials were gradually developed, and in this section, we will follow the approach of the Kiritsis group (Gursoy et al. 2008; Gursoy and Kiritsis 2008); for review see Gursoy et al. (2011). They have developed a systematic approach toward building a gravity dual model to QCD. The fields in the action include the (tensor) metric, (scalar) dilaton, and (pseudoscalar) axion in the common Lagrangian. One more scalar field, the "tachyon", is added later to model the chiral symmetry breaking and the quark condensate.

Let us start with the pure YM theory: for its holographic dual, one uses generic form of the (five-dimensional) action, with the bulk and boundary terms:

$$S = -M_p^3 N_c^2 \int d^5x \sqrt{g}\left[R - \frac{4}{3}(\partial\Phi)^2 + V(\Phi)\right] + 2M_p^3 N_c^2 \int_{\partial M} d^4x \sqrt{h}K$$

(19.22)

Here R and K are bulk and boundary curvature; the factors in front correspond to five-dimensional Newton constant $G_5 = 1/(16\pi M_p^3 N_c^2)$ small in the large N_c limit. The exponent of the dilaton $exp(\Phi) = \lambda$ is identified with running 't Hooft coupling $N_c g_{YM}^2$.

What needs to be carefully chosen is the potential $V(\Phi)$, as it incorporates the beta function of the theory. These details can be found in the original works. A clever trick allows to incorporate the beta function (which demands first-order differential equation) with Lagrangian formalism which demands second-order ones.

There are basically two types of the solutions, with qualitatively different metric behavior, without or with the black hole. Aiming at thermal physics of quark-gluon plasma, we start with Euclidean time which, with slight abuse of notations, we still call t. In the former case, the following simple ansatz is used ($m = 1, 2, 3$):

$$ds^2 = b_0^2(r)\left(dt^2 + dr^2 + dx_m dx^m\right)$$

(19.23)

In the latter,

$$ds^2 = b^2(r)\left[f(r)dt^2 + \frac{dr^2}{f(r)} + dx_m dx^m\right]$$

(19.24)

the warping factor $f(r)$ has zero at horizon location $f(r_h) = 0$, and only $r < r_h$ are considered. These two geometries correspond to low-T and high-T phases, as calculation of their free energy shows. The obtained temperature dependence of the thermodynamical quantities—energy and entropy densities e, s and pressure p—is compared in Fig. 19.3 with the YM lattice data (dots). In spite of the model assuming large N_c limit and the lattice data at physical $N_c = 3$, the agreement is rather good.

The phase transition is of the first order, and as, e.g., is the case for elementary Van-der-Waals case, there is the third solution (here called "the small black hole") which is unstable and corresponds to free energy maximum between the two minima.

We will not go into discussion of kinetic quantities, the diffusion constant, the viscosities, the jet quenching parameter, etc.—which can also be calculated in this model.

The next extension of the model is adding the pseudoscalar *axion* field a: this allows to fix negative parity glueballs and also to calculate the topological susceptibility, but we will not discuss it. Finally, one should add quarks to construct

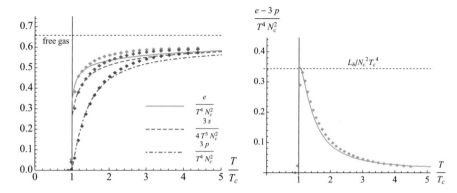

Fig. 19.3 Temperature dependence of the dimensionless combinations, normalized so that they all have large-T limit $\pi^2/15$ (dashed horizontal line). The second plot is the so-called trace anomaly combination, the measure of the non-conformal behavior

really holographic model for QCD. The topic has its history which we do not go to but just jump to the paper (Jarvinen and Kiritsis 2012) in which quark-generating D_4-anti D_4 branes are not considered in the "probe approximation."[4] In this work, the so-called *Veneziano limit* is taken in which the ratio of the number of flavors to the number of colors

$$x_{Veneziano} \equiv \frac{N_f}{N_c} \qquad (19.25)$$

is kept fixed in the $N_c, N_f \to \infty$ limit. Its value is from zero to the maximal value $x_{max} = 11/2$ at which the asymptotic freedom is lost.

The model action is supplied with the fermionic part

$$S_f = -x_{Veneziano} N_p^3 N_c^2 \int d^5 x \, V(\lambda, \tau) \sqrt{det(g_{\mu\nu} + h(\lambda)\partial_\mu \tau \partial_\nu \tau^+)} \qquad (19.26)$$

containing the "tachyon" field τ.

One interesting aspect of this model is existence of critical value x_c, approximately near 4, separating "conformal window" at $x_{max} > x > x_c$, in which the system flows in IR into the zero of the beta function, and the more common regime at $x < x_c$ in which the tachion value at the boundary is related to the nonzero quark condensate. The solution to EOM is numerical and highly technical: details can be found in the original papers, so let me only present some results.

After the background field configuration is established, one can study quantum deviations from it. Quadratic terms, as usual, give the excitation spectrum of the

---

[4] We remind the reader that in real-world QGP, quark-antiquark degrees of freedom are twice more important than those of gluons.

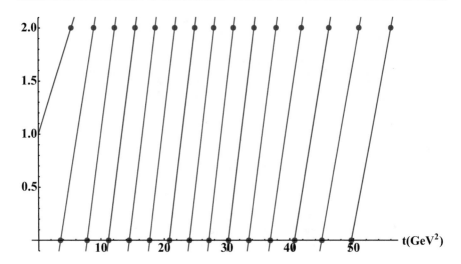

Fig. 19.4 The 2^+ and 0^+ glueball masses on the Regge plot $J - M^2$, calculated from improved holographic QCD in the Veneziano limit at $x_{Veneziano} = 1$ by Iatrakis et al. (2015)

fields involved. For example, excitations of the graviton and dilaton fields give the spectrum of tensor and scalar glueballs. Connecting those one can obtain the lower parts of multiple "daughters" Regge trajectories, as shown in Fig. 19.4.

Meson scalar gives its own set of excitations. More precisely, the dilaton and tachyon excitations (scalar glueballs and scalar mesons) get mixed together. I will discuss this mixing, following Iatrakis et al. (2015), because it was important for understanding interaction strength of the QCD strings, and (perhaps more importantly) it is the part of AdS/QCD I was involved myself.

As a brief historic introduction, let me just state that in hadronic spectroscopy, one can hardly find more confused subject than scalar mesons. Experimentally, there are more resonances than the simple $\bar{q}q$ mesons may exist. Clearly there should be scalar glueballs and four-quark mesons as well, all mixed up in some proportions. AdS/QCD is at least a self-consistent model, providing a well-defined method to calculate these mixings. How it works is explained in Fig. 19.5.

Completing this section, let me note that expanding the action to terms quadratic in all fields deviations from the background is not the only thing one can do. Nothing (except technical complexity) can stop anyone to expand the action further, e.g., to the third order, defining the interaction constants between all these states.

The question is not only the magnitude of the coupling but the index structure of the corresponding vertices as well. For example, as shown by Anderson et al. (2014), the index structure of triple vertex of two tensor fields and *pseudoscalar*

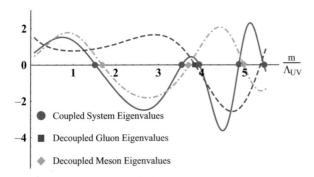

Fig. 19.5 The determinant of the UV boundary value of two linearly independent solutions of the scalar fluctuation equations, versus the mass parameter in Λ_{UV} units. The red solid line with five zeroes indicates values of the lowest scalars from full Hamiltonian. Blue dashed lines and squares indicate unmixed glueball masses, and green dash-dotted curve and green diamonds indicate unmixed meson masses. The lowest state—sigma meson $f_0(500)$—shifts little, but the second meson is close to the first glueball and their mixing is more serious. In fact these two states collide at $x_{Veneziano} \approx 1.5$, but including mixing one observe classic "avoided crossings" of the levels

0^- production can be determined uniquely:

$$V_{hho^-} \sim \epsilon_{\alpha\beta\gamma\delta} q_1^\alpha q_2^\beta h^{\gamma\sigma} h^{\delta\sigma} \tag{19.27}$$

Indeed, the 2^+ Pomeron is identified with symmetric tensor $h^{\mu\nu}$, which cannot be directly paired with an antisymmetric epsilon symbol. Two momenta of the Pomerons, q_1, q_2 need to be also included for obvious reasons. If transverse part of those is parallel, $\vec{q}_1^\perp \sim \vec{q}_2^\perp$, the vertex vanishes. In general, it is predicted to be proportional to $\sin(\phi_{12})$ (the azimuthal angle between $\vec{q}_1^\perp, \vec{q}_2^\perp$), and the data is in agreement with this prediction.

Unfortunately, the index structure of triple vertex of three tensor fields V_{hh2^+} is not unique; it may be written in several possible forms. The holography, however, suggests the correct one: if all objects are holographically related to gravitons, one can try to mimic the Einstein-Hilbert action, expanded to the third order in $h^{\mu\nu}$ perturbation.[5] This idea by Iatrakis et al. (2016) was compared to the experimentally observed Pomeron-Pomeron-tensors (PPT) vertex. It was shown that the Einstein-Hilbert vertex indeed works for tensor glueball $f_2(2300)$, but not for ordinary tensor mesons like $f_2(1200)$.

---

[5]Other evidences that the Pomeron should be accompanied by a tensor polarization were given by Ewerz et al. (2016).

19.6   QCD Strings and Multi-String "Spaghetti"

In the holographic QCD, there are bulk strings, moving in a potential created by
nontrivial profile of the gravitational metrics as well as the profile of the dilaton. The
combined potential is shown in Fig. 19.6 from Iatrakis et al. (2015). The bulk string
is pointlike, but since it is at some distance from the boundary, its hologram has
finite size, the radius of the QCD string. The oscillations of the bulk string around
the minimum of the potential are perceived at the boundary as the "breathing mode"
of the QCD string.

The string interaction with gravity and dilaton means that they can interact
via exchanges of tensor and scalar glueballs. (Recall their masses displayed in
Fig. 19.4 in the preceding section.) The problem is both are rather heavy, and the
resulting interaction range $V \sim exp\left(- M_{glueball}r\right)$ is very short. This is where
the glueball-meson mixing becomes relevant: it allows the string interact via sigma
meson ($f_0(500)$ in particle data book now) exchange. In chapter on QCD strings,
we discussed it and found that this idea fits the lattice data rather well.

Multi-string configurations are produced in high-energy collisions. Recall that
we discussed that Pomeron exchange leads to production of (at least) two strings.
The collisions of the proton with Pb nucleus at LHC, depending on impact
parameter, include a range of string number from 2 to about 30 at central collision.
As the string ends—the through-going quarks—fly along the beam directions, one
finds string in "spaghetti" situation (see Fig. 19.7).

In the transverse "plane" (which is in holography three-dimensional, $\vec{x}_\perp, z$),
these strings have no extension and are shown as just points. Weak attraction which
string has to each other only becomes important when the density of strings is
sufficiently large. The study of collective multi-string collapse has been studied by

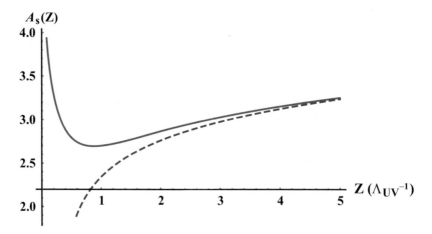

Fig. 19.6 The red solid line shows effective potential for a string in the bulk, as a function of
holographic potential z. The minimum corresponds to location of a string in equilibrium

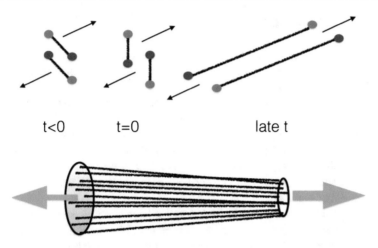

Fig. 19.7 Upper part displays production of two strings in a Pomeron, a result of color exchange of the incoming color dipoles. The lower part is a sketch explaining the "spaghetti" multi-string configuration

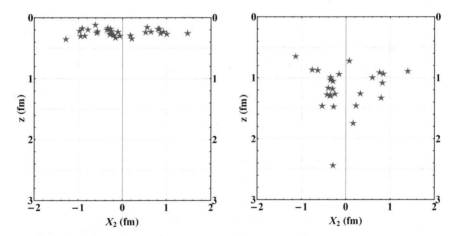

Fig. 19.8 Early and late time locations of 30 interacting parallel strings, in a plane transverse coordinate x_2—holographic coordinate z. The set is falling into IR under its common weight

Iatrakis et al. (2015), using their EOM in the transverse plane. Two snapshots, at early and late time of some particular configuration, are shown in Fig. 19.8. One can see that they become closer to each other, as a result of attraction. Also, they all fall into curved z direction. On the boundary, this motion is seen as coalescing into some common "fireball."

19.7   Brief Summary

- The so-called "top-down" holography focuses on case in which gauge-string duality can be proven. One may start with AdS/CFT and then deform the setting in a certain way. The opposite "down-up" approach is to *assume* such correspondence, model certain bulk Lagrangian, solve the problem, and then calculate its hologram on the boundary (where our gauge fields live).
- Witten and Sakai-Sugimoto models are examples of the first kind. They started from D_4 branes and then add D_8 ones, at which fundamental fermions (quarks) live. Certain geometry models vector and axial mesons, their interactions, etc.
- Instanton solution in gauge theory can be used in a bulk, with vector meson fields, as a model of baryons.
- Defining metrics with certain action for dilaton, cutting space off by some "walls," one finds background with confinement. The Coulomb law is appended by linear potential. Spectroscopy takes the form of well-defined Regge trajectories, including the Pomeron.
- One can discuss another kind of wall, from UV side, separating weak and strong coupling domain. This wall is the location of the "instanton liquid."
- Holographic QCD models describe QCD spectroscopy and QGP thermodynamics well, provided the so-called Veneziano limit (19.26) is kept. We discussed example of scalar mesons and glueballs and their mixing.
- Strings sit in effective potential with a minimum at certain distance from the boundary. This distance holographically defines the width of QCD flux tubes. One can calculate interaction between flux tube and compare it to lattice and phenomenology. We also discussed multi-string dynamics, including cases with system imploding into a common QGP fireball.

References

Anderson, N., Domokos, S.K., Harvey, J.A., Mann, N.: Central production of η and η? via double Pomeron exchange in the Sakai-Sugimoto model. Phys. Rev. **D90**(8), 086010 (2014)

Bigazzi, F., Caddeo, A., Cotrone, A.L., Di Vecchia, P., Marzolla, A.: The holographic QCD axion. J. High Energy Phys. **2019**(12), 56 (2019)

Erlich, J., Katz, E., Son, D.T., Stephanov, M.A.: QCD and a holographic model of hadrons. Phys. Rev. Lett. **95**, 261602 (2005)

Ewerz, C., Lebiedowicz, P., Nachtmann, O., Szczurek, A.: Helicity in proton-proton elastic scattering and the spin structure of the pomeron. Phys. Lett. **B763**, 382–387 (2016)

Faccioli, P., Schwenk, A., Shuryak, E.V.: Instanton contribution to the pion and proton electromagnetic form-factors at Q**2 approximately greater than 1-GeV**2. Fizika **B13**, 193–200 (2004)

Frankfurt, L., Miller, G.A., Strikman, M.: Coherent nuclear diffractive production of mini - jets: Illuminating color transparency. Phys Lett **B304**, 1–7 (1993)

Gursoy, U., Kiritsis, E.: Exploring improved holographic theories for QCD: Part I. J. High Energy Phys. **20008**(02), 032 (2008). arXiv:0707.1324

Gursoy, U., Kiritsis, E., Nitti, F.: Exploring improved holographic theories for QCD: Part II. J. High Energy Phys. **2008**(02), 019 (2008)

Gursoy, U., Kiritsis, E., Mazzanti, L., Michalogiorgakis, G., Nitti, F.: Improved holographic QCD. Lect. Notes Phys. **828**, 79–146 (2011)

Hata, H., Sakai, T., Sugimoto, S., Yamato, S.: Baryons from instantons in holographic QCD. Prog. Theor. Phys. **117**, 1157 (2007)

Iatrakis, I., Ramamurti, A., Shuryak, E.: Collective string interactions in AdS/QCD and high-multiplicity pA collisions. Phys. Rev. **D92**(1), 014011 (2015). arXiv:1503.04759

Iatrakis, I., Ramamurti, A., Shuryak, E.: Pomeron interactions from the Einstein-Hilbert action. Phys. Rev. **D94**(4), 045005 (2016)

Jarvinen, M., Kiritsis, E.: Holographic models for QCD in the Veneziano limit. J. High Energy Phys. **2012**(03), 002 (2012)

Karch, A., Katz, E., Son, D.T., Stephanov, M.A.: Linear confinement and AdS/QCD. Phys. Rev. **D74**, 015005 (2006a)

Polchinski, J., Strassler, M.J.: Hard scattering and gauge/string duality. Phys. Rev. Lett. **88**, 031601 (2002)

Rho, M., Sin, S.-J., Zahed, I.: Dense QCD: a holographic dyonic salt. Phys. Lett. **B689**, 23–27 (2010)

Sakai, T., Sugimoto, S.: Low energy hadron physics in holographic QCD. Prog. Theor. Phys. **113**, 843–882 (2005)

Schafer, T.: Euclidean correlation functions in a holographic model of QCD. Phys. Rev. **D77**, 126010 (2008)

Schafer, T., Shuryak, E.V.: Implications of the ALEPH tau lepton decay data for perturbative and nonperturbative QCD. Phys. Rev. Lett. **86**, 3973–3976 (2001a)

Shuryak, E.: Building a 'holographic dual' to QCD in the AdS(5): Instantons and confinement (2006). arXiv:hep-th/0605219

Shuryak, E.: A 'Domain Wall' Scenario for the AdS/QCD (2007). arXiv:0711.0004

Shuryak, E.V., Verbaarschot, J.J.M.: Chiral symmetry breaking and correlations in the instanton liquid. Nucl. Phys. **B341**, 1–26 (1990)

Witten, E.: Anti-de sitter space, thermal phase transition, and confinement in gauge theories. Adv. Theor. Math. Phys. **2**, 505–532 (1998b). [,89(1998)]

Summary

20

At the end, let me summarize again the main content of these lecture notes.

20.1 Semiclassical Theory

The semiclassical theory, in quantum mechanics and QFTs, is the main technical method on which the applications were based. It always implies existence of some parameter due to which the action of certain classical solutions is considered large $S \gg \hbar$. These solutions generally correspond to extrema of the action (or path integrand $\sim exp(-S[x(\tau)])$) in imaginary (Euclidean) time, the maxima or sometimes saddles.

The contributions of those extrema to the path integral can be then calculated by writing the path (in QM) or field configuration (in QFT) as classical+quantum fluctuation $\phi = \phi_{cl} + \delta\phi$ substituting it to the action and reading from it the *Green functions* (propagators)—inverse to the operator of order $O(\delta\phi^2)$—and *vertices* of higher orders in $\delta\phi$. One can then use standard Feynman diagrams to calculate the effects of these fluctuations to the desired order.

The simplest example of classical Euclidean paths are *fluctons*, for which this procedure is very straightforward. In quantum mechanics, they are dependent on the point x_0 (in QFTs on the field configuration $\phi_0(x)$) for which the density matrix $P(x_0)$ is evaluated. Semiclassical theory requires the corresponding action be large $S(x_0) \gg \hbar$.

We also discussed quantum-mechanical *instantons*, the tunneling events in quantum mechanics, and *monopoles, instantonss and instanton-dyons* in QFTs. In all those cases, large action was due to presumed small coupling appearing in denominator $S \sim 1/g^2 \gg 1$.

In all those cases, there are *bosonic zero modes* induced by symmetries of the solutions, like displacements, scale change, rotations, etc. Zero modes correspond

477
E. Shuryak, *Nonperturbative Topological Phenomena in QCD and Related Theories*, Lecture Notes in Physics 977,
https://doi.org/10.1007/978-3-030-62990-8_20

to non-Gaussian integrals: those over *collective coordinates* were defined. A general
new feature is appearance of the delta function ensuring that quantum fluctuations
are orthogonal to all zero modes. The Jacobian generated by this delta function leads
to new Feynman diagrams, not coming from the Lagrangian.

While the amplitudes for *quantum-mechanical instantons* has been calculated for
a number of models up to three loops, for the QFTs and *gauge field instantons*, the
semiclassical theory has only been carried to the one-loop (determinant) order.

(In one case, the static magnetic balls known as *sphalerons*, there is a mode
corresponding to instability of these solutions. The classical solution corresponding
to sphaleron decay, in Minkowski space-time, has also been discussed.)

After (i) single-soliton semiclassical amplitudes are established, the next steps
are:

(ii) calculation of the soliton interactions;
(iii) formulation of the partition function of their ensemble;
(iv) analytic or numerical integration over all collective coordinates.

The interaction can be of *classical* or *quantum* origin. In general, extrema
of integrals are connected by certain "functional lines," called "streamlines" or
"Lefschetz thimbles," [1] being solutions of the gradient flow equation containing the
force $\partial S / \partial \phi$. We met three examples of those: for instanton-anti-instanton pairs
in QM and gauge theory and for instanton-dyon-antidyon pairs. In the dyon-dyon
case (when classical interaction is absent), we also studied the quantum one-loop
interaction, generated by the determinant in two-soliton background.

Significant role in our discussion has been played by the *fermionic zero modes*.
For monopoles, those are just bound states of fermions; for QFT instantons, those
lead to 't Hooft effective Lagrangian, in QCD with $2N_f$ fermion legs and in
electroweak sector with nine quarks and three lepton left-handed legs. Fermion-
induced forces between instantons can be considered as diagrams with this effective
Lagrangian.

The semiclassical theory defines amplitudes in a form of *transseries*, complementing perturbative series by terms of the order of

$$e^{-\frac{const}{g^2}} \ (instanton \ series \ in \ g)$$

---

[1]In many cases, for the usual integrals as well as the QM and QFT settings, it may require
complexification of the paths/configurations. The notion of "thimbles" explains and generalizes
the famous "Stokes phenomena" known for the ordinary integrals and special functions, and it is
related to the phase transitions in the partition functions in QFT and statistical mechanics settings.
All of them can be seen as thimble reconnections, happening at some values of the parameters.
Theorems by Lefschetz help us to explain when thimbles can or cannot cross, but in practice their
geometry is often quite complex and is studied on case-by-case basis.

as well as terms with exponent and powers of $\log(1/g^2)$ coming from the "stream-lines."In QM applications, certain *resurgence relations* are known, relating series in powers of g for all these terms. In QFT case, such relations are not (yet?) known.

20.2 The QCD Vacuum and Instantons

I was asked many times the following question:*Why should the semiclassical theory of topological solitons be studied, in view of the fact that lattice gauge theories anyway include automatically all of these objects ?* Indeed, the lattice simulations do reproduce hadronic spectrum, (Euclidean) correlation function equations of state of hot matter, and many other observables.

The answer is that the degrees of freedom associated with the topological objects are orders of magnitude more important than zillions of gauge degrees of freedom involved in these simulations. We discussed various versions of "cooling," removing a lot of noise, and revealing the underlying topological structure above.

But probably the most striking example is the appearance of topology-induced 't Hooft effective Lagrangian, violating some of the symmetries which the original QCD Lagrangian has.

The topological solitons have zero modes, which are also topological in nature, and thus cannot be moved by any small deformations. So, ignoring the perturbative noise, they generate effective multi-fermion operators. Without understanding of the topology and the corresponding the index theorems, and these zero modes themselves, it would be impossible to figure out how interaction between quarks inside hadrons works.

Moreover, the topology-induced interactions between quarks are not just some curios addition to pQCD, but they are in fact the strongest forces shaping the observed hadronic spectrum. It is this interaction which makes pions π near-massless and σ quite light,

$$m_\pi^2 \approx 0, \quad m_\sigma^2 \sim 0.2 \, \text{GeV}^2$$

while η' and spin-0 isospin-1 a_0 meson (we also called by its old name δ) are nearly as heavy as the nucleon,

$$m_{\eta'}^2 \sim m_\delta^2 \sim 1 \, \text{GeV}^2$$

We systematically studied the QCD point-to-point correlation functions in the corresponding chapter and had shown how their splitting develops as a function of the distance between the operators, "probing" the QCD vacuum, and have seen that they are indeed stem from the topology-induced effective 't Hooft Lagrangian.

Moreover, the distances at which such nonperturbative phenomena turn out to be rather small indicated that these topologically nontrivial gauge fields are rather strong. They are even more important for spin-zero gluonic operators, dominating the perturbative effects (and thus limiting any pQCD applications) for momentum transfer $Q^2 < 10 \, \text{GeV}^2$.

We have shown above that the gauge topology is key to understanding of the $SU(N_f)$ and $U(1)_a$ chiral symmetries. Lattice practitioners know well the Casher-Banks relation, relating the density of the Dirac eigenvalues near zero to the quark condensate. Much less widely known is the notion of the *zero modes zone* (ZMZ), a thin layer of Dirac eigenvalues made out of collectivized zero modes. All textbooks repeat the standard notion of the light quark masses m_u, m_d, m_s, as being much smaller compared to some strong coupling scale Λ_{QCD} and thus justifying a Taylor expansion in their powers. In reality, many masses used on the lattice are actually comparable to the ZMZ width.

As a result, lattice practitioners continue to be surprised by large deviations from linear chiral perturbation theory, for quark masses they routinely use. The reason for that is an observation—quite elementary in terms of instanton ensembles and going back to the early 1980s—that the hopping matrix elements for quark jumping between topological solitons, and thus the width of the ZMZ, are of the order of only 20–30 MeV, an order of magnitude smaller than Λ_{QCD}.

The reason the ZMZ has small width is also of the topological origin. No matter how small or large are perturbations of the solitons, their topology remains integer-valued, and their zero modes remain *unperturbed*, at zero Dirac eigenvalues. The only effect which does perturb them is a presence of another topological anti-soliton nearby.

Finally, let us talk about some practical effects of topology. Modern super-computers simulate path integrals in QCD-like gauge theories with millions of variables: and yet the results—such as correlation functions, hadronic masses, and other properties—are still subject of large fluctuations, from configuration to configuration. Also some observables—e.g., mean topological charge $< Q >$—may show nonzero values, in contradiction to CP invariance of the theory. The reason for both phenomena is the fact that local update algorithms used are notoriously inefficient in updating the topology of the configurations. This can in principle be improved by new algorithms able to identify/create/destroy the topological objects and appropriately update their collective coordinates. Furthermore, even the largest simulated volumes contain only $O(10)$ topological solitons, not a very large number.

20.3   Magnetic Monopoles and the Near-T_c QCD Matter as a Dual Plasma

In one dimension, the topological solutions are *kinks* also known as the *domain walls*. In two Dimensions, one has *vortices* or *flux tubes*. Their mutual linking also creates rich topological structures, e.g., in turbulent flows or Sun plasma. Such structure also underlines the *confining QCD strings* and multi-string configurations produced in high-energy collisions.

Yet in these lectures, we jumped over those right to the three-dimensional soliton, the *magnetic monopole*. While monopoles were never found in electrodynamics, in the non-Abelian gauge theories, they were discovered by t' Hooft and Polyakov. Recall that the setting was the Georgi-Glashow model, possessing not only the gauge field but also the *adjoint scalar*.

Since adjoint scalars naturally appear in *extended supersymmetric theories* $\mathcal{N}$ =2,4, the monopoles were extensively studied in that setting. Recall that supersymmetry forbids appearance of nonzero vacuum energy; therefore vacua with different VEVs of the scalars are all degenerate and form the so-called moduli space. Seiberg and Witten have found that in the $\mathcal{N}$=2 theory at weak coupling, monopoles are heavier than "electric" $W^{\pm}$ particles, but at certain points, the coupling goes to infinity and monopole (or dyon) mass becomes zero. The renormalization group flow connects those two limits smoothly. So, the same theory appears as weakly coupled non-Abelian theory with asymptotic freedom in one end of the moduli space and as weakly coupled dual (magnetic) electrodynamics with monopoles. The Dirac condition—product of electric times the magnetic coupling must be an integer—is preserved everywhere.

Fermions (gluinoes and quarks) have bound state with the monopoles. If SUSY is unbroken, those make certain supermultiplets. This statement has been explicitly verified semiclassically. The $\mathcal{N}$=4 theory was the most important case: it was found to be *electric-magnetic self-dual*. This fact, without any perturbative diagrams calculated, explains why it must have *zero* beta function: it should be the same for g and 1/g. (Quarks in QCD also should be able to be bound to monopoles: these issues are however not yet studied in any details.)

The monopoles in pure gauge and QCD-like theories were studied extensively by lattice numerical simulations. We know that their density increases toward T_c, and at $T < T_c$, they form Bose-Einstein condensate.

We discussed phenomenology of the hadronic matter near the QCD phase transition, its thermo and kinetic properties, related to heavy-ion collision experiments at RHIC and LHC colliders. While the equation of state predicted by the lattice was well confirmed by hydro explosions, the subsequent discoveries yield rather unexpected values of the kinetic coefficients, such as entropy-density-to-sheer-viscosity s/η ratio, heavy quark diffusion coefficient D/T, and jet quenching parameter $\hat{q}/T^3$. We argue further that all of them have rather peculiar T-dependence, indicating extremely small mean free path of the matter constituents, especially between T_c and $2T_c$. This transition is the deconfinement transition.

Rescattering between constituents is proportional to their densities. Yet the (dimensionless) densities of quarks and gluons $n_{q,g}/T^3$ rapidly decrease as temperature goes down to critical value T_c, due to confinement. The only density which is *peaked* at T_c is that of the magnetic monopoles n_m. This argument, and more detailed calculations we discussed in that chapter, indicates that significant density of the particle-monopoles is the most probable cause of the unusual sQGP kinetics observed experimentally.

In short, QGP near T_c seems to be a "dual plasma," containing a density of monopoles comparable to that of "electric" objects, quarks, and gluons. A number of theoretical methods were used subsequently to study such dual plasmas, from classical molecular dynamics to quantum cross section and kinetic theory, to quantum path integral Monte Carlo (PIMC). Those reproduce correlation functions of monopoles observed on the lattice in detail.

20.4   Instanton-Dyons, Deconfinement and Chiral Restoration Phase Transitions

Another version of a semiclassical theory at $T \sim T_c$ is based on other classical solutions, the *instanton-dyons*.

Incorporation of *nonzero VEV of the Polyakov line*—called holonomy— leads to a shift from studies of instantons to studies of their constituents: *instanton-dyons* or instanton-monopoles. Recent papers on their ensembles, done by a variety of methods, we discussed in the previous chapter lead to very significant advances.

Unlike instantons, these objects have three different set of charges, *topological, magnetic, and electric*, therefore back-reacting on the holonomy potential. The calculations showed that at sufficiently large density of the dyons, the minimum of the free energy *shifts to confining value of the holonomy*, at which the mean value of the Polyakov line vanishes, $< P >= 0$. This creates the so-called symmetric phase, in which all types of the dyons obtain equal actions and densities.

Unlike instantons, the dyons possess magnetic charges, and thus their ensemble generates *the magnetic screening mass*.[2] While we have not discussed it above, let me just mention that it clearly indicates a transition from electric (QGP) to magnetic plasma, as the coupling grows with decreasing temperature.

Further remarkable findings were obtained using a very sensitive tool, a deformation of the QCD-like theories into those with *nonzero flavor holonomies* θ_f, also known as imaginary chemical potentials, or modified periodicity phases. The so-called Roberge-Weiss transitions, originally postulated at high T only on perturbative grounds, are not confirmed to be related with the "hopping" of quark zero modes from one type of dyon to the next. On the way from ordinary QCD to the "most democratic" version called "$Z(N_c)$ QCD" (with isotropic distribution over these flavor holonomies), one observes dramatic modifications of the phase transition. While the deconfinement transition is strengthened, from crossover to strong first-order transition, the chiral symmetry restoration transition is weakened instead, finally disappearing at any finite temperature.

These findings complement another kind of "deformed QCD," in which quarks are substituted by adjoint gluinoes periodically compactified on Matsubara circle. The theory with *one* such gluino, $\mathcal{N}=1$ SYM ($N_a = 1$), preserves both confinement and chiral symmetry breaking all the way to very small circle (very high "T"), eliminating phase transitions. The theory with *two* such gluinoes shows even more spectacular behavior, with three phase transitions in the Polyakov line and

---

[2]Recall that perturbative polarization tensor does not generate it (Shuryak 1978): but, according to lattice data, in the near-T_c region, it even surpasses the electric mass.

four subsequent phases (confined-deconfined another deconfined-re-confined), all of them in the chirally broken phase!

One should stress, that flavor holonomies constitute very "soft" modification of the theory: the fields and Lagrangian remain the same, only periodicity condition on the Matsubara circle is changed. All perturbative physics and RG flow of the theory remain unchanged—and yet these phase transitions are changed dramatically. It would be impossible to explain it using theory of instantons alone: the number of their fermionic zero modes is defined solely by the topological charge and is insensitive to these holonomies. While those changes appear natural in the framework of the instanton-dyon theory, so far no other known explanation of them exists. It is a significant challenge now to any other model of the deconfinement and chiral transitions to explain these phenomena. Let us even speculate that we are now perhaps approaching "the point of no return," at which the mechanism of QCD phase transitions we discussed in this book is going to be finalized.

We also discuss lattice studies based on the so-called "fermion method" to study underlying topology. It is unmistakably found to be the instanton-dyons. The semiclassical model can be extensively checked on the lattice: so far we see unexpectedly high accuracy of its predictions, at least for the lowest Dirac eigenstates.

20.5   The "Poisson Duality" Between the Monopole-Based and the Instanton-Dyon-Based Descriptions

The physics behind this duality can be explained in very general terms. For a description of a thermal system of any particles, one can adopt two well-known strategies, which produce two different forms of the partition function.[3] Needless to say, they both describe the same system and thus must be identical. While to show its equivalence explicitly is often difficult, as they are simple in two opposite limits, of low and high T, it was in fact possible in certain highly symmetric examples, e.g., extended supersymmetry.

The standard *approach 1* is to find all the states of the system and perform the usual statistical sum $Tr[exp(-\hat{H}/T)]$. It is working best at low temperatures, where only some lowest states need to be included.

The *approach 2* is to go to Euclidean formulation and evaluate the partition sum using the path integral over the paths periodic in the Matsubara time. Those can be classified by their winding number (also known as BEC cluster number). This approach works best at high T or small Matsubara circumference $\beta = 1/T$, in which case the paths with zero winding number dominate.

In order to see how these two approaches work for monopoles, one needs a setting in which both the monopoles and instanton-dyons are well-defined semiclassical objects. Convenient settings thus include theories with extended supersymmetry,

---

[3]One may call them the Hamiltonian and the Lagrangian approaches, respectively.

because those have scalar fields and 't Hooft-Polyakov monopoles. Both of these theories attracted a lot of theorist's attention in the 1990s: some of that will be useful for our current goal, which can be formulated as *getting some understanding on interrelation between the physics of particle-monopoles and instanton-dyons.*

In Sect. 15.1.3, we had discussed QCD with two (or more) adjoint gluinos. By adding appropriate number of scalars, those theories can be upgraded to theories with extended supersymmetry. Specifically, adding one complex scalar a to the $N_a = 2$ theory, one gets $\mathcal{N}=2$ SYM, and by adding six scalars to the $N_a = 4$ theory, one gets $\mathcal{N}=4$ SYM. This has been done in literature, first for $\mathcal{N}=4$ SYM by N.Dorey and collaborators (Dorey 2001; Dorey and Parnachev 2001; Chen et al. 2010) and then also for $\mathcal{N}=2$ SYM by Poppitz and Unsal (2011).

What is common to all those papers is that they start by compactifying one of the dimensions to a circle S^1 of a circumference β. In contrast to the thermal theory, however, the gluino fields are assumed to be *periodic* on this circle, and therefore the supersymmetry is *not* going to be broken. We will still call this direction the zeroth one. Furthermore, the compactification allows one to define two holonomies, the Polyakov loop with the A_0 field and the magnetic field with a dual magnetic potential. Dorey et al. call those ω and σ, respectively, and their VEVs can be considered two main parameters of the settings, together with β. In order to make the discussion simpler, one assumes the minimal number of colors $N_c = 2$, in which there is only one diagonal generator τ^3, breaking $SU(2) \rightarrow u(1)$, so these VEVs ω and σ are just parameters.

On top of that, the $\mathcal{N}=4$ theory is discussed[4] in the so-called Coulomb branch, which means that, on top of the holonomies ω and σ, one—or more of six available—scalar is also assumed to have a nonzero VEV, called ϕ, for the same color-diagonal component. This of course leads to monopoles with the action

$$S_m = \left(\frac{4\pi}{g^2}\right) \sqrt{\beta^2 |\phi|^2 + |\omega - 2\pi n|^2} \qquad (20.1)$$

including the contribution from the scalar VEV ϕ, electric holonomy ω, and the winding number of the path in the S^1 circle n. The $n = 0$ term is what we called above the M-type instanton-dyon and the $n = -1$ the L dyon, and higher n corresponds to the paths with stronger time-dependent twists. We would not derive the partition function[5] but just present the resulting expression

$$Z_{inst} = \sum_{k=1}^{\infty} \sum_{n=-\infty}^{\infty} \left(\frac{\beta}{g^2}\right)^9 \frac{k^6}{(\beta M)^3} exp\left[ik\sigma - \beta kM - \frac{kM}{2\phi^2\beta}(\omega - 2\pi n)^2\right] \qquad (20.2)$$

---

[4]Dorey et al also discussed two brane constructions corresponding to two Poisson-dual formulations, which we would not discuss here.

[5]We simplify it here a bit, compared to the original paper, by putting one more external parameter of the setting, the CP-odd θ angle, to zero.

where $M = (4\pi\phi/g^2)$, the BPS monopole mass without holonomies, and thus the second term in exponent is interpreted as just the Boltzmann factor. The index k is the magnetic charge of the configuration. Note that it appears in the exponent times the magnetic holonomy σ, and since k is an integer, the expression is periodic in σ with the 2π period, as it should. The second index n is the winding number of the path in the β circle. Note that all values of n need to be included in the sum, because Z should be periodic in the electric holonomy ω as well as in σ. Finally note that the last term of the exponent has unusual position of β (or "temperature" in the numerator): the sum over n therefore converges better at small β (high T) limit.

Now we switch to other description, which is better convergent in the opposite case, of large β and low T. It operates with states of motion of monopoles in its four collective coordinates. Three of those are locations of the monopole and are included in trivial way. The fourth collective coordinate is the angle α of monopole color rotation which preserves the holonomy, $\Omega = exp(i\alpha\tau^3)$ defined on *another* circle S^1. Therefore, the problem includes a "quantum rotator." As was explained by Julia and Zee, the corresponding integer angular momentum q is nothing but the electric charge of the rotating monopole. The partition function looks in this approach as follows:

$$Z_{mono} = \sum_{k=1}^{\infty} \sum_{q=-\infty}^{\infty} \left(\frac{\beta}{g^2}\right)^8 \frac{k^{11/2}}{\beta^{3/2}M^{5/2}} exp\left(ik\sigma - iq\omega - \beta kM - \frac{\beta\phi^2 q^2}{2kM}\right)$$

(20.3)

Now both holonomies appear with appropriate integers, so the periodicity in σ, ω is as required. The last term in exponent is due to the kinetic energy of the rotation, with angular momentum squared and the rest being nothing else but the momentum of inertia of the monopole (in denominator). Note also that β (or temperature) is in this term in the usual thermal position, so that the sum in Z_{mono} is more suppressed at higher β or lower temperature.

We have copied those expressions from the original work in order to demonstrate the key statement of this section, pointed out by Dorey and called the *Poisson duality*: the two seemingly different expressions, addressing seemingly different motions in two different circles, do in fact lead to the *same* partition function:

$$Z_{inst} = Z_{mono}$$

Indeed, performing the sum over q in the latter expression—the discrete Fourier transform of a Gaussian—one gets the so-called *periodic Gaussian* given by the sum over n in the former expression.

In fact, both expressions generate the same function. Let us understand why it is so. The first exponents (outside of the sums) are the same for obvious reasons: the masses of the particle-monopole and instanton-dyon are the same, as so are their magnetic charge—thus $e^{i\sigma}$. To understand what is happening with the sums, let us simplify the issue to the simplest problem possible, in this case a particle moving in a circle. Thermal Euclidean time theory is thus defined on two circles $S^1 \times S^1$.

The low-T theory starts with defining the spectrum of excitations. As usual, since there is no dependence on the position on the circle, the angular momentum l is conserved (commutes with the Hamiltonian), and the excited states are numerated by it. The spectrum of a rotator is $E_l = l^2/2mR^2$ where m is the particle mass and R is the circle radius. Furthermore, if there is a magnetic flux Φ through the circle, so that our particle gets the Aharonov-Bohm phase, the spectrum shifts to $E_l = (l - e\Phi)^2/2mR^2$. The partition function is $Z = \sum_l exp(-E_l/T)$.

The dual description at high T describes paths of the particle in terms of how many times it is "winding" around the circle. At high T, the thermal circle is very short, so most of the periodic paths would be approximately time-independent. But as the thermal circle gets longer, there appear paths in which a particle rotates by additional $2\pi n_w$ times around the spatial circle. It is not difficult to calculate the action for such paths and get another representation for Z, better converging at high T. Since both descriptions correspond to the same quantum mechanical problem, one should not be surprised that both of them give the same Z.

The next step was to apply this Poisson duality to QCD, calculating semiclassical sum with all winding numbers into the monopole-like identical sum. What was found is that in the QCD case, the monopole action is $S \sim \log(1/g^2) \sim \log(\log(T))$, and thus the density

$$exp(-S) \sim \frac{1}{\log(T)^2}$$

This explains long-known lattice data on the monopole density. It also tells us that in pure gauge and QCD-like theories, the monopoles are *not* classical objects(!)

20.6   Holography and QCD Strings

On the theory side, interest to electric-magnetic duality has been superseded by the holographic dualities of the AdS/CFT type , at the end of the 1990s . Originally found for conformal $\mathcal{N}$=4 SYM, it has been extended, by softly broken supersymmetry, to $\mathcal{N}$=2 and so on. The Seiberg-Witten elliptic curve has been reformulated in terms of certain brane constructions. While the top-down holography extension to $\mathcal{N}$=1 and to non-supersymmetric ($\mathcal{N}$=0) theories meets with problems, multiple *down-to-top* models, commonly known as AdS/QCD, were developed. As we have discussed in the corresponding chapter, those describe rather well many aspects of hadronic spectroscopy, thermal equation of state, and even kinetic quantities like viscosity.

The holographic approach also revitalized discussion of *QCD strings*, now appeared just as holograms of the fundamental strings in the ten-dimensional "bulk." New way of looking at Reggeons and especially Pomerons came about. As an example, consider collapse of multi-string configurations created in collisions.

Let me also remind that the instanton-dyons were co-discovered using a particular brane construction by Lee and Lu (1998) at the same time as the explicit solution in terms of gauge fields was found by Kraan and van Baal (1998). And yet, to my test, too little has been done in this direction from the holographic models.

References

Chen, H.-Y., Dorey, N., Petunin, K.: Wall crossing and instantons in compactified gauge theory. J. High Energy Phys. **06**, 024 (2010). arXiv:1004.0703

Dorey, N.: Instantons, compactification and S-duality in N=4 SUSY Yang-Mills theory. 1. J. High Energy Phys. **04**, 008 (2001). arXiv:hep-th/0010115

Dorey, N., Parnachev, A.: Instantons, compactification and S duality in N = 4 SUSY Yang-Mills theory. 2. J. High Energy Phys. **08**, 059 (2001). arXiv:hep-th/0011202

Kraan, T.C., van Baal, P.: Monopole constituents inside SU(n) calorons. Phys. Lett. B **435**, 389–395 (1998). arXiv:hep-th/9806034

Lee, K.-M., Lu, C.-H.: SU(2) calorons and magnetic monopoles. Phys. Rev. D **58**, 025011 (1998). arXiv:hep-th/9802108

Poppitz, E., Unsal, M.: Seiberg-Witten and 'Polyakov-like' magnetic bion confinements are continuously connected. J. High Energy Phys. **07**, 082 (2011). arXiv:1105.3969

Shuryak, E.V.: Theory of Hadronic plasma. Sov. Phys. J. Exp. Theor. Phys. **47**, 212–219 (1978) [Zh. Eksp. Teor. Fiz.**74**, 408(1978)]

Conventions for Fields in Euclidean vs Minkowskian Space-Time

A

A.1 The Gauge Fields

The QED/QCD gauge part of the Lagrangians is

$$S = -\frac{1}{4} \int d^4x \, \left(G^a_{\mu\nu}\right)^2$$

where the QCD field

$$G^a_{\mu\nu} = \partial_\mu A^a_\nu - \partial_\nu A^a_\mu + g f^{abc} A^b_\mu A^c_\nu$$

where f^{abc} are structure constant of the $SU(N_c)$ Lee algebra (for SU(2), it is ϵ_{abc}). Alternative form of the gauge fields is a matrix notation, in which the generator is included together with fields $A_\mu \equiv A^a_\mu T^a$: then the second term is the commutator.

In pQCD, one uses the so-called *perturbative definition* in which the coupling constant is explicitly written in the nonlinear terms, while the kinetic terms are free of it. We will use the nonperturbative definition, used in lattice and instanton studies, which is obtained by an inclusion of g into $\tilde{A} = gA$ and $\tilde{G} = gG$ so that g no longer appears in front of the nonlinear terms but is placed instead in front of the action $S = \frac{-1}{4g^2} \int d^4x \, \tilde{G}^2$.

The transition to Euclidean time is done by

$$A^M_0 = i A^E_4 \qquad A^M_m = -A^E_m$$

© The Author(s), under exclusive license to Springer Nature Switzerland AG 2021
E. Shuryak, *Nonperturbative Topological Phenomena in QCD and Related Theories*, Lecture Notes in Physics 977,
https://doi.org/10.1007/978-3-030-62990-8

Note the minus sign on spatial components, different from what happens with coordinates themselves. This is done in order not to modify covariant derivatives, so that both E and M read as

$$iD_\mu = i\partial_\mu + \frac{g}{2}A_\mu^a t^a$$

A.2   Fermionic Path Integrals

Introduction of fermion fields into the path integral needs special definitions appropriate for Grassmannian (anti-commuting) variables. Those have been defined in a classic work (Berezin 1975) and is discussed in any modern textbook on QFTs. Additional subtleties appear with its transformation into the Euclidean space-time, in which $\bar{q}$ and q must be treated as independent variables.

I have to explain in what sense the integral over ψ should be defined. It is a fermionic variable, not just the ordinary field. They are called Grassmann variables, or anti-commuting ones $\chi_1\chi_2 = -\chi_2\chi_1$. In particular, for such variables, one has $\chi^2 = 0$, which represents the Pauli principle. These variables have funny rules for the integrals. They are given essentially by two basic integrals:

$$\int d\chi = 0; \qquad\qquad \int \chi d\chi = 1 \qquad\qquad (A.1)$$

One can then derive the following formula

$$\int exp\left(-\chi_k^* M_{k,l}\chi_l\right) d\chi_1^* d\chi_1 \cdots d\chi_N^* d\chi_N = det M \qquad (A.2)$$

which can be proved in the eigenvector basis by decomposition of the exponent. Note that the ordinary Gaussian integral with N variables of such type is equal to

$$\int exp\left(-\sum_{k,l}\phi_k M_{k,l}\phi_l/2\right) d\phi_1 d \cdots d\phi_N = \frac{(2\pi)^{N/2}}{\sqrt{det M}} \qquad (A.3)$$

and the determinant stands in the denominator. So, if one of the eigenvalues is zero, the fermionic integral vanishes, while in the corresponding bosonic one diverges.

Note also in passing that one fermionic integral can compensate two bosonic ones, if the eigenvalue spectra happen to be equal. This comment is important for supersymmetric theories, in which fermionic and bosonic integrals do indeed compensate each other, in the vacuum energy and many other observables.

Fortunately, all of the above can be bypassed because a general integration over the fermions in QCD can be made in a simple form because those appear in the action only *linearly*. The schematic master formula for it is

$$\int D\bar{q}\,Dq\,e^{\bar{q}Mq} = det(M) \tag{A.4}$$

where M is in general a matrix in all fermionic indices and also possibly a differential operator.[1] In QCD, $M = i D_\mu \gamma_\mu$ is the Dirac operator, with the color matrix in covariant derivative obviously in fundamental representation. The final comment is, as the operator $i\hat{D}$ is Hermitian, its eigenvalues λ are all real. However, one may still ask how it happens that at nonzero mass (entering with i in the Euclidean formulation)

$$\prod_f det\left[i\hat{D}(A_\mu(x)) + im_f\right]$$

and non-positive λ, the ratio of the determinants happen to be positive (otherwise one cannot use probability language). We noticed that the former Dirac operator is Hermitian; therefore its eigenvalues λ are real. Due to chiral symmetry, they go in pairs, and therefore we have always

$$\prod_{\lambda>0}(\lambda + im)(-\lambda + im) = -\prod_{\lambda>0}\left(m^2 + \lambda^2\right),$$

so this factor is real too and even has definite sign and can be made positive. This is important, as we want to prescribe to it the meaning of the probability of occurrence of the corresponding configuration in the vacuum.

A.3   Quark Fields

We denote quark fields as ψ or q, usually omitting but implying their spinor index $\alpha = 1..4$, color $i = 1..N_c$, and flavor $f = 1..N_f$. The QED/QCD Lagrangian is

$$S = \int d^4x\,\bar{q}(i\gamma_\mu D_\mu - m)q$$

The transition to Euclidean time is done by

$$q^E = q^M, \quad \bar{q}^M = -i\bar{q}^E,$$

---

[1] To prove it, imagine that the operator is diagonalized, the exponent is expanded, and the "Pauli principle" in the form $q^2 = 0$ holds.

and the gamma-matrices change by

$$\gamma_4^E = \gamma_0^M, \ \gamma_m^E = -i\gamma_m^M,$$

We remind that the anti-commutators are

$$\left\{\gamma_\mu^M, \gamma_\nu^M\right\} = 2g_{\mu\nu} \qquad \left\{\gamma_\mu^E, \gamma_\nu^E\right\} = 2\delta_{\mu\nu}$$

Another often used notation is a slash or a hat, indicating a convolution of a 4-vector with the gamma matrices. For example, the relations just described can be written using these notations as

$$\hat{a}\hat{b} + \hat{b}\hat{a} = (ab)$$

where a_μ, b_μ are any 4-vectors.

Finally, the Euclidean fermionic action looks like

$$S^E = -iS^M = \int d^4x\,\bar{q}(-i\gamma_\mu D_\mu - im)q$$

Let me explicitly mention the definitions used here:

$$\psi_e = \psi_M, \ \bar{\psi}_M = -i\bar{\psi}_E, \ \gamma_E^0 = \gamma_M^0, \ \gamma_E^m = -i\gamma_E^m$$

We get

$$S_E^f = \int d^4x\,\bar{\psi}_E(i\hat{D} + im)\psi$$

thus the complete partition function of QCD is

$$Z = \int DA_\mu(x)D\bar{\psi}D\psi\,exp\left(-S_E - S_E^f\right)$$

Perturbative QCD

B

B.1   Renormalization Group and Asymptotic Freedom

A systematic approach using continuous renormalization group (RG) has been developed originally by Gell-Mann and Low (1954) and describes the renormalization of the charge as a function of (momentum) scale. They introduced the so-called beta function defined as the charge *derivative* over the scale magnitude at which it is defined

$$\frac{\partial g}{\partial \log \mu} \equiv \beta(g) \tag{B.1}$$

If the beta function is known, this equation can be integrated:

$$\log\left(\frac{\mu}{\mu_0}\right) = \int_{g_0}^{g} \frac{dg'}{\beta(g')} \tag{B.2}$$

For small coupling values, the beta function can be determined perturbatively, and its standard definition includes the so-called first and second beta function coefficients

$$\beta(g) = -b \frac{g^3}{16\pi^2} - b' \frac{g^5}{\left(16\pi^2\right)^2} + \ldots \tag{B.3}$$

which are computed from one- and two-loop diagrams. Let us consider a number of examples with different beta functions (see Fig.B.1).

Example B.1 A very special class of QFTs are *conformal* (CFT). In those theories, $\beta(g) = 0$ and the coupling is independent of the scale. The famous example in four dimensions is $\mathcal{N} = 4$ supersymmetric gluodynamics. It has four kinds of gluinoes

© The Author(s), under exclusive license to Springer Nature Switzerland AG 2021   493
E. Shuryak, *Nonperturbative Topological Phenomena in QCD
and Related Theories*, Lecture Notes in Physics 977,
https://doi.org/10.1007/978-3-030-62990-8

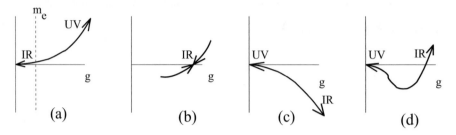

Fig. B.1 Schematic behavior of the beta functions in the four examples considered: (**a**) QED, (**b**) second-order phase transition, (**c**) QCD and (**d**) QCD with sufficiently large number of light flavors. IR and UV with arrows indicate direction of the charge motion at small and large momenta, respectively

and six scalars, and their contributions to the beta function cancel that of the gluons, order by order. This theory was studied in the 1990s and was historically central for a discovery of the so-called AdS/CFT duality.

Example B.2 Quantum electrodynamics (QED) is the first QFT studied since the 1950s. In one loop, the so-called vacuum polarization diagram leads to the following answer:

$$\beta(e) = \frac{e^3}{12\pi^2} \tag{B.4}$$

The solution for the charge vs scale is therefore

$$e^2(\mu) = \frac{e^2(\mu_0)}{1 - \left(e^2(\mu_0)/6\pi^2\right) log(\mu/\mu_0)} \tag{B.5}$$

and when μ grows (the charge is measured deeper inside the electron, at distances $\sim 1/\mu$), it appears larger. Eventually one finds zero of the denominator, the so-called Landau pole, where the formula tells us the charge is infinite. Before taking this answer at its face value, one should be reminded that the formula has applicability limitations, and already when the coupling becomes of the order 1, we cannot any more trust any perturbative results. What actually happens in QED at such small distances remains unclear. QED might be the first QFT studied, but by no means, it is a well-defined theory: it lacks any nonperturbative definition. Therefore, one cannot answer, or even in principle formulate how to answer, many questions.

Example B.3 The second-order phase transitions or scalar theories in three dimensions. In this case, the beta function changes sign and has a zero at g^*, the "fixed point," so when the coupling reaches this value, its "running" stops. Assuming that close to its zero it can be approximated by a linear function $\beta(g) = a(g - g^*)$, one finds power-like dependence near the fixed point:

$$g(\mu) - g^* = (g(\mu_0) - g^*)(\mu/\mu_0)^a \tag{B.6}$$

Example B.4 QCD-like gauge theories have the so-called ultraviolet fixed point, near which the coupling is small.

$$\beta(g) = -b\frac{g^3}{16\pi^2} - b'\frac{g^5}{(16\pi^2)^2} + \dots \tag{B.7}$$

with *positive*

$$b = (11/3)N_c - (2/3)N_f. \tag{B.8}$$

where N_c, N_f are the number of colors and quark flavors; both can be taken to be 3 in QCD. Positivity of the first coefficient of RG beta function behavior is also known as "asymptotic freedom." Its first published version (for $N_c = 2, N_f = 0$) was obtained in Khriplovich (1969), using Coulomb gauge quantization. In 1973 (Gross and Wilczek 1973; Politzer 1973) have not only obtained this result in covariant gauges, but, more importantly, they related this feature with the experimentally observed (in DIS at SLAC) weakly interacting pointlike quarks, suggesting that the fundamental theory of strong interactions, quantum chromodynamics, describes it.
    Integrating the beta function, one has the explicit asymptotic freedom formula

$$g^2(\mu) = \frac{g_{\mu 0}^2}{1 + b\left(g_{\mu 0}^2/8\pi^2\right)log(\mu/\mu 0)} \tag{B.9}$$

and at large μ (small distances), the charge goes to *zero*. There is less and less color charge deep inside the quarks.
    The expression can be written as

$$g^2(L) = \frac{8\pi^2}{[(11/3)N_c - (2/3)N_f]log[1/L\Lambda_{QCD}]} \tag{B.10}$$

where the constant $g(a)$ is traded for a dimensional one, Λ_{QCD}, giving to the strong interaction theory its natural scale.

Example B.5 The QCD-like theories close to the $b = 0$ line have both infrared and ultraviolet fixed points (Banks and Zaks 1982). The expression for the second (two-loop) coefficient is

$$b' = (34/3)N_c^2 - (13/3)N_cN_f + (N_f/N_c) \tag{B.11}$$

and so when $b = 0$ (e.g., in a theory with $N_c = 3$ colors and $N_f = 33/2$ flavors), b' is *negative*. Putting expressions for b, b' to the definition of the beta function, one finds that it possesses a zero (fixed point) at

$$\frac{g_*^2}{16\pi^2} = |b/b'| \tag{B.12}$$

Note that it is small if b/b' is small; in this case the charge is *always* small. Therefore hadrons cannot exist, and the correlation functions have power-like, rather than exponential, decrease with the distance.

Studies of this "conformal regime," at zero and nonzero temperatures, require to handle many quark species, and therefore they only recently became available to lattice practitioners, due to advances in computing. There are evidences that QCD with $N_c = 3$ colors and $N_f = 12$ flavors is already in this regime. We will not discuss this interesting subject any more, since it is too close to the cutting-edge of supercomputer ability, at the time of this writing.

For supersymmetric theories, it has been suggested that rather than calculating beta function in vacuum, one can rather find it from calculation of the instanton amplitude. We have discussed the so-called NSVZ beta function in Sect. 6.2.4.

B.2 Gross-Pisarski-Yaffe One-Loop Free Energy for Nonzero Holonomy

The notations used in their paper (Gross et al. 1981) for the fourth component of the gauge potential on the Matsubara circle, of circumference β, are

$$A_4 = \frac{2\pi}{\beta} diag \left(\frac{q}{2}\right) \tag{B.13}$$

so their variables $q^i, i = 1 \ldots N_c$ are related to the phase fractions we use just by $\mu_i = q^i/2$. Note further that they use $[q]_+ = [q]_{mod2} - 1$, $[q]_- = [q + 1]_{mod2} - 1$ so that $[q]_\pm \in [-1, 1]$.

The results for the fundamental and adjoint color fermionic determinants come from generic sums

$$log \, det_\pm \left[(\partial_\mu + q\xi_\mu)^2 \right] = 2Re \int \frac{d^3k V_3}{(2\pi)^3} log \left(1 \mp e^{-\beta|k|+ikq} \right)$$

$$= -\frac{\pi^2 V_3}{\beta^3} \left(\frac{1}{45} - \frac{1}{24} \left(1 - [q]_\pm^2 \right)^2 \right) \tag{B.14}$$

where $\pm$ are for periodic and antiperiodic boundary conditions. Using that for periodic adjoints (gluons and periodically compactified gluinoes), one then gets

$$log\, det\, \slashed{D} = -\frac{\pi^2 V_3}{\beta^3} \left(\frac{N_c^2 - 1}{45} - \frac{1}{6}tr \left[\left(log\frac{P}{i\pi} \right) \left(1 - log\frac{P}{i\pi} \right) \right]^2 \right) \qquad (B.15)$$

while for quarks—antiperiodic fundamental charges—it is

$$log\, det \left(-D^2 \right) = -\frac{2\pi^2 V_3}{\beta^3} \left(\frac{N_c}{45} - \frac{1}{24}tr \left[1 - \left(log\frac{P}{i\pi} \right)^2 \right]^2 \right) \qquad (B.16)$$

This last expression should be rotated by complex chemical potential angle θ if it is nonzero, as we discussed in the section devoted to Roberge-Weiss symmetry.

Instanton-Related Formulae

C

C.1   Instanton Gauge Potential

We use the following conventions for Euclidean gauge fields: the gauge potential is $A_\mu = A_\mu^a \frac{\lambda^a}{2}$, where the $SU(N)$ generators are normalized according to $\mathrm{tr}[\lambda^a, \lambda^b] = 2\delta^{ab}$. The covariant derivative is given by $D_\mu = \partial_\mu - I A_\mu$, and the field strength tensor is

$$F_{\mu\nu} = [D_\mu, D_\nu] = \partial_\mu A_\nu - \partial_\nu A_\mu + [A_\mu, A_\nu]. \tag{C.1}$$

In our conventions, the coupling constant is absorbed into the gauge fields. Standard perturbative notation corresponds to the replacement $A_\mu \to g A_\mu$. The single-instanton solution in regular gauge is given by

$$A_\mu^a = \frac{2\eta_{a\mu\nu} x_\nu}{x^2 + \rho^2}, \tag{C.2}$$

and the corresponding field strength is

$$G_{\mu\nu}^a = \frac{4\eta_{a\mu\nu}\rho^2}{(x^2 + \rho^2)^2}, \tag{C.3}$$

$$\left(G_{\mu\nu}^a\right)^2 = \frac{192\rho^4}{(x^2 + \rho^2)^4}. \tag{C.4}$$

© The Author(s), under exclusive license to Springer Nature Switzerland AG 2021
E. Shuryak, *Nonperturbative Topological Phenomena in QCD and Related Theories*, Lecture Notes in Physics 977,
https://doi.org/10.1007/978-3-030-62990-8

The gauge potential and field strength in singular gauge are

$$A_\mu^a = \frac{2\bar{\eta}_{a\mu\nu}x_\nu\rho^2}{x^2(x^2+\rho^2)}, \tag{C.5}$$

$$G_{\mu\nu}^a = -\frac{4\rho^2}{(x^2+\rho^2)^2}\left(\bar{\eta}_{a\mu\nu} - 2\bar{\eta}_{a\mu\alpha}\frac{x_\alpha x_\nu}{x^2} - 2\bar{\eta}_{a\alpha\nu}\frac{x_\mu x_\alpha}{x^2}\right). \tag{C.6}$$

Finally, an n-instanton solution in singular gauge is given by

$$A_\mu^a = \bar{\eta}_{a\mu\nu}\partial_\nu \log \Pi(x), \tag{C.7}$$

$$\Pi(x) = 1 + \sum_{i=1}^{n} \frac{\rho_i^2}{(x-z_i)^2}. \tag{C.8}$$

Note that all instantons have the same color orientation. For a construction that gives the most general n-instanton solution (Atiyah et al. 1978).

C.2   Fermion Zero Modes and Overlap Integrals

In singular gauge, the zero mode wave function $i\slashed{D}\phi_0 = 0$ is given by

$$\phi_{av} = \frac{1}{2\sqrt{2}\pi\rho}\sqrt{\Pi}\left[\slashed{\partial}\left(\frac{\Phi}{\Pi}\right)\right]_{\nu\mu} U_{ab}\epsilon_{vb}, \tag{C.9}$$

where $\Phi = \Pi - 1$. For the single-instanton solution, we get

$$\phi_{av}(x) = \frac{\rho}{\pi}\frac{1}{(x^2+\rho^2)^{3/2}}\left(\frac{1-\gamma_5}{2}\right)\frac{\slashed{x}}{\sqrt{x^2}}U_{ab}\epsilon_{vb}. \tag{C.10}$$

The instanton-instanton zero-mode density matrices are

$$\phi_I(x)_{i\alpha}\phi_J^\dagger(y)_{j\beta} = \frac{1}{8}\varphi_I(x)\varphi_J(y)\left(\slashed{x}\gamma_\mu\gamma_\nu\slashed{y}\frac{1-\gamma_5}{2}\right)_{ij}\otimes\left(U_I\tau_\mu^-\tau_\nu^+U_J\right)_{\alpha\beta}, \tag{C.11}$$

$$\phi_I(x)_{i\alpha}\phi_A^\dagger(y)_{j\beta} = -\frac{i}{2}\varphi_I(x)\varphi_A(y)\left(\slashed{x}\gamma_\mu\slashed{y}\frac{1-\gamma_5}{2}\right)_{ij}\otimes\left(U_I\tau_\mu^-U_A^\dagger\right)_{\alpha\beta}, \tag{C.12}$$

$$\phi_A(x)_{i\alpha}\phi_I^\dagger(y)_{j\beta} = \frac{i}{2}\varphi_A(x)\varphi_I(y)\left(\slashed{x}\gamma_\mu\slashed{y}\frac{1+\gamma_5}{2}\right)_{ij}\otimes\left(U_A\tau_\mu^+U_I^\dagger\right)_{\alpha\beta}, \tag{C.13}$$

with

$$\varphi(x) = \frac{\rho}{\pi} \frac{1}{\sqrt{x^2}(x^2 + \rho^2)^{3/2}}. \tag{C.14}$$

The overlap matrix element is given by

$$T_{AI} = \int d^4x \, \phi_A^\dagger(x - z_A) i \slashed{D} \phi_I(x - z_I)$$

$$= r_\mu \, \text{Tr} \left(U_I \tau_\mu^- U_A^\dagger \right) \frac{1}{2\pi^2 r} \frac{d}{dr} M(r), \tag{C.15}$$

with

$$M(r) = \frac{1}{r} \int_0^\infty dp \, p^2 |\varphi(p)|^2 J_1(pr). \tag{C.16}$$

The Fourier transform of zero-mode profile is given by

$$\varphi(p) = \pi\rho^2 \frac{d}{dx} \left(I_0(x)K_0(x) - I_1(x)K_1(x) \right) \Big|_{x = \frac{p\rho}{2}}. \tag{C.17}$$

C.3   Group Integration and Fierz Transformations

In order to perform averages over the color group, we need the following integrands over the invariant $SU(3)$ measure:

$$\int dU \, U_{ij} U_{kl}^\dagger = \frac{1}{N_c} \delta_{jk}\delta_{li},$$

$$\int dU \, U_{ij} U_{kl}^\dagger U_{mn} U_{op}^\dagger = \frac{1}{N_c^2} \delta_{jk}\delta_{li}\delta_{no}\delta_{mp} + \frac{1}{4(N_c^2 - 1)} (\lambda^a)_{kj} (\lambda^b)_{il} (\lambda^a)_{on} (\lambda^b)_{mp}$$

The so-called Fierz transformations are, in general, identities allowing to rewrite multi-fermion operators in different forms. Four-fermion operators written as product of two "neutral" brackets (meaning all indices, Dirac, color, and flavor, convoluted inside a bracket) transformed into a sum of different brackets. Since a given quark can be convoluted with three others, there are three possible "channels" (corresponding to three Mandelstam kinematical variables s, t, u.) Their generic form is

$$\sum_A \Gamma_{ij}^A \Gamma_{kl}^A = \sum_B C_B \Gamma_{ik}^B \Gamma_{jl}^B$$

where indices i, j, k, l (no sums) are quark variables, and indices A, B number all possible matrices, with C_B being some coefficients of the Fierz identity.

Since we mostly deal with QCD quarks with three colors, let us start with Fierz relations for $SU(3)$ spinor generators

$$T^a = \frac{1}{2}\lambda^a, \quad a = 1..8$$

where λ^a are eight traceless Gell-Mann matrices. If $\Gamma^a = t^a$ and $i, j, k, l = 1, 2, 3$ color indices, the corresponding relations follow from the identity:

$$t^a_{ik}t^a_{jl} = \frac{1}{2}\left(\delta_{il}\delta_{kj} - \frac{1}{3}\delta_{ik}\delta_{jl}\right) \tag{C.18}$$

In original papers and textbooks, validity of this and other identities is usually proven by doing multiple convolutions, e.g., with $\delta_{ik}\delta_{jl}$ or $\delta_{ij}\delta_{kl}$ plus some additional considerations which force us to think too much. Nowadays one can check them in seconds, directly using Mathematica. Indeed, the l.h.s. and r.h.s. are 4-index tables, with just $3^4 = 81$ components, so there is no problem to calculate them *all* for both sides. One indeed finds that all 81 of them *are* the same, namely, (in Mathematica table notations with i, j, k, l order)

$$\{\{\{\{1/3, 0, 0\}, \{0, -(1/6), 0\}, \{0, 0, -(1/6)\}\}, \{\{0, 0, 0\}, \{1/2, 0, 0\},$$
$$\{0, 0, 0\}\}, \{\{0, 0, 0\}, \{0, 0, 0\}, \{1/2, 0, 0\}\}\},$$

$$\{\{\{0, 1/2, 0\}, \{0, 0, 0\}, \{0, 0, 0\}\}, \{\{-(1/6), 0, 0\}, \{0, 1/3, 0\},$$
$$\{0, 0, -(1/6)\}\}, \{\{0, 0, 0\}, \{0, 0, 0\}, \{0, 1/2, 0\}\}\},$$

$$\{\{\{0, 0, 1/2\}, \{0, 0, 0\}, \{0, 0, 0\}\}, \{\{0, 0, 0\}, \{0, 0, 1/2\}, \{0, 0, 0\}\},$$
$$\{\{-(1/6), 0, 0\}, \{0, -(1/6), 0\}, \{0, 0, 1/3\}\}\}\}$$

The somewhat simplified form of such relations is often used, in which matrices are mentioned without explicit indices. For example, the relation given above can be rewritten using such notations as

$$\left(\hat{1} \bigotimes \hat{1}\right)_{direct} = \left(2t^a \bigotimes t^a + \frac{1}{3}\hat{1} \bigotimes \hat{1}\right)_{cross} \tag{C.19}$$

Another relation of the type is

$$\left(t^a \bigotimes t^a\right)_{direct} = \left(-\frac{1}{3}t^a \bigotimes t^a + \frac{4}{3}\hat{1} \bigotimes \hat{1}\right)_{cross} \tag{C.20}$$

and using both one can transfer any expression containing color convolution from one channel to the next.

Let me add that so far no assumptions about open four indices were made: but sometimes their symmetry is known, simplifying relations further. For example, if the cross channel combines two quarks (diquark) in the baryon, we know that in this case, only *antisymmetric* color indices are allowed. Then in the r.h.s. out of nine matrices t^a, $\hat{1}$, only *three* ones, which are antisymmetric $t^a_{anti} = t^2, t^4, t^7$ produce nonzero contributions.

Quark isospin matrices are related to $SU(2)$ generators, Pauli matrices σ^a, $a = 1, 2, 3$ divided by two. Of course, those coincide with the first three Gell-Mann generators. Here are corresponding relation for Pauli matrices:

$$\left(\hat{1} \otimes \hat{1}\right)_{direct} = \left(\frac{1}{2}\hat{1} \otimes \hat{1} + \frac{1}{2}\sigma^a \otimes \sigma^a\right)_{cross} \tag{C.21}$$

$$\left(\sigma^a \otimes \sigma^a\right)_{direct} = \left(\frac{3}{2}\hat{1} \otimes \hat{1} - \frac{1}{2}\sigma^a \otimes \sigma^a\right)_{cross} \tag{C.22}$$

In diquark channel with isospin zero, and *antisymmetric* indices in cross channel, there remains only one antisymmetric Pauli matrix σ^2, so r.h.s. in this case is simply proportional to $(ud - du)$ flavor combination. It is of course the same as the wave function for two spins 1/2 added into total spin zero.

Similarly, one can define the matrix of Fierz transformation for all 16 Dirac matrices. Defining symbolically five products of quark bilinears,

$$S = 1 \otimes 1, \ V = \gamma_\mu \otimes \gamma_\mu, \ A = \gamma_5\gamma_\mu \otimes \gamma_5\gamma_\mu, \ T = \sigma_{\mu\nu} \otimes \sigma_{\mu\nu}, \ P = \gamma_5 \otimes \gamma_5$$

one can perform Fierz transformation from direct channel (s) to cross channel (u) (coupling antiquark to another quark) and find the following matrix of Fierz transformations:

$$\begin{pmatrix} S \\ V \\ T \\ A \\ P \end{pmatrix}_{direct} = \begin{pmatrix} 1/4 & 1/4 & -1/8 & -1/4 & 1/4 \\ 1 & -1/2 & 0 & -1/2 & -1 \\ -3 & 0 & -1/2 & 0 & -3 \\ -1 & -1/2 & 0 & -1/2 & 1 \\ 1/4 & -1/4 & -1/8 & 1/4 & 1/4 \end{pmatrix} \begin{pmatrix} S \\ V \\ T \\ A \\ P \end{pmatrix}_{cross} \tag{C.23}$$

Some Special Theories

D

D.1   Gauge Theory with the Exceptional Group G_2

Interest to this particular theory is related with the question of whether the Z_N symmetry of the $SU(N)$ gauge theories is or is not related to confinement. (The answer is *not at all*.)

The construction starts with the $SO(7)$ group, which has 21 generators and rank 3. As for any SO group, its 7×7 real matrices satisfy

$$det\,\Omega = 1, \quad \Omega^{-1} = \Omega^T \tag{D.1}$$

G_2 is its subgroup which additionally satisfies seven more relations

$$T_{abc} = T_{def}\,\Omega_{da}\Omega_{eb}\Omega_{fc} \tag{D.2}$$

where T is an antisymmetric tensor such that

$$T_{123} = T_{176} = T_{145} = T_{257} = T_{246} = T_{347} = T_{365} = 1 \tag{D.3}$$

This leaves us 14 generators listed, e.g., in Cossu et al. (2007). The rank of G_2 is 2, so, like in SU(3), a nonzero Polyakov line leaves two massless U(1)s to make the Abelian monopole charges. In fact SU(3) is the largest subgroup of the G_2, with its eight generators. The rest can be constructed via six SU(2) subgroups: and such construction is explicitly used in lattice updates.

Yet the center of this SU(3) does not commute with extra generators of SU(2)s, leaving G_2 without a nontrivial center (only the unit matrix commutes with all generators). Therefore there is no "spontaneous breaking of the center symmetry" in the deconfined phase!

© The Author(s), under exclusive license to Springer Nature Switzerland AG 2021
E. Shuryak, *Nonperturbative Topological Phenomena in QCD and Related Theories*, Lecture Notes in Physics 977,
https://doi.org/10.1007/978-3-030-62990-8

D.2 $\mathcal{N} = 2$ SYM and SQCD, and their Seiberg-Witten Solution

The $\mathcal{N} = 1$ and corresponding SUSY QCD theories will be discussed only in one chapter, on instanton-dyons: this is how the puzzle of quark condensate was resolved. In general, these theories have properties similar to ordinary QCD.

Therefore we just jump to $\mathcal{N} = 2$ and then $\mathcal{N} = 4$ theories, in which theoretical progress was more impressive.[1]

D.2.1   The Field Content and RG Flows

Let us start with the **field content** of those theories. The $\mathcal{N} = 2$ gluodynamics or super-Yang-Mills (SYM) theory has gluons (spin 1), two real gluinoes λ, χ (spin 1/2), and a complex scalar (spin 0) which we will call a. Each of them has two degrees of freedom, thus four bosonic and four fermionic ones.

The $\mathcal{N} = 2$ QCD is a theory with additional matter supermultiplets of structure ψ_f, ϕ_f with spin 1/2 and 0, respectively. We will call N_f the number of Dirac quarks, as in QCD, or $2N_f$ Majorana ones.

There are two different Higgsing possible, defining "branches" of these theories. If $< a > \neq 0$, $< \phi > = 0$, Higgsing is like in Georgi-Glashow model, with massless photon/photino multiplet: this branch is called the "Coulomb branch." However if both $< a > \neq 0$, $< \phi > \neq 0$, Higgsing is like in the Weinberg-Salam model, with all gluons massive: this is called the "Higgs branch."

The **coupling renormalization** in these theories is done only via the one-loop beta function, with the coefficient

$$b = 4 - N_f \tag{D.4}$$

while two-loop and higher coefficients vanish. The explanation for that was given in the instanton chapter, see NSVZ beta function.

Starting with $N_f = 0$ or $\mathcal{N} = 2$ SYM, one can add 1, 2 or 3 quark–squark multiplets, obtaining QCD-like theories with running coupling.

At the opposite end, at $N_f = 4$ QCD, one finds zero beta function and is thus a conformal theory: we will not discuss it.

D.2.2   The Moduli

Supersymmetry also requires that for *any* value of v, the vacuum energy remains zero: thus there is a whole *manifold of non-equivalent vacua*, known as *moduli space*, labeled by a complex number $u = Tr(\phi^2)$. All properties of the system

---

[1]This fits general mathematical expectations: the more symmetry the problem has, the easier it is to solve. Unfortunately, real world around us is not that symmetric as one would like.

Fig. D.1 The map of the
moduli space according to
Seiberg-Witten solution

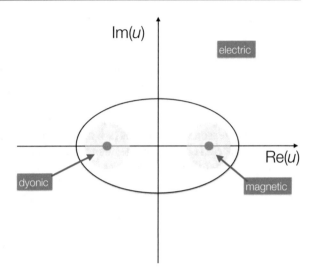

are expressed as derivatives of one fundamental holomorphic function $F(u)$, in particular the effective charge and the theta angle are combined into a variable τ which is given by its second derivative:

$$\tau(u) = \frac{\theta}{2\pi} + \frac{4\pi i}{e^2(u)} = \frac{\partial^2 F(u)}{\partial u^2} \qquad (D.5)$$

We will return to the approximate form of this function in connection with its instanton-based description.

The map of the moduli space is schematically given in Fig. D.1. There are three distinct patches:

(1) At large values of $|u| \to \infty$, there is a "perturbative patch," in which the coupling $e^2(|u|)/4\pi \ll 1$ is weak. It is dominated by electric particles—gluons, gluinoes, and Higgses—with small masses $O(ev)$, which determine the beta function. Monopoles have large masses $4\pi v/e^2(|u|)$ there and can be treated semiclassically.

(2) A "magnetic patch" around the $u = \Lambda^2$ point, in which the coupling is infinitely strong $e^2 \to \infty$; the monopole mass goes to zero as well as the magnetic charge $g \sim 1/e \to 0$.

(3) A "dyonic patch" around the $u = -\Lambda^2$ point, in which a dyon (particle with electric and magnetic charges both being 1 gets massless.

D.2.3   Singularities for $\mathcal{N} = 2$ QCD

Let us now focus on the next case, $N_f = 3$. Here are the limiting cases discussed by Seiberg-Witten. Suppose first the quark/squark triplet has large mass m. This means that one needs large value of the VEV $< a >$ to cancel it; the structure of the $< a >$ plane is as follows: two original singular points, inherited from $N_f = 0$ or $\mathcal{N} = 2$ SYM, plus a triply degenerate singularity at large $< a >$. As $m_f \to 0$, one by one, one may reason what happens (see details in the second Seiberg-Witten paper). When the process is complete, they came up with the following unusual structure: a 4-degenerate zero where monopoles ($n_m = 1, n_e = 0$) get massless, plus another single zero where some particle with $n_m = 2, n_e = 1$ gets massless. The former quartet of monopoles includes 3 monopole+occupied zero mode by one quark plus perhaps the old unoccupied monopole (the singularity existing before). The latter $n_m = 2, n_e = 1$ state is some kind of di-monopole bound state bound by quarks: its possible nature has been discussed at the end of the monopole chapter.

D.3 $\mathcal{N} = 4$ Super-Yang-Mills Theory

As the name suggests, it has four supersymmetries, and thus fermions χ_i have ("flavor" or R-charge) index $i = 1..4$, on top of assumed adjoint color index. This makes eight fermionic degrees of freedom, compensated by two polarizations of the gauge field and six complex scalars ϕ_{ij} in the symmetric tensor representation of the R-charge. The usual cancellations between bosonic and fermionic loops follow.

This remarkable theory has been called "a harmonic oscillator of the twenty first century".[2] As follows from NSVZ beta function, derived from instanton above, in this case, bosonic and fermionic parts of beta function cancel *to all loops*, and thus it is a super-conformal theory. The charge keeps its value, independent of the scale.

It is due to this remarkable high symmetry of this theory (and to J. Maldacena, of course) that we now have AdS/CFT holographic correspondence.

We already discussed another derivation of this fact from electric-magnetic duality, in the monopole chapter. The monopole dressed by all types of gluinoes make spin-1, spin1/2, and spin-0 objects, which together form precisely a $\mathcal{N} = 4$ supermultiplet. So, its magnetic theory, with charge $1/g$, is the same as the electric theory, with charge g. Beta functions must be the same for electric and magnetic theory, up to a sign: the only possible solution is that it is identically zero!

---

[2]This saying is attributed to D. Gross, but I do not know if it is indeed true.

AdS/CFT Correspondence

E

E.1 Black Holes and Branes

The original Schwarzschild solution for the usual black hole in asymptotically flat $3 + 1$ dimensions has a metric tensor (in spherical coordinates)

$$ds^2 = g_{\mu\nu}dx^\mu dx^\nu = -(1 - r_h/r)dt^2 + \frac{dr^2}{(1 - r_h/r)} + r^2 d\Omega^2 \tag{E.1}$$

and the horizon radius (in fulll units) is $r_h = 2G_N M/c^2$, containing the mass M, the Newton constant G_N, and speed of light c.

E.2 Colors and the Brane Stack, the Road to AdS/CFT

The ten-dimensional solution

$$ds^2 = \frac{-dt^2 + dx_1^2 + dx_2^2 + dx_3^2}{\sqrt{1 + L^4/r^4}} + \sqrt{1 + L^4/r^4}\left(dr^2 + r^2 d\Omega_5^2\right) \tag{E.2}$$

AdS_5 limit: the "near-horizon region," at $r \ll L$, when 1 in both roots can be ignored and the metric splits into noninteracting $AdS_5 \times S_5$. Using a new coordinate $z = L^2/r$, we get it into the "standard AdS_5 form" used below:

$$ds^2 = \frac{-dt^2 + dx_1^2 + dx_2^2 + dx_3^2 + dz^2}{z^2} \tag{E.3}$$

The Maldacena relations between the gauge coupling, the AdS radius L and the string tension α' (which comes from the total mass of the brane set):

© The Author(s), under exclusive license to Springer Nature Switzerland AG 2021
E. Shuryak, *Nonperturbative Topological Phenomena in QCD
and Related Theories*, Lecture Notes in Physics 977,
https://doi.org/10.1007/978-3-030-62990-8

$$L^4 = g^2 N_c (\alpha')^2 = \lambda (\alpha')^2 \tag{E.4}$$

AdS_5 in six-dimensional Global coordinates, standard and Poincare coordinates
the AdS_5 space can be described by the equation

$$-\left(X^{-1}\right)^2 - \left(X^0\right)^2 + \left(X^1\right)^2 + \left(X^2\right)^2 + \left(X^3\right)^2 + \left(X^4\right)^2 = -L^2 \tag{E.5}$$

in the six-dimensional space $X^{-1}..X^4$. "Standard" coordinates

$$X^{-1} = \sqrt{L^2 + \rho^2} cos\left(\frac{\tau}{L}\right), \quad X^0 = \sqrt{L^2 + \rho^2} sin\left(\frac{\tau}{L}\right), \quad X^i = \rho \Omega^i \tag{E.6}$$

The last term contains the coordinates of the three-dimensional unit sphere, with the
standard line element:

$$d\Omega^2 = d\chi^2 + sin^2\chi \left(d\theta^2 + sin^2\theta d\phi^2\right) \tag{E.7}$$

Poincare coordinates, defined by ($i = 1..3$)

$$X^{-1} = \frac{z}{2}\left(1 + \frac{L^2 + \vec{x}^2 - t^2}{z^2}\right), \quad X^0 = L\frac{t}{z}, \tag{E.8}$$

$$X^i = L\frac{x^i}{z}, \quad X^4 = \frac{z}{2}\left(-1 + \frac{L^2 - \vec{x}^2 + t^2}{z^2}\right)$$

Relation between the two sets

$$\frac{z}{L} = \left[\sqrt{1 + \frac{\rho^2}{L^2}} cos\left(\frac{\tau}{L}\right) + \frac{\rho}{L} cos\chi\right]^{-1}, \tag{E.9}$$

$$t = z\sqrt{1 + \frac{\rho^2}{L^2}} sin\left(\frac{\tau}{L}\right), \quad \vec{x} = z\frac{\rho}{L} sin\chi \vec{\Omega}$$

The metric tensor in both sets are related

$$ds^2 = -dt^2 + (dx^i)^2 = W^2\left(-d\tau^2 + L^2 d\Omega^2\right) \tag{E.10}$$

where the so-called *conformal factor* is

$$W^2 = \frac{1}{(cos(\tau/L) + cos(\chi))^2} = \frac{t^2}{L^2} + \frac{1}{4}\left(1 + \frac{r^2}{L^2} - \frac{t^2}{L^2}\right)^2 \qquad (E.11)$$

E.3   Propagators in AdS_5

The Laplace-Beltrami operator is in general

$$\Delta = \frac{1}{\sqrt{g}}\partial_\mu\sqrt{g}g^{\mu\nu}\partial_\nu \qquad (E.12)$$

and the equation for massive propagator is obtained by simply adding the mass term

$$\Delta D(x, x') - m^2 D(x, x') = \frac{1}{\sqrt{g}}\delta(x - x') \qquad (E.13)$$

where the mass can be negative but not the r.h.s. of

$$\nu^2 = m^2 + d^2/4 > 0 \qquad (E.14)$$

The main object one needs for holography is Euclidean bulk-to-boundary propagators: those were constructed by Witten (1998a) in the following way. The propagator is a scalar field of a point charge, placed somewhere in the space under consideration. Clearly the simplest case is when we put this charge at the *origin* of the space, $z = \infty$. In this case, no dependence on $x^i, i = 1..d$ is expected by symmetry, and so one can keep only the z part of the Laplacian. Considering the massless case for simplicity, we only have one term in the equation

$$\frac{d}{dz}z^{-d+1}\frac{d}{dz}D(z) = 0 \qquad (E.15)$$

and the only solution which vanishes at z=0 is

$$D(z) \sim z^d \qquad (E.16)$$

It is singular at $z = \infty$, as of course it should, as there is a source there. (One can show it is exactly the right delta function.)

The next step is to use conformal invariance and move the source to a place we actually want it to be, namely, at the boundary. This is done by the conformal transformation

$$x^i \rightarrow \frac{x^i}{z^2 + \sum_{j=1}^{d}(x^j)^2}, \quad z \rightarrow \frac{z}{z^2 + \sum_{j=1}^{d}(x^j)^2} \qquad (E.17)$$

which moves the source to the point $z = 0$, $x^i = 0$ and changes the solution to

$$D \sim \left[\frac{z}{z^2 + \sum_{j=1}^{d} (x^j)^2} \right]^d \tag{E.18}$$

All that remains is to shift the point to arbitrary location by $x^i \rightarrow x^i - x'^i$ and finding the normalization constant.

For more complete discussion of propagators, we suggest (Danielsson et al. 1999): we will not copy its beginning with a concise description of what vacuum one should select, in AdS curved space; yet it is worth reading.

The basic geometry needs the distance $d(X, X')$ between the two points which is defined as the line element integrated along the geodesic. In this space, it is

$$cosh(d(X, X')) = 1 + \frac{z^2 + z'^2 + (X^0 - X'^0)^2 \dots (X^3 - X'^3)^2}{zz'} \tag{E.19}$$

(see that it makes sense along each of the axes).

AdS_5 corresponds to spatial dimension $d = 2m = 4$, and the solution for massive bulk-to-bulk Euclidean propagator is

$$D = \frac{1}{8\pi^2 sinh(u)} \frac{d}{du} \frac{e^{vu}}{sinh(u)} \tag{E.20}$$

where we use $u = d(X, X')$ defined above. Note that the function is more complicated than just the power of the distance: that was a large-distance tail, which is the only one needed for bulk-to-boundary applications.

Classical applications sometimes need retarded bulk propagators, which are not obtainable from the Euclidean version by an analytic continuation. Those are only nonzero for real-time-like distance v

$$cos(v) = 1 - \frac{t^2 - r^2 - (z - z')^2}{2zz'} \tag{E.21}$$

and we only give a simple answer for integer v

$$D = -\frac{cos(vv)}{2\pi sin(v)} \tag{E.22}$$

E.4   Nonzero Temperatures in Holography

Global AdS-Schwarzschild black hole, GAdSBH, is a charged five-dimensional black hole. The Einstein equation with a particular cosmological term

$$R_{ab} + \frac{4}{L^2} g_{ab} = 0 \tag{E.23}$$

the metric

$$ds^2 = -f d\tau^2 + \frac{d\rho^2}{f} + \rho^2 d\Omega^2 \quad f = 1 - \frac{\rho_0^2}{\rho^2} + \frac{\rho^2}{L^2} \tag{E.24}$$

where τ, ρ, Ω have the same meaning as in "standard" AdS coordinates above. Parameters ρ_0, L are two parameters related to the mass

$$M = \frac{3\pi^2 \rho_0^2}{\kappa^2}, \tag{E.25}$$

and charge of the black hole.
    The Hawking temperature

$$T = \frac{\rho_h}{\pi L^2} \left(1 + \frac{L^2}{2\rho_h^2} \right) \tag{E.26}$$

and the Bekenstein entropy

$$S = \frac{4\pi^3 \rho_h^3}{\kappa^2}, \tag{E.27}$$

define thermodynamical variables on the brane. In this case the horizon radius, the upper root of $g_{00} = f = 0$, is

$$\rho_h/L = \sqrt{\sqrt{(1 + 4\rho_0^2/L^2 - 1)}/2} \tag{E.28}$$

κ is the five-dimensional Newton constant, which is related to L and the brane number N via

$$\frac{L^3}{\kappa^2} = \left(\frac{N}{2\pi} \right)^2 \tag{E.29}$$

Note that if $N \gg 1$ is large, then the space parameter L is large compared to the five-dimensional Plank scale, and thus gravity is classical.

Bibliography

Atiyah, M.F., Hitchin, N.J., Drinfeld, V.G., Manin, Yu.I.: Construction of instantons. Phys. Lett. **A65**, 185–187 (1978). [133(1978)]

Banks, T., Zaks, A.: On the phase structure of vector-like gauge theories with Massless Fermions. Nucl. Phys. **B196**, 189–204 (1982)

Berezin, F.A.: General concept of quantization. Commun. Math. Phys. **40**, 153–174 (1975)

Cossu, G., D'Elia, M., Di Giacomo, A., Lucini, B., Pica, C.: G(2) gauge theory at finite temperature. JHEP **10**, 100 (2007)

Danielsson, U.H., Keski-Vakkuri, E., Kruczenski, M.: Vacua, propagators, and holographic probes in AdS / CFT. JHEP **01**, 002 (1999)

Gell-Mann, M., Low, F.E.: Quantum electrodynamics at small distances. Phys. Rev. **95**, 1300–1312 (1954)

Gross, D.J., Wilczek, F.: Asymptotically free gauge theories-I. Phys. Rev. **D8**, 3633–3652 (1973)

Gross, D.J., Pisarski, R.D., Yaffe, L.G.: QCD and instantons at finite temperature. Rev. Mod. Phys. **53**, 43 (1981)

Khriplovich, I.B.: Green's functions in theories with non-abelian gauge group. Sov. J. Nucl. Phys. **10**, 235–242 (1969). [Yad. Fiz.10,409(1969)]

Politzer, H.D.: Reliable perturbative results for strong interactions? Phys. Rev. Lett. **30**, 1346–1349 (1973). [274(1973)]

Witten, E.: Anti-de Sitter space and holography. Adv. Theor. Math. Phys. **2**, 253–291 (1998a)

© The Author(s), under exclusive license to Springer Nature Switzerland AG 2021   515
E. Shuryak, *Nonperturbative Topological Phenomena in QCD
and Related Theories*, Lecture Notes in Physics 977,
https://doi.org/10.1007/978-3-030-62990-8

Index

© The Author(s), under exclusive license to Springer Nature Switzerland AG 2021   517
E. Shuryak, *Nonperturbative Topological Phenomena in QCD*
and Related Theories, Lecture Notes in Physics 977,
https://doi.org/10.1007/978-3-030-62990-8

Printed in the United States
by Baker & Taylor Publisher Services